Environmental Literacy from A to Z for Students and Educators

A quick reference guide for students who want to write original research papers, educators who want to develop cross-discipline lessons and lectures, and everyone else who wants to keep informed about our environment

by

H. Steven Dashefsky

"Everyone is entitled to their own opinion, but not their own facts."

Daniel Patrick Moynihan
(former U.S. Senator from New York)

Once again, for Lindsay and Kim

Sci-Tec

Simplifying Science and Technology
Ridgefield, Connecticut
www.E-Literacy.com

Copyright © 2013 H. Steven Dashefsky
All rights reserved.
ISBN: 149496788X
ISBN 13: 9781494967888

Library of Congress Control Number: 2014900966
CreateSpace Independent Publishing Platform, North Charleston, SC

Contents

How to Use This Book vi

Quick Start Suggestions for Students Writing Papers or Reports vii

Quick Start Suggestions for Educators Developing
  Lesson Plans or Lecture Notes ix

Sample Category Lists xi

Crowd Suggestions xv

A Note to the Reader about Cross-Referenced Italicized Words xvi

A to Z entry listings 1-743

Source Material 745

About the Author 750

Acknowledgments 751

Further Information 752

How to Use This Book

People have their own interests, and finding a book tailored to an individual's needs is difficult. <u>Environmental Literacy from A to Z</u> offers an alternative way to read about our environment. It is designed to flow with your interests, whatever they might be, opening new doors each step of the way. Reading this book should be like a free-form learning experience, with each topic an open invitation to the next.

The best way to begin reading this book is by flipping through the pages. When an entry catches your eye, read it and notice the italicized words, which are themselves entries. These words will lead to other entries to help you delve deeper into the topic, to provide background material for a better understanding, or to take you to unexpected but related topics. Let each entry guide you to the next, creating your own custom, on-demand reading and learning experience.

This book can be used for a high school or college course.

- Students should read "Quick Start Suggestions for Students Writing Papers or Reports."
- Educators should read "Quick Start Suggestions for Educators Developing Lesson Plans or Lecture Notes."
- Everyday environmentalists could start by looking up specific entries, looking at the Sample Category Lists section, or just skimming the entries.

Quick Start Suggestions for Students Writing Papers or Reports

To make the best use of this book, follow these steps:

1. Skim the book for an entry that interests you and note it.
2. Follow any italicized cross-reference that also interests you. Note it and repeat this process until you have built your topic outline.
3. Go back to each entry in your outline, in whatever you consider the logical sequence, and write your paper or report.

Here are a few examples.

If you start with the entry *e-waste* (discarded smartphones and computers), it will lead you to cross-references such as *e-waste recycling* and *e-waste and you*, both of which are about the problems this type of waste causes and what you can do about it. But *e-waste* can also lead you to unexpected entries such as *urban miners*, *gold fingers*, and the *Basel Convention*, all of which explain how your e-waste contributes to international social injustice when poor people in poor nations are made sick by America's computer waste.

Another example might be to start with the entry on labeling *cosmetics*, which can lead you to entries about the *cosmetics and animal cruelty* issue. However, you could follow other references to take you in a totally different direction about toxic substances in cosmetics, such as *lipsticks and toxic substances* and how to avoid them. This can then lead you to environmental toxic substances in general with entries such as *breast milk and toxic substances*, *toxic cocktail*, and *body burden*.

And finally, to round out these examples, you could start with an entry about *green blogs* and end up writing about how *green microblogs* (*tweets*) have been used to educate millions of Chinese people about their horrible *air pollution* problems. And, depending on the length of your paper, you could follow up with any of the dozens of entries about air pollutants and what's being done (or not done) about them. Or take a different route and write about the use of technology as an environmentalist's tool for change.

These three topic outlines might look like:

e-waste

e-waste recycling and disposal
   programs

e-waste product life cycle

e-waste and you

urban miners

gold fingers

Basel Convention

cosmetics. labeling

lipstick and toxic substances

nanoparticles in cosmetics and
   personal care products

breast milk and toxic
   substances

toxic cocktail

body burden

blogs, green

microblogs, green

apps, eco-

social media, environmental

crowd research, scientific

tweetstorm, Rio+20

tweetstorm, Red Panda on the
   loose

As you can see, each entry can turn into a paper about any number of topics. The book includes over 2000 entries, most of which have numerous cross-references that can lead you to countless combinations—resulting in topics you probably would have never thought about. Let the entries and your imagination lead the way.

~

The next section is "Quick Start Suggestions for Educators." Of course you can use that information to help in your student research, but the main purpose there is to help educators incorporate environmental topics into their particular courses. Educators will focus on one specific topic, such as *environmental philosophers and writers*. This will help educators develop a class about what they specifically want to teach, but it won't lead you to the creative topic outlines as explained above.

~

Environmental Literacy Crowd Suggestions

See the "Crowd Suggestions" page for information about having your topic outline included in the next release of this book, in its print or e-book version, and getting a discount for you or your friends.

Quick Start Suggestions for Educators Developing Lesson Plans or Lecture Notes

Many schools and colleges are trying to integrate environmental topics across their curriculums. As an educator, you may be looking for ways to weave these topics into your particular discipline. For those with some expertise in environmental science, this is not difficult. But if your specialty lies elsewhere, identifying relationships between these seemingly disparate areas may be difficult. That is where this book can help.

Take a look at the "Quick Start Suggestions for Students" in the previous section. The assumption is that the students are not sure what they want to write about when they begin their research. As an educator, however, you know what you are trying to accomplish from the start: to relate your course to some environmental or sustainability issue or topic.

You are the expert in your particular field. This book is designed to help you realize relationships to the environment you might not have previously thought about. The book gives you information about that environmental issue for you to use with your expertise. Here are some suggestions about how to accomplish this:

See the "Sample Category Lists" section that follows. These are lists of related entries from a few different fields to help you link an environmental issue to your course.

Look for cross-over entries that might relate to your field. For example, numerous entries in "People of Importance" can be discussed not only for the person's expertise, such as writing or painting, but also for a link to an environmental issue. Consider *John Muir's* friendship with *Theodore Roosevelt,* as it relates to the *Hetch Hetchy* environmental debate; *Rachel Carson's* crusade against *pesticides;* or *Thomas Cole* and the *Hudson River School art movement* in the formative years prior to *environmentalism,* just to mention a few.

Environmental ethics, The Land Ethic, Native American environmentalism, primitivism, NIMBY and *YIMBY FAP,* and many others can be used for an ethics or sociology course. Over two dozen entries about food can be used for a health course, from introductory entries such as *organic farming and food* to unique topics such as *microgreens and urban gardening;* to the dark side with entries on *pesticide residue on food, nanoparticles in food,* and *food dyes,* and finally to debatable issues such as *genetically modified organisms you routinely eat.*

The *environmental movement,* ranging from *Richard Nixon* to *Earthfirst!* and *eco-terrorism,* can enhance a social studies or humanities course. Dozens of treaties, conventions, and summits about global agreements (and disagreements) can be used for a government course. Dozens of entries on the *economy vs. environment* debate and various alternative economic models are appropriate for a business or economics course. Historical perspectives can be found throughout this book in dozens of entries, including differences between the European Union's *precautionary principle* approach and the United States' *industry self-regulation* approach.

These topic outlines might look like:

Muir, John	organic farming and food	economy vs. environment
preservation	conventional farming	economics and sustainable
Roosevelt, Theodore	conservation farming	development
national parks	community-supported agriculture	Beyond GDP
Yosemite	urban gardens	corporate social responsibility
Grand Canyon	heirloom plants	steady-state economics
Hetch Hetchy	pesticide residues on food	top ten green global companies
environmental movement	nanoparticles on food	climate capitalism
	circle of poison	CERES
	food eco-labels	decoupling
	organic food labels	

With over 2000 entries, the possibilities are endless. The following Sample Category Lists just scratch the surface of entries available within this book that might tie in with your course.

~

Environmental Literacy Crowd Suggestions

See the "Crowd Suggestions" page for information about having your topic outline included in the next release of this book, in its print or e-book version, and getting a discount for you or your colleagues.

Sample Category Lists

The following lists are meant to help you, as an educator, find entries that relate to your courses. They include many (but not all) of the entries about one particular category. You can link these entries to topics in your field of expertise. These lists represent only about 10 percent of all entries, so if you don't find something here, the book will give you plenty more to choose from.

PEOPLE OF IMPORTANCE

Abby, Edward
Adams, Ansel
Audubon, John James
Bartram, William
Bennett, Hugh Hammond
Berry, Thomas
Berry, Wendell
Bierstadt, Albert
Boulding, Kenneth
Brockovich, Erin
Brower, David R.
Brown, R. Lester
Brundtland, Gro Harlem
Bullard, Robert D.
Burroughs, John
Carson, Rachel
Catlin, George
Chief Seattle
Cole, Thomas
Commoner, Barry
Cousteau, Jacques Yves
Daly, Herman E.
Darling, J. N. "Ding"
Darwin, Charles
Dillard, Annie
Douglas, Marjory Stoneman
Douglas, William O.
Dubos, René

Ehrlich, Paul R.
Eiseley, Loren
Emerson, Ralph Waldo
Fossey, Dian
Fuller, Buckminster
Gibbs, Lois
Goodall, Jane
Gore, Al
Hardin, Garret
Hayes, Denis
Hill, Julia "Butterfly"
Leopold, Aldo
Lovins, Amory B.
Maathi, Wangari
Malthus, Thomas
Maraniss, Linda
Marsh, George Perkins
Mather, Stephen
McKibben, Bill
Mendes, "Chico"
Merchant, Carolyn
Moran, Thomas
Mowat, Farley
Muir, John
Nixon, Richard M.
Odum, Eugene
Orr, David
Pinchot, Gifford
Porter, Elliot

Quammen, David
Roosevelt, Theodore
Seeger, Pete
Silkwood, Karen
Thoreau, Henry David
Ward, Barbara
White, Gilbert
Whitman, Walt
Wilson, E. O.

CLIMATE CHANGE

archeologists, ice-patch
arctic ice, changes in
asthma and the environment
Bangladesh and sea rise
belching cows
carbon budget
carbon markets
carbon sequestration storage
CIA and climate change
climate change and biodiversity
climate change and human health
climate change and marine life
climate change deniers
climate change legislation
Climategate
Convention on Climate Change,
    The United Nations
coral bleaching

deforestation and climate change

extreme weather and climate change

Green Climate Fund

greenhouse effect

hockey-stick graph

Intergovernmental Panel on Climate Change

karst, carbonate

Keeling Curve

Kyoto Protocol

megafloods

oysters and climate change

seeds and climate change

peatlands

Petermann Glacier ice detachment

rice paddies and methane

ton of carbon dioxide look like? What does a

HUMAN POPULATION

carrying capacity

China's One-Child Program

demographic transition, theory of

doubling time in human populations

human population throughout history

population ecology

population growth, limits of human

Population Reference Bureau

push-pull hypothesis

K-strategists

J-curve

species on Earth

United Nations Population Fund

hunger, world

water shortages, global

SPECIES IN CRISIS

amphibians in decline

Amphibian Ark project

biodiversity, loss of

Biodiversity Treaty

captive breeding

charismatic megafauna

colony collapse disorder

debt-for-nature swaps

de-extinction

dodo bird

doomsday seed vault

endangered and threatened species lists

Endangered Species Act, U.S.

extinct in the wild

extinction and extraterrestrial impact

extinction, mistaken

extinction, newly identified species and

extinction, sixth mass

god squad

invasive species

elephants

ivory trade

Lazarus taxon

Lonesome George

migratory bird protection

poaching

polar bear trade

Red List

rhinoceroses and their horns

seafood eco-ratings

species recently gone extinct, ten

suburbia and biodiversity

Ugly Animal Preservation Society

whale and commercial ship collisions

whaling

white-nose syndrome

wildlife trade and trafficking

FOOD: FOR BETTER OR FOR WORSE

high-impact agriculture

monocultures

factory farms

egg farms

fish populations and food

overfishing

green revolution, the

pesticides residues on food

synthetic fertilizers

pollination crisis

super bugs

superweeds

organic farming and food

community-supported agriculture

conservation agriculture

farmers' markets, fake

heirloom plants

Organic Consumers Association

organic food labels

regenerative farming

sugar, sugar alternatives, and the environment

vertical farms

Dirty Dozen – fruits and vegetables

farm-to-table movement

farm-to-school movement

food dye

food eco-labels

food miles

food waste

fregan

Good Samaritan Food Donation
    Act

grain production and use

hunger, world

locavore

meat production

meat, eating

nanoparticles in food

Non-GMO Project Verified label

rice, contaminated

seasonal foods

slow food movement

tomatoes – tasteless but red

vegan and vegetarian diets and
    the environment

vineyards and assisted migration

wine, organic and biodynamic

WWOOF

genetically modified organisms
    that we routinely eat

genetically modified organisms in
    your food, the five most common

BUSINESS / ECONOMICS

economy vs. environment

Beyond GDP

cap-and-trade

CERES

climate capitalism

companies, top ten green global

corporate acquisitions of small
    green companies

corporate social responsibility

cost/benefit analysis

decoupling

economics and sustainable
    development

economics, ecological

economics, environmental

ecopreneur

extended product
    responsibility

externalizing costs

Genuine Progress Indicator

gross national happiness

growthmania

integrated bottom line

triple bottom line

investing, green

microfinance, green

microlending

nature capital

polluter pays principle

Production Tax Credit

shareholders, corporate

steady-state economics

waste-conscious product
    development

CONSUMERISM

consumer, green

dry cleaning

cosmetics

ecotourism

ecotravel

flat screen televisions, green

flip-flops and preconsumer waste

green fatigue

green marketing

green scam

greenwash

holiday waste

hotels, green

lipstick and toxic substances

mycowood

nanoparticles in cosmetics and
    personal products

planned obsolescence

Skin Deep, EWG's

sneakers, green

stonewashed jeans

blue-jean pollution

textile industry

toilet paper, eco-

tourism, ten best countries for eco-

wine bottle cork debate

CITY LIFE

urbanization and urban growth

Alliance for a Paving Moratorium

brownfields

deicing roads

eco-cities

gardens, rain

greenways

interim use

lawns

megacities

eco-cities

right-to-dry laws

smart cities

urban farms

urban forests

urban gardens

urban sprawl

white roofs

zoning, land use

mass transit

bicycles and bicycling

indoor air pollution

indoor ecology

mattresses, recycling

laundry and the environment

TECHNOLOGY

apps, eco-
biomaterials
biomimetic
bio-ore
bioplastics
bioremediation
biosensors
blogs, green
microblogs, green
cloud computing and
   the environment
crowd funding, green
crowd research, scientific
crowdsourcing, eco-
data centers, energy
   consumption of
DNA sequencing technology
e-waste and you
genome
geoengineering
hyperaccumulators
Landsat
nanoparticles
nanotechnology
robotic bees
satellite mapping
social media, environmental
space debris cleanup
synthetic biology
terraforming
tweets, eco-
tweetstorm, Rio+20

RANDOM THOUGHTS

body burden
toxic cocktail
China, People's Republic of

Dust Bowl
explosive remnants of war
Spaceship Earth
zoos, future
"Land Ethic, The"
baseball stadiums, green
coffee, environmental impact
dissecting frogs in schools
ecological footprint
eco-terrorism
news literacy
precautionary principle
primitivism
Sagebrush Rebellion
SLAPP
sustainababble
voluntourism
suburbia and biodiversity
energy poverty
water shortages, global
unburnable carbon
Mining Law of 1872
eco-cities
smart cities
smart energy
light bulb - recent history and
   legislation
nuclear waste dilemma, the global
animals experiencing grief
ant slaves
blobfish
bumblebee decline
circus elephant ban
communications between plants
drones, conservation
elephant ivory, dating
flagship species
honeybee sperm bank

sixth sense in animals
sonar testing, underwater
whale earwax
zombie bees
ship ballast and invasive species
Rigs-to-Reefs Program
fracking
energy-saving apps and programs
tidal power
sugarcane to biofuel
waste-to-energy power plants
light bulb technologies, domestic
overfishing
farm animal cruelty legislation
rendering
recycling
downcycling
upcycling
municipal solid waste
landfills
toilet-to-tap
oil spills in history, ten worst
oil spills in U.S.
pipeline spills
dam removal
fracking
wind farms
fish populations and food
aquaculture
ghost fishing
genetically modified
   organisms - debate about
plastic pollution
garbage patch, marine

Crowd Suggestions

You might have heard of crowd funding and crowd research. This is an attempt at crowd suggestions. If you submit a suggestion about this book, such as a new entry, new cross-reference, or—better yet—your own topic outline you used for a class, you will receive a 20 percent discount code toward the purchase of the next updated edition of this book or a 10 percent discount on the e-book for you or your friends to use. If we use your topic outline, your name will be included with the outline, if you wish. (Updates are released each semester.) The discount also applies for any corrections you submit, such as incorrect or missing cross-references or anything else that might need fixing.

Send your suggestions to: suggestions@E-Literacy.com. You will receive your discount code via a reply message.

A Note to the Reader about Cross-Referenced Italicized Words

Italicized words within an entry are cross-references to other entries, but they may not be exact. For example:

- They might be listed in the inverse, such as *commercial fishing* is listed as *fishing, commercial*.
- The full entry name might not be included, such as *online petition sites* is listed as *online petition sites and environmentalism*.
- And finally, many entries are better known by their acronyms or abbreviations and are listed as such. For example, Bisphenol A is best known and listed as *BPA*, just as confined area feeding operations is listed as *CAFO* and the Organization of Petroleum Exporting Countries is listed as *OPEC*.

Additional cross-references are placed in [*square brackets*] at the end of many entries. Website addresses are placed in {curly brackets} at the end of some entries.

ℬ

Abby, Edward

(1927–1989) Abby was a novelist and writer, but best known as an outspoken *environmentalist* who has reached almost cult status. His books include <u>Desert Solitaire</u> and <u>The Monkey Wrench Gang</u>, which provided the backdrop for the formation of the radical group *Earth First!* His beliefs are well-depicted in a quote from <u>Desert Solitaire</u>: "There is a cloud on my horizon. A small dark cloud no bigger than my hand. Its name is Progress." A collection of his journals written between 1951 and 1989 is titled <u>Confessions of a Barbarian</u>. [*monkeywrenching; ecotage; deep ecology; environmental movement*]

abiota

An *ecosystem* consists of two major components interacting with one another. The *biota* is the living component (plants and animals), and the abiota is the nonliving component, which includes *soil*, *water*, and *air*. [*biosphere*]

abyssal ecosystem

An *ecosystem* that exists deep within the ocean where no light penetrates. Organisms at this depth depend on organic debris (dead plants and animals) that drifts down from higher levels where light is visible. [*marine ecosystem; benthic organisms; giant squid in native habitat*]

acid rain (wet deposition)

When *fossil fuels* such as *coal*, *oil*, and *natural gas* are burned, many substances are emitted into the air. *Sulfur dioxide*, *nitrogen compounds*, and *particulates* are three such substances, and all are considered primary air pollutants responsible in part for *air pollution*. These substances travel through the air and react with each other in the presence of sunlight to form *secondary pollutants*, such as *sulfuric* and nitric acids. When these acids fall to Earth with rain, it is commonly called acid rain.

However, because these acids also come to the Earth's surface in the form of snow, fog, dew, or small droplets, the phrase "acid deposition" or "wet deposition" is often used. (There is also "dry deposition," when the same acids accumulate on dust and *particulate matter* and fall to Earth in a dry form, causing the same harm as wet deposition.)

Because these secondary pollutants float and are carried by winds, acid deposition often occurs far from its source. For example, the northeastern United States has high concentrations of acid rain, but most of it is produced by power and industrial plants in the Midwest.

Normal rain is slightly acidic, having a *pH* of about 5.6. Average rainfall in most of New England and adjoining parts of Canada is around 5.0. Mountain tops in New Hampshire have recorded rains with a pH of 2.1, about the same acidity as lemon juice. Overall, however, the acidity has improved significantly because of the *Clean Air Act*.

The most apparent damage caused by acid deposition is the destruction of statues, which crumble from the acids, but the most serious effects are less noticeable. Studies show acid deposition at levels below 5.1 kill fish and destroy *aquatic ecosystems* because most organisms have narrow pH *tolerance ranges*.

Acid deposition can weaken and kill trees and stunt the growth of crops. It can also harm lake and pond *ecosystems*.

China has become one of the worst producers of acid rain because of their reliance on *coal*-burning *electric power plants*. By 2025 they are on track to produce more sulfur dioxide than the United States, Canada, and Japan combined.

acid deposition

See *acid rain.*

acid mine drainage

During *strip mining*, sulfur within *coal* can become exposed to the elements and form (with the help of *bacteria*) *sulfuric acid*. This sulfuric acid can seep into streams or lakes and damage *aquatic ecosystems*. Acid mine drainage results in *water pollution* and possibly *groundwater pollution*.

acoustic niche

As you walk through the woods or fields, you hear sounds—some would say the sounds of nature. All of the sounds within a *habitat* are called the *soundscape* of that habitat. But a soundscape can further be studied by listening to the sounds made by individuals within that habitat.

The acoustic niche specifically refers to the sounds a particular organism makes as part of its niche. (It is similar to an *ecological niche* but only refers to sound.) Different species use different calls and frequencies to be heard among the din. When an *invasive species* moves into a new area, some of the native species might be forced to change their calls—whether they be mating or warning—so as to be heard among the similar sounds of the invaders. There is even a field of science called audio ecology.

Environmentalists, especially birders, record certain calls and then play them back to attract that species. [*apps, eco-; beehive sound research; jazz, Earth; noise pollution and control*]

active solar-heating systems

Active solar-heating systems use solar panels mounted on a roof. The panels collect and concentrate the sun's energy in a series of tubes that contain an antifreeze solution or pumped air. As the antifreeze or air heats up, it is pumped into an insulated storage tank. Fans, controlled by a thermostat, distribute the stored heat through conventional air ducts into the building.

Most areas in the United States receive enough sunlight to heat a home at least 60 percent of the year in this manner. In areas that don't receive enough sunlight, backup conventional heating systems are needed.

Domestic hot water can also be created with active solar water-heating systems. More than one million buildings in the United States get their hot water via this method. Sixty-five percent of all domestic water in Israel is supplied by active solar hot-water systems. [*solar power; baseball stadiums, green; eco-cities; photovoltaic cell film; solar charging of electronic devices; solar glitter*]

acute toxicity

The harmful effect a substance has on an organism shortly after exposure to that substance. By definition, "shortly" means less than 96 hours from contact or ingestion. The harmful effect might be an illness, burns, or death. [*toxic waste; chronic toxicity; hazardous waste*]

Adams, Ansel

(1902–1984) Ansel Adams is America's best known and most loved photographer, whose majestic, large-format, black-and-white photos captured the wide-open expanses of the Western landscape. One look at his work makes it is no surprise that he was a true *environmentalist* and even activist—his love of the wilderness is visible in every picture. He grew up visiting *Yosemite* and worked there several years; later Yosemite became one of his favorite subjects, with iconic photos of El Capitan and Half

Dome. He joined the *Sierra Club* and served on their Board of Directors. Adams' interest in environmental issues led him to meet with *Rachel Carson* to discuss her work.

As his fame grew—as well as the cost of his fine art—he decided to share his work with everyone by selecting a few of his best photographs and offering them as affordable poster prints. Later, he had many more selected to create his still well-known and popular calendars.

Americans learned to appreciate nature and the landscape from artists such as those within the *Hudson River School* movement and writers such as *John Burroughs*, all of whom used the country's Northeast and especially the Hudson Valley as their subjects. Ansel Adams' photographs and *John Muir's* writings, both of the wild and beautiful Sierra Nevada Mountains, continued and enhanced this appreciation as the country moved west. [*Porter, Eliot*]

adaptation

Any trait that makes an organism more suitable to survive in its *environment*. These traits can be physical, physiological, or behavioral. Adaptations are a result of *natural selection*. [*genetically modified organisms—resistance in*]

aerobic organisms

Organisms that require oxygen to survive, as opposed to *anaerobic organisms*, which live without oxygen.

aerobiology

The study of organisms, such as *bacteria* and *algae*, or reproductive cells, such as spores and *pollen*, that float freely through the air. [*high-altitude ecosystems; ecosystems; extremophiles*]

aerosols

An aerosol is a suspension of a liquid or solid particles suspended in a gas. Aerosols can be created naturally, such as the blast of a volcano that spews out aerosol or by a strong desert wind that kicks up an aerosol. Or it can be a manufactured (anthropogenic) aerosol, such as a can of hairspray or spray paint. *Coal*-burning *electric power plants* emit aerosols of *sulfur dioxide* into the air. Aerosols in the *atmosphere* can have a significant impact on climate. When Mount Pinatubo erupted in 1991, the plume of ash (an aerosol) cooled global temperatures by 1.8 degrees F for a year. Because manufactured aerosols can affect climate, there have been discussions about injecting aerosols into the atmosphere as a *geo-engineering* plan to slow *climate change*. [*air pollution*]

aesthetic pollution

Aesthetic pollution includes odor, visual, and noise pollution. Aesthetic pollution is difficult to define. Just as beauty is in the eye of the beholder, so too is what someone considers to be environmentally offensive. For example, some odors may offend almost anyone, while others may be unpleasant to only a few. The matter is complicated by the fact that people who constantly smell a certain odor can become oblivious to it after a while. Aesthetic pollution is defined by the person subjected to it, so what is done to resolve it depends on those identifying it.

However, when the general public finds some form of aesthetic pollution offensive, action is often taken. In some cases, ordinances have been passed to prevent *noise pollution*, *malodor pollution*, and *visual pollution* such as billboards.

afforestation

The process of planting large numbers of trees in areas that did not previously contain trees. This is often done to replace forested areas lost to *deforestation*. The new forested areas are intended to increase natural *carbon capture sequestration* and lost *biodiversity*. Studies have shown that even though most forested areas created by afforestation do not provide the same *habitat* as natural forests, they do act as *carbon sinks* and can improve overall biodiversity in the area. [*forest health, global*]

age distribution

An important aspect of any *population* is its age distribution. Populations are divided into three groups of individuals: prereproductive juveniles, reproductive individuals, and postreproductive individuals. In the wild, most stable populations have more juveniles than reproductive individuals, and more reproductive individuals than postreproductive individuals. This occurs because most *species* (*insects* and small animals) have high *mortality* throughout their lives as a result of predation (being eaten) or disease. Large numbers of young are produced, but only a small percentage make it to the reproductive stage and fewer still to the postreproductive stage. For example, less than 25 percent of young cottontail rabbits survive to sexual maturity and many insect species have fewer than 5 percent survive to reproduce. [*population ecology; r-strategists; K-strategists*]

age distribution in human populations

Age distribution in human *populations* can be divided into the same three groups as described in *age distribution* for wild populations. However, the number of individuals found in each of the three groups differs.

When the majority of the population is in the preproductive stage, as in *Kenya* and many *less developed countries*, a long-term, rapid population increase is expected. This is called an "expansive" population profile. If a large percentage of the population is in the reproductive stage, a short-term baby boom is expected. This occurred in the United States between the late 1940s and mid 1960s and was called the *baby boom*. Today, the boom has passed but a similar percentage of population remains in the reproductive and prereproductive stages, resulting in slow growth. If a small percentage of the population is in the preproductive stage, such as in Italy and *Germany*, the population profile becomes "constrictive," resulting in slower or even negative growth. [*population explosion, human; population growth, limits of human*]

Agenda 21

One of five documents developed at the United Nations *Conference on Environment and Development*, commonly called the Earth Summit, in Rio de Janeiro in June 1992; it is the largest of the five documents, containing more than 900 pages in 40 chapters. Its objective is to strive for *sustainable development* in the 21st century. It is a suggested action plan for how countries, and our planet as a whole, should move forward. It is a nonbinding agreement, meaning no one is actually committed to following the suggested plan.

One part of the plan that has received much attention is the suggestion that economic growth should be accompanied with environmental protections. It encourages *mass transit* and *smart growth* of urban areas that could house lots of people with minimal environmental impact.

Some of the main points discussed in this document are poverty, changing consumer patterns, population, human health, policy-making for sustainable development, protecting the *atmosphere*, *hazardous wastes*, safeguarding the oceans' resources, promoting environmental awareness, and who will pay for all of the above.

Twenty years later, Rio+20 (the *United Nations Conference on Sustainable Development*) followed up the work done at this conference. [*Agenda 21 conspiracy*]

Agenda 21 conspiracy

A couple of environmental topics have become targets of a few high-profile extreme anti-environmentalists. Their outrageous and often time ludicrous statements, articles, and books get a great deal of air time but provide few if any actual facts. *Agenda 21* is one of these topics (*climate change* the other). They assert that *Agenda 21* is a conspiracy perpetrated by the *United Nations* to end the American way of life.

Some of this is based on the document's call for *smart growth* that includes building environmentally *smart cities* and using *mass transit* instead of cars. Smart growth is based on science and could help reduce further degradation of our environment even as our population grows. Conspiracy seekers have converted this suggestion into a document designed to take away our cars, our homes, and our freedom—quite a stretch from reality.

People who espouse these assertions have obviously never read the document. It is unfortunate that some people choose to waste time listening to falsehoods instead of spending time learning the facts. The same is true about climate change when people refuse to listen to the *science* involved. [*climate change deniers*]

Agent Orange and Vietnam War

Agent Orange was a commonly used mixture of two *herbicides,* (2,4-D) and (2,4,5-T). The drums containing this substance often had bright orange stripes, providing its name. During the Vietnam War, over six billion gallons of Agent Orange was dumped in Southeast Asia to defoliate areas to expose the enemy. From 1965 to 1970, when its use was discontinued, about 50,000 U.S. military personnel and an unknown number of Vietnamese were exposed to the substance.

Agent Orange was found to be contaminated with the highly toxic substance *dioxin.* Exposure to the contaminated herbicide was suspected as the probable cause of illnesses that developed in many war veterans. In 1984, court action resulted in manufacturers compensating a quarter of a million of the victims named in the suit.

In 2012, the United States agreed to begin a four-year program to clean up the remaining Agent Orange *toxic waste* sites in Vietnam at a cost of 43 million dollars. [*explosive remnants of war; hazardous waste, military*]

ag-gag bills

See *farm animal cruelty legislation.*

aggressive mimicry

A form of *mimicry* in which a *predator* mimics a nonpredatory organism. This tactic deceives the *prey* into dropping its guard, making it vulnerable to attack.

agricultural pollution

The release of pollutants into the environment by farming or ranching practices. It can be caused by agrochemicals such as *pesticides*, *herbicides*, and *synthetic fertilizers* or by *runoff* of waste products or excessive nutrients from feedlots, resulting in *nutrient enrichment.* When cultivation techniques cause *topsoil* erosion, this is also considered a form of agricultural pollution. [*high-impact agriculture; conservation agriculture; factory farms; CAFO*]

agriculture

The growing of plants and raising of animals for human use. It can be as simple as *subsistence farming,* where people raise their chickens and goats and pigs along with a few crops in a small, family-run *agroecosystem.* Or it can be a larger operation that still uses eco-friendly techniques such as *organic farming* or *conservation agriculture.* At the other end of the spectrum, *high-impact agriculture* is one of the primary causes of environmental degradation. It is also called mechanized or industrial farming and is the most common farming practice.

agriculture, high-impact

High-impact agriculture or farming (also called mechanized or industrial farming) is so called because it uses techniques that have a significant negative impact on the *environment.* These farms grow large *monocultures* that require vast amounts of *pesticides* to kill *pests, herbicides* to kill *weeds, synthetic fertilizers* to replace lost *nutrients,* and *conventional tilling* techniques that result in *soil erosion.*

It is also true that this form of agriculture, developed during the *Green Revolution,* is the primary reason there is food for most of the seven billion plus people on our planet. *Conservation agriculture* and *organic farming* produce far less environmental harm but so far have not been scaled up to meet these needs. [*agroecology*]

agroecology

The science that integrates both *agriculture* and *ecology* with the goal to develop sustainable practices—those that do not harm *topsoil,* water, or organisms within the immediate *ecosystem.* It uses methods such as *integrated pest management, conservation agriculture,* and improved *irrigation* methods. [*low-input agriculture; sustainable agriculture; regenerative agriculture; agroforestry*]

agroforestry

The practice of cultivating both crops and trees within the same area. This practice benefits both types of plants by offering a more natural *ecosystem* that causes far less harm to the region than *high-impact agriculture.* Many less developed nations continue to use this form of farming, although in recent years it has lost ground to modern *monoculture* farms that cause more environmental harm. [*permaculture; integrated multi-trophic aquaculture; agroecology*]

agrofuels

See *biofuels.*

air

See *atmosphere*.

air conditioning (A/C)

A/C was created in the early 1900s to cool equipment—not people. In a few years, however, it became popular to cool people in places such as movies and stores. Today, A/C is ubiquitous and considered a must in all developed nations and even in many developing nations. From an economic perspective, it is known to improve productivity. From a public health perspective, it saves lives, because heat can and often does kill, especially during heat waves.

But the amount of energy used to cool buildings increases our *ecological footprint* significantly, and *global warming* will likely make things worse. A/C use is expected to increase dramatically over the next few decades. The United States uses more energy just for A/C than the African continent uses for everything put together. *China*, however, will take the lead away from the United States by 2020 as the biggest A/C user. In 2009, 100 million homes in the United States had air conditioners. In China, 50 million units were sold in just one year, 2010.

The good news is that A/C technologies continue to improve from an environmental perspective. Advances in technologies should make them run on 30 percent to 50 percent less energy in the near future. Also, the *Montreal Protocol* helped stop the use of the *ozone-depleting gas*, Freon, thus reducing that particular harm.

On the negative side, however, many of the developing nations generate electricity by using *coal*. As people in countries such as China become more affluent, one of the first luxuries they get is an air conditioner. This increases electrical demand, resulting in more coal use and its accompanying pollution.

Another problem associated with A/C is improper building design. The A/C units in closed buildings create heat canyons, where the hot air from within the buildings is ventilated outside. This makes adjacent buildings hotter, so they require more A/C, producing a vicious cycle. Buildings no longer have courtyards or verandas so people cannot go outside where air circulates and remains cooler.

Green buildings can offer some common-sense design improvements to reduce these problems. Natural designs, such as light outside colors, trees, and other methods of shading all reduce A/C use. Proper insulation, hi-tech windows, and water sprayed on roofs to cool by evaporation also help. [*heating, ventilation, and A/C systems; energy use*]

airplane pollution

Planes account for a great deal of air pollution and an inordinate portion of *carbon emissions*. One round-trip flight from New York to San Francisco produces over two tons of *carbon dioxide* per person on board the plane. You would have to drive 100 miles per week, every week, all year, in a midsize car to produce the same two tons.

The European Union passed, in 2012, the world's first and still only mandatory program curbing aviation emissions. This new directive should remove enough carbon to equal removing 30 million cars off the road each year. [*cruise-ship pollution; ship pollution; air pollution; ton of carbon dioxide look like?, What does a*]

air pollution

The five primary air pollutants are *carbon monoxide, hydrocarbons, nitrogen compounds, particulate matter*, and *sulfur dioxide*. The major source of the first three pollutants is the *automobile,* which is called a mobile source of air pollution. Burning *fossil fuels* such as *coal* and *oil* to generate electricity also contribute to air pollution and are called stationary sources.

In addition to the primary pollutants, *secondary air pollutants* form when the primary pollutants react with each other in the presence of sunlight. *Ground-level ozone* and *lead* produced by automobile emissions also play a role in air pollution.

When pollutants are released into the air, they are mixed, diluted, and circulated around the globe. Densely populated areas produce large amounts of air pollutants in small regions, making it difficult to dilute and more dangerous to breath. Local weather conditions create periods of intense air pollution in urban areas, such as *thermal inversions* and *dust domes*. Global wind currents move air across the Earth's surface, accumulating air pollutants as it goes. In the United States, westerly winds carry air pollutants from west to east, further increasing the pollutant concentration.

Until recently, air pollution meant only outdoor air. Today, studies have found *indoor air pollution* to be a serious concern, with a different set of sources and associated problems. [*airplane pollution; ship pollution; seaport-related air pollution; air pollution in Beijing, China; atmospheric brown cloud; brick kilns; brown cloud; Clean Air Act; electric vehicles; ozone, ground-level; ground-level ozone, ten U.S. cities with the worst; hazardous air pollutants; automobile; natural gas as an energy source; Persian Gulf War pollution; textile industry; tire fires; toxic pollution; urbanization and urban growth; wood-burning cookstoves, primitive*]

air pollution in Beijing, China

China's capital, Beijing, has some of the worst *air pollution* in the world. In January 2013, the city had a number of days that broke the hazardous level and was reported as beyond measurement. This resulted in a frenzy of activity on *microblogs* and *social media* about the *environment*.

Air pollution in Beijing, primarily caused by their use of *coal*-burning *electric power plants*, is estimated to reduce the life expectancy of its citizens by five years.

alar

A substance sprayed on apples to modify their growth. It postpones the fruit's dropping from the tree, which enhances the apple's color and shape and extends its storage life. In 1989 a public panic arose when it was announced that alar might be *carcinogenic*. This was especially worrisome to parents, because children drink large volumes of apple juice. That year, the manufacturer voluntarily withdrew the product from use on fruit, and it was subsequently banned from use on apples. It is still legally used on some ornamental plants.

The risks involved were highly debatable and typify some of the problems involved in environmental *risk assessment*. The facts show that alar causes cancer in laboratory animals, but it was not established to cause cancer in humans. Few chemicals, however, can be directly linked to cancer in humans. [*genetically modified organisms you routinely eat; apples, antibiotics on organic*]

alarm pheromone

A chemical substance released by members of a species to warn other members of the same species of danger. Commonly used by *insects* such as ants and aphids. [*pheromones; semiochemicals; sex pheromones; biological control*]

Alaska pipeline

Properly called the Trans-Alaska Pipeline, it was approved by Congress in 1973 and completed in 1977. It extends a little more than 800 miles, from Prudhoe Bay within the *Arctic* Circle in northern Alaska to Port Valdez in the south of Alaska, crossing more than 800 rivers and streams. It is about five feet in diameter and carries about 2 million barrels of oil per day. Because the pipeline passes through the *tundra*, with its *permafrost*, environmentalists have long been concerned about the potential for harm. To protect the permafrost, about half of the line is elevated above ground. The pipeline has not caused the harm many originally feared. Concerns are now more about the fact that it is about 35 years old.

Today, the pipeline transports less *oil* than at first, because of a decline in production. To replace this lower yield, the oil industry has increased its efforts to open the *Arctic National Wildlife Refuge* for development.

The pipeline has been a huge financial bonanza for the region. Ever since the pipeline opened, it has financed much of the region with a $350 million annual budget. Alaskans pay no state income taxes and even get a check averaging $5000 per family each year. [*Keystone XL pipeline; pipeline oil spills; oil spills in U.S.; oil spills in history, worst ten*]

albedo

When energy from the sun enters the Earth's *atmosphere*, one of three things happens. Roughly 35 percent is reflected away by dust particles and clouds. This reflected energy is called albedo. About 15 percent of the energy is absorbed by the atmosphere and the remaining 50 percent reaches the Earth and is called insolation. [*greenhouse effect*]

aldrin

One of the *persistent organic pollutants*.

algae

Algae are primitive aquatic plants ranging from microscopic single-celled organisms to large multi-celled plants such as seaweed. Algae are of great importance in *aquatic ecosystems* because it fills the role of the *autotrophs*. However, their importance extends well beyond aquatic ecosystems. It is estimated that about 80 percent of the oxygen in our *atmosphere* comes from algae in the *oceans*.

Algae are harvested for food in many Asian nations and are a common harvest in *aquaculture* farming today. Some algae products are used in processed foods. For example, kelp (a common marine algae) provides alginic acid, which is used to add consistency to ice cream and puddings. There are over 20,000 species worldwide. [*algal bloom; aufwuch; biofuels, second generation*]

algal bloom

A sudden and dramatic increase in the density of *phytoplankton* in a body of water. Algal blooms can occur during *natural eutrophication,* but they have become a human-induced phenomenon associated with *nutrient enrichment*, and the resulting *cultural eutrophication*, and can result in *environmental hypoxia* and even *dead zones* in bodies of water.

On rare occasions, they have produced large enough amounts of hydrogen sulfide gas to be dangerous to animals and people. In France, a large number of wild boar were found dead on a beach, and the cause was found to be hydrogen sulfide poisoning. A truck driver died while removing the dead *algae* from the scene. On occasion, algal blooms have caused beaches to be closed for safety reasons.

However, the primary concern about algal blooms is not about these unusual events. It is about the environmental harm caused to local ecosystems. [*manatee deaths and algae blooms*]

algalculture
See *biofuels, second-generation*.

alien species
An alien species, also called *exotic species*, is any species that is not native, or indigenous to an area. It typically arrives in a new area with the help of humans, either as an accident, such as being carried within the ballast of a boat, or with intent, such as being introduced and released to control pests. When alien species overtake a *habitat* and force out native species, they are often called *invasive species*. [*ship ballast and invasive species*]

Allen's Principle
The concept that warm-blooded animals in cold climates have shorter appendages, such as ears and tails, than in warm climates. Less surface area (on the shorter appendages) results in less heat loss. For example, northern species of rabbits have shorter ears than their southern counterparts. [*adaption; natural selection*]

alley cropping
Alley cropping is a *soil conservation* method in which crops are planted in rows (alleys) between other rows of trees or shrubs. This reduces *soil erosion*. The trees or shrubs can be harvested along with the crop, for fruit or wood. [*conservation agriculture*]

allergens, natural
When a person is sensitive to contact with certain natural substances (as opposed to synthetic substances), it is commonly called a natural allergen. These substances are typically proteins and can include plant *pollen*, mold spores, feathers, or fur. [*aerobiology; asthma and the environment; climate change and human health; pollination*]

Alliance for a Paving Moratorium
The goal of this organization is to halt the environmental, social, and economic damage caused by endless road building. Members believe a paving moratorium would limit the spread of the population, redirect investment to inner cities, and revitalize

the economy. They try to save *wetlands*, farms, and *forests* from becoming paved in the name of progress. [*urbanization and urban growth; combined sewer systems; gardens, rain; pointless pollution; runoff; eco-cities*] {culturechange.org/apm_page.htm}

alligators

Two species of this freshwater reptile exist, the American and Chinese alligators. They are considered *apex consumers* and are a *keystone species*.

In the 1960s, the American alligator was overhunted. Its population was seriously reduced, and it was placed on the *endangered species* list of the *U.S. Fish and Wildlife Service.* Its numbers have since recovered and it was removed from the list in 1987. Most of the alligator products purchased in the United States today come from alligators raised on farms.

As a result of overhunting and of *habitat loss*—primarily from conversion of *wetlands* to *rice paddies*—the Chinese alligator is near extinction and considered a *critically endangered species* on the *IUCN Red List*.

All Natural labels

See *food eco-labels*.

alluvium

Accumulations of particles such as *sand* and silt that are carried downriver and deposited along river banks or at the mouth of the river, in areas such as deltas and *floodplains*. Regions with large amounts of alluvium deposits are considered some of the world's most fertile. [*deposition; dredging; sedimentation; soil texture; wetlands*]

alpine tundra

See *tundra*.

alternative energy sources

Fuels that can replace our dependence on *fossil fuels* (*oil, natural gas,* and *coal*) are considered alternative energy sources. Alternatives are necessary, because the combustion of fossil fuels—especially from our *automobiles* and *electric power plants*—have side effects that damage our environment, including *air pollution, acid rain, and global warming*. In addition, fossil fuels are finite and will someday become exhausted, probably within a few hundred years, even though this has become a hotly debated topic with newfound *reserves*.

Alternative energy sources include *nuclear* and *renewable energy*. Renewable energy sources are considered inexhaustible even if we continually use them. They include *solar power, wind power, hydroelectric power, geothermal energy*, and *biomass energy*.

alternative fuels

Fuels used as alternatives to *gasoline* that cause less harm to the environment. They are being used throughout the world but represent only a small portion of the total fuels used to run vehicles. Most are considered *biofuels* and called a form of *biomass energy*.

Methanol is often mixed with gasoline—for example, M85 means it is 85 percent methanol and 15 percent gasoline—and can be produced from *natural gas* or *biomass. Ethanol*, E85, is produced from corn, sugar, or other forms of biomass. (Ethanol is also called *bioethanol*.) Other alternative fuels include compressed or *liquid natural gas,* and someday we might have hydrogen-fueled cars. [*automobile propulsion systems, alternative*]

Amazon River basin

The Amazon river basin encompasses about four and a half million square miles, which is almost 5 percent of the Earth's land surface. Most of it is *tropical rainforest*. Two-thirds of it is in *Brazil* and the remainder in Bolivia, Columbia, and seven other countries.

The *biodiversity* found in this basin is staggering. Thirty percent of all plant species are believed to exist there. Many activists over the past decades have worked to protect this region and its people, including *Chico Mendes,* who was assassinated in 1988.

The biggest threat to the region is *deforestation,* primarily caused by *clearcutting*, which continues at a frightening pace to this day. Estimates are that about 240,000 square miles have been destroyed since 1980, due to the conversion of forests to pastures for grazing of cattle and to farmland.

American Grassfed label

See *food eco-labels*.

American Zoo and Aquarium Association

Accredits *zoos*, aquariums, and wildlife parks in North America with the purpose to advance *conservation, environmental education,* recreation, and scientific study. Most recently it stresses *wildlife* conservation and *habitat* protection. [*zoos, captive breeding programs in; Zoo, Frozen; zoos, future; touchpools; nature centers*] {aza.org}

ammonia as a car fuel

When ammonia is burned, it produces water vapor and nitrogen (the most common element in the atmosphere). If it could be used as an automobile fuel, it would be a clean *alternative fuel* to *gasoline*. The problem has been producing the ammonia. Conventionally, ammonia is created by burning a *fossil fuel* such as *coal* and getting the hydrogen released to combine with nitrogen, forming the ammonia.

However, this is an energy-intensive and dirty process, and even though clean-burning ammonia would replace a fossil fuel (gasoline for cars), producing the ammonia still requires burning a fossil fuel. So there is not much gained by this process.

However, new methods being studied include releasing the hydrogen not from a fossil fuel but from water to combine with the nitrogen. Plus, these methods use *wind power* to run the entire process. A few ammonia fuel test vehicles are on the road now. If this can be scaled up to be economically feasible, it might be an entirely new form of a clean, natural vehicle fuel. [*automobile propulsion systems, alternative; natural gas vehicles; fuel cell vehicles, hydrogen*]

Amoco Cadiz oil spill

To date, this remains the largest *oil spill* involving an *oil tanker*, spilling more than eight times that of the *Exxon Valdez*. It occurred in 1978 off the coast of Brittany, France, where it grounded and broke up, spilling more than 220,000 tons of oil. *Oil* was deposited on about 200 miles of coastline and as deep as a foot. Sea birds died en masse, *food webs* were compromised, and much of the *commercial fishing* industry in the region was curtailed because the catch was unfit for consumption. To this day, sheltered areas near the spill still contain signs of it. [*oil spills in history, ten worst*]

Amphibian Ark (AArk) project

A group trying to save the world's 6000+ species of frogs and salamanders from being wiped out by the *chytrid fungus* infection and other threats. They are asking *zoos*, aquariums, and *nature centers* to collect these organisms, ensure they are fungus-free, and then protect them until this fungal epidemic can be resolved. At that time, they would be released back into the wild. [*amphibian decline; biodiversity, loss of*] {amphibianark.org}

amphibian decline

Amphibians include frogs, toads, salamanders, and other organisms. They are the most *threatened* group of *vertebrates* on Earth. Close to 200 species are believed to have gone *extinct* since 1970—not to be seen again except preserved in museums.

Today, more than 30 percent of all amphibian species are declining in numbers. Large-scale die-offs are occurring, caused by infections such as the *chytrid fungus* that kills entire local populations of some frog species. *Parasites* are also thought to contribute to their demise, but humans appear to be the primary culprit.

Frogs have not done well cohabitating with humans. Housing developments, parking lots, *urbanization,* and industrialization have all reduced frog *habitats.* *Pesticides* and *herbicides* such as *atrazine* reduce the male frog's ability to reproduce. Even recreational fishing hastens their demise when lakes are stocked with fish that decimate tadpole populations. It is also thought that rapid fluctuations in temperature probably increase the likelihood of fungal infections, and this will be exacerbated by *climate change.*

This reduction in amphibian populations and species loss causes significant harm to *ecosystems.* Amphibians play a major role in many *food chains* because they are an important food source for larger *vertebrates*, such as snakes, birds, and *mammals.* Ongoing research suggests that natural compounds found in some frog species have been shown to fight HIV and some forms of cancer. [*biodiversity, loss of; Amphibian Ark project; indicator species*]

anaerobic organisms

Organisms that do not require oxygen to survive, as opposed to *aerobic organisms*, which do. Some *bacteria* are anaerobic and are responsible for the decomposition of organic matter (dead plants and animals). [*methane gas; food webs*]

ancient forests

Forests that have never been harvested and therefore contain ancient trees—some 700 or more years old. A typical stand of trees in an ancient forest contains 250-year-old trees with trunks more than 20 feet in diameter. Unless we are prepared to wait 700 or more years, we will never see these original forests again.

Globally, it is estimated that about seven percent of the Earth's lands contain ancient forests, but about 25 million acres are cut down each year.

In the United States, almost all of the ancient forests that once existed on private lands have been logged. Most that remain in the United States are found in *national forests* and parks. The few remaining ancient forests in the United States—also called *old-growth forests*—are found in the Cascade Range of northern California, western Oregon and Washington, and southeast Alaska. About 2.3 million acres remain, but fewer than one million acres are designated as wilderness areas and therefore protected from logging. Over the past few years, the U.S. *Forest Service* has allowed tens of thousands of acres of trees more than 200 years old to be cut down annually.

These trees are mainly cedar, Douglas fir, western hemlock, and Sitka spruce. Most environmentalists believe all remaining ancient forests should be protected and preserved as natural monuments. [*dual-use policy; redwood forest destruction; Hill, Julia "Butterfly"*]

Animal Damage Control program (ADC)

In 1931 Congress created the Animal Damage Control program within the *Department of Interior*. Its purpose is to destroy "animals injurious to *agriculture*, horticulture, *forestry*, animal husbandry, wild game animals, fur-bearing animals, and birds." ADC hunters and trappers kill more than four million animals each year to fulfill this purpose. About half a million of those killed are *mammals,* including bears, beavers, deer, and mountain lions. The agency calls this killing "wildlife services." It costs the U.S. taxpayers about $120 million dollars a year to conduct this slaughter of wildlife for the sake of protecting livestock. Agents perform this "control" with many methods, including shooting animals from low-flying planes and setting poison bait.

Most environmentalists believe the ADC to be a misdirected, misguided attempt to resolve animal/human conflicts simply by killing without any scientific basis or purpose. The ADC's budget is vastly higher than any damages perpetrated by the *predators* being killed. It's been more than 80 years since this law was enacted and many believe the time has come to stop it. Alternatives have been found in some locations, such as Marin County, California, where they stopped using the federal agency and began using a community-based program called the Marin County Strategic Plan for the Protection of Livestock and Wildlife. [*buffalo; wildlife management; wildlife management; Wildlife Conservation Society; culling; wildlife refuge; multiple use policy*]

animal law

Animal protection laws, anti-cruelty laws, or the *Endangered Species Act*, used to defend animals in court. In these cases, you can think of the animal as the defendant in the case. Many of these laws can carry felony convictions. Because the animals are obviously not paying a lawyer, the attorney fees are typically paid by an animal protection *NGO* or are possibly NGO staff attorneys. Many people want animal law to become mainstream, similar to *environmental law*. [*farm animal cruelty legislation; greyhound dogs; circus elephant ban; cosmetics, cruelty-free; vivisection; pollution police; law and the environment*]

animal manure

Animal manure is used as an *organic fertilizer*. It adds *nitrogen* to the soil, improves *soil texture*, and encourages the growth of beneficial *soil organisms*. Animal manure use has diminished in the United States because animals are no longer raised on the same farms that grow crops and transporting animal manure long distances is expensive. Because farms no longer use animal manure or other organic fertilizers, they depend on *synthetic fertilizer*, which does not improve soil texture and has a host of environmental problems associated with it. [*biogas; CAFO; composting, large-scale; green manure; night soil - biosolids*]

animal rights movement and animal welfare

The animal rights movement believes people should protect all non-human life from human exploitation and abuse. The movement, as most typically defined, believes animals should be treated humanely and not subjected to pain, suffering, or death at the hands of humans. This includes the killing of domesticated animals such as livestock. The movement began in the 19th century and continues today by many groups, but most notably PETA (People for the Ethical Treatment of Animals).

Animal welfare, as opposed to animal rights, only pertains to the humane treatment of animals. For example, animal welfare is not against the use of livestock for food but requires humane slaughter of the animals so they do not suffer. [*cosmetics, cruelty-free; farm animal cruelty legislation; primate research; vivisection; circus elephant ban; sealing; zoos; illegal trade and trafficking; poaching; World Wildlife Fund; Xerces Society; NGOs; biodiversity, loss of; greyhound dogs*] {leapingbunny.org}

animals experiencing grief

In the past, the scientific consensus has been that only humans grieve. Any thought that animals do so was considered nothing more than *anthropocentric* sentimentality and unscientific. Recent research is beginning to show that possibly animals, other than humans might express grief, including *whales*, *dolphins*, great apes such as *gorillas*, *elephants*, and even some farm animals—and, as many of us have always suspected, some of our pets. The research has focused on emotions being elicited when a relative or close companion of the animal dies. [*sixth sense in animals; smell sensitivity in albatross; stereoscopic smell in moles; bumblebee senses; camouflage, squid; communications between plants; hummingbird and flight; whistle names, bottlenose dolphin*]

annual plants

Plants that live their entire lives, from germination through seed production and death, in one year. [*perennial plants; succession, primary terrestrial; pollination*]

Antarctica

Antarctica is the coldest, windiest, and highest continent on the planet, encompassing ten percent of the Earth's land surface. It is covered in ice sheets averaging 6500 feet thick but reaching more than 15,000 feet. It contains about 75 percent of all the world's *freshwater*. Temperatures can drop to –128 degrees F and winds reach 180 miles per hour.

One hundred million birds breed each year in Antarctica and 100 species of fish and *mammals*—such as porpoises, *dolphins,* and *whales*—live in its harsh *habitat*. Although not usually thought of as dry, it receives less precipitation than the *Sahara desert*. (Its waters are locked up in the frozen ice.) Antarctica is considered by many to be the last great *wilderness* on the planet. [*migration; flyways; arctic; Antarctic Treaty; Arctic ice, changes in; Third Pole, The*]

Antarctic Treaty

This was the first attempt to protect the *Antarctic*, enacted in 1961 and amended in 1991 (when it became commonly called the Madrid Protocol). The treaty promoted scientific research and attempted to maintain peaceful relations among all parties active in the huge region. The Madrid Protocol suspends oil exploration in the region for 50 years.

The treaty includes three documents: the Protocol on Environmental Protection to the Antarctic Treaty, the Convention on the Conservation of Antarctic Seals, and the Convention on the Conservation of Antarctic Marine Living Resources.

The debate will be ongoing over whether to use this fragile, pristine *ecosystem* as an international world park, as some have suggested, or to extract its vast resources. But the Madrid Protocol took a huge step forward in attempting to keep the region an environmentally safe haven. [*oil recovery or exploration; arctic drilling*]

anthromes

Also called anthropogenic biomes, this is a new way to study *biomes*. Most *ecology* textbooks teach about traditional biomes or land zones such as *tropical rainforests*, deciduous *forests*, *taiga*, *tundra*, and *deserts*. However, if you take a look at some of these regions today, they look nothing like a traditional textbook biome. Today, much of the land consists of cropland where we grow our crops, rangeland where we raise our livestock, and urban areas where we live. Some of these regions have remained the same, such as deserts and taiga, but in general, biomes are now thought of as the traditional classification in need of a more modern classification—now called anthromes.

For example, some areas traditionally classified as temperate deciduous forest biomes are now classified as *urban* or rice village anthromes because they are inhabited primarily by cities or *rice paddies*, respectively.

Some of the new classifications are residential irrigated cropland, residential rain-fed cropland, remote cropland, urban, and mixed settlement. Still others include the term mosaic, such as woodland mosaic, indicating that only small portions of the original habitat (woodland) remain and are scattered within a larger anthrome, such as urban. This is called *habitat fragmentation*.

anthropocene geological epoch

Geologists categorize periods of the Earth's geological past as epochs. There have been 41 official geological epochs, the latest being the Holocene Epoch, which includes time from the end of the last *Ice Age*—about 12,000 years ago—to today. This is the epoch when humans built civilizations and we came to be what we are today. It is considered a period of stability that provided an opportunity for our species to evolve and thrive.

Some scientists think we should consider officially assigning a new epoch for today—the Anthropocene Epoch, meaning the Age of Man. The idea is that our planet, from the *Industrial Revolution* onward, is today manipulated by man in almost every way and therefore, worthy of a new designation.

The eco-theologian *Thomas Berry* provided an excellent argument for why this designation is appropriate. (He professed that we were entering a new Ecozoic Era, which is a much longer period of time than an epoch, but his logic is the same.) He stated, "What is happening was unthinkable in ages gone by. We now control forces that once controlled us, or, more precisely, the Earth process that formerly administered the Earth directly is now accomplishing this task in and through the human as its conscious agent. Once a creature of earthly providence, we are now extensively in control of this providence."

Others believe humans are a passing stage and our impact will eventually, at least in *geological time scales*, disappear. [*natural selection*]

anthropocentric

Interpreting the actions of organisms in terms of human values. For example, "The bird must be disappointed that it didn't catch the bug." It is also a belief that people are more important than all other forms of life and that other life is subservient to people; if something exists, it must be there for our use. This contrasts the *biocentric* belief. [*ethics, environmental*]

anthropogenic stress

The effect human intervention has on other organisms and the *environment*. Although interaction between organisms is universal, the impact of human intervention is usually extraordinary in scope and magnitude. We inflict uniquely human-caused stress on the planet. The most common use of this phrase is about anthropogenic *greenhouse gases*, the gases—mostly *carbon dioxide*—that human activities emit as opposed to the ones emitted by nature.

But the word can be used for any human-induced stress such as those that cause *air pollution*, *habitat destruction*, *deforestation*, and *climate change*.

anthropologists, environmental

Scientists who study the importance of culture, religion, society, and *ecology* in combination. [*philosophers and writers, environmental; political ecologist; sociobiology*]

antibiotics and the human microbiome

Antibiotics are one of the greatest medical advances of all time and have saved untold millions. However, they are not without risks, and these risks are gaining attention. Recent studies about the *human microbiome* have helped clarify the negative impact antibiotics can have on our bodies. Broad spectrum antibiotics—those that kill a wide range of *bacteria*—can be compared to *pesticides* for the body. As *Rachel Carson* said, pesticides should be called *biocides*, because they don't just kill pests; they kill all types of organisms. Broad spectrum antibiotics not only wipe out bad bacteria but also good bacteria; and we are just now learning how important they are to our well-being. [*antibiotics for livestock*]

antibiotics for livestock

About 80 percent of all antibiotics sold in the United States are not used on humans—they are not even used on any organism that is sick. Instead, they are used on *factory farms* and *CAFOs* for animals that are not sick. Antibiotics are routinely used because these animals are reared in a manner that often makes them ill if antibiotics are not administered. Wide use of preventive antibiotics has resulted in new antibiotic-resistant strains of *bacteria* often called *superbugs,* which are becoming a major health concern. Denmark has banned most antibiotics from factory farms. The practice began in the 1950s, when farmers began placing livestock in confined, crowded, unsanitary conditions conducive to disease.

The growth of these superbugs is considered by many to be out of control. The number of resistant bacteria (superbugs) found on poultry, for example, has increased tenfold in just the past few years.

Attempts have been made to control this use to protect the nation's public health, but the *Food and Drug Administration*—which regulates drugs—appears more concerned about protecting the livestock industry than protecting humans. The FDA believes the industry can self-regulate themselves. The results of self-regulation has been basically. . .nothing. A number of bills have been presented in Congress to get tough on limiting antibiotic use on livestock. However, at the close of 2013, it appears that some meaningful action might actually take place.

In addition to this problem is a lack of available data. Many people in the medical profession are concerned that there are no requirements for these factory farms to report their antibiotic use. This leaves researchers with little data to study—something the industry is perfectly happy about. [*antibiotics and the human microbiome*]

Antiquities Act

A U.S. federal law enacted in 1906 and formally known as "An Act for the Preservation of American Antiquities."

It was signed into law by President *Theodore Roosevelt* and gives the President authority to protect lands by placing them under control of the federal government by executive order—completely bypassing input from Congress. For this reason, its use is often controversial. It was used with great flourish first by Theodore Roosevelt and is still used to this day by Barack Obama.

The Act was originally intended to let a President protect certain valuable public natural areas to be designated as U.S. National Monuments. This is opposed to the more formal process of getting Congressional approval, which results in the creation of a U.S. *National Park*. However, more recently it has been used to designate both monuments and parks. (A park's primary purpose is to set aside a large tract of land for scenic or recreational purposes, while a monument is for historical, archaeological, or scientific value and can be of any size.)

The Act was first used by Theodore Roosevelt in 1906 to designate Devils Tower National Monument in Wyoming and has been used many times by all but three presidents (*Nixon*, Reagan, and George H.W. Bush). The *Grand Canyon* and the Statue of Liberty became landmarks in this way. In 2013, President Obama—frustrated with getting approval from Congress—designated five new national monuments with the Antiquities Act.

The Act is used as part of a very informal method of rating how environmentally friendly presidents have been throughout the years. The method was created by

Bruce Babbitt, former President Clinton's Interior Secretary. It compares two values. The amount of land set aside and protected by a president, using measures such as the Antiquities Act, is compared to how much land a president opened up to fossil fuel development. For example, President George W. Bush and President Clinton both "protected" roughly one acre for each acre "opened" to development. but President Obama so far is running at about one acre protected to about 2.5 for development. [*Arctic National Wildlife Refuge; Biosphere Reserves; Blueways, National; Grand Canyon; Great Barrier Reef; Yellowstone National Park; Yosemite National Park; Nature Conservancy; national parks, global*]

ant slaves

In some New England states you can find two species of ants with a strange relationship. In summer, the "master" ant, from the genus Protomognathus, attacks the brood of the "slave" ants, of the genus Temnothorax, and brings the eggs back to their own nest. The slave ants hatch and spend their lives catering to their masters. They are responsible for feeding and caring for the master's brood (young) throughout their lives.

Recent research shows that the slave ants retaliate; they often stop feeding or cleaning the young master ants and occasionally even attack and dismember them. Researchers found that mortality of the young masters with slave care is about 60 percent, but in colonies without slave care, only about 20 percent of their young die. [*dulosis; predator/prey relationships; symbiotic relationships; decapitating flies; follicular mites; sixth sense in animals; zombie bees*]

apex consumers and ecosystems

If you look at an *energy pyramid*, it is composed of layers, called *trophic levels*. The *producers* form the wide bottom layer. As you go up the pyramid, the narrowing layers consist of the primary *consumers*, secondary consumers, and so on, until you get to the top point of the pyramid, where apex consumers are found. These might be lions, sharks, eagles, or others, depending on the *ecosystem*.

Until recently, the scientific consensus had been that the bottom trophic-layer species play the most important role in creating a stable ecosystem. Some scientists have now come to believe apex consumers might play the most important role in an ecosystem and therefore *conservationists* should spend time protecting these species rather than those at lower levels. If this is correct, ecosystems might be in more trouble than we thought, because apex consumers have been the most affected by human intervention and are the most *endangered species*. [*predator/prey relationships; flagship species; biodiversity, loss of*]

aphotic zone

That portion of the ocean that receives no light. [*oceanic zone ecosystems; marine ecosystems*]

aposematic coloration

Coloration or structures on an organism that warn a *predator* of danger. For example, the distinct stripe on a skunk is aposematic. [*mimicry; directive coloration*]

Appalachian Trail

The Appalachian Trail is considered by many to be the most famous hiking trail in the United States. It runs almost 2200 miles from Maine to Georgia. It is on federal or state protected lands or along right-of-ways for more than 99 percent of its length. Annually, more than 4000 volunteers maintain the trail. The trail follows the ridge-line of the Appalachian Mountains and extends, with few exceptions, almost entirely through *wilderness*.

Stories have been written about this trail and the experiences people have gained along the way. The popular author Bill Bryson wrote a funny, entertaining book about two guys trekking this trail, called <u>A Walk in the Woods</u>. [*National Park and Wilderness Preservation System; outdoor recreation; Blueways, National; green-ways; Outward Bound USA; Student Conservation Association*]

apples, antibiotics in organic

If you buy *organic* apples or pears, you would expect them to be *antibiotic* free, because that is part of the definition of organic produce. However, you would be wrong. Apples and pears are susceptible to a bacterial infection called fire blight that can quickly wipe out the fruit and even kill entire trees. Two antibiotics kill this blight, saving the tree and making everyone happy—except the *consumer*.

In 2002, when *organic labeling* was just getting started, the *Food and Drug Administration* gave an exclusion for these two antibiotics for use on apples and pears. The exclusion would be good until some form of organic control of the blight could be established. That's why it's most likely on your apples and pears now. This exclusion expires in 2014, but because no one has come up with an alternative that does not require these antibiotics, organic growers are asking for an extension of the exclusion. [*organic farming and food*]

appliance recycling

There are many appliance recycling centers in the United States, Canada, and else-where. They disassemble old refrigerators and recycle various parts and substances.

The coolants are siphoned and stored, and metal and glass are separated and shipped to the appropriate recycling centers. *Toxic substances* are sent to special *incineration* centers.

Most are privately owned and contracted by utility companies that want to reduce the number of old, inefficient refrigerators still plugged-in and drawing power. It also fosters good will from their customers. [*refrigerator recycling; refrigerator doors; energy, ways to save residential; automobile recycling; mattresses, recycling; cigarette butt litter and recycling; downcycling; upcycling; motor oil recycling; paper recycling; recycling metals; plastic recycling; e-waste recycling*]

applied ecology

A division of *ecology* that deals with environmental problems directly affecting our society. This discipline tries to identify existing and potential problems, separate them from imagined or unfounded problems, and offer possible solutions. Applied ecology presents facts and theories that are the tools to protect and possibly save our planet from ourselves. [*anthropologist, environmental; cost/benefit analysis; risk assessment, environmental; risk management, environmental*]

apps, eco- (software applications)

Countless software applications have an environmental twist. Ranging from simple to sophisticated, some apps simply enhance printed material by using multimedia. For example, there are apps that provide the most up-to-date information about *national parks,* with pictures, videos, and sounds. Other apps identify birds by sight and sound and also play certain calls to attract birds. Some apps use GPS to locate things such as the closest *Superfund* site or indicate *air pollution* factors such as *ozone* and *particulate matter* in your location. Still others link your home's electricity rates with your usage patterns, all provided by your utility company, perhaps explaining what all those *watts and kilowatt hours* on your bill mean and showing you how to save money, in real time.

At the more hi-tech end are eco-apps that work with *biosensors*. Once available, they will monitor the environment—for example, air quality—and your vital signs—for example, respiration rate—to see if the air might be causing your body stress and warning you if so. As the saying goes, there's an app for that. [*biotechnology; best available technology; geoengineering; blogs, green; microblogs, green; crowd sourcing, eco-; online petition sites and environmentalism; robotic bees; social media, environmental*]

aquaculture

The cultivation of fish, shellfish, or aquatic plants in controlled *marine* or *freshwater* environments. Only ten years ago, the quantity of *seafood* supplied by aquaculture was negligible. However, in 2013, people consumed more seafood raised via aquaculture than wild caught for the first time.

Aquaculture has been practiced since ancient times, but only recently has *overfishing* caused a dramatic decline in the availability of wild seafood, resulting in an economic necessity to grow and harvest seafood artificially. Most of the aquaculture harvest pertains to fish and shellfish, but plants are also important. Seaweed, a type of *algae,* is especially important in Asian countries, where it is a popular food.

Many see aquaculture as the best way to feed billions of people in coming decades. As farmable land becomes scarce, we are turning to the seas that cover two-thirds of our planet as an almost limitless habitat to cultivate food.

As aquaculture becomes more prevalent, it is beginning to cause the same environmental problems found with other forms of agriculture, such as *nutrient enrichment.* Uneaten food and wastes that collect in the water cause *cultural eutrophication,* with all of its accompanying problems. Like livestock in *CAFO*, fish in closed, dense populations are less resistant to disease, and serious outbreaks have wiped out some aquaculture farms. Finally, just like livestock, aquaculture farmers use large amounts of *antibiotics* on these fish in efforts to prevent disease rather than cure it. The threats of *superbugs*, resistant to antibiotics, resulted in the Chilean government banning antibiotics from fish farms in 2010.

Efforts have begun to make aquaculture more sustainable by reintroducing practices such as raising many different species on the same farm (called *integrated multi-trophic aquaculture*). For example, many rice farmers also raise fish and even some aquatic plants in the same *rice paddy*, similar to efforts to make land-based farms more sustainable. [*shrimp farms, aquaculture; tuna aquaculture; hydroponic aquaculture; fish populations and food; fishing, commercial; forage fish; genetically modified organisms-basics about; aquaculture-producing countries, ten biggest; Aquaculture Stewardship Alliance; biofuels, second-generation; vertical farms*]

aquaculture-producing countries, ten biggest

The largest *aquaculture* producers in the world are *China*, Japan, *India*, Chile, Vietnam, *Indonesia*, Thailand, *Bangladesh*, Korea, and the Philippines. [*Aquaculture Stewardship Alliance*]

Aquaculture Stewardship Alliance

This *NGO* was founded in 2009 by the *World Wildlife Fund* and the Dutch Sustainable Trade Initiative to develop and manage global standards for responsible *aquaculture*. They work with aquaculture producers, processors, retail and foodservice companies, scientists, and conservation groups to promote responsibly farmed seafood. The

Alliance manages an *eco-certification* and an *eco-label* program for specific species, including salmon, tilapia, and some shellfish. These standards address issues such as natural *habitat conservation, water* quality, and the fish feed used in these farms. [*aquaculture-producing countries, ten biggest*] {asc-aqua.org}

aquatic ecosystems

About 71 percent of our planet's surface is covered with water, providing *habitats* for many aquatic *ecosystems*. Five factors dictate what kind of ecosystem can exist in water: 1) salinity (the concentration of dissolved salts), 2) depth of sunlight penetration, 3) amount of dissolved oxygen, 4) availability of *nutrients*, and 5) water temperature. Bodies of water with high concentrations of dissolved salts are called *marine ecosystems*, and those with low levels are *freshwater ecosystems*.

aquifers

More than 90 percent of the Earth's *freshwater* is stored in underground aquifers. This water is commonly called *groundwater*. Aquifers are not actual bodies of water but are large areas of permeable rock, gravel, or *sand* saturated with water much like a soaked sponge. They provide drinking water to about two billion people worldwide and irrigate about 40 percent of the world's *agriculture*. In the United States, about half of all drinking and *irrigation* water comes from underground aquifers. Aquifers can cover a few square miles or up to thousands of square miles, such as the *Ogallala aquifer* that stretches from South Dakota to Texas.

There are two types of aquifers: unconfined and confined. Unconfined aquifers are located near the surface and are replenished with water directly from the surface in a process called *infiltration*. The water in these aquifers is not under pressure, so a well requires pumps to draw the water up.

Confined aquifers are found deeper in the soil and have an impermeable layer of rock above, so water cannot simply infiltrate down to them. Instead, water can only enter in some areas where the impermeable layer does not block access, called recharge areas. These recharge areas can be miles away from portions of the aquifer. Because water within confined aquifers is enclosed, it is under pressure and moves slowly, usually just a few inches per day. Wells that tap confined aquifers draw water by using the natural pressure that exists within the aquifer.

Both the quantity and quality of groundwater in aquifers have been affected by humans, resulting in *groundwater pollution and depletion*. The amount of groundwater extracted in the United States jumped from 30 billion to more than 85 billion gallons per day between 1950 and 2013. In 35 out of the 48 contiguous states, more groundwater is being extracted than can be naturally replenished. This depletion, called *groundwater mining*, has resulted in water shortages in many portions of the country.

Pesticides, *fertilizers*, leaking *septic tanks*, and *toxic waste* from *landfills* are contaminating aquifers across the country. Numerous abandoned *hazardous waste* disposal sites and the practice of injecting *toxic wastes* into deep underground wells are also contaminating aquifers.

Other problems also exist. For example, aquifers near coasts have a problem called *saltwater intrusion,* and groundwater mining can result in the land settling, called subsidence, or even collapsing and causing *sinkholes*. [*water shortage, global; land grabs*]

Aral Sea

Water diversion redirects water from one area to another, usually for the purpose of irrigating crops. The Aral Sea, in Soviet Central Asia, was the fourth largest freshwater lake in the world until the 1920s. That's when the Soviet government decided to divert the lake and its incoming rivers for *irrigation* to increase the country's cotton crop and later for *hydroelectric power*. An irrigation canal diverted the water 800 miles away to farmland.

During the late 1930s, the cotton crop was a success, but the Aral Sea was doomed. The Aral Sea shrunk to one-tenth its original size. Where shoreline towns were, now dry wastelands exist 25 miles from the water's edge. In 1987, the water level had dropped so low that a barrier of land appeared, dividing the sea into three smaller lakes. The fishing trade was minimal, and so were the crops that once grew alongside the banks. The catch that once totaled about 50,000 tons a year yielded only about 50 tons by the late 1980s. The remaining water continued to become more polluted and more saline.

In the early 1990s, after the Soviet Union's demise, the newly formed state nations that surround the Aral Sea basin signed a series of agreements to manage and protect the area. In 2005 a dike was created in the northern end of the lake to increase its depth. That portion of the lake has made an amazing recovery and now supports about 18,000 tons of catch annually.

The Aral Sea is an example of how unplanned growth and development can result in ecological disaster. Some use this as a preview of what could happen if the *Great Lakes* are ever used to divert water to the American west, where water shortages get worse each year. [*water shortages, global; land grabs; waterlogging and soil salinization; comprehensive water management; Kenya water find; Colorado River*]

Arbor Day Foundation

This *NGO* concentrates on *conservation* and *environmental education* pertaining to trees and *arboriculture*. It was founded in 1972, a hundred years after the first Arbor Day observance in the 19th century. The Foundation is the largest nonprofit membership organization dedicated to planting trees. They have more than one million members and programs that include Tree City, Tree Campus, Tree Line, Replant Our National Forests, Rain Forest Rescue, Community Tree Recovery, and a huge Volunteer Center Program. [*forests; deforestation; afforestation; reforestation*] {arborday.org}

arboricide

A chemical that kills trees. [*herbicides; pesticides*]

arboriculture

The cultivation of trees. [*forests; deforestation; Arbor Day Foundation; afforestation; reforestation; deforestation and climate change*]

archeologists, ice-patch

Archeologists study how humans lived in the distant past. They do so primarily by searching for artifacts, such as early stone tools or vessels.

Ice-patch archeologist is an informal name given to a new type of archeologist studying something that could only recently be studied. As ice sheets melt across the globe as a result of *climate change*, large expanses of land are appearing that have not seen the light of day for thousands of years. Finding new regions (patches) to look for artifacts is an exciting opportunity for archeologists, and some have made this their specialty. In 2010, what is believed to be the oldest such artifact found under these conditions was lying on the ground near *Yellowstone National Park*. It is believed to be a 10,400-year-old wooden dart shaft. It even has marks left by its Stone Age maker. [*Arctic ice, changes in; glacial lake outbursts; glaciers, mountain; melt ponds; permafrost, Siberian; Petermann Glacier ice detachment*]

arctic

The arctic is the region surrounding the North Pole and consists of *ocean*, polar ice, and *tundra*. The Arctic Circle, which demarks the region, is roughly the area where *forests* end and tundra begins. The following countries claim territory within this region: Canada, *Greenland* (Denmark), Russia, the United States, Iceland, Norway, Sweden, and Finland.

The first pollution spotted in the arctic was back in the 1950s and was called arctic haze. This was later found to be *aerosols* of sulfate and *black carbon* particles that traveled from industries in Europe and Russia. Animals living in the arctic have *pesticides* and other toxic substances such as *PCBs* in their bodies via *bioaccumulation*, demonstrating how no part of our planet is safe from environmental problems.

The first environmental protection proposed for this area was initiated in the early 1970s, with an agreement to protect polar bears. During the 1980s, many transnational treaties were signed to protect the arctic environment through the auspices of an *NGO* called the International Arctic Science Committee, with members from eight countries. By the 1990s the *Arctic Council* was formed to promote *sustainable development* in the arctic and assess changes in the *arctic ice* sheets as a result of *climate change*. [*Arctic Council; arctic drilling*]

Arctic Council

A group of eight nations that, in 1996, got together to consider how to manage the region within the *Arctic* Circle. Their original objective was to promote *sustainable development* within the region. They also help guide research and conservation efforts. In 2004, they published a report called the Arctic Climate Impact Assessment that predicts dire impact to the region as a result of *climate change*. One of their most recent actions was to sign a binding agreement on how to respond to *oil spills* in the region. Member states are Russia, Norway, Iceland, Denmark, Canada, the United States, Sweden, and Finland.

Only states that have territory within the region can have full member status, but others have been allowed to join with permanent observer status. There has been a rush to join this club now that melting ice caps have opened up the area to shipping, fishing, and oil and gas exploration. Twenty-six nations are on permanent observer status. [*arctic drilling; arctic ice, changes in; Arctic National Wildlife Refuge*]

arctic drilling

The Alaskan Outer Continental Shelf lies just north of Alaska, within the *Arctic* Circle. It is an untouched, beautiful region, but appears poised for change. The U.S. government is considering whether to let oil and gas companies begin exploration in the region. Many experts believe drilling in the arctic will be far more dangerous and difficult than in the Gulf of Mexico.

Reports clearly state that the *ecosystem* would suffer from normal operations but be devastated by an *oil spill* on or near the ice. These reports repeatedly state that more information is needed to know what the impacts might be. But even with these warnings, U.S. plans have moved ahead, albeit with starts and stops.

A battle has been raging over this proposal. The Coast Guard Commandant at one point stated the country was woefully unprepared to respond to a major spill in the arctic. The Interior Secretary, however, said he believed company claims that, in event of a disaster, a new oil-spill containment device could clean up at least 90 percent of any oil spilled. But these statements were overly optimistic.

In 2012, Shell started test drilling, with the disastrous result that their drilling rig, the Kulluk, broke free and ran aground. Also, an experimental containment dome was damaged during testing. As a result of these failures, the Interior Secretary changed his tune and put future arctic drilling on hold. However, oil companies are determined to begin drilling at some point in the future, and the government has a history of accommodating them. [*oil recovery or extraction; oil spills in history, oil spills in U.S.; pipeline spills; Arctic National Wildlife Refuge*]

Arctic ice, changes in

The *Arctic* ice is found in three types of formations: ice over water, ice sheets over large tracts of land, and *glaciers.*

Climate change is warming the region at twice the pace as the rest of the planet, causing the Arctic ice to recede. Arctic ice sheets normally recede each summer, but in recent years, the ice has been doing so for a much greater distance for a longer time. During the summer of 2012, it receded to the smallest coverage ever recorded; so much so it even surprised scientists studying the region.

And the ice is receding earlier in the year, as well. A couple of years ago, most scientists thought the summer sheet ice would be totally gone by the end of this century. However, now a large group of scientists believe summer ice will be nonexistent before 2040 if things continue on the same path. The sheet ice coverage now is about half of what it was in the 1960s.

In addition, glaciers over Greenland are retreating and have dropped in area by about 20 percent since the 1960s. The land beneath that is *permafrost*, is also shrinking in size.

These changes have major environmental implications. Increased *sea levels* will impact millions of people and make some island nations uninhabitable. *Biodiversity* will suffer at both ends of *food chains*. At the bottom, the copepods that make up much of the *zooplankton* and the *algae* that make up most of the *phytoplankton* will suffer because water from melting doesn't mix well with sea water, impacting both organisms' life cycles. At the top of the food chain, *polar bears* and walruses depend on the ice to survive.

The economic implications are also large because it would open shipping lanes for longer periods of time, such as the Northern Sea Route that cuts a trip from *China* to *Germany* by a third (about 4000 miles), as well as shortening the Northwest Passage.

These improved sea routes will most likely hasten offshore drilling in an environment that can ill afford it.

Many scientists believe the Arctic could reach a *tipping point* that would accelerate the melting even further. This tipping point is based on the fact that ice reflects about 85 percent of sunlight. If the sunlight is reflected back into space, it doesn't get a chance to warm up the region. As snow and ice melt, the ice becomes less reflective, and when it's gone, the remaining water and the ground below reflect far less sunlight. Less reflected light means more sunlight hitting the land and warming it up. This becomes a positive feedback loop—or what some would simply call a vicious cycle.

The only reason more people are not shocked at what is happening in the Arctic is that so few people live there and see it occurring. [*melt ponds; glacial lake outbursts; glaciers, mountain; melt ponds; permafrost, Siberian; Petermann Glacier ice detachment; Third Pole, The*]

Arctic National Wildlife Refuge

A 19-million-acre refuge located in northeast Alaska and containing vast numbers of wildlife, including more than 200 animal species. Millions of birds nest and breed in this refuge annually. It is visited each year by a herd of more than 180,000 caribou that use it as a calving ground. It was protected in 1980 and is the most northern park run by the *U.S. National Park Service.*

A portion of this refuge, called the northern slope, has been open to resource extraction since the park's inception. However, there is an ongoing battle to protect a remaining portion of the refuge, called the coast plain. Much to the dismay of environmentalists, Congress has been considering, since the 1990s, to open the coast plain portion of this refuge to oil exploration and development. Even though the battle goes back and forth in the United States, the Canadian government made its decision a while ago by protecting its entire portion of the coastal plain. [*arctic; arctic drilling; oil recovery or extraction; oil spills in history, oil spills in U.S.*]

Argentina

Argentina contains the Patagonian desert (fifth largest in the world), the Andes Mountains, subtropical *rainforests,* and scrubland. The nation cultivates a great deal of *agriculture* and is a major exporter of *oil, natural gas,* and such metals as gold, zinc, and copper. The country has many environmental *NGOs,* including a *Greenpeace* office; The Wildlife Foundation, which helps protect the country's *biodiversity;* and the *Environmental Defense Fund.* The government includes a federal agency called the Federal Council of the Environment, with the mission of protecting their environment.

To their credit, Argentina has 29 *national parks*, 13 UNESCO *World Biosphere Reserves,* and eight *World Heritage* sites.

argumentum ad ignorantiam (argument from ignorance)

Also translated to mean "appeal to ignorance," this refers to an illogical reasoning many people use when they don't know or understand the facts about something. If no explanation is satisfactory, they assume something else, no matter how outlandish it might be, must be true—even if there are no facts to back it up. This logic, or illogic, is how many conspiracy theories originate, how aliens are responsible for many things we don't understand, and countless other things we simply do not have scientific answers for—yet.

The expression was first used by John Locke, but more contemporaries have helped clarify the meaning. Carl Sagan's belief was that "science is saying, in the absence of evidence, we must withhold judgment." Bertrand Russell said, "If you can't find out whether [a thing] is true or whether it isn't, you should suspend judgment." The best statement might be from Neil deGrasse Tyson, who said, while speaking about UFOs, "Remember what the U stands for. . . Unidentified. . . . If you don't know what it is, that's where your conversation should stop, [rather] than say: It must be. . . anything. That's what argument from ignorance is." [*Agenda 21 conspiracy; climate change deniers; environmental literacy; environmental education; pig-gate; science; pseudoscience; opinions and science*]

arid

Pertains to *habitats* that receive less than 25 cm (10 inches) of precipitation annually and the evaporation exceeds the amount of precipitation. For example, a desert is an arid *biome*. [*desert; desertification; anthrome*]

artesian springs

Most standing bodies of water, such as ponds and lakes, are fed by water supplied from *surface waters* such as *rivers* and streams. However, some get their water from natural underground *freshwater* springs that bubble to the surface and create the ponds and lakes. These are called artesian springs. In many cases, you would only know if the body of water is fed by an artesian spring if you took a trip all around the lake and saw no streams or other incoming surface water.

Some places—Florida, for example—have numerous artesian springs that, like most *freshwater ecosystems*, are in environmental jeopardy. Natural drought in the region is partly to blame, but much of the problem comes from overconsumption of the *groundwater aquifers* that feed the springs. The water is pumped to supply the long-running population boom in Florida and their ever-present sprinklers as well as *irrigating agriculture*. The result is a dropping water table in the groundwater aquifers, a process called *groundwater mining*.

In addition, the water remaining in the springs is often affected by *cultural eutrophication, nutrient enrichment* that causes *algal blooms*, and the loss of beautiful *ecosystems* and their accompanying *biodiversity*.

artesian wells

Wells that do not require a pump to draw water to the surface. Artesian well water comes from *confined aquifers*, which are under natural pressure. [*drinking water; water conservation; water grabs; artesian springs*]

asbestos

Asbestos has been used since ancient times for its strength, flexibility, and fire resistance. It is also waterproof and sound resistant. All of these advantages made it a popular building material and insulation during the 1950s through the 1980s. And contrary to popular belief, it is not banned in the United States and is still commonly used, although in a slightly different form.

Asbestos is dangerous when fibers are released and become airborne. That is why old asbestos insulation, which deteriorates, poses a serious health threat. Cutting, scraping, or sanding materials containing asbestos also releases these fibers and poses a danger. Asbestos fibers, once inhaled, become lodged in the lungs—probably for life. Asbestos is called the "silent killer" because it remains in the lungs for decades before causing disease. The *Environmental Protection Agency* (EPA) declared the substance a *hazardous air pollutant* as far back as the 1970s. These fibers have been conclusively linked to scarring lung tissue and mesothelioma, a rare lung cancer.

After all of this information, you would think the product is no longer used. Although the EPA initiated a ban on the material during the 1980s, the manufacturers successfully challenged the ban in the courts and it was overturned. Today, asbestos is used in about 3000 products in the United States, including home construction materials such as insulation and flooring, as well as in the textile industry and in *automobile* parts such as brake linings.

Most of today's products contain a form of asbestos called chrysotile, which the industry considers to be less dangerous than the old form, called amphibole.

Asbestos removal is a job for professionals and should not be attempted as a do-it-yourself project. Most states require asbestos-removal professionals to be licensed. Your state environmental protection agency could refer you to licensed professionals.

[persistent organic pollutants; toxic pollution; couches and indoor pollution; indoor air pollution; indoor ecology]

aseptic containers

Aseptic containers, also called juice or drink boxes, are popular alternatives to conventional beverage containers. They have many advantages for consumers, but most *environmentalists* feel they also have disadvantages. As advantages, they are sterile, resist breakage, and don't have to be refrigerated. They have very little superfluous *product packaging* material, so they minimize the amount of waste produced. A full container is composed of only 4 percent packaging materials and 96 percent drink. That is about as good as a packaged product gets.

The major problem with these products is that they are not readily recyclable. These boxes contain layers of paper (70 percent), *polyethylene plastic* (24 percent), and *aluminum* (6 percent). *Recycling* programs for these boxes are few, and establishing markets for the recycled materials is difficult because the material cannot be reused to make new aseptic containers. A few recycling programs *downcycle* them into paper products such as writing paper.

There is also concern that this product harms advances made in *recycling glass* and *plastic* beverage bottles. The market for these recycled materials already exists because they can be used to remanufacture more beverage bottles and many other products.

asexual reproduction

Reproduction that involves only one parent and does not involve sex cells, as opposed to *sexual reproduction*. Asexual reproduction can occur by simple division, called fission, which occurs with *bacteria*. *Fungi*, however, reproduce asexually by producing spores. [*pollination; hermaphrodites; parthenogenesis; traumatic insemination*]

ash, fly and bottom

Ash is a pollutant composed of *particulate matter* that forms during high-temperature combustion as in an incinerator. When it goes up the chimney it is called *fly ash*. When it settles at the bottom of the combustion chamber, it is called *bottom ash*. [*incineration; incineration problems; municipal solid waste disposal*]

ash tree demise

Ash trees are used for baseball bats, hardwood flooring, tool handles, kitchen cabinets, plus many other products. Today, the ash is under attack. In 2002, the Emerald Ash Borer— a pretty, iridescent-green *beetle*—was discovered in Michigan. It probably arrived on U.S. shores onboard a ship from Asia that docked in Detroit. It has already killed tens of millions of trees in Canada and 17 U.S. states.

The beetle eggs are laid in the tree's bark and the larvae consume the wood, destroying its ability to transport water and nutrients and killing the tree. The beetles can fly but cover far greater distances within firewood as it is transported on trucks. This is why many states now have bans on bringing firewood in from other states. Research is ongoing to find ways to stop the threat, including the potential use of natural *predators* and other *biological control* measures. [*invasive species; forests; timber, DNA sequencing of; coast redwood*]

Asian carp

One of our most notorious *invasive species,* these fish have become famous because of their habit of jumping as much as 20 feet up and out of the water when agitated by boat noise. Even though they appear almost comical as they leap, they could be dangerous if they make a direct hit on someone in the boat. But this jumping feat is not what makes them problematic as an invasive species.

Asian carp eat *phytoplankton* and other *algae* in such vast quantities as to wreak havoc with local *ecosystems.* These fish typically grow to 40 pounds but can grow to 100 pounds. They spawn many times a year, starting at only a couple of years of age and continuing until their death, which can be up to 20 years. They simply overwhelm the *habitat*, out-eat the native fish, and destroy both the *ecosystem* and the local economies.

Of seven Asian carp species now found in U.S. waters, four are especially problematic. They are the bighead, silver, black, and grass carp and all belong to the minnow family. They were imported from southeast Asia to the United States in the 1970s by catfish farmers who thought they would keep fish ponds clean. But flooding during the 1990s released many into the Mississippi river and the rest is history. Young fish have a few *predators,* but large adults have none other than man.

They have invaded the northern Mississippi River and are threatening the *Great Lakes.* Many efforts are underway to prevent the fish from making their way into the Great Lake system where the seven-billion-dollar commercial sport-fishing industry would be threatened. Electric barriers on the Illinois River, a controversial poisoning of a six-mile stretch of the river near Chicago in 2009 that killed 90 tons of fish—only a few of which were carp, and many other attempts have failed. Future efforts at control include sound barriers, *pheromones*, and bubble barriers. Even a complete permanent basin separation has been discussed.

Some environmental groups and a few states have taken the U.S. Army Corp of Engineers to court urging for more aggressive action to prevent these carp from entering the Great Lakes. This species has the potential to be one of the worst invasive species in the United States if it makes its way into the Great Lakes. [*biodiversity, loss of*]

assisted migration

Plants grow in an area because the conditions are suitable for that species. So what happens when temperatures rise and that plant can no longer survive? Under natural situations, the temperature increase would be slow and the plants might adapt (by *natural selection*) to the new conditions over time or gradually *migrate* to another location. Under human-created situations such as *climate change*—which is happening far too fast for adaptation or natural migration to occur—the plants would most likely not grow well or might die off completely.

But what if the plant was your country's most important crop and vitally important to the economy? This is the situation in British Columbia, Canada, where climate change is threatening the survival of timber, their most important product. They have begun a large-scale assisted-migration program, moving 250,000 larch seedlings. The tree's native *range* is in danger of becoming too warm, so they're moving the seedlings 200 miles away to an area they believe will become more suitable as the temperature rises.

So Canadians are helping the plants migrate at an artificial rate to accommodate the rapidly changing temperature. This proactive human intervention might become one of the methods commonly used to accommodate climate change. [*vineyards and assisted migration; climate change and biodiversity; climate change and marine life; natural selection*]

asthma and the environment

Many environmental factors contribute to asthma. *Climate change* will most likely exacerbate asthma, as well as produce other public health issues. Higher pollen counts will arrive with an earlier spring, *ground-level ozone* in urban areas will increase, dust storms from advancing *desertification* will be more frequent, and additional wildfires will produce more *particulate matter*.

Other environmental problems that contribute to asthma symptoms include exposure to *phthalates* that can build up in household dust, *BPA* found in plastic bottles, *VOCs* in everything from household cleaning fluids to air fresheners, and increased use of household pesticides. Even natural substances contribute to *indoor pollution*—for example, dust mites in bedding, mold in bathrooms, and insect or mouse droppings. [*climate change and human health*]

Aswan High Dam

Completed in 1970 at a cost of about one billion dollars and located on the Nile River, in Aswan, Egypt, this dam creates a reservoir, now called Lake Nasser, more than 330 miles long and 10 miles wide and stores water that *irrigates* about seven million acres of dry land. It was built to regulate floods and generate electricity for the city of Cairo. The artificial lake relocated 90,000 people prior to construction and destroyed ancient ruins or required them to be relocated. The dam generates about 12 billion *kWh* annually, has controlled flooding, and is considered by proponents to be of great economic value. However, most environmentalists condemn large-scale dam projects. Even though the Aswan is no longer one of the *ten largest dams* in the world, it has numerous environmental issues.

Prior to the dam, the Nile flooded each year and all surrounding farmland was blessed with a new load of fresh *nutrients*. After the dam, flooding ceased and the farmlands began to fail, so farmers initiated large-scale use of *synthetic fertilizer*. Prior to the dam, the river water picked up nutrients and deposited them in the Mediterranean Sea, where they nourished the *food chain* for the sardine fisheries and many other harvests. After the dam, most of the fisheries downstream collapsed. Before the dam, schistosomiasis, a deadly disease, was already a problem, but after the dam, it increased significantly because of changes in the populations of disease-carrying snails.

The Aswan High Dam is often used as an example of environmental problems caused by large-scale hydroelectric projects. Those building newer, larger dams, often called megaprojects—such as the *Three Gorges Dam* in *China*—appear to have learned little from the early knowledge gained at the Aswan Dam. [*dam removal; dam, hydroelectric; dams, ten largest; hydropower, traditional; hydropower, nontraditional; pumped-storage hydropower; run-of-river hydropower*]

atmosphere

The mixture of gases, commonly called the air, that envelopes Earth. Excluding moisture, it is composed of about 78 percent nitrogen in the form of N_2, 21 percent oxygen in the form of O_2, a little less than 1 percent argon, slightly less than 0.04 percent *carbon dioxide*, and 0.0002 percent *methane*. Other trace elements are also found in the atmosphere, including *ozone* as O_3 and *nitrous oxide* as N_2O. It also contains varying amounts of *particulate matter*. Carbon dioxide and the trace elements are typically measured not as a percentage but as parts per million (ppm) for scientific studies. For example, the exact amount of carbon dioxide in the atmosphere (currently just below 400 ppm) is critically important and closely measured because of its impact on *climate change*.

The lowest portion of the atmosphere is called the *troposphere*, where weather is produced, the climate controlled, and *air pollution* found. It continues roughly seven miles up into the atmosphere. The next higher level is the *stratosphere*, which starts at seven miles and goes to about 30 miles, where commercial airlines fly (at the lower end) and the ozone layer (*stratospheric ozone*) is found. [*biosphere; air pollution*]

atmospheric brown cloud

This is not really a cloud as much as a haze that settles over large regions. It is caused by *air pollution* consisting primarily of *aerosols* such as *fly ash* and *black carbon*. Most of the pollutants found in this cloud are caused by burning *fossil fuels* in *coal*-burning *electric power plants* and *automobiles*. The clouds can extend from the ground up to two miles into the *atmosphere*. They are much larger and longer lasting than *photochemical smog*, which is localized over cities. Atmospheric brown clouds both absorb solar radiation and scatter it; therefore, its exact relation to *climate change* remains under study.

Although the link to the climate is unsure, the link to public health is not. Brown clouds, like many forms of air pollution, cause health issues, especially pertaining to the lungs and respiratory system. These clouds can be long-lasting, so the health issues are of greater concern than many other forms of air pollution.

atmospheric rivers

Discovered in the late 1990s, these are bands of water vapor about 200 miles wide, thousands of miles long, and floating about a mile high. They form in the tropics above oceans and act like conveyor belts, carrying moisture either north or south, heading toward the poles. They can carry vast amounts of water vapor, the equivalent of 10 Mississippi Rivers. When these rivers reach landfall, they can cause a series of storms, lasting days or weeks. Some are called the pineapple express because they come from the Hawaiian Islands neighborhood. Scientists theorize that these rivers might be responsible for *megafloods*.

Nine small atmospheric rivers annually find their way to the U.S. West Coast. The good news is that scientists believe these rivers can be used to predict oncoming storms well in advance. [*climate change*]

atrazine

The active ingredient in some *herbicides* that acts as an endocrine *hormone disruptor* and is known to feminize male frogs, making them unable to mate. Many are concerned about the effect this substance might have on humans as it makes its way into the environment. [*body burden; toxic cocktail; toxic pollution; flame retardants, brominated*]

Audubon, John James

(1785–1851) Audubon is one of America's best known artists, famous for his paintings of birds and other wildlife. He was born in Haiti, raised and educated in France, and then moved to America where he perfected his art. He was a trained ornithologist and accomplished writer. He is most famous for his large portfolio book collections of his paintings, including <u>The Birds of America</u> and <u>Quadrupeds of North America</u>.

Audubon's paintings, along with those of other painters of the period, such as those in the *Hudson River School art movement* and nature writers such as *John Burroughs* and *John Muir*, are often credited for spreading an appreciation for nature in the formative years of America. The *National Audubon Society* was named for him.

aufwuch

A *community* of plants and animals living on or around a submerged surface, such as a rock or plant stem, in a lake or pond. The *dominant* organisms are usually *algae*, with many *insects* living in close association. [*aquatic ecosystem*]

Australia

Australia has environmental protection agencies at many government levels. Active *NGOs* include the *Wilderness Society* and the Australian Conservation Foundation. The country has had active *green parties* for decades, including at the national level.

One of the country's most serious environmental issues is *biodiversity* loss, primarily due to *invasive species*. The country was a major advocate of the *Convention on Biological Diversity*. Another serious issue is water—or lack of it. Australia is the driest inhabited continent, so water shortages have long been a problem. The country has passed many regional and national agreements in efforts to manage these water issues.

Australia, like the United States, did not sign the *Kyoto Protocol* to curb *greenhouse gases* back in 1997, but unlike the United States, it finally did ratify it in 2007 and created the country's own Department of Climate Change.

autecology

The study of individual organisms or a single *species*. It concentrates on how an organism's characteristics allow it to survive (or not survive) in certain *habitats*. Besides studying an organism's anatomy, it also uses sophisticated instrumentation to analyze relations between an organism and its *environment* at the molecular and chemical level.

An obvious topic of study would be why and how one organism lives in *fresh water* while another lives in salt water. Other studies resulted in discovering that plants in differing habitats use different types of *photosynthesis*. *C3 plants* (which photosynthesize a 3-carbon molecule) are found in all *aquatic* and most *terrestrial* habitats. *C4 plants* (which photosynthesize a 4-carbon molecule) are only found in hot, arid environments.

autism studies and environmental factors

Recent studies by highly regarded institutions suggest a possible link between rising autism rates and environmental exposure to *air pollutants* and *pesticides* prior to pregnancy. In one study from the Harvard School of Public Health, *air pollution* from road traffic was studied; another, from University of California, Davis, studied insecticides used in homes. There is a long way to go before drawing any conclusions from this research. [*toxic cocktail; body burden; breast milk and toxic substances; medical ecology; nanoparticles and human health; puberty and the environment, early*]

automobile

Americans—and most everyone else—love their cars. The automobile and the infrastructure (roads and bridges) to support them are in part responsible for the growth of nations and for the environmental degradation of them as well. In 2012, there were more than one billion cars and commercial vehicles on our planet—each one emitting *greenhouse gases* and causing *air pollution*. Even though pollution-control devices such as *catalytic converters* significantly reduce these emissions, the overall amount spewed into our *atmosphere* continues to rise, because the number of these vehicles continues to increase. Efforts to reduce emissions rely on increased miles per gallon, technological advances that reduce emissions, and increased use of *mass transit*.

Most nations are improving their miles per gallon averages. Japan and the European Union have the best, followed by South Korea and *China*. In 2013, President Obama announced new U.S. *CAFÉ* (corporate average fuel economy) goals for 2025; however, many people consider these goals weak. For example, Japan will have met these same goals ten years earlier.

Technologies such as the catalytic converter have made vehicles much cleaner than in the past. Promising alternative *automobile propulsion systems*, such as *hybrid cars* and *electric cars,* offer some hope for reducing our dependency on *fossil fuels*.

However, even with dramatic increases in miles per gallon and other technological improvements, the biggest problem remains the total number of cars on our planet and the total miles driven. China, alone, went from ten million cars in the year 2000 to 73 million in 2011. As developed nations become more affluent, one of the first things people do is buy a car. These numbers make it difficult to find something positive about this environmental issue.

There are some signs that Americans might be getting tired of the expense that comes with a car and the problems they cause. Nineteen-year-olds are a good bellwether, and the number of them getting a license dropped from 87 percent in 1983 to 70 percent in 2010. Also, Americans are keeping their old cars far longer than they used to, with an average age of more than 11 years. The best alternative to our automobiles is readily available mass transit, at least in urbanized areas. [*transportation; greenest cars, top ten; urbanization and urban growth; rail transportation; eco-cities*]

automobile propulsion systems, alternative

An *automobile's* propulsion system, also called powertrain system, is the new term used to describe the type of engine powering a vehicle. Until recently, the only system of importance has been the internal combustion engine running on *gasoline* or diesel fuel. But things are changing.

Compared to 25 years ago, the United States has made some advances in lowering auto *emissions*, a major contributor to *air pollution* and *global warming*. Unfortunately, even though less now comes out of our car's tailpipes, there are many more tailpipes on the road today. Therefore, the auto emissions problem continues to get worse—not better. The U.S. *transportation* sector uses one million more barrels of *oil* per day today than it did in 1973. Federal and state laws demanding improved fuel efficiency (called *CAFÉ*) have forced car manufacturers to look for alternatives to conventional gasoline and diesel automobiles.

Alternative fuels and *electric vehicles* offer hope to further reduce emissions. *Compressed natural gas* (CNG) and alcohol alternatives such as *methanol* and *ethanol* can also reduce emissions. Even though electric vehicles produce zero emissions, they only displace the source of *pollution* because electric power plants must still generate the electricity to charge these cars. However, this will still result in substantial emission reductions.

Some of these options will not become mainstream any time soon because fuels such as methanol and natural gas require a distribution network so you can "fill up" somewhere—something not likely to occur for years.

No particular alternative propulsion system will probably take the market by storm—at least for a while. Instead there will likely be a mix of technologies on the road before one wins the day. The mix will probably include traditional gasoline and diesel internal combustion engines; *hybrid vehicles* with a combination of electric and internal combustion engines; plug-in hybrids with an internal combustion engine; plug-in capable, all *electric vehicles*; and probably natural gas vehicles running on *compressed natural gas* (CNG) or *liquid natural gas* (LNG) because of the abundance of *natural gas*. And we should not discount the always hopeful but continually experimental hydrogen fuel cell car.

We have a long way to go before any of this happens because—at the moment—the vast majority of vehicles continue to have internal combustion engines running on gas or diesel. Hybrids of any sort have less than two percent of the total market, and electric vehicles—even though getting lots of press—still don't even show up on the charts. [*ammonia as a car fuel; natural gas vehicles;, fuel cell vehicles, hydrogen*]

automobile recycling

Each year about 14 million *automobiles* are discarded just in the United States. Executives at the big automakers have come to realize that many consumers want to buy "green" cars, meaning they will take into consideration how much of a car has been and can be recycled, along with its gas mileage and other features.

About 70 percent of a typical car is steel and can be extricated when junked by using shredders, magnets, and other devices. This metal is valuable and has a viable *recycling* market. Almost all cars today are composed of a significant portion of recycled steel.

About 75 percent of the metal in cars today is recycled and used by steelmakers to make new products, including new cars. An industry called mini-mills removes the metals from autos (and large *appliances*) and prepares them for the steel manufacturers.

The remaining 30 percent of the car is called *fluff*—various types of plastics, glass, and other materials. This is far more difficult to recycle or reuse because it is a mixed bag of small materials. Efforts are being made to standardize some of these materials—such as the *plastics*—to make them easier to recycle. Some automakers recycle rubber car bumpers into new car parts. Some companies have built auto "disassembly plants" to take apart and recycle car components. [*greenest cars, top ten; appliance recycling*]

autonomous automobiles

Cars that drive themselves. The driver just plugs in the destination and the car does the rest. Google was the first company to get a legal permit to road-test such a car, and Audi the second, both in California. Some people believe these cars will be ready to go to market by 2018, but others say not until the early 2020s. Either way, it is closer than we think.

At the moment, few people seem concerned about backlash from those who drive cars in the routine manner. It might be frightening to see cars without anyone in control. But frightening or not, the first accident caused by a driverless car, and the ensuing lawsuits might slow down the acceptance of this new technology.

Many cars on the road today have autonomous aspects, such as self-parking and emergency maneuver features. It is probable that we'll see a gradual rollout of autonomous features, leading up to a fully autonomous automobile. [*intelligent vehicle highway systems*]

autotrophs

Organisms are categorized according to how they obtain energy and nutrients to survive. Autotrophs (also called producers) are capable of using *inorganic* substances to synthesize *nutrients* to live. (As opposed to *heterotrophs* that must obtain their energy source and nutrients by eating other organisms.)

There are two types of autotrophs, depending on the type of inorganic substances they use. Photoautotrophs (the green plants) use *carbon dioxide* and *water* during the process of *photosynthesis*. Chemoautotrophs (almost all of which are types of *bacteria*) use *methane*, ammonia, or hydrogen sulfide during the process of chemosynthesis.

The photoautotrophs use radiant energy from the sun—during photosynthesis—to build complex organic molecules in the form of sugars. These sugars provide the energy for the plants to survive as well as most other forms of life that consume the plants (heterotrophs).

The chemoautotrophs are organisms that use inorganic chemicals directly from the Earth to build complex organic molecules for energy and nutrients. Some of these unique organisms get inorganic chemicals from *hydrothermal ocean vents* deep within the ocean. They are the bottom rung of *food chains* in these *habitats,* fed upon by a variety of unusual organisms, such as tube worms and white crabs (the heterotrophs in these habitats). Other chemoautotrophs include *bacteria* capable of *nitrogen fixation*—converting free nitrogen from the air into organic *nitrogen compounds* that can then be used by plants and animals.

baby boom

Following the end of World War II—in 1944—U.S. births began a surge that continued through 1964. Everyone born between those years is considered to have been born during the baby boom. The birth rate (number of births per 1000 population) went as high as 3.7 in 1957 from a low of 2.1 in 1937. This boom skewed the *age distribution* of the U.S. population by adding 75 million individuals in a short period of time. Today, 35 percent of all adults in the United States are baby boomers. As these individuals grow older, the age distribution has been dramatically shifting. The median age of the U.S. population in 1970 was 29, but today it is close to 38. By 2015, 45 percent of all Americans will be 50 or older. [*population explosion, human; population growth, limits of human; human population throughout history; age distribution in human populations*]

bacteria

Bacteria are single celled, microscopic organisms found in almost all environments in vast numbers. They reproduce *asexually* by simply dividing (fission).

They are incredibly diverse. Blue-green algae (actually *cyanobacteria*) are *autotrophs* containing *chlorophyll*, therefore responsible for *photosynthesis*, and are at the foundation of many *food chains*. Other types of bacteria play a vital role in the *nitrogen cycle* because they are some of the few organisms capable of using the free nitrogen in the air and converting it into organic forms of nitrogen that other organisms can use. Still others play an important role in *detritus food webs* by decomposing organic matter (dead plants and animals) and returning the chemicals to the soil for reuse. Many are *symbiotic*, living in close association with other organisms. Humans have billions of bacteria living in their guts—called the *human microbiome*.

Most bacteria require oxygen to survive and are therefore called *aerobic*. Others, however, survive without oxygen and are called *anaerobic*. Some of these anaerobic bacteria form the basis for *chemosynthesis*. [*kingdoms and domains*]

bacteriophage

See *hyperparasite*.

Bakken shale bed

A geological formation covering about 200,000 square miles beneath North Dakota, Montana, and parts of Canada, that has been found to be rich in *oil* and *natural gas*. This geological discovery—along with *fracking* and horizontal drilling technologies—has put North Dakota second only to Texas in oil production and ahead of Alaska in just a few years. Beginning in 2006, this formation was producing more than 660,000 barrels a day. Currently, about 8000 wells are in operation, with estimates that it could end up at about 50,000. Estimates project that in more than a 20-year period possibly 14 billion barrels of oil may have been removed. All this in addition to the natural gas fracking frenzy that is also in progress because of the Bakken shale bed.

This dramatic change in *oil reserves* has changed forecasts about *peak oil* and has caused concerns that any progress toward *renewable energy* sources will be thwarted or at least delayed because of less expensive oil and gas and the potential for U.S. *energy independence*. [*shale gale, the*]

bald eagle

A species of eagle found only in North America, in every state except Hawaii. Half of all these birds reside in Alaska, where they thrive on salmon runs. The eagle became an environmental icon in the 1990s when the *pesticide DDT* threatened its existence and it was placed on the *endangered species* list. When the United States was founded, bald eagles ranged from the east coast of Canada to Florida in vast numbers. But with DDT and other *pesticides*, along with the process of *bioaccumulation*, their numbers dwindled down by the early 1960s to only about 450 nesting pairs in the contiguous 48 states. The pesticides had caused the birds to produce eggs with thin shells that often could not survive.

Thanks to a ban on DDT in 1972 and other restoration programs, bald eagles have made a comeback and were upgraded to the *threatened list*. Currently, they are doing well, with more than 10,000 nesting pairs in the lower 48 states. [*endangered and threatened species lists; apex consumer and ecosystem*]

bamboo

Bamboo is popular as a green alternative to wood. Even though people call it a wood, it is actually a grass that possesses many wood traits, allowing it to be used for everything from fishing rods and furniture to floors and doors. Its primary advantage is that it grows very quickly. Compare building a table from oak, which takes well over 100 years to mature, to one of bamboo, which takes about three years to reach full maturity. This speed of growth lends itself to *sustainable development*.

The global market for bamboo is about $7 billion per year and expected to grow to $17 billion within a few years because of its growing popularity. Most bamboo comes from *China* and Vietnam. At the moment, there are no bamboo farms in the United States, but some are on the drawing board, with the first one planned for Alabama in the near future. [*sustainable forests; forest management, global; mycowood; tree farms*]

bananas, Panama disease on

Panama disease is a viral infection that only grows on one variety of banana, called Cavendish. But this one variety is the most commonly grown commercial banana. Panama disease is indicative of a global problem—*monocultures. High-impact agriculture* typically grows one variety of crop in huge numbers, called monocultures. This lack of diversity and uniformity of plants allows pests or infections to run rampant. This disease is wiping out entire crops of bananas in *Australia* and New Zealand, and now there are concerns about crops in Latin America.

This is the only banana grown in these regions, so the entire industry is jeopardized by this one infection. Some in the business are calling for more diversity in the varieties of bananas commercially grown. The lack of diversity in high-impact agriculture causes the same problems found in natural ecosystems when *biodiversity* is lost. [*heirloom plants; fire blight on apples and pears*]

Bangladesh and sea rise

Bangladesh could be the poster child country for *climate change* and the accompanying *sea level rise*. What the world does or does not do about climate change will probably impact the people of this nation more than most. It is bordered by *India*, Myanmar, and the Bay of Bengal. It is in a subtropical zone that is very wet, flat, and fertile and is also one of the most densely populated countries. Most people are poor and farm small plots of land to grow rice.

The land is just above sea level; every year, floods wash away much of their coastal farms, so people must start over annually. If climate change occurs as predicted and sea levels rise as expected, about 30 percent of what is now Bangladesh will be under water. How these people will survive or where they will go is not known. They are resourceful and are trying new methods, such as using new varieties of rice that can be harvested more quickly. They are also using what appears to be a cross between *crop rotation* and *integrated multilevel aquaculture,* such as growing rice before the monsoon season and then stocking their flooded fields with small fish and crabs to also be sold.

Even though most people might find it simply frustrating to hear about the lack of action resulting from major international summits such as *Rio+20*, the people of countries such as Bangladesh react with fear.

barbless hooks

When fisherman release their catch in an effort to save the fish, the likelihood of the fish's survival is dramatically improved when barbless hooks—or hooks that have their barbs flattened—are used. Throwing back a fish that dies because of injuries received from the hook accomplishes nothing. Many trout streams in the United States have mandatory *catch-and-release* programs requiring the use of barbless hooks. [*fishing, commercial; bycatch; overfishing; pirate fishing; turtle excluder device*]

barrel of oil

A barrel of *oil* contains (is equal to) 42 gallons. [*oil recovery or extraction*]

Bartram, William

(1739–1823) A botanist who spent 4 years, starting in 1773, travelling 5000 miles through the southern United States, documenting nature as he saw it unfold. He also wrote about *Native American* cultures. His book, Travels, is considered one of the finest of its time and credited with helping to popularize botany. His writings were read by Europeans wanting to learn about the new American landscape. [*philosophers and writers, environmental*]

baseball stadiums, green

Many baseball stadiums are making green statements that their fans appear to appreciate. They typically can accomplish these improvements when building a new stadium or upgrading an old one. There are numerous examples.

The Miami Marlins' new stadium is the first-ever retractable roof building that is certified *LEED* Gold. The Boston Red Sox celebrated Fenway Park's 100th birthday in 2012 by installing rooftop *solar panels* behind home plate, cutting energy use in the stadium by 37 percent. St. Louis's Busch Stadium now also has solar panels.

New York's Yankee Stadium installed more efficient light fixtures that cut 200,000 pounds of *carbon emissions* for every night game, the equivalent of planting one tree per pitch. The Cleveland Indians installed a large *wind turbine* to contribute power to their stadium, and fans can join in producing extra power by using stationary bikes that tie in to the turbine. The Seattle Mariners' recently renovated SafeCo Field is exemplary. They now save about $400,000 per year in energy costs and keep 1000 tons of *municipal solid waste* out of *landfills*.

The stadium of the Washington Nationals just received LEED certification as well, with low-flow urinals and recycled *toilet paper*, a scoreboard with *LED* lights, solar panels on the sky bridge that links the parking lots to the stadium, two *light-rail* stations and a few bus lines within minutes of the stadium, electric chargers for *electric vehicles,* and—possibly most important—two mascots who use on-field antics to educate the fans about the environment, named Kid Compost and Captain *Plastic.* [*NFL stadiums, green; NBA stadiums, green; universities, top ten green; cities, ten best American; green gym movement, green; bicmaqiunas*]

Basel Convention

The Basel Convention on the Control of Transboundary Movements of Hazardous Wastes and Their Disposal is the full name of this international agreement, signed in 1989 and effective in 1992. Its purpose is to control the movement and disposal of *hazardous waste* products, such as *e-waste* from cell phones and other high-tech devices. This waste is produced in one country, usually a developed nation, and then transported and disposed of in another country, usually a poor nation in need of revenue. The practice became common during the 1980s, and this treaty was the first major effort to develop principles of *environmental justice.*

In 2008, 170 countries ratified the agreement. The only country in the *OECD* that did not was the United States. In 1995 the wording of this agreement was strengthened, but even with this change, it is not a ban but an attempt to set limits on how much of this type of waste can be transported and how. It has not become an international law, but many of the signatories adhere to it in good faith. [*e-waste and you; e-waste product life cycle; e-waste recycling and disposal programs; beryllium in computers*]

baseload power

Both *water power* and *geothermal power* are energy sources that can generate power 24/7 and are said to produce baseload power. That means it can add power to the *power grid* all of the time, as opposed to *renewable energy* sources such as *solar* and *wind* that can only generate power at certain times, when the wind is blowing or the sun is shining. These forms of power require *time shifting* to balance out the availability. [*alternative energy sources*]

Batesian Mimicry

A form of *mimicry* in which an edible (nonpoisonous) organism resembles another species that is poisonous. This adaptation protects the edible organism by tricking a *predator* into thinking it is harmful to eat. For example, there is a nonpoisonous snake that resembles the highly poisonous coral snake. Predators leave both alone. [*Müllerian mimicry*]

bathroom water-saving techniques

Saving water is more important today than ever before. *Water shortages* occur more often in more places, *aquifer* levels are dropping, and *sewage treatment facilities* are often overwhelmed. More than half of all water use in a typical U.S. home occurs in the bathroom, so it is a good place to start reducing your *ecological footprint.*

Toilets are the number-one culprits, using about one-third of all residential water. Fixing a toilet that constantly "runs" can save 200 gallons of water per day. Using a screw-on sink-faucet aerator reduces the volume of water used, without reducing the pressure.

Old toilets use a lot more water than newer ones. If they were made before 1980, they probably use between five and seven gallons of water per flush. If made from the 1980s to the 1990s, they use three or four gallons, and the most recent toilets use 1.6 gallons or less.

But you don't have to replace a toilet to save water. Placing a large bottle or jug in the tank reduces the amount of water used for each flush.

Even better than low-flush toilets are dual-flush toilets, very popular in Europe, where you choose between two quantities of water volume, depending on how much is needed during the flush. But again, you don't need to buy a new toilet for this feature; you can convert any toilet into a dual-flush system with a converter you can purchase at a plumbing store.

If you really want to minimize your bathroom's ecological footprint, *composting* toilets use little water and biodegrade the waste without a septic system.

Showerheads also make a huge difference in water use. The popular high-pressure and rain-style showerheads are nice but use lots of water. Many new state laws require low-flow showerheads, with rates less than 2.5 gallons per minute, but you can find some that run at 1.75 per minute. Some people avoid low-flow showerheads because they think water pressure will be reduced, but a good quality low-flow head should not do that.

Finally, there are plumbing fixtures that fill the toilet tank with *gray water* from your sink (after washing your hands, for example). These devices are very efficient and save a great deal of water. [*energy-saving apps and programs; energy, ways to save residential; tub bath vs. shower*]

bathyal zone

The level within the open oceans too deep for *photosynthesis* to occur but close enough to the surface for some light to filter through. This region between the *euphotic zone* (where photosynthesis occurs) and the *abyssal zone* (where no light penetrates at all) is often called the twilight zone. It ranges from 660 to 5000 feet in depth. [*oceanic zone ecosystems; marine ecosystems*]

bats

Bats are the only *mammals* that can truly fly. It is a large order, containing more than 850 species. They are divided into two suborders called megabats and microbats.

Bats are found almost everywhere on our globe except the *Arctic*. Most are nocturnal and use echolocation to sense their surroundings. Bats play an important role in many *ecosystems*. Many microbats are voracious, consuming huge numbers of *insects* and keeping them in check. Some megabats are fruit eaters and help *pollinate* and distribute seeds. These *eco-services* they provide far outweigh the bad rap they get because of the few species, called vampire bats, that feed on the blood of cattle and other animals.

Some species of bats are on the *endangered species* list. Many are threatened by the *white nose syndrome* disease that is decimating many populations. The loss of a large numbers of bats within an ecosystem has serious ramifications because the numbers of insects dramatically increases.

batteries

Hundreds of billions of batteries are used each year. In the United States alone, about three billion so-called flashlight batteries are purchased each year. But there are many types for many purposes; everything from the batteries in your flashlight, to those found in small electrical devices and the battery in your car. There are disposable batteries and rechargeable batteries. But they all create the same environmental problems that need be understood.

All batteries contain *heavy metals* and other *toxic* substances, such as *lead*, *cadmium*, and *mercury*. If these are not disposed of properly, they can contaminate *soil* and *groundwater* as they leach out of *landfills.*

Batteries are not likely to go away anytime soon, so *recycling* programs and proper disposal are important. Most batteries are recyclable and have existing viable markets. In some cases these recycling programs are mandated. For example, many states require *automobile batteries* to be returned and recycled. Other programs are voluntary, such as return programs for common flashlight batteries.

The worst thing a consumer can do is to simply throw out an old battery in the trash, where it becomes part of the *waste stream* and ends up as *municipal solid waste* in our landfills, which are not designed to handle these dangerous substances.

batteries, automobile

Electric vehicles have finally arrived and are no longer rare to see on the roads. This is possible because of advances in the batteries that provide the energy in lieu of *fossil fuels* (in the form of *gasoline*). These batteries have transitioned from lead-acid to nickel-metal-hydride and are now moving to lithium-ion.

Even though electric vehicles are environmentally friendly—when compared with gas-driven vehicles—they are not without their problems. These batteries still contain *toxic* substances that must be disposed of in some way when the cars reach their end of life. *Battery disposal* is currently an environmental problem that must be addressed.

The toxicity of the types of batteries, however, is improving. *Lead*-acid is the most toxic and least efficient of the three. This is the typical battery found in all cars and is even found in the Prius, but only as the battery to run electrical components such as the headlights and radio.

Nickel-metal-hydride is not as toxic as lead. The substance used in these batteries is possibly *carcinogenic,* and mining for nickel is an environmental issue. For now, these are the batteries of choice for most hybrid electric cars such as the Prius and many others.

Lithium-ion batteries are the least toxic of the three and the most efficient, but also the most expensive. This is what powers the Nissan Leaf, Chevrolet Volt, and Tesla Roadster—all more expensive than cars using nickel. The costs should be coming down and the future appears to be lithium-ion batteries. Honda states that forthcoming Civics will use this type of battery. [*battery disposal in Mexico, automobile*]

battery disposal in Mexico, automobile

In 2008, new laws in the United States made it more costly to *recycle* materials containing *hazardous wastes* such as *lead*. This resulted in many recycling firms moving to Mexico, where they have few health and environmental regulations. Companies in the United States now send about 20 percent of our dead, used vehicle batteries to Mexican *recycling* facilities, making the recycling process less expensive but more dangerous for those working at the facilities and for those living near them. It is also more harmful to the environment, because the lead is typically released, contaminating the soil and water surrounding these facilities.

Vehicle batteries are recycled for two reasons. First, they contain a lot of lead, which has become more valuable because of its use in high-tech structures such as *wind turbines* and cell-phone towers. Second, recycling is less expensive than disposal, because products containing lead are considered a *hazardous waste* and must be sent to special landfills. [*Basel Convention; environmental justice; gold fingers; e-waste*]

beach water quality and closings, U.S.

When you go to the beach for a day in the sun, you are probably not thinking about environmental issues. But the quality of the water you jump into is certainly of concern. Like so many issues, water quality is only in the news when something is wrong, such as when beaches are closed because of bacterial contamination from stormwater *runoff* and raw *sewage*. The United States had more than 23,000 such closings in 2011.

One of the main causes for these frequent beach closings is the *combined sewer systems* used by nearby cities. After a heavy rain, these outdated systems can release untreated wastewater into the waters near beaches.

The *Environmental Protection Agency* (EPA) sets federal water quality standards for the maximum amount of bacterial contamination they deem acceptable in beach waters. The *Natural Resources Defense Council* (NRDC) states these current levels mean the EPA finds it acceptable for 1 in every 28 beachgoers to become sick with a gastrointestinal illness.

For some positive information, here is the NRDC's 2013 data on the seven best beaches for clean water in the United States: Gulf Shores Public Beach (Alabama), San Clemente State Beach (California), Dewey Beach (Delaware), Ocean City (Maryland), Bay City State Recreation Area (Michigan), 13th Street South Beach (Minnesota), and Hampton Beach State Park (New Hampshire). [*water pollution*]

bedbugs and remedies

Bedbugs were a huge public health problem back in the 1950s until the advent of the *pesticide DDT*, which was sprayed with abandon. Wallpaper was even infused with DDT to keep the critters away. Although bedbugs had been almost completely eliminated in the United States for many decades, they were never well controlled elsewhere in the world. The recent resurgence is thought to be caused by *insects* becoming resistant to DDT in parts of the world where the insect prevails and the pesticide is still commonly used. It is thought that resistant individuals made their way back to the United States, thanks to modern transportation. Bedbugs can live for six months without food, so they make perfect long-term stowaways. Some people, however, believe the resurgence comes from poultry farms where bedbugs are commonly found.

How they made a comeback in the late 1990s really doesn't matter if your home becomes infested. Until this latest invasion, few entomologists even specialized in the study of bedbugs. They are tough little invaders. They are small, shy, nocturnal, and hard to find until their numbers make them hard to avoid. A single female can produce enough young to infest an entire building. They are also odd; they reproduce by using traumatic insemination, meaning the male stabs the female directly through the abdomen to deposit sperm.

Some people are using an old method of bedbug control. In eastern Europe during the early 20th century, people with bedbugs placed bean leaves on the ground at night. When awakening, the leaves would be filled with dead bedbugs that appeared to be stuck to the leaves. The leaves and bugs where disposed of and all was well. Bean leaves have small projections that impaled the bugs. Today scientists are trying to copy nature by mimicking the natural method. [*pesticides, biological control; biological control methods*]

beehive sound research

Recent research shows that the buzzing sound produced within a hive can be indicative of the overall health of its residents. Using sensitive equipment, the subtle differences are thought to indicate whether a hive is prone to *colony collapse disorder*, which is ravaging bee hives on a global scale. Unfortunately, at this time, even if this technique can be used to detect if a hive is in trouble, there is no known cure for stopping the collapse. [*neonicotinoid pesticide and honeybees; soundscape; bumblebee senses; honeybee sperm bank; zombie bees*]

beetlecam

Picture a small (about three inches long) remote-controlled robot, looking like a dune-buggy with a webcam mounted on top. It is being used to video-record wildlife in Africa. The beetle-like device drives up to lions, *elephants*, and wildebeest, getting great views a photographer would never be able to get. [*biosensors; nanotechnology; robotic bees*]

beetles

Beetles belong to a specific order of *insects* called Coleoptera. There are more species of beetles on our planet than any other form of life. Of the roughly 1.8 million described species on Earth, about 350,000 of them are beetles. [*biodiversity*]

belching cows

See *methane gas.*

Bennett, Hugh Hammond

(1881–1960) Bennett is called the father of *soil conservation.* He was the first to realize the damage caused to soils by cultivated fields and the likelihood and consequences of *soil erosion.* His predictions were proven with the *Dust Bowl* of the 1930s. In response to the Dust Bowl, Congress created the *Soil Conservation Service,* which is part of the *Department of Agriculture.* Bennett was its first director.

benthic organisms

Organisms that live on or near the bottom of bodies of water—in particular, *marine ecosystems.* This includes plants that are rooted to the sediment when the water is shallow enough and animals such as crabs, lobsters, clams, and aquatic worms. Fish such as skates, rays, and flounder are considered benthic because they spend much of their time at the sea floor, often in hiding from *predators.* Also, sponges, sea anemones, and barnacles are sessile and permanently attached to the sea floor. [*oceanic zone ecosystems; abyssal ecosystems; squid in native habitat*]

Bering Sea

The extreme northern portion of the Pacific Ocean, which separates the United States from Russia. The Sea is of great economic importance because more than 50 percent of all U.S. fish production comes from this area. The global economic value of this region is in the billions of dollars each year. As so often happens, the bounty has been *overfished,* and seafood populations are greatly stressed and in decline. Quotas and various schemes to limit catch have been put in place to control the damage. [*fishing, commercial; fish populations and food; pirate fishing*]

beryllium in computers

One of the *rare earth metals,* this element is used to build circuit boards found in most high-tech devices, from smartphones, to computers, to televisions and GPS. It is added to other elements—creating alloys—to improve how they perform.

Beryllium is both *acutely* and *chronically toxic* to humans and is classified as a class 1 *carcinogen.* Using products that contain the element is okay, but the people who extract it from the Earth or come in contact with it during the manufacturing process of these devices are at a significant risk. There is even a disease called chronic beryllium disease, found in people who are in constant contact with the dust often found at these sites. The disease primarily damages the lungs.

Once beryllium is disposed, it becomes an environmental problem. Like other rare earth metals, it should be considered a *hazardous waste*, requiring special disposal facilities. However, many people simply dump their high-tech devices in the regular trash, so they end up in *landfills* or *incinerators*, where they cause *air pollution* or possible *groundwater pollution*. Many *e-waste recycling programs* are available that everyone should check out. [e-waste; e-waste and you]

Berry, Thomas

(1914–2009) Thomas Berry was an author, historian, and eco-theologian—one who ties spirituality with *ecology*. He has many wonderful books, but his most well-known is The Dream of the Earth. Berry writes about how we live today in a "Technozoic era," where humans try to control nature and overcome its limitations and how we must move into a new "Ecozoic era," where humans build a sustainable relationship with the universe. This new relationship between humans and nature he calls the New Story. [*philosophers and writers, environmental*]

Berry, Wendell

(1934–) Wendell Berry is a poet, writer, farmer, and *environmental philosopher*. He has been a longtime proponent of *conservation* and especially *sustainable agriculture*. His books, The Gift of Good Land and The Unsettling of America, emphasize human stewardship of Earth. He places great value on traditional farms and farmers and worries about the loss of the family farm to *high-impact agriculture*.

best available technology (BAT)

The most state-of-the-art technology available for a particular industry; environmental BAT technology causes the least harm to the environment. Using BAT doesn't mean a technology is nonpolluting. It only means it is the best that can be done with existing technology. However, when standards require the use of BAT, typically pollutants have decreased substantially. [*biotechnology; biosensors; primitive technologies; robotic bees; satellite mapping; synthetic biology*]

Beyond GDP (gross domestic product)

Beyond GDP is a relatively new concept that urges governments to look at indicators of economic health other than the traditional *GDP* (gross domestic product). This is developing in many ways and includes considerations for *natural capital*, social equality, *eco-services*, and human happiness. It addresses the need for a *decoupling* of economic growth as the only indicator of "progress." It uses new models such as *steady-state economics* and new indicators such as the *Genuine Progress Indicator* and green net national production. [*CERES; climate capitalism; corporate social responsibility; economics and sustainable development; economy vs. environment; integrated bottom line; polluter pays principle*] {ec.europa.eu/environment/beyond_gdp}

Bhopal

Bhopal, *India* (population at the time of 850,000), was the scene of the world's worst industrial accident, which happened in December of 1984. Thirty tons of a gas (methyl isocyanate) used to manufacture *carbamate pesticides* leaked from a storage facility at a Union Carbide plant. It created a cloud covering a 23-square-mile area over the city for a little less than an hour. Estimates state that about 2000 people died instantly, but in the aftermath, well over 3500 people died and 300,000 were injured. To this day, estimates vary widely.

Even though the plant had a good safety rating previous to the accident, many people believe the disaster could have been prevented by an investment of about one million dollars for the proper safety equipment. In 1989, the company settled out of court to pay victims $470 million dollars in total; an amount many believe to be inadequate considering the scope of the disaster. Many countries have since passed stricter laws controlling the production of toxic chemicals. [*toxic pollution; toxic waste; Toxic Substances Control Act; hazardous waste*]

bicmaqiunas

In Guatemala, an *NGO* called Maya Pedal provides a device called bicmaquinas that allows residents to sit on a bike-like seat with pedals. The pedal power is hooked up to perform work such as pumping water up from a well or running grinders or threshers to prepare food. These devices are very popular and extremely functional in many poor nations. Similar types of devices are also being used in some green gyms in the United States. [*frugal science; gym movement, green*]

bicycles and bicycling

Alternative fuels must be found to reduce our dependence on *fossil fuels*, but the best alternative is to use a mode of *transportation* that replaces the gasoline-powered automobile completely. *Electric vehicles* will offer a partial solution in the future, and *mass transit* is a viable alternative today in many cities. But in many parts of the world, the alternative form of transportation is the bicycle. Roughly 100 million bikes are manufactured each year globally. No one is sure how many bikes exist at any given time, but a reasonable estimate appears to be about one billion.

Bicycling is not a form of transportation found only in poor nations, as many people think. *China*, Japan, Denmark, and The Netherlands, as well as many other nations, rely heavily on pedal power. Bicycles don't cause *air pollution*, they relieve traffic congestion, and they offer healthy exercise.

In the United States, only a small percentage of people, in a few regions, use bikes for anything other than sport or pleasure, meaning they don't replace the car for most trips. Few people commute to work on bicycles in the United States, as opposed to some of these other nations, where it is a common form of commuting.

Why do some nationalities embrace the concept of cycling as a means of transportation but others don't take it seriously? Pro-bicycling countries encourage cycling by providing extensive bike lanes and roads separated from auto traffic with protective barriers. They have nonauto-zoned areas within cities and plentiful bike-parking facilities. Although Americans enjoy riding, as indicated by the numbers who do it for fun, we are limited by the lack of available bike lanes and facilities, making riding impractical—if not unsafe—for commuting.

Even though the nation as a whole has not embraced the concept, many cities do—as indicated by many publications that offer lists of the *ten best cities for bicycling*. Changes hint that bike riding has the potential to become an important alternative to the automobile in the United States. The number of people who commuted to work on bikes increased from 2000 to 2009 by almost 60 percent, albeit still a small percentage of the total. Many cities also have *bicycle-sharing programs.*

The *Rails-to-Trails Conservancy* is an organization that works with railroad companies and municipalities to acquire and convert abandoned tracks into bike lanes. These trails become immensely popular as soon as they open, with trails near cities becoming daily commuting routes. The popularity of these new bike lanes show that many more people would ride to work if they had safe and comfortable places to do so.

Less than two percent of a typical city's federal *transportation* monies is used for bikes and *walking* infrastructure. If cities would use more of these federal dollars to build and encourage bikes and walking, not only would it be better for the environment but also for more immediate health reasons. Those cities that encourage biking and walking have the lowest levels of obesity, high blood pressure, and diabetes. [*urbanization and urban growth; e-bikes*]

bicycle-sharing programs

Programs where a large number of bikes are provided at central locations across a city, at little cost. The bikes can be removed from the stations and later returned to the same or any other station. If the locations are properly set up, people can use these bikes not just for recreation but for commuting to work or school. People can *walk* to a nearby station, check out a bike and ride it to a station near work or school, reversing the process at the end of the day.

Successful *bicycle*-sharing programs have been in existence in many countries for a long time. The idea, however, is just now becoming popular in the United States. Both New York City and Washington, DC, have unveiled new programs. As one would expect in large cities, there have been some startup problems, but for the most part they are considered successes.

Paris has had large bike-sharing programs for many years. Their program has four times as many bikes as any program in the states. Paris was used as a model by New York—both on what to do and what not to do. For example, 80 percent of the original Parisian bikes were stolen or damaged, so New York started off with a more robust form of security in place.

In Washington, bike sharing started back in 2008 but has been recently upgraded. The program, called Capital Bikeshare, has well over 1000 bikes available in more than 100 metro stations. The bike stations themselves are *carbon neutral* because they use *solar power*. Moscow recently started a bike-share program, and many other cities in many nations are studying the idea. [*bicycles and bicycling; bicycling cities, ten best; e-bikes; car-pooling; mass transit; rail transportation; ride-sharing*]

bicycling cities, ten best

Bicycling.com, Walkscore.com, and *E Magazine* all publish their own lists of best cities for *bicycles*. The criteria differ slightly but include such things as number and types of bike lanes, ability to commute to downtown areas, *bicycle-sharing* programs, and many others.

Bicycling.com: 1) Portland, OR; 2) Minneapolis, MN; 3) Boulder, CO; 4) Washington, DC; 5) Chicago, IL; 6) Madison, WI; 7) New York, NY; 8) San Francisco, CA; 9) Eugene, OR; and 10) Seattle, WA.

Walkscore.com: 1) Portland, OR; 2) San Francisco, CA; 3) Denver, CO; 4) Philadelphia, PA; 5) Boston, MA; 6) Washington, DC; 7) Seattle, WA; 8) Tucson, AZ; 9) New York, NY; and 10) Chicago, IL.

E Magazine: 1) San Francisco, CA; 2) Portland, OR; 3) Minneapolis, MN; 4) Boulder, CO; and 5) Madison, WI.

Bierstadt, Albert

(1830–1902) Bierstadt was born in *Germany*, moved to America early, and began painting natural landscapes. Even though he is considered one of the *Hudson River School art movement's* finest artists, he travelled west and painted many Rocky Mountain landscapes as well as the New York area scenes.

big ag

Another term for *high-impact agriculture* and *factory farms*.

Billion Acts of Green campaign

See *Earth Day.*

bioaccumulation

A process where substances build up in an organism's body. For example, many *nutrients,* such as vitamins and minerals, go through a process of bioaccumulation within our bodies, as they should. However, this natural process has a dark side, because many *toxic substances* such as *pesticides* also accumulate in our bodies.

Pesticides on or in plants are eaten and then absorbed into an animal's fatty tissue, where they remain. This ongoing bioaccumulation of pesticides (or their breakdown products) and other *toxic substances* such as *PCBs* within an animal's body occurs routinely. These accumulated substances can harm that particular animal or be passed along food chains to many other organisms, including humans, in a process called *biological amplification.* As they collect in humans they become part of our *body burden.*

biocentric

Contrary to *anthropocentric*, biocentric means "life centered." This is a belief or viewpoint that all forms of life have value and must be given the right to exist. [*ethics, environmental; primitivism; "Land Ethic, The"; ecological footprint; Native American environmentalism; philosophers and writers, environmental*]

biochar

A form of *biomass energy* that uses agricultural waste that would otherwise be disposed of. A process called pyrolysis converts the waste into this charcoal-like product that can be used for fuel. Because biochar does not require trees—the most common source for biomass fuel—it does not contribute to *deforestation.*

Some scientists have suggested using biochar as a method of *carbon sequestration storage.* The idea is to create the biochar but not burn it as a fuel. Instead, bury it underground, where it will act as a *carbon sink.* Carbon stored within the biochar would take a very long time to decompose. Storing carbon this way removes it from the natural *carbon cycle* and keeps it out of the *atmosphere* so it cannot contribute to *climate change.* [*karst, carbonate; forams; geoengineering; metal organic frameworks; regenerative agriculture*]

biochemical conversion, biofuels

Biofuels—a form of *biomass energy*—can be produced by either *thermochemical conversion* or biochemical conversion. The latter uses microbes such as *yeasts* or *bacteria* that live with little or no oxygen and feed on the *biomass*.

The most common method of biochemical conversion uses yeasts to ferment the carbohydrates and sugars in corn, sugar cane, wheat, or other crops to produce alcohol fuels such as *bioethanol* fuel, also just called ethanol. The next most common method uses fats and vegetable oils to produce *biodiesel* fuel.

The third method uses microbes to produce a mixture of *methane* gas and *carbon dioxide* called *biogas,* which is then used as fuel. This process occurs in nature but can be controlled in *methane digesters*. These devices use plant matter and animal wastes to produce methane, which is collected and used for heating and cooking. *China*, *India*, and Korea have tens of thousands of these digesters in use.

Landfills naturally contain these bacteria and therefore generate methane gas. Many landfills have pipes that collect the gas so it can be used as fuel. Experimentally, animal wastes from large feedlots and *sewage treatment* plant sludge are also being used to produce biogas for fuel. [*alternative fuels*]

biochemical oxygen demand (BOD)

Also called biological oxygen demand, this is a commonly used measure of water quality that works as follows: At any time, plants and animals are dying in a body of water. This *organic* matter is decomposed by microorganisms, especially *bacteria* that use oxygen (as we do). The more organic matter available to be decomposed, the more bacteria is present and the more oxygen used, meaning there is a high BOD. If, on the other hand, the water contains little organic matter to be decomposed, fewer bacteria are present and less oxygen is used, meaning the BOD is low.

From the perspective of an organism living in a pond or lake, a low BOD is good, because plenty of oxygen is available. This is typically true in a clear, clean body of water, because the amount of decomposing organic matter is balanced within the *ecosystem* and the BOD is relatively low, so all of the organisms within the ecosystem have plenty of oxygen.

However, when *synthetic fertilizers* or raw *sewage* enters bodies of water (a process called *nutrient enrichment*), the population of *algae* explodes, called *algal blooms*. When this large population of algae dies and decomposes, it is eaten by bacteria that then have their own population explosion. These vast numbers of bacteria use up unusually large amounts of oxygen, so the BOD is very high. This results in very little oxygen available for all the other organisms in the ecosystem and the health of the entire aquatic community is in danger and it can collapse. [*eutrophication, cultural*]

biocide

A chemical that is dangerous to all life. For example, most *pesticides* do not just kill pests (the *target organism*); they kill most organisms and therefore should be called biocides. [*Carson, Rachel*]

biodegradable

The ability of a substance or product to naturally break down into the basic elements or compounds so they can be reused as *nutrients* by plants. This decomposition occurs when *insects, fungi, bacteria,* and other microbes feed on the substance. These types of organisms are called *decomposers,* or saprophytes.

Organic matter—such as dead plants and animals and their waste products—biodegrades quickly in nature. This is why forests and other habitats are not littered knee-deep in dead plants and animal carcasses. Virtually everything created naturally is biodegradable. It is only manufactured products such as plastics that do not biodegrade readily—if at all.

Many manufactured products today are sold as biodegradable. Some of these boasts are true but many are not. For example, some *plastics* claim to be biodegradable, but virtually none of the plastic in the marketplace will ever degrade. You can check if a product is truly biodegradable by seeing if it has an appropriate *eco-label* or *eco-certification*.

People are working on biodegradable consumer products that will easily decompose without causing pollution. Suggestions have even been made to create devices that simply dissolve in water or can be *composted*. Imagine instead of throwing out your two-year-old cell phone, or even bringing it in to be recycled, you just throw it in your compost pile and use it to grow plants next summer.

Most of today's manufactured products do not break down at all or break down slowly, so they must be disposed of in some way such as *landfills* and *incinerators*. [*municipal solid waste disposal*]

biodiesel fuel

Biodiesel is a *biofuel* created from *feedstock* such as animal fats, vegetable oils, soy, rapeseed, mustard, flax, sunflower, *palm oil*, hemp, and *algae*, among others (as opposed to regular diesel fuel that is refined from petroleum). More than 19 billion liters of biodiesel fuel was created in 2010. It is used in various blends with regular gasoline, much like bioethanol. Pure biodiesel, called B100, produces the fewest pollutants of all diesel fuels, but it is typically used as blends. For example, *Argentina* has a B7 mandate (7 percent) and *Brazil* uses a B5 (5 percent) blend. Biodiesel is the most common biofuel in Europe and has been gaining popularity in the United States, as well. [*alternative fuels*]

biodiversity

A term made popular by *Edward O. Wilson* in his 1988 book <u>Biodiversity</u>. Biodiversity refers to the vast range of plants and animals on Earth and implies the importance of all. Today, somewhere between 1.2 and 1.8 million *species* are identified.

Scientists disagree on how many species actually exist if all were to be identified. Estimates range from 3 million to as many as 100 million. A recent study has narrowed this range by using a mathematical formula based on the numbers of species we already know to exist within each group (such as family and genus). This study projects there to be 8.7 million species on Earth (excluding bacteria). Of these, 6.5 are thought to live on land and 2.2 in the oceans. This is the newest estimate, but it won't be the final one because many scientists believe this number is far too low.

Organisms are found everywhere. *Habitats* include *fresh water*, *marine* waters, *brackish* waters, the *soil*, and the air. Organisms are found from the *arctic* to *deserts* and everywhere in between. The process of *natural selection* appears to have filled virtually every conceivable space on our planet with life.

Possibly, the most important aspect of biodiversity is not identifying every species but protecting them. The only way to know where we should put our efforts is to know which species are *endangered*. This is a burgeoning aspect of the life sciences and it now uses everything from traditional field scientists, walking through the wilds studying species, to high-tech global position imaging.

What we learn about our Earth's biodiversity is being applied to the developing study of *eco-services*; a field that helps us assess the value of nature's resources. Further recognition of the importance of biodiversity is seen by the creation of the Intergovernmental Panel on Biodiversity and Ecosystem Services, a counterpart of the *Intergovernmental Panel on Climate Change*. [*biodiversity, loss of; biodiversity hotspots*]

biodiversity, loss of

When people speak of the loss of *biodiversity*, they are referring to the exceptionally large numbers of species forced to the brink of extinction because of human activities. Species becoming *extinct* is not a new phenomenon and was happening long before humans ever roamed the planet, but the speed with which organisms are being lost is a major concern.

There are many facets to this loss and why we should care. One is the intrinsic value of every form of life. Many people believe humans don't have the right to force any organism into extinction. Another facet is the impact on an *ecosystem*. Losing a single species, such as a *keystone species,* can damage an entire ecosystem, and even slight changes often have major effects on the entire system.

More tangible effects of reduced biodiversity include the loss of potentially useful substances, such as medicines, crops, and fibers. A report by the *United Nations Convention on Biological Diversity* states that "the livelihoods and food security of hundreds of millions of people" will be threatened if the trend in biodiversity loss continues.

Perhaps you are not concerned about the survival of other forms of life, but we are dependent on these other species for our food supply, air quality, climate control, water purification, *pollination*, *erosion* prevention, and medicines. Potential cures for cancers and other deadly diseases are lost when plants become extinct before they can be studied. Some estimate as many as ten percent of all plants have medicinal value. These are all called *eco-services* and must be assigned value.

The numbers are rather frightening. According to the *International Union for Conservation of Nature's Red List,* more than 20 percent of all known mammals face extinction, more than 30 percent of all known amphibians face extinction, and more than ten percent of all known birds face extinction. This comes to about 20,000 species of plants and animals at a high risk of extinction. Many scientists believe the situation is dire and that we are approaching *tipping points*, when the damage done might be irreversible. This loss of species is being called by many, the *sixth mass extinction*.

The causes are many, but *habitat loss* leads the list. *Deforestation*, *desertification,* and *urbanization* all contribute to the problem of habitat loss. It is estimated that when a habitat is reduced in size by 10 percent, 50 percent of the species within that habitat are lost. *Tropical rainforests* containing about half of all known species are being destroyed at a staggering rate. Exploitation of resources is another major cause and includes *overfishing* and hunting, both legal and illegal. *Water pollution* from agricultural nutrient *runoff* is another cause. More recent contributors include *invasive species* and the ubiquitous *climate change*.

One of the most practical ways to prevent further loss of our Earth's biodiversity is to reduce the habitat loss by setting aside protected areas. Many environmental *NGOs*, such as the *Nature Conservancy,* are at the forefront of this activity. Also the federal government must play an important role in protecting the biodiversity of public lands and balance the preservation of natural resources against exploiting them. *Marine ecosystems* are beginning to be protected with *marine protected areas*.

On a global scale, the *United Nations Convention on Biological Diversity (CBD)* and the *United Nations Environment Programme* work to preserve biodiversity.

The illegal international *wildlife trade and trafficking* of species, such as the *ivory trade,* also contributes to this loss. The *Convention on International Trade in Endangered Species of Wild Fauna and Flora (CITES)* and the *Convention on Wetlands of International Importance* both work to protect biodiversity on a global scale. [*biodiversity loss, prevention of; zoos, captive breeding programs in; turtle excluder device; dodo bird; passenger pigeon; Lonesome George*]

biodiversity hotspots

In the past, scientists concentrated on *mammals*, birds, and plants as the primary indicators of where to look for regions that have—or should have—the most *biodiversity*. These so-called hotspots were considered to be in need of the most help to prevent *biodiversity loss*. These larger animals have always been considered indicators that help scientists identify biodiversity hot spots.

Recent studies, however, indicate that more effort should go to "following the *insects*." By doing so, scientists are discovering that biodiversity hotspots are not always where they thought, meaning we are missing some important regions in need of protection. By studying ant populations, for example, scientists have calculated that Togo and Burundi each have almost 40 previously unknown species of ant genera associated with complex *food webs* never before studied. Both countries have lost about half of their forests in the past few decades, so these regions should be considered biodiversity hotspots in need of more study and protection. [*biodiversity loss, prevention of*]

biodiversity loss, prevention of

Biodiversity loss is preventable and a great deal of research goes into this field. However, measures are also being put into place for the unfortunate reality of what to do when a species is on the brink of *extinction* or has become extinct. This includes *captive breeding programs* in *zoos* and wildlife preserves where people maintain the only remaining individuals of a species. The San Diego Zoo has its own *Frozen Zoo* where the DNA of thousands of animal species are stored. This type of preparation is not only for animals but also includes *seed banks* that store plant seeds in case a species becomes extinct.

Another futuristic attempt to save *endangered species* involves the use of stem cells. Work is being done to help resolve one of the major problems with captured breeding programs—the fact that many endangered species will not mate when in captivity, making these efforts futile. It has been more than a dozen years since the northern white *rhinoceros*, believed to be *extinct in the wild*, has given birth in captivity. Some scientists are using stem cells in an effort to create egg and sperm cells. This would allow them to create fertilized eggs to be implanted in surrogate mothers. [*biodiversity, loss of; zoos, captive breeding programs; turtle excluder device*]

biodiversity, increased disease with a loss in

Recent research shows that a reduction in *biodiversity* within an *ecosystem* often results in greater risk of a disease outbreak within that ecosystem. Greater biodiversity appears to reduce the ability of disease to be transmitted between individuals.

For example, white-footed mice play an important role in transmitting *Lyme disease*. The mice are not harmed, but they transmit the disease to most of the ticks that feed on them, which then pass it along to other species, including humans. Opossums, on the other hand, play an important role in reducing the spread of the disease. They do not transmit the disease readily to ticks that feed on them—actually they eat most of the ticks that try to feed on them, thereby reducing the spread of the disease. Suburbia fragments woodlands to a size that supports mice populations but diminishes opossum numbers. The result is an increase in Lyme disease, because an important part of the food web is lost—the opossum. [*biodiversity, loss of*]

Biodiversity Treaty

See *United Nations Treaty on Biodiversity*.

bioethanol

Ethanol that is produced from *biomass*, as opposed to *fossil fuels*, is called bioethanol. Even though the process to create bioethanol and *ethanol* are different, there is little difference in the chemical structure of the two, so you often hear people use the terms bioethanol and ethanol interchangeably. Bioethanol is also called ethyl alcohol or grain alcohol (this is the same alcohol as found in alcoholic drinks).

Bioethanol is a *biofuel* created by the fermentation of sugar or grain crops or from plant and *animal wastes* and used as an alternative to *gasoline*. More than 86 billion liters of ethanol fuel was created in 2010. The United States is the leader in bioethanol production. Corn is the primary feedstock used to produce bioethanol in the United States. Most U.S. cars can run on E15 (a blend of 15 percent ethanol and 85 percent gasoline), although most gasoline contains less than this amount. (When mixed with gasoline, it is sometimes called gasohol.)

Brazil uses sugarcane to produce its ethanol, and most cars there now run on *flex fuel*, which can be any ratio of ethanol to gasoline, depending on the price of each. [*alternative fuels*]

biofouling

The growth and colonization of plants and animals on submerged surfaces. Biofouling occurs on ship hulls, buoys, wharfs, and almost any other marine surface. *Aquaculture* nets are also affected by biofouling. When these nets become colonized, nutrients and oxygen cannot get in and waste products cannot get out of the nets, resulting in the death of their inhabitants.

Prior to 1987, organo-metal paints were applied to surfaces to prevent biofouling. However, these substances were found harmful to the environment and banned. A new generation of antifoulants are nontoxic and make it physically—as opposed to chemically—difficult for organisms to attach themselves to surfaces or nets.

Invasive species are known to make their way to new locations via biofouling. They attach to a ship hull or get into the *ship ballast* in one port and are then released or break off in another port.

biofuels

One type of *biomass energy* source. Some believe biofuels could replace one-third of the U.S. consumption of *fossil fuels,* dramatically cutting *greenhouse gas* emissions. One of the most popular uses of biofuels is to replace *automobile gasoline.* It is estimated that biofuels can meet 25 percent of the global *transportation* fuel needs by 2050.

The original methods for developing biofuels are now called first-generation biofuels. They are produced in three ways: 1) *biochemical conversion*, which uses organisms such as yeasts and *bacteria* to create fuels such as *bioethanol, biodiesel,* and *biogas*; 2) *thermochemical conversion,* which uses heat to create fuels such as *syngas* and *methanol*; and 3) using refined *plant-oils* as fuel.

We have come to realize that even though these biofuels have several advantages over fossil fuels, they are problematic because growing crops for fuel competes with growing crops for food. On many occasions, food prices have increased because of this competition. *Lester Brown* noted that the amount of corn needed to make enough ethanol to fill an SUV with fuel just one time could feed a hungry person for an entire year.

In addition to this dilemma, many of these first-generation biofuels have lower energy content than fossil fuels, meaning they are not economically feasible on a large scale. Scientists have created *second-generation biofuels* to address these concerns.

biofuels, second-generation

Also called advanced *biofuels*, these new types of biofuels reduce the problems associated with first-generation biofuels. The problems include a fuel vs. food issue, because crops grown for fuel compete with crops grown for food. Also, most current biofuels have less energy content then *fossil fuels*, meaning they are not economically competitive. Second-generation biofuels address both of these issues.

The most promising is *cellulosic biomass,* which uses parts of plants such as stalks that cannot be digested by humans and therefore do not compete with food. These indigestible substances—cellulose and lignin—compose the woody portion of a plant. Others types of cellulosic biomass include sawdust and citrus peels, which are considered waste products and would otherwise be discarded. In addition to alleviating the fuel vs. food issue, cellulosic ethanol—the fuel produced by cellulosic biomass—has a greater energy content then *bioethanol*—the most common first-generation biofuel. This makes it a much more efficient fuel that can compete with fossil fuels.

You are not filling up your car yet with cellulosic ethanol because research continues to look for ways to economically produce it. It is difficult to break down cellulose and convert it into fuel. This is, in part, why people don't eat it—we cannot digest it. Ruminating livestock such as cows can break down cellulose—but they have four stomachs, nothing like ours. Termites also have this ability thanks to *bacteria* in their guts.

Research is ongoing on many fronts. Some scientists are creating cellulosic ethanol by using *fungi* or *GMO (genetically modified organism) bacteria* that have been modified so they can break down cellulose. For example, a GMO bacteria is converting switch grass—a nonedible, easily grown plant—into cellulosic ethanol.

Other research uses seaweed—a type of *algae.* They have created a new form of agriculture called algaculture to grow large quantities of these algae. Algae are not typically used as a food crop in most regions so do not compete for food resources. This research also uses GMO bacteria to break down a cellulosic-like component in the algae called alginate. The first large-scale algaculture biofuel farm is being built in Columbus, New Mexico. The company hopes to make about 100 barrels of oil a day from this facility, expected to open in 2018. The company has financial backing from Bill Gates and others.

Possibly the best answer to the food vs. fuel issue is to create biofuels not from plant biomass at all, but from waste. Preliminary plans are being made to open a waste-to-biofuel facility in Reno, Nevada, that would convert more than 150,000 tons of *municipal solid wastes* into biofuels to be sold at competitive prices.

biogas
Biogas is a type of *biofuel* that is usually a mixture of 60 percent *methane* and 40 percent *carbon dioxide*, which is produced by *anaerobic bacteria* feeding on *biomass* such

as plants and livestock manure. Biogas can be produced under controlled conditions in *methane digesters* and then used as fuel.

Perhaps the biggest potential for biogas is to use animal manure from livestock in *CAFO* (confined area feedlot operations*)*. This manure is a waste problem most farms must pay to have removed. Instead, the manure is burned in digesters to produce biogas, which is then sold. Some farms have become energy independent by using their own manure to create their own fuel to run their own equipment.

It is also created naturally in *landfills*, where it is often collected and used for fuel.

biogeochemical cycles

The cycling of chemicals essential to life (*nutrients*) between the *abiotic* (nonliving) and *biotic* (living) parts of the *biosphere*. Chemicals are taken in from the *soil*, *water*, or *air* by organisms and used as an energy source or for growth. Once in the organism, they are transformed into biologically active substances. Some of the elements involved in biogeochemical cycles are carbon, nitrogen, sulfur, and phosphorus. These cycles can be categorized as being either gaseous or sedimentary. Gaseous biogeochemical cycles occur rapidly—usually taking from a few hours to a few days—as opposed to sedimentary cycles, which take thousands or millions of years to complete.

Gaseous cycles move chemicals back and forth between the air or water and organisms; examples include the oxygen and *nitrogen cycles*. Sedimentary cycles include the solid Earth in the cycle; the sulfur and *phosphorus cycles* are examples. Both gaseous and sedimentary cycles are being modified by human intervention. [*biogeochemical cycles, human intervention in*]

biogeochemical cycles, human intervention in

Human (*anthropogenic*) intervention affects the normal flow of *biogeochemical cycles*. These cycles are the foundation of all life on Earth. As an old saying goes, "It's not nice to mess with mother-nature," and this certainly pertains to interfering with these cycles. The importance of biogeochemical cycles and the ramifications of human interference become apparent if we look at the recently defined *safe operating space* of our planet.

Scientists believe humans must stay within certain boundaries if our planet is to remain a safe, sustaining place for future generations. They identified nine boundaries we must not cross if the planet is to remain safe. Two of these nine planetary boundaries pertain to biogeochemical cycles: changes made to the *carbon cycle*, resulting in *climate change,* and changes made to the *nitrogen cycle* and *phosphorus cycle*. Only two of these nine boundaries are considered to have already been crossed and they both pertain to these cycles.

Climate change is the first boundary crossed and it is because of changes we have made to the carbon cycle. The carbon cycle has been modified by extensive burning of *fossil fuels* as the world's primary energy source. Burning fossil fuels, like all organic matter, releases *carbon dioxide* into the *atmosphere*. It is a *greenhouse gas* and responsible, in part, for climate change. In addition to burning fossil fuels, vast tracts of *forests* are burned (*deforestation* and *slash-and-burn cultivation*), adding still more carbon dioxide. Not only does forest-burning add carbon dioxide to the atmosphere, but also the destruction of trees and other plant life means they can no longer remove carbon dioxide from the *atmosphere*.

The *nitrogen cycle* is the other boundary crossed. It has also been modified by human intervention, removing free nitrogen from the atmosphere and adding large amounts of fixed nitrogen into the soil and water *ecosystems*. Most of this occurs because of *high-impact agriculture* and its addiction to *synthetic fertilizers,* the manufacturing of which converts the free nitrogen into fixed forms such as ammonia. Much of these fertilizers, loaded with nitrogen, never make it into a plant's roots where they belong but end up running off of the farms and entering bodies of water, resulting in a process called *nutrient enrichment.*

Also adding fixed nitrogen into the cycle are *factory farms* where large numbers of livestock are raised. They produce vast amounts of animal wastes in concentrated areas (*CAFOs*), contributing high concentrations of nitrogen that add to the problem. Nutrient enrichment collects fixed nitrogen from all of these sources, negatively impacting local ecosystems by causing *cultural eutrophication* in ponds and lakes and also in much larger ecosystems such as oceans, where they create large *dead zones* in the oceans.

In addition, burning fossil fuels not only adds carbon dioxide to the atmosphere, as mentioned earlier, but also creates and releases *nitrogen oxides* that react with water vapor and come back to the soil as *acid precipitation,* adding still more fixed nitrogen. All of these additional inputs of fixed nitrogen entering the natural nitrogen cycle have caused scientists to consider the problem beyond the scope of what our planet can sustain and beyond the safe operating system.

The *phosphorus cycle* has also been modified by the extensive use of synthetic fertilizers and the dumping of wastes into bodies of water. This cycle has not been identified as having passed the safe operating space limits, although it is well on its way to do so.

biogeochemistry

Biogeochemistry is primarily concerned with how elements such as carbon, nitrogen, and phosphorus are used by organisms and the impact this has on the chemical composition of the Earth. [*carbon cycle; nitrogen cycle; phosphorus cycle*]

biological amplification

Organisms that feed on plants that have been contaminated with *pesticides* or other *toxic* chemicals often accumulate these substances in their tissues—a process called *bioaccumulation*. If these animals are eaten by *predators*, these substances are passed along to those organisms, as well. As organisms higher up the *food chain* feed on contaminated individuals, the toxic substances increase dramatically in a process called biological amplification. (Also called biological magnification.)

Biological amplification has been documented in many food chains with many toxic substances, but the best known studies were done on *DDT* in the Long Island Sound of New York. Water in the Sound was found to contain 0.000003 ppm (parts per million) of DDT. The plankton living in the water, by the process of bioaccumulation, had a concentration of .04 ppm. Small fish feeding on the *plankton* had accumulated a concentration of about 0.3 ppm. Larger fish that preyed on the small fish had levels of 2.0 ppm. Predatory birds, such as the osprey, that fed on these fish had extreme levels of 25.0 ppm. This is ten million times greater than originally found in the water!

Biological amplification can affect any organism. Humans feed at the high end of food chains, so significant concentrations of many toxic substances have been found in our bodies, often called our *body burden*.

biological control

Before synthetic *pesticides* became the standard method of controlling *insect* pests, people successfully used natural control methods. These natural methods have been advanced with science and technology and are now viable ways of controlling pests without chemicals. Biological control is one of the few types of pesticides that only kill the *target organism* (meaning the actual pest) and not all forms of life.

Biological control uses populations of *parasites*, *predators*, and *pathogens* to control pests. Successful implementation of these methods goes back more than 100 years when the cottony cushion scale—an insect pest in California—was successfully controlled by releasing the vedalia *beetle*, a predator that fed on the scale insect. Hundreds of biological control success stories have been documented throughout the world.

One of the most common biological control methods uses *bacteria* called *BT* (*Bacillus thuringiensis*). These bacteria come in different strains, each targeted for a different pest. Some strains kill mosquitoes, others are used to control the caterpillars of leaf-eating moths such as the gypsy moth, and there are many others. BT is now a *genetically modified organism* and routinely used on *high-impact agriculture* to control pests.

Introducing large numbers of ladybird beetles (ladybugs) is another common biological control. The larval stage of the beetle devours enormous numbers of aphids and scale insects. Many small parasitic wasps (harmless to humans) parasitize immature forms of pest insects, killing them before they mature. Some small wasps, for example, lay eggs in the caterpillar of the tobacco hornworm. The young wasps hatch within their host (the hornworm caterpillar) and feed on the caterpillar's insides, killing it. The well-fed wasps then emerge from the carcass.

Biological control and the broader method of *integrated pest management* remain underutilized but are highly effective. [*biological control methods*]

biological control methods

Three types of *biological control* are used: importation, conservation, and augmentation. Importation involves identifying, in some remote area, a natural enemy of the pest. The natural enemy is then imported and released in the region containing the pest. The imported insect must be able to survive in the targeted area. Once released, these natural enemies thrive on their own, destroying the pest. In the United States, the alfalfa weevil has been successfully controlled by using this method.

Conservation involves finding natural enemies of the targeted pest locally. Once a natural enemy is found, techniques are implemented to help it prosper and flourish. This includes discontinuing use of any chemicals harmful to the natural enemy and using cultural control methods, such as plowing and *irrigation* techniques that would aid the natural enemy's ability to thrive.

Augmentation involves introducing large numbers of *parasites* or *predators* of the pest. These natural enemies are mass-reared specifically for this purpose. This differs from importation and conservation in that the natural enemy isn't expected to reproduce and survive indefinitely as a deterrent. Small parasitic wasps are often used for augmentation control. They lay their eggs in the larval stages of the pest, preventing it from maturing. [*biological pesticides; pesticides; invasive species; conservation agriculture*]

biological magnification

See *biological amplification*.

biological oxygen demand (BOD)

See *biochemical oxygen demand.*

biological pesticides

A biological *pesticide* is a type of *natural pesticide* used as an alternative to synthetic pesticides. Synthetic pesticides pollute the environment, contaminate our *groundwater*, and can become concentrated in our foods by *biological amplification*. In addition, many *insect* pests have become resistant to these synthetic chemicals by the process of *natural selection.*

Biological pesticides use natural enemies including microbes such as *bacteria, viruses*, protozoans, and *fungi,* that attack and destroy certain pests (mainly insects). They have two important advantages over synthetic pesticides. First, they cause far less harm to the environment—if any. Second, they typically kill only the *target organism.* As *Rachel Carson* said in <u>Silent Spring</u>, synthetic pesticides should be called *biocides* because they do not distinguish between harmful and helpful organisms—they kill all forms of life. These biological pesticides are *predators, parasites,* or disease-causing organisms, so they kill only the host species they attack—the pest.

The sale of biological pesticides has increased substantially over the past decade but remains a tiny percentage of the total pesticides applied. [*biological control methods; conservation agriculture; nanopesticides; pesticide dangers*]

bioluminescence

Some organisms, such as the firefly and numerous deep-sea marine life, can convert chemical energy into light. This light is often called cold light because very little heat is produced. Only 2 percent of the energy used to generate bioluminescence is lost as heat compared with a typical *incandescent* light bulb, which loses over 96 percent of its energy as heat and is therefore hot to touch. [*light bulb technologies; biomimetics*]

biomass

The dry weight of any organism. It is commonly used in two ways. First, biomass is used to describe the quantity of certain groups of organisms (called *trophic levels*) that exist in an *ecosystem.* For example, the biomass of *producers* (green plants) in an ecosystem is far greater than the biomass of *primary consumers.* In other words, by weight, a typical ecosystem includes more plants than animals.

Biomass is also used as an energy source. For example, burning wood and converting plants into *biofuels* to be burned later are ways of using *biomass energy.* [*alternative energy sources*]

biomass direct combustion

Biomass direct combustion is one of three ways of using *biomass energy*. It refers to burning *biomass* to create heat, steam, or electricity. Almost any type of biomass can be burned, the most common being wood. Only about five percent of U.S. homes use wood as their primary source of heat; however, in *less developed countries*, wood and animal manure supply about half of all energy needs.

Wood and animal manure are typically burned to create heat but also to generate electricity. Small wood-burning *electric power plants* exist, most of them owned by independent power producers associated with the forest industry. Most of these power plants use waste wood produced in lumber and paper mills as the biomass, because it is cheaper to burn it than to dispose of it.

Other types of biomass are now commonly used for direct combustion. *Energy plantations* cultivate fast-growing trees, shrubs, and grasses specifically to provide biomass for energy. *Municipal solid waste* (MSW) is occasionally burned in *waste-to-energy power plants* instead of going into *landfills*. This power is typically used to provide electricity for the local surrounding area.

Biomass direct combustion has environmental advantages. Substituting biomass for *fossil fuels* reduces emissions of *carbon dioxide*, a *greenhouse gas*, but only if the plants burned are replanted. These new plants will continue to remove carbon dioxide from the atmosphere (during *photosynthesis*), something fossil fuels cannot do.

Using waste products such as sawdust from wood mills or agricultural waste such as bagasse (leftover sugar cane) keeps these materials out of landfills and is put to good use. Most of this biomass is waste material so it does not compete with actual food crops.

However, there are also some disadvantages. *Air pollution* from biomass direct combustion is an important concern. Biomass combustion produces large amounts of *particulate matter* that contributes to air pollution. If municipal solid waste is burned, it might also contain high levels of toxic substances and therefore create a *hazardous waste* disposal problem. Establishing large energy farms means more environmentally damaging *monocultures* and accompanying *pesticides*. In addition, using land to grow energy plants competes with food crops.

With all of this said, there is still evidence that burning biomass produces far less overall greenhouse gases than *coal*.

Although some good applications exist for biomass direct combustion, converting biomass into *biofuels* shows the most promise as a viable alternative to fossil fuels.

biomass energy

Many *alternative energy sources* can replace *fossil fuels*. *Biomass energy* is one of these alternatives. The three major forms of biomass energy are *biomass direct combustion*, where *biomass* such as wood, animal manure, or waste material is burned directly for heat or to create electricity; *biofuels*, where biomass is converted into a fuel that is then burned for energy; and *plant oil fuel*, where oils naturally produced by some plants are refined and used as fuel.

Biomass energy sources only provided about four percent of the total energy used in the United States in 2010. Almost half of that was from wood or wood-waste products such as wood chips and sawdust; a little more than 40 percent came from biofuels, and a little more than ten percent was from *municipal solid waste*.

The Biomass Users Network (BUN), founded in 1985, has more than forty country members that exchange technologies and ideas to promote the use of biomass energy.

biome

Terrestrial regions of the Earth are divided into large *ecosystems* called biomes, each with distinct combinations of climate, geology, and relatively stable collections of organisms. The two most important factors that determine the types of plants and animals found in any biome are temperature and precipitation. Authorities differ on the number of different kinds of biomes; descriptions include as few as six or as many as twelve. Eight biomes are listed here: 1) desert, 2) *tundra*, 3) *grassland*, 4) *savanna*, 5) woodland, 6) taiga, 7) *temperate deciduous forest*, and 8) *tropical rainforest*.

Biomes are found at corresponding latitudes and altitudes because they both can produce the same type of environment. For example, tundra is found in subpolar regions (latitudes near the poles) and also in alpine regions (high altitudes well above the tree line).

Today, many people believe biomes should be replaced with new descriptions called *anthromes*.

biomaterials

It seems like almost everything is made of *plastics*—a *petroleum* product. Someday petroleum and other *fossil fuels* will become scarce and too expensive to be used to make plastic toys and other "stuff." When this happens, we will have to find alternative materials to make all of this stuff. The answer is to look for natural materials to replace petroleum-based plastics—what many are calling biomaterials. Which organisms are the most likely to replace plastics is unknown at this time, but a great deal of research is being done with *fungi*. Time will tell.

Biomaterials might make a grand entrance before petroleum products make their exit. Many people believe that even though fossil fuels are a nonrenewable resource, they will be around for a very long time. However, even if this turns out to be true, biomaterials might have to replace plastic products anyway. Plastics are one of our biggest environmental waste problems because they do not decompose and many are toxic. So whether we run out of petroleum and can no longer afford to use it to make our stuff—or if we decide plastics are a waste problem that cannot continue—biomaterials will probably make their entrance in the not too distant future. [*biomimetic; bio-ore; bioplastics; biotechnology; nanoparticles; nanotechnology*]

biomimetic

Inspired by nature. What could be designed any better than something designed by nature? If you take the process of *natural selection* and combine that with a few million years, sooner or later nature comes up with the most efficient and eloquent forms and designs. Humans still cannot match most of nature's designs such as the flying abilities of a dragonfly. However, we have always learned from nature, going back thousands of years, such as mimicking the natural material silk, first introduced to the world by *insects* but now synthetically manufactured by humans.

But today we have taken this mimicry to the next level—developing products that borrow nature's incredible engineering skills. These new naturally designed hi-tech products are often called biomimetic products.

For example, some of the beautiful colors on butterfly wings are not created by the typical pigments we are accustomed to but by a crystalline structure that breaks up light like a prism. Some companies use a manufactured version of this structure to brighten the colors of cell phone displays and reduce energy requirements.

Other examples include copying the ability of spiders to make their webs reflect ultraviolet (UV) light. (Birds can see UV light, so they don't fly into the spider's web, destroying them.) A company has created windows embedded with materials similar to what the spiders produce naturally. This makes the glass panes visible to birds so they are less likely to fly into the office building's windows.

A beetle's ability to capture morning fog in a desert has been turned into a biomimetic product. The beetle's entire body is covered in a waxy substance that repels water. It also has tiny bumps in an unusual arrangement on its body. Droplets of water form on these bumps and are repelled by the wax so they flow down the insect's back and into its mouth. A company has copied this natural design to create biomimetic devices that capture moisture from the morning air to collect and provide fresh water for indigenous people living in the desert.

Biomimetics is not limited to copying the physical structure of organisms; it can include mimicking behavior as well. Studies about ant behavior analyze the best way for people to evacuate a building during an emergency. Biomimetic examples are endless and the potential enormous. [*bio-ore; bioplastics; biotechnology; nanoparticles; nanotechnology; biomaterials*]

bio-ore

Bioremediation uses plants and animals to cleanse contaminated environments. For example, some plants can absorb high concentrations of contaminates such as *heavy metals* from the *soil*. The soil is cleansed of the contaminate because it all ends up in the plants. The plants are called bio-ores, because theoretically they can then be processed like a mineral ore, to retrieve the metal, which can be recycled for other purposes. [*hyperaccumulators; phytoremediation; remediation of hazardous waste; Rocky Mountain Arsenal*]

Biophilia Hypothesis

A hypothesis put forth by *E. O. Wilson* in his 1984 book <u>Biophilia</u>. It means "love of life." He proposes that humans have an innate connection with all forms of life and that we have a need to be connected to all life. This might be demonstrated by the positive impact animals can have on people with disorders such as autism. [*nature-deficit disorder; zoos; national parks (global); nature centers; Gaia Hypothesis*]

biopiracy

The use of biological materials, such as medicinal plant extracts or genetic cell lines, obtained by people usually in more advanced nations without appropriate compensation to those in less advanced nations that provided these materials, often occurring without the consent or even knowledge of these people. In many cases, the materials pirated are considered "traditional knowledge"—for example, a plant that has been harvested and used for hundreds of years within a less developed nation. If a company from another (probably more developed nation) procures the plant, decodes the genetic structure, and then patents it for commercial gain, this would be considered an act of biopiracy. However, few international agreements pertain to this type of activity, and when they do, they are difficult, if not impossible to enforce in a court of law. [*environmental justice; environmental ethics; genetically modified organisms; DNA sequencing technology; primitive technologies*]

bioplastics

Plastics are made from petroleum, which is a *fossil fuel*. Bioplastic refers to plastic-like materials produced from living organisms. For example, a company is producing a bioplastic from fungi—a mushroom, to be specific—that has properties similar to *PS* (polystyrene). It will be used as a packing material or to manufacture *flip-flops* (both typically made from real plastic).

This bioplastic uses the mycelium of a mushroom, which looks like very fine roots. The mycelium is grown in a form or mold and can assume any shape needed. (Much like plastic does.) The mycelium produce their own glue-like substance that binds them together so the material becomes firm within the mold, and the end result is very similar to regular plastic. Products made from bioplastic are *biodegradable,* so there is no waste.

Other bioplastic companies are using chicken feathers, *algae,* and soy. These bioplastics cost at least ten percent more than regular plastics, but when the *externalized cost* of *landfills* and *plastic pollution* in the oceans are figured in, bioplastics offer a great cost savings to the planet. The possibilities seem endless and the industry is just getting started. Look for bioplastic products soon in a store near you. [*biomaterials; biomimetics; bio-ore; biotechnology*]

bioregions

National borders are probably the most important type of border. Bioregions, however, are areas distinguished by natural borders such as where water meets the land or where forest meets grasslands. Some people believe environmental policy would be much better served if it was based on bioregions—not regions delineated by national borders. Although this might be a logical idea, it certainly is not a practical one in this day and age. [*biomes; anthromes; Holdridge Life Zone System; zones of life*]

bioremediation

The process of using organisms to detoxify, absorb, or otherwise render harmless *hazardous* or *toxic wastes* found in the environment. *Bacteria* and plants are being used to clean up these wastes found in water and the soil.

One recent example of bioremediation occurred after the *BP Deepwater Horizon oil spill*. The oil spilled during this disaster contained large amounts of *methane*, a *greenhouse gas*. Many species of bacteria found in the ocean consume *methane* gas, so scientists wanted to see if these bacteria would remove the excess methane released during the spill. The studies showed that these bacteria consumed almost all of the excess methane gas shortly after the spill, demonstrating the power of bioremediation. Much of the oil itself was also consumed by oil-eating bacteria and *fungi*.

Another example of bioremediation is the cleaning up of *radioactivity* after the *Chernobyl* nuclear power plant disaster. *Fungi* and *algae* were used to clean up the radioactive soil and water around the site, and algae and bacteria were used to clean up radioactivity from sludge found in nearby swimming pools. [*hyperaccumulators; phytoremediation; remediation of hazardous waste; Rocky Mountain Arsenal*]

biosensors

High-tech sensors that are worn to detect changes in a person's body or in the environment. Once a change is detected, the sensor causes some sort of indicator device to inform the person of the change or to record the event.

For example, long distance runners or cyclists can wear a temporary tattoo that detects changes in the person's sweat indicating the presence of lactic acid. (This means the person is no longer metabolizing *aerobically* and is beginning to use glucose stored in muscles, a form of *anaerobic* respiration—commonly called "hitting the wall"). A tiny device attached to the tattoo indicates the change.

Biosensors can also work by detecting changes that affect one's health, such as *air pollution* biosensors worn by people with severe *asthma*. [*biotechnology; biomaterials; nanotechnology; robotic bees; synthetic biology*]

biosolids and wastewater

Sewage treatment facilities separate solids from the liquid portion of sewage. The solid waste, called *sludge* at the end of the process, is then usually dumped in bodies of water, causing *water pollution* and *dead zones*. Some municipalities are trying to reduce this damage by converting the sludge into a usable substance called biosolids. They do this by allowing microbes, both *aerobic* and *anaerobic*, to stabilize the sludge, dramatically reducing both smell and pathogens. The end result is a marketable product similar to *compost* that can be used as an *organic fertilizer*. Some cities sell it and others provide it free to residents for use on their gardens or *lawns*. In addition, the anaerobic portion of the stabilizing process produces methane gas used to generate electricity in *waste-to-energy* facilities.

biosphere

The biosphere is the relatively small portion of our Earth that contains life. The term was first coined in 1927 by a Russian scientist, Vladimir Vernadsky. Organisms are typically found in the lower portion of the *atmosphere* (troposphere), on and just below the surface of the land (*lithosphere*), and within bodies of water (*hydrosphere*) and the *sediment* below. This places the vast majority of life no more than a few inches below and no more than a few hundred feet above the Earth's land and waters.

However, in recent years numerous exceptions have been discovered in extreme environments. These organisms, called *extremophiles*, have been discovered deep within the solid Earth, at great depths of the oceans and high into the Earth's atmosphere. [*noosphere; biosphere, shadow; aerobiology*]

biosphere, shadow
See *weird life*.

Biosphere 2
Biosphere 2 started as a privately financed business enterprise and was a cross between a grand scale science project and an entertainment business. (It is called 2 because 1 is planet Earth.) The 3.15-acre greenhouse facility in the Arizona desert, 20 miles north of Tucson, originally cost $150 million dollars to build. The artificial habitat contains thousands of species of plants and animals and was designed to replicate many natural *biomes*, including a *rainforest*, *savannah*, ocean *marsh*, and *desert*.

In September 1991, four men and four women were sealed inside the facility with the intent of performing ongoing environmental experiments in a closed environment for two years. The *wilderness* areas would naturally *recycle* gases and purify the waters. Waste would be recycled as *fertilizer* and a small farm would grow and produce the Biospherians' food.

The private enterprise was a failure and the facility abandoned until Columbia University took it over to conduct research. They found it too expensive to run and stopped operations in 2003. In 2007 it was taken over by the University of Arizona to conduct *climate change* research. The university has hired scientists to work specifically in the facility. Plans have it slated for large-scale improvements and additions. The hope is that a facility of this scale can perform research on climate change few others can duplicate. [*Biosphere Reserves*] {b2science.org}

Biosphere Reserves
Protected areas around the globe set aside by the *United Nations Man and the Biosphere Program*. These areas represent unique or threatened *ecosystems*. The actual area being protected is surrounded by a human buffer zone, which is meant to nurture its continued success through some form of *sustainable development* such as *ecotourism*, research, or education. There are 621 such reserves currently set aside in 117 countries. [*marine protected areas; marine reserves; national parks (global); World Heritage Sites*] {unesco.org/new/en/natural-sciences/environment}

biota

The living component of an *ecosystem*. The biota consists of the *flora* (plants) and *fauna* (animals). [*biosphere*]

biotechnology

Biotechnology combines biology and technology (often shortened to biotech). It attempts to solve problems by using organisms or biological principles. *Selective breeding* is one of the oldest forms of biotechnology and is responsible for vast improvements in *agriculture* and livestock, such as cattle. When organisms are bred selectively, they are still considered natural.

However, in recent years a new aspect of biotech has become prevalent and highly controversial. It includes genetic engineering—the manipulation of genes to create new varieties of plants or animals, now commonly called *genetically modified organisms* (GMOs). These organisms can produce more or "better" versions of existing crops, such as plants resistant to pests or those that remain fresh longer in storage. In some cases, an organism's genes are modified; in other cases, genes from one organism are placed into another—called transgenic organisms. In these and similar GMO organisms, they are actually manufactured organisms that would never have been created naturally.

But biotechnology is also used for many purposes besides breeding and GMOs. For example, many organisms are routinely used in the manufacture of products such as natural inks as opposed to petroleum-based ink for your printer. [*biosensors; biomimetics; biomaterials; bio-ore; bioplastics; geoengineering; nanotechnology; robotic bees; satellite mapping*]

bird flu virus

Also called avian influenza, this is a viral respiratory infection contracted by people from poultry. Various strains have been found in the past few years. The latest strain (identified in 2013) is called H7N9. They have all been localized in Asia, especially in China, Japan, and Taiwan. These strains of virus are exceptionally lethal. As of April 2013, more than 100 people contracted the H7N9 strain of the disease, of which at least 20 died. Ten years ago, the H5N1 strain was first reported and killed more than half of the 600+ people who contracted it.

The virus appears to jump from poultry to people, a *zoonotic disease*, but the fear is that it could mutate into a form that could jump from person to person, with the potential to result in a pandemic. To date, this is not believed to have occurred. [*spillover effect*]

bisphenol A
See *BPA*

bisphenol AF
See *BPAF*.

black carbon
See *soot*.

black finger of death fungus
This *fungus* is being used to control an invasive weed called cheatgrass. The fungus looks like a tiny little Mohawk haircut growing on the seeds found on the top of each cheatgrass stem. About 60 million acres of land has been overrun by cheatgrass, and researchers are looking for any way possible to stop the onslaught, including this fungus. [*invasive species; biological control; biological control methods; biological pesticides*]

black liquor
A waste product created when wood is converted into pulp (which is then used to manufacture paper). The waste is collected and used by the paper mill industry as a fuel source to run their facilities. [*cogeneration; waste-to-energy power plants; alternative energy sources*]

black yogurt
This term is used by seafarers for the bunker fuel used for commercial ships. It is composed of the thick, yogurt-like, black residue left over from refining oil. It contains a large amount of sulfur and is considered one of the dirtiest fuels used in industry. [*ship pollution; cruise-ship pollution*]

blobfish
Beauty is in the eye of the beholder, but most people would probably agree this must be one of the ugliest fish in the world—and it might be on its way to *extinction*. Blobfish are found in deep waters off the coasts of *Australia* and New Zealand. Even though they are inedible, deep-water trawlers bring them up as *bycatch* with crabs and lobsters, which is reducing their numbers. [*Ugly Animal Preservation Society; charismatic megafauna; flagship species; biodiversity, loss of*]

blogs, green

Blog is short for "web log," which is a journal someone or some organization writes and posts online for anyone to see. Many excellent blogs focus on environmental issues. Because they can be updated at a moment's notice, blogs are one of the best ways to stay on top of environmental issues. Some are independent, written by a solo writer, while others are linked to newspapers or other media outlets. Two excellent environmental blogs are *DOT Earth;* a <u>New York Times</u> blog and *Yale Environment 360.* Both blogs will keep you up-to-date on important green issues.

Microblogging, often just called tweeting in the United States, has been used successfully in *China* to raise public awareness about *air pollution* and has given rise to environmental activism in that country. [*social media, environmental; crowd funding, green; crowd research, scientific; crowdsourcing, eco-*]

Blue Angel Program

See *eco-labeling.*

blue baby syndrome

Nutrient enrichment is a serious environmental problem that has a health issue associated with it. Exposure to high levels of nitrates in *drinking water*, often caused by nutrient enrichment, is linked to a disease called blue baby syndrome. This can be lethal because it blocks the blood's ability to carry oxygen, thus resulting in the baby's blue appearance. [*autism studies and environmental factors; breast milk and toxic substances; geomedicine; medical ecology; nanoparticles and human health; puberty and the environment, early*]

bluefin tuna

See *longlines, commercial fishing.*

blue-jean pollution

Blue jeans are dyed and that dye contains *heavy metals* and many other *organic pollutants,* some of which are known *carcinogens*. The problem is that most jeans are made in one place—Xintang, Guangzhou, China—called the Blue Jean Capital of the World. If you produce 200 million pairs of blue jeans per year from one site and are allowed to dump your contaminated waste water into a river, it's not surprising there's going to be a problem. The Pearl River runs through this city and is often colored black from the dye. The river is an important source of water and food for the people downstream. [*water pollution; industrial water pollution; rivers in decline; water grabs; water shortages, global*]

Blue Marble photographs, original and updated

If one photograph can be given credit for helping launch the *environmental movement*, it is the famous Blue Marble, taken by the crew of the Apollo 17 spacecraft on December 7, 1972, as it circled the planet. The iconic image showed us we are truly aboard *Spaceship Earth*, flying through emptiness. It raised awareness about our environment's vulnerabilities and our own.

On January 4, 2012, NASA released a new, updated version of the Blue Marble with far better resolution, creating an even more striking view of Earth from space. It was taken by a high-tech camera that uses infrared imaging, onboard one of NASA's newer satellites called Suomi NPP, from an altitude of more than 500 miles. [*Earthrise photograph; Ansel Adams*] {nasa.gov/multimedia/imagegallery/image_feature_2159.html}

bluesign eco-label

See *textile industry*.

Blueways, National

This is a new White House initiative designed to help conserve *wildlife habitats* and preserve or enhance recreational opportunities in major river systems across the United States. The Connecticut River and its watershed received the first such designation in 2013. The river runs over 400 miles south from the Canadian border, dividing Vermont and New Hampshire, through Massachusetts, and into Connecticut, emptying into the Long Island Sound and the Atlantic Ocean. [*national parks, global; National Park Service*]

body burden

Describes all of the chemicals our bodies have absorbed or ingested, which now reside within us. In most cases, these chemicals are found within our fatty tissue but can also be found in bones or elsewhere. Most of these chemicals are typically not found in nature, so they do not belong in our bodies. Everything from *pesticides* and *herbicides* to *PCBs* and *BPA* are included in your body burden. Mother's *breast milk* is contaminated with many of these chemicals, so a person's body burden begins developing from birth. [*toxic cocktail; geomedicine; medical ecology; nanoparticles and human health; puberty and the environment, early*]

bog

Typically an intermediate step in *ecological succession*. It is a *wetland* that accumulates a large volume of *peat,* which is acidic and contains a great deal of decomposing plant matter. Grasses, sedges, and moss are the typical flora, with *insects*, reptiles, *amphibians,* and waterfowl, the fauna.

In many cases, bogs become filled with *sediment,* providing a *habitat* for shrubs and trees and then reach a *climax community* as a *forest ecosystem.*

Boone and Crockett Club

The Boone and Crockett Club was founded by *Theodore Roosevelt* in 1887. Its purpose is to protect *wildlife habitats* and ensure hunting is practiced in a responsible way. This club helped save *Yellowstone National Park* from development and pushed legislation that founded the National Forest System and the *National Park Service.* By creating the Rules of Fair Chase for responsible hunting, the Boone and Crockett Club serves as one of the nation's advocates of hunter's rights. {boone-crockett.org}

boreal forest

See *taiga.*

borehole disposal

See *nuclear waste disposal.*

botany

The division of biology involving the study of plants.

bottle bills

First attempts to initiate *recycling* have usually been in the form of bottle bills. This legislation, typically passed at the state level, enforces a deposit on bottles so they are returned (for a deposit) and then recycled. The cost for these programs is typically paid by the manufacturers. Oregon was the first state to implement a bottle bill, back in 1972. Estimates show that bottle *litter* drops drastically after these bills become law. So far, nine states (plus Guam) have bottle bills, with about a dozen pending passage. One state (Delaware) repealed a bottle bill that had previously been passed in favor of what they call a universal recycling bill. This might be a sign of things to come. As cities and states mandate the recycling of many types of waste materials, bottle bills might duplicate these new laws. However, most environmentalists believe bottle bills are a simple, effective way to keep litter down and recycling up.

The ten states with bottle bills are California, Connecticut, Hawaii, Iowa, Maine, Massachusetts, Michigan, New York, Oregon, and Vermont. (*disposal fee; cap-and-trade; internalizing costs*]

bottled water

Bottled water is very big business globally. In the United States alone, it is worth about 22 billion dollars a year. It takes 17 million barrels of *oil* to make all of those *plastic* bottles each year. When you think about it—it is a business that takes a product that is free in most developed nations, packages it in energy-consuming, nonbiodegradable plastic bottles that will never decompose, ships it all over the world, and sells it at inflated prices. (Of course, because it would have been free, any price is an inflated price.) In many cases, it is the same water that comes out of a resident's faucet. The amount of energy and accompanying environmental damage that goes into making it, shipping it, and then disposing of it is staggering when you think that it only provides a few minutes of use.

From an environmental perspective, the energy used to create and ship this product is large, but the most noticeable and possibly the most harmful aspect is the disposal problem. Most of these bottles are made from *PET* plastic. Manufacturers have reduced the amount of plastic in each bottle significantly over the past few years and *recycling* of PET products is one of the more successful *plastic recycling* programs. Even so, fewer than one in three bottles are recycled. And those other two add up to a lot of bottles sitting in *landfills*. More than one million tons of plastic bottles end up in U.S. landfills each year. The simple, obvious solution is to get a reusable water bottle and save yourself money and reduce your *ecological footprint*, big time. [*plastic pollution; plastic recycling; economy vs. environment*]

bottom ash

The *incineration* of *municipal solid waste* produces *fly ash*, which is emitted up the smokestack into the *atmosphere*. What is left are the charred remains that do not go up the stack, called bottom ash. Bottom ash must be disposed of either in *landfills* or by being re-incinerated. Because it often contains concentrated amounts of *toxic substances*, it is usually handled as a *hazardous waste*.

bottom trawling, deep sea

Bottom trawling is a method of fishing that is so destructive to the environment it is called *strip-mining* of the ocean floor. One or two boats drag a large, heavy net (the trawl) along the bottom of the sea bed, trapping everything it runs over. Only a small percentage of the fish caught are the intended catch. Most of what is pulled in is *bycatch*—unintended species of little commercial value, including starfish and sponges. A single pass of a trawl removes up to 20 percent of the entire sea floor life.

In the United States, the *National Oceanic and Atmospheric Administration* banned bottom trawling off the Pacific coast in 2006 and has placed many restrictions on the remaining U.S. coastline. Some countries have bans or some form of restrictions.

Outside of a nation's coastal waters, considered the *exclusive economic zone* and controlled by each nation, there is no international ban against bottom trawling on the high seas. In 2006, an effort was made in the United Nations to create such a ban, but so far it has failed. Countries, including Iceland and Japan, that have large fleets of bottom trawlers derailed a planned moratorium.

However, in 2007 an international agreement was created to protect much of the South Pacific region—about 25 percent of all international waters—from this destructive fishing method. [*fishing, commercial; fish populations and food; seafood eco-ratings; driftnets; factory ship; ghost fishing; gillnets, commercial fishing; illegal, unreported, and unregulated fishing; longlines, commercial fishing; overfishing; purse seine nets; aquaculture*]

Boulding, Kenneth

(1910–1993) A U.S. economist credited with coining the phrase *Spaceship Earth*, alluding to how our planet and all onboard are within a closed system, with limited resources that must be used wisely. He argued that unlimited *population growth* is not sustainable considering our finite resources. He stated that all "first-class passengers" on our spaceship will suffer the same consequences of a depleted spaceship as do the "second-class passengers." He was one of the first to promote a sustainable economy as opposed to our *throwaway economy*. He also wrote about the abolishment of war to improve the prospects of our spaceship. [*philosophers and writers, environmental; conspicuous consumption; the tragedy of the commons*]

bovine growth hormone (BGH)

In *factory farms*, milk cows are often dosed with this hormone to increase milk production. BGH can result in mastitis (a painful inflammation of the udder) as well as many other serious health risks to the cows. BGH is not used in *organic* milk, and many regular milk brands have stopped using it because of consumer opposition. Even though it is still found in some milk in the United States it has been banned by the European Union and Canada. [*organic beef; Organic Consumers Association; organic farming and food; food eco-labels; organic food labels*]

BPA

Short for Bisphenol A. This synthetic petrochemical product is the main ingredient in a common type of *plastic*—polycarbonate plastic. It has been used to make baby bottles, water bottles, dental sealants, compact discs, canned foods, and numerous other products. The chemical can *leach* out of the plastic into food or directly into our bodies. It is also used to print receipts on most of the thermal cash-register printing machines, where it can rub off and be absorbed through skin.

It has been so prevalent that trace amounts are showing up in *drinking water* and the *environment* in general. BPA has been found in most people's urine and has become part of our *body burden* of a *toxic cocktail* that most of us now carry with us in our bodies.

Research shows that BPA is an endocrine *hormone disruptor*—a substance that mimics the female hormone estrogen. The petrochemical industry insists the chemical is harmless, but hundreds of independent research studies have linked it to cancer, low sperm count, and early onset of puberty.

Canada was the first to take action against this chemical by providing a list of all products containing the substance. The European Union has recently banned the substance. In the United States, the *Food and Drug Administration* banned BPA in baby bottles, children's cups, and a few other products, but not elsewhere. Today many plastic products are labeled as BPA-free on a voluntary basis.

BPA, like thousands of other substances, was grandfathered in when *Toxic Substance Control Act* legislation was passed back in 1976, so it was never properly tested and has not been to this day. [*puberty and the environment, early; BPAF*]

BPAF

Short for Bisphenol AF. A substance similar to *BPA* but more recently discovered. It has most of the same characteristics and harmful attributes as BPA, but it is considered more dangerous because of the slightly different molecular structure. It too is an endocrine *hormone disruptor*. Like BPA, it is found in polycarbonate plastic but also in epoxy resins, optical fibers, components that process foods and drugs, and some electronic equipment, among others.

Just as the public found out about the dangers of BPA, we will probably also learn the hard way about the dangers of BPAF—after it is already in our bodies. These substances get into our environment and our bodies because of weak legislation, such as the *Toxic Substances Control Act* of 1976, which has never been updated. This logic is contrary to that of many other countries, such as those in the European Union, that follow the *precautionary principle* and only allow substances into the marketplace after they have been proven to be safe.

Bp fungus

See *chytrid fungus*.

brackish water

Water that has some degree of salinity (salt concentration) between fresh and salt water. *Estuaries* contain brackish waters because they are typically found at the mouth of a river and the salinity changes with the tides. Organisms that live in brackish water must have a wide *tolerance range* for salinity, something quite unusual. In tropical regions, *mangrove forests* are the best known of these brackish habitats. Mangrove trees are among the few plants capable of surviving and thriving in these waters. In temperate regions, salt *marshes* fill the same niche. Both of these *ecosystems* are exceptionally fragile because of the limited number of species that can inhabit them, and they are therefore sensitive to environmental stress. [*aquatic ecosystems; marine ecosystems*]

BP Deepwater Horizon oil spill

A deep sea oil well blew out about 50 miles from the Louisiana coast on April 20, 2010, causing a large explosion on the Deepwater Horizon Drilling platform that stood above this well. Eleven workers were killed in the blast. At the ocean floor, the 5000-foot pipe that connected the well to the drilling platform snapped. A safety device called a blowout preventer that was supposed to shut down the flow in just this situation did not function.

Over the next few months, almost five million barrels of crude oil spewed into the Gulf of Mexico before it could be capped. It is the largest *oil spill in U.S.* history (for contrast, the *Exxon Valdez* oil tanker spilled less than 250,000 barrels). The company spent billions of dollars cleaning up the immediate effects of the vast spill.

Shortly after the spill, large oil plumes were observed in the Gulf of Mexico, many miles wide and long and up to 300 feet thick. The plumes were floating in the middle of the water column—not on the bottom. The exact damage to *ecosystems* from these plumes could not be ascertained, but most agree it cannot be good. Also, oil was found resting on the sea surface. A recent report, three years after the event, shows a serious negative long-term impact. Some of the results include above-average *dolphin* deaths and more than 1700 *sea turtles* found stranded—far more than normal.

After the incident there was an outcry for stricter regulations on deep water drilling and safety. There was a brief moratorium, but few meaningful new laws put in place, and deep sea drilling is expected to continue to expand. [*Oil Pollution Control Act; oil pollution in the Persian Gulf; oil recovery or extraction; oil spills in history, ten worst; pipeline spills*]

Brazil

Much of the *Amazon tropical rain forests* and river reside in Brazil, meaning survival of one of the world's largest and most productive *ecosystems* depends in large part on this country's government and people. The country's interest in protecting the environment has been cyclical.

The Amazon was in the news for many years in the late 1980s when *deforestation* reached its worst, and activist *Chico Mendes* was murdered. However, Brazil then dramatically turned the tide against deforestation. Most of the loss resulted from clearing land for cattle ranching and agriculture. But in more recent times, the Brazilian government was credited for its efforts to protect their environment. Deforestation is at its lowest since 1988, and Brazil now protects more than 50 percent of the Amazon within its borders. They have made progress on many fronts, including programs such as the *governments of the forest*. In 2002 they established the Amazon Protected Areas Programme, which has protected millions of acres.

Brazil leads the world in the use of *biofuels*. Beginning back in the 1970s, the government made a commitment to a national *bioethanol* program. They became the first country to convert virtually all of their own *automobile* fuel to a biofuel (from sugar cane).

Recent policy government changes in Brazil are once again concerning *environmentalists* who fear the country might be relaxing its environmental protection regulations. A debate about changes to their most important piece of legislation—called The Forest Code—might signal this change once final votes are completed. Also of concern is *Belo Mondo*, one of the world's largest *hydropower dams,* being built in Brazil.

Environmental *NGOs* are popular and successful in Brazil. For example, the Projeto TAMAR group protects *sea turtles*, Pro-Natura supports sustainability, and Conservation International and the Amazon Institute for Environmental Research work to protect the Amazon. Umbrella NGOs have formed to help coordinate the many groups, including the Social Movement Forum and the Brazilian Network of Civil Organizations.

breast milk and toxic substances

The degree to which toxic substances have permeated our environment can be seen in studies showing the existence of these substances in most women's breast milk. The following toxic substances have all been found to occur in mother's milk: *pesticides* such as *DDT*, *heavy metals* such as *lead* and *mercury*, and industrial byproducts such as *dioxin*. These substances enter the human body when we eat foods that contain low levels of these toxic substances, or they are absorbed directly through the skin.

Almost everyone now carries some level of toxic substances, called our *body burden*. [*autism studies; blue baby syndrome; medical ecology; nanoparticles and human health; puberty and the environment, early; toxic cocktail*]

breeder reactors

Breeder reactors and conventional *nuclear reactors* both generate *nuclear power* by using *nuclear fission*. Breeder reactors, however, have significant differences. During nuclear fission in conventional reactors, the fuel rod is used up and must be replaced. This also occurs in breeder reactors, but new radioactive fuel is constantly being created. In this process, a form of uranium that cannot be used as fuel (nonfissionable) becomes usable (fissionable) while the reactor is in operation. This produces a continuous supply of nuclear fuel for nuclear power and helps alleviate the *nuclear waste disposal* issues. The latest incarnation of breeder reactor is called the liquid metal fast-breeder reactor (LMFBR).

Nuclear power generated from breeder reactors has all the same disadvantages as from conventional nuclear reactors. Breeder reactors, however, are considered more dangerous, because the chain reaction is more difficult to control and the potential for explosions is greater when compared with conventional reactors.

But perhaps the biggest issue facing these reactors and the reason they remain few is that they have the potential to create weapons-grade plutonium. This means countries claiming to be producing radioactive materials for peaceful purposes such as power generation might actually be planning to develop *nuclear weapons*. For these reasons, breeder reactors in the United States, United Kingdom, Germany, and Canada are not in operation. Russia, China, and Japan have working breeder reactors, and both Russia and *India* have at least one under construction. [*nuclear waste disposal; nuclear waste dilemma, the global; weapon dumps, abandoned; plutonium pit disposal*]

brick kilns

In *India*, there are over 100,000 brick kilns turning out 200 billion bricks per year. The sheer scale of this industry draws attention to the *air pollution* it generates. *Soot* is the primary pollutant, which—when mixed with other local pollutants such as *diesel fuel* and *ozone*—produces *atmospheric brown clouds* that blanket the city. These clouds cause respiratory problems and other health issues for local residents as well as contributing to *climate change*.

Suggested improvements include expensive, modern designed kilns that burn *coal* more cleanly. Their high costs limit the likelihood of this having an impact. Some startup companies have shown that simple changes to the existing kiln's design can result in significant improvement. Simply adding more smokestack vents and changing the pattern of bricks to be fired reduces soot emissions by 60 percent and at very little extra cost. [*energy poverty; particulate matter; wood-burning cookstoves, primitive; primitive technologies*]

bridge fuel

Natural gas has been called a bridge fuel—a fuel that can transition the world from *fossil fuels* to *renewable energy*. Natural gas is considered the cleanest of the fossil fuels and produces the least amount of *carbon dioxide*, so switching from other fossil fuels would in theory reduce carbon emissions, thereby slowing *climate change*. It would provide a bridge until the world made a total switch to renewable energy that produces no carbon emissions.

The natural gas boom in the United States, caused by *fracking*, has made the bridge fuel concept more complicated. Many people are using the natural gas boom as an excuse to postpone the transition to renewable energy, saying natural gas is "good" for the environment. In fact, natural gas is a *fossil fuel* and therefore contributes to climate change.

Environmentalists have long agreed natural gas was a good bridge fuel, but that was based on two factors that have changed. First, the thought was that the natural gas would come from *conventional natural gas extraction* methods—not *unconventional* methods such as fracking that produce far greater environmental harm. The second was the assumption that there simply wasn't a great deal of natural gas available and we had passed the point of *peak oil*—meaning fossil fuels would become scarce in the not too distant future. But we now know natural gas supplies can last for many decades, and there is even talk about *energy independence* in the United States—something unheard of a just a few years ago. With these two new factors, many environmental groups are rethinking the original logic of the natural gas bridge fuel.

If natural gas prices remain low in the United States, renewable energy sources cannot compete, resulting in less investment and development and thereby slowing the transition from fossil fuels toward renewable sources.

In addition, some recent research suggests the use of natural gas as a bridge fuel will not accomplish its goal of reducing *global warming*. The reason is that natural gas is mostly methane, a far more potent greenhouse gas than carbon dioxide. Leaks or venting would release at least some methane from these natural gas wells, causing almost as much harm as burning other fossil fuels, such as coal, and releasing carbon dioxide.

Brockovich, Erin

(1960–) Like many, Brockovich never set out to become an environmental activist. But while working as an assistant in a law firm, her personal interests led to an investigation of a utility company that had polluted a local town's groundwater for years, jeopardizing the health of the entire community and making many ill. Her tenacity led to a $333 million settlement paid to the residents of the town of Hinkley, California, in 1996.

Her story became famous when it was made into a movie in 2000 and she was played by Julia Roberts. Today she remains a consumer advocate, running a research and consulting firm involved in a range of environmental and health issues. Her latest target is a birth control procedure many consider to be dangerous. [*environmental justice; Lois Gibbs, Linda Maraniss; Mendes, "Chico"; Hill, Julia "Butterfly"; Silkwood, Karen*]

Brower, David R.

(1912–2000) David Brower has been called a modern day *John Muir*. He was the director of the *Sierra Club* (founded by Muir) from 1952 to 1969 when the membership grew from 2000 to 77,000. Brower founded the *Friends of the Earth* in 1969 and the *League of Conservation Voters* in 1970. Later, in 1982, he founded the *Earth Island Institute*. He was nominated for the Nobel Peace Prize twice during the 1970s and received numerous honorary degrees.

He helped prevent the construction of the Echo Park Dam in Dinosaur National Park, much the way Muir fought to protect *Hetch Hetchy* and proposed (along with others) the creation of United Nations *World Heritage Sites* designed to protect worldwide *biodiversity* and *ecosystems* in some of the world's most unique environments. Brower's autobiography was published in two volumes, For Earth's Sake: The Life and Times of David Brower and Work in Progress.

brown cloud

Brown clouds refer to local *air pollution* conditions such as *photochemical smog*, typically caused by *thermal inversions*. This form of air pollution is different from another type with a similar name—*atmospheric brown clouds*. [*industrial smog; ground-level ozone*]

Brown, R. Lester

(1934–) Brown has been one of the most prominent leaders of the *environmental movement*, primarily through his writings and publications. The Washington Post called Lester Brown "one of the world's most influential thinkers," and The Telegraph of Calcutta called him "the guru of the environmental movement."

He began his career as a farmer but started his public life with the U.S. *Department of Agriculture*. He was president of the highly respected *Worldwatch Institute,* which he founded in 1974 with the help of the Rockefeller Brothers Fund. This research institute is devoted to the analysis of global environmental issues. Ten years after the Institute was established, Brown began to publish the annual State of the World reports, which are translated into all the world's major languages and—along with its supplement, called Vital Signs, which began publication in 1993—has now achieved semiofficial status.

He later left Worldwatch, and in 2001 founded the Earth Policy Institute "to provide a vision and a road map for achieving an environmentally sustainable economy." He now writes the annual Plan B reports and many other excellent publications. His books present the facts about how our present economy must transition to a sustainable economy for our survival.

He is the author of more than 50 books and the recipient of many awards, including the United Nations Environment Prize, Humanist of the Year, and the Robert Rodale Lecture Award.

brownfields

These fields usually refer to industrial property that has (or is believed to have) some form of environmental contamination in the *soil, surface waters*, or *groundwater* within the property. It might contain *oil* products, *heavy metals*, or other *hazardous wastes*. Because *CERCLA* (Comprehensive Environmental Response, Compensation, and Liability Act of 1980) could result in fines for any company purchasing these brownfields, they often remain vacant, ugly, dangerous eyesores in a community. In the 1990s some state and local governments created new laws to allow these sites to be purchased, cleaned up, and built on without fear of legal settlements against them.

Some states even provide tax incentives and credits to encourage development in these areas. States with brownfield incentives include Alabama, Connecticut, Delaware, Florida, Iowa, Kentucky, Louisiana, Maryland, Massachusetts, Michigan, Mississippi, New Hampshire, New York, Ohio, Pennsylvania, South Carolina, Tennessee, and Wisconsin. [*schools and brownfields; interim use; smart cities; urbanization and urban growth; transition towns*]

Brundtland, Gro Harlem

(1939–) Considered one of the most important world leaders of the *environmental movement* during the 1980s and 1990s. In Norway, she was the Minister of the

Environment in 1974 and became Prime Minister in 1981, where she remained for ten years. She is best known (environmentally) as the chairperson of the *World Commission on Environment and Development* from 1983 to 1987.

In 1992 Brundtland was instrumental in the formation of the *United Nations Conference on Environment and Development* and the principal author of the conference report, <u>*Our Common Future*</u>. The report is commonly called the Brundtland Report and is considered a pivotal point in environmental history. From 1998 to 2003, she served as the Director-General of the World Health Organization and is given credit for refocusing the organization's efforts on links between the environment and health.

Bt (Bacillus thuringiensis)

Bt is short for the scientific name of a species of *bacteria*, Bacillus thuringiensis. These bacteria produce a substance that is toxic to many *insects*. In the 1970s, scientists began using Bt as a *natural pesticide*, or what is called a *biological control* method. It was often used as part of an *integrated pest management* program to control insect pests naturally. This was a great advance in pest control, because it worked well and did not require dangerous synthetic *pesticides* that wreaked havoc on the *environment*. Bt, like many natural pesticides, only kills specific, targeted pests, as opposed to most synthetic pesticides that don't know a pest from any other organism and tend to kill (or sicken) all forms of life.

At first the actual bacteria were sprayed on plants. Later, the natural toxin was determined and that was sprayed on crops to control pests (without the need for the entire bacteria). Then in the early 1990s, with the advent of *synthetic biology* and the use of *genetically modified organisms* (GMOs), scientists figured out how to insert the gene from the Bt bacteria that produced the toxin directly into certain crop plants such as cotton. The result was a new variety of cotton that was resistant to certain pests. It was so successful that almost all cotton grown today is this genetically modified type. Many crops now have Bt varieties, such as Bt eggplant and Bt corn. Much of the discussion about *genetically modified organism labeling* is about Bt crops.

Over time, however, the Bt crops have run into problems similar to other crops—the insect pests become resistant. New research shows techniques that make resistance to Bt crops highly unlikely. For example, if a farmer grows a conventional crop near the Bt crop, insects will spend some of the time in areas without the toxin and therefore will not build up resistance.

BTU (british thermal units)

This is the standard measure for the amount of heat available in a fuel. Technically, it is the amount of heat (energy) required to raise the temperature of one pound of water by one degree Fahrenheit. This translates to roughly one BTU, being equal to the energy released when one wooden matchstick is lit. One gallon of *gasoline* contains a little over 114,000 BTUs of energy. [*energy (fuels)*]

bubonic plague

Commonly called simply the plague, it is believed to be responsible for three historical pandemics. Best known is the Black Death, which occurred from 1347 to 1352 and killed somewhere between one-third and one-half of the entire population of Europe. [*antibiotics for livestock; bird flu virus; geomedicine; medical ecology; nanoparticles and human health; spillover effect; superbugs*]

buffalo (bison)

Lewis and Clark wrote of buffalo herds covering the landscape as far as the eye could see. Buffalo are North America's largest land animal, and their vast herds were icons of the great West. An adult male can easily weigh a ton and stand six feet high at the shoulders.

Their numbers in the early 1800s was estimated to be 65 million. They were found primarily from the Rockies to the Mississippi. Considered one of the clearest examples of human impact on the wild in the United States, their numbers dwindled in the late 1800s, and buffalo were all but wiped out by hunting for food and sport. By 1890, only about 1000 individuals remained and most of them were in protected herds in *Yellowstone National Park*.

In 2012, 63 buffalo were relocated from the park to a northern Montana reservation in an effort to reestablish a wild herd. At this time only three free herds exist outside of Yellowstone, with about 5000 head in total. Unfortunately, they are not welcome. Local ranchers sued, stating they could spread disease to their cattle and would compete for the same grazing lands. Ranchers pay $22 per head for private grazing land but, thanks to public subsidies, only $1.35 per head on public lands. The ranchers don't want wild buffalo competing with their privately owned cattle on America's public lands.

The state of Montana is considering legislation that would allow hunting of buffalo on public lands, with the likelihood of wiping out these wild herds and leaving only the herds in Yellowstone. Proponents of the buffalo release believe there is room for both cattle and buffalo on public land. [*biodiversity, loss of; captive breeding; CITES; endangered and threatened species lists; extinct in the wild; extinction, sixth mass*]

bug

Although often used to describe any *insect*, true bugs belong to a single order called Hemiptera. Included in this group are many insects that feed on plants by sucking the juices out of them, such as milkweed bugs. The so-called ladybug is actually a *beetle*, which is a different order of insects. [*biological control; biological pesticides; natural pesticides; sex pheromones*]

building-related illness (BRI)

Building-related illnesses involve specific identified diseases that can be linked to a building's environment, such as Legionnaire's Disease (caused by *bacteria* that make the *air conditioning* system their home) or nausea caused by the *outgassing* from building materials. Building-related illness disappears once the cause has been removed from the building. [*indoor air pollution; indoor ecology; humidifier fever; couches and indoor pollutants*]

buildings, green

The U.S. *Environmental Protection Agency* (EPA) states that "green buildings use energy, water and other resources more efficiently; protect occupant health. . .and reduce waste, pollution, and environmental degradation." The process of creating a green building "is the practice of creating structures and using processes that are environmentally responsible and resource-efficient throughout a building's life-cycle from siting to design, construction, operation, maintenance, renovation, and deconstruction."

Energy *Star* (created by the EPA) and *LEED* certification (by the *U.S. Green Buildings Council*) are the recognized leaders in *eco-certification programs* that define and rate green buildings.

Green buildings (also called sustainable buildings or green construction) provide savings to both the environment and the owner. They cause less environmental degradation throughout the building's lifecycle, and improved energy efficiency saves the owner on energy costs throughout the life of the building. Estimates vary, but even though a green building has higher upfront costs, the owner usually saves ten times those additional upfront costs over the life of the building. [*homes, gree; indoor ecology*]

Bullard, Robert D.

(1946–) Bullard is considered the father of *environmental justice*. In the late 1970s, working as an environmental sociologist, he studied the siting of *garbage* dumps in poor neighborhoods and concluded this was a form of systemic injustice. He wrote a book in the 1990s titled <u>Dumping in Dixie</u>, which many believe to be the first indepth study of the concept now called environmental justice. He worked for the Clinton administration, helping create legislation that mandated federal agencies to take environmental justice into consideration as they developed new programs.

He is currently a Dean at Texas Southern University, remains an advocate of this cause, and has written more than a dozen books on the subject. [*environmental movement; Women's Earth Alliance*]

bumblebee decline

Even though bumblebees don't produce honey, they are important pollinators and therefore, important components of *ecosystems*. (Tomatoes cannot be pollinated by honeybees, so you can thank bumblebees instead.) Similar to *honeybees*, bumblebees are in decline but for different reasons. Bumblebees nest in the ground, and with more and more pavement and buildings in urban areas, there is less and less *habitat* for their nests. They also appear to be attacked by *parasites* that take their toll on bumblebee populations.

Another issue recently discovered shows that bumblebees prefer diverse flower habitats, as opposed to dense patches of the same species of flower—something also common in urban settings. The solution is to provide more open, natural space where these bees can find more nesting sites and more *biodiversity*.

The combination of a decline in both bumblebees and honeybees is causing a pollination crisis. [*colony collapse disorder; honeybee sperm bank; honeybees, pollination, and agriculture; Monarch butterfly decline; neonicotinoid pesticide and honeybees; Alliance for a Paving Moratorium*]

bumblebee senses

Recent research shows that *bumblebees* not only see color into the ultraviolet end of the spectrum, but also can sense electric fields in plants. Flowers have a slight negative charge. The bees can sense this charge and are attracted to the flower. But there is more. When bees fly, the friction of the air against their bodies gives them a slight positive charge. So, when a bee lands on a negatively charged flower, the positively charged bee attracts the pollen onto the bee with little or no effort.

Even more amazing is the fact that when the bee lands on the flower, the positive and negative charges cancel out one another. Even after the bee leaves the flower, the charge on the flower is not restored for over a minute. If another bee flies near this flower, it knows a bee has recently visited and removed much of the pollen, so it won't bother to land. [*beehive sound research; beetlecam; communications between plants; hummingbirds and flight; sixth sense in animals; smell sensitivity in albatross*]

Bureau of Land Management (BLM)

An agency within the U.S. *Department of the Interior* that manages about 250 million acres of America's *public lands* and another 300 million acres of mineral reserve lands used specifically for *mineral exploitation*. Most of these lands are in the western states and Alaska.

The public lands are mandated for *multiple use* management, meaning they are not *wilderness* to only be protected but to be used, whether for recreation, *mining*, cattle *grazing*, or other activities. Their stated mission is "to sustain the health, diversity and productivity of the public lands for the use and enjoyment of present and future generations."

Because the BLM must manage contradictory activities such as camping, hunting, hiking, and birding along with *logging*, mining, and *fracking*, it has always had a contentious battle within its management. And because BLM offices are managed by the federal government, their emphasis—*preservation* or extraction—changes with the administration currently in office. [*dual-use policy; wise-use movement; multiple use policy*]

buried lakes, antarctic

About 250 lakes are sandwiched beneath *Antarctica's* ice sheets and the land mass below. Most of these subglacial lakes have never been seen by humans. In 2012, after drilling down about one-half mile through ice into a subglacial lake called Lake Whillans, Russian scientists found *bacterial* life in the water. Previously, scientists drilled down over two miles into Lake Vostok, but those studies remain inconclusive as to whether life exists in that lake. Some scientists believe we will soon find new forms of life in these newly discovered buried lakes. [*extremophiles; arctic; tolerance range; high-altitude ecosystems; hydrothermal vent ecosystems; ice ages; glacial lake outbursts*]

Burroughs, John

(1837–1921) Burroughs was the most popular American nature writer of his time. (And his contemporaries were *Thoreau* and *Muir*, among others.) Although he is not a well-known name today, he was the equivalent of rock star in his day. His books were immensely popular. Many were adapted to become standard reading in schools across the country. When he and his friend President *Roosevelt* went on a train tour together, it was reported that Burroughs commanded more attention than the President.

Burroughs was friends with industry titans such as Firestone, Ford, and Edison. He was also a good friend of John Muir; they were often called "the two Johnnys." Both Burroughs and Muir personally helped convince Theodore Roosevelt to set aside vast swaths of land as *national parks*.

But Burroughs' most important contribution was his pastoral, descriptive, and educational essays about nature around his home in the Hudson Valley of New York

State. He wrote about the prevalent *anthropocentric* view and urged man to respect nature at all levels. It is unfortunate that today's readers rarely come across his writings—they are worth finding and reading. [*philosophers and writers, environmental; Hudson River School Art; preservation*]

bycatch

Some methods of harvesting fish and shrimp result in capturing unwanted species, called bycatch. The bycatch is thrown back into the water, usually dead or injured. Hundreds of millions of pounds of bycatch are killed annually. Common bycatch include *dolphins* trapped in *purse seine nets, sea turtles* trapped in shrimp trawler nets, numerous species of fish and *mammals* trapped in *driftnets* and *gillnets*, and seabirds caught on *longlines. Bottom trawling*, where a huge net is dragged along the bottom of the sea floor, devastates the entire *ecosystem*, bringing up virtually anything and everything.

The estimates of the quantity of bycatch are staggering. Tens of millions of *sharks* and rays die as the outcome of bycatch each year. Hundreds of thousands of *whales*, dolphins, and porpoises also die each year. About 100,000 albatrosses are killed by longlines. But possibly the worst bycatch offender is bottom trawling—also called *strip-mining* of the ocean floor. Along with the target fish, a single pass of a bottom trawler removes up to 20 percent of the entire seafloor plant and animal life. [*turtle excluder device; overfishing*]

c3 plants

Plants growing in aquatic and most terrestrial environments build molecules of sugar containing three carbon atoms and are therefore called C3 plants. Examples include rice, wheat, oranges, and grapes. [*photosynthesis; carbon cycle; C4 plants; biogeochemistry; respiration, cellular*]

c4 plants

Some plants growing in hot, dry regions produce a four-carbon sugar molecule instead of the usual three-carbon molecule and are therefore called C4 plants. Examples include crabgrass, corn, and sugarcane. [*photosynthesis; carbon cycle; C3 plants; biogeochemistry; respiration, cellular*]

cacti

Members of this plant family are well suited for *desert* and near-desert *environments*. The most obvious adaption of cacti is their modified leaves, which form needles. Less surface area means less water loss and they also protect the plant. The thick, fleshy body and stems store water. There are about 2000 species, found mostly in the American southwest, Mexico, and Central and South America. About 20 species are at risk of *extinction*, mostly because of over-collecting by humans and damage by grazing animals. [*natural selection; biomes; tolerance range*]

cadmium

Remember nickel-cadmium or "ni-cad" rechargeable *batteries*? Cadmium is an element used in many products, including these batteries, less expensive *solar panels*, electronic switches, and other devices. It is resistant to corrosion and conducts electricity well. However, cadmium can cause *chronic* and *acute* illness in people working with it during the manufacture of these products—typically by inhalation or direct contact.

Products containing cadmium should be disposed of as a *hazardous waste,* but many people simply throw these batteries out with the *garbage.* The cadmium thus becomes part of the *municipal solid waste,* ends up in *landfills* or *incinerators,* and makes its way into the *soil*, the *groundwater,* or the air. Its use is being phased out in many countries and most vigorously in Europe.

CAFÉ (corporate average fuel economy)

A law created by Congress in 1975 designed to reduce U.S. *energy gasoline* consumption by requiring *automobile* manufacturers to increase the fuel economy of their cars and light trucks. This results in fewer *fossil fuel* pollutants and *carbon emissions* spewing into the *atmosphere* and saves everyone money at the pump. It is managed by the National Highway and Traffic Safety Administration, which sets these standards. The averages have been increasing in recent years, but the United States remains far behind many other nations. [*alternative fuels; biofuels; bioethanol; flex-fuel engines; synfuels*]

CAFO (concentrated animal feeding operations)

Also called confined animal feeding operations, CAFOs are a type of *factory farm.*

In these operations, livestock (cattle, poultry, or pigs) are kept and raised in confined areas. The entire operation—the animals themselves, their feed, their manure and urine— takes place in a small area. The feed is delivered to the animals rather than the animals grazing in pastures or on rangeland. The *EPA* defines an operation as a CAFO if the animals are confined for least 45 days out of a 12-month period and there's no grass or other vegetation in the confinement area.

These tight quarters and overall conditions are responsible for serious degradation to the environment as well as health issues for the general public and are viewed by many as inhumane treatment of the animals.

CAFOs produce vast amounts of animal wastes—well beyond what natural processes can decompose. In most cases, the waste is simply dumped into lagoons or open pits. These concentrated wastes contain *nitrogen, phosphorus, pathogens, heavy metals, hormones,* and ammonia, among other substances. Mismanagement of these wastes results in *nutrient enrichment, algal blooms*, fish kills, and basic *water pollution* in bodies of water. It can also result in the contamination of *drinking water* supplies.

But other environmental issues go well beyond the local environment, issues that affect our planet and all onboard. Feeding livestock uses more than 33 percent of all the crops grown. Many people believe this should be used to feed people directly—not livestock. Also, clearing land for livestock is a major cause of *deforestation* in the *Amazon* and South America. And finally, farm animals release 18 percent of all *greenhouse gases* by *belching* and flatulence.

Nearby residents often have respiratory issues and show other ill effects from these nearby wastes. However, a larger public health issue is the use of *antibiotics for livestock,* believed to be the primary cause of resistant *superbugs* now infecting people.

Then there are the animals themselves. Many people believe CAFOs are inhumane treatment of animals. These conditions often cause disease and injury in the animals. Many people say they became *vegetarians* because of the inhumane treatment of livestock and the entire industry.

There are still other issues of concern. Just as agriculture depends on only three grain crops, livestock relies on very few species. Only two cow breeds comprise 97 percent of the entire U.S. dairy herd. Depending on so few species increases the chance of large-scale shortages if something compromises these animals' ability to survive, such as *climate change*. What would happen to milk supplies if something happened to these two breeds?

Conservation agriculture, sustainable agriculture, permaculture and *agroecology* are alternatives to CAFOs, factory farms, *and high-impact agriculture*. They are all methods of feeding people but do so in entirely different ways. [*egg farms; dairy cow output*]

Cage-free label

See *food eco-labels.*

California condor

North America's largest bird, with a wingspan of almost ten feet. It gained notoriety when this huge bird came close to *extinction*. Through the 1800s, it was abundant in western United States and some of Mexico and Canada. By 1900, however, its numbers were in decline because of *habitat loss, pesticides,* and hunting. In 1930, a sanctuary was created near Los Angeles to try to save the six remaining wild condors. When most died, the last remaining wild bird was captured in 1987 and taken to the San Diego Zoo, where it joined 26 captive birds. The captive birds' numbers slowly increased, and in 1991 a few were released back into the wild. Today there are about 350 birds in total, with about 150 living in the wild. [*biodiversity, loss of; captive breeding; CITES; de-extinction; dodo bird; endangered and threatened species lists; extinct in the wild; extinction, mistaken; Lonesome George; Tasmanian tiger*]

caliology

The study of animal homes, such as burrows, nests, and hives. [*habitat; habitat loss; colony collapse disorder; cryptozoology; giant squid in native habitat; rookery; white-nose syndrome*]

camouflage, squid

Many species of squid can disappear in their *habitat*. They can change their skin color quickly with changes in *pH*, allowing them to blend into their background. Researchers are creating synthetic versions of this squid camouflage for the U.S. military. [*biomimetic; bioluminescence; giant squid in native habitat; whistle names, bottle-nose dolphin*]

campus drilling for fossil fuels

Fracking is one of the most contentious environmental issues raging in the United States. Both *NIMBY* (not in my backyard) and *YIMBY FAP* (yes in my backyard for a price) advocates can be found at either end of the issue. But many college campuses, in spite of their usual reputation as being environmentally conscious, are taking the money. Many colleges are financially stressed and cannot refuse offers of significant income. More than a dozen colleges now allow *natural gas* or *oil drilling* on campus. Many people believe the health problems associated with *fossil fuel mining* are not worth the profits, while others simply don't think mining is conducive to learning. However, it appears this trend will continue unless students decide to go elsewhere, negating the profit side of the balance sheet. At the moment, colleges in Indiana, Pennsylvania, West Virginia, and Texas either are already involved or plan to do so.

campus quad farms

The *farm-to-table movement* has spread far and wide. Some colleges have joined the movement by converting their quads or other open spaces from *lawns* (with all the environmental problems associated with them) into small farms, called quad farms, where students grow produce or other plants. The food is used in campus cafeterias, allowing students to become true *locavores*, or it is offered to food banks. These unique small farms turn an environmental negative—the lawn—into a positive—a food source.

Some colleges use the campus farm as a service-learning facility, a way to fulfill a physical exercise requirement or volunteer work. At the University of Massachusetts in Amherst, their three such farm plots produce at least 1000 pounds of food each year, used by the food service and a campus *farmers' market.* More than 1200 volunteers work these farms. [*farm-to-school movement; food miles; Meatout Mondays; seasonal foods; slow food movement; WWOOF*]

cancer alley

This location is often used as an example of environmental racism in the United States. This roughly hundred-mile stretch between New Orleans and Baton Rouge, Louisiana, has high rates of cancers, birth defects, and miscarriages. Over 140 industrial plants exist in the area, spewing forth all sorts of pollutants into the air and water. The fact that it is a poor area, 90 percent black and home to almost 30 percent of the nation's entire petrochemical production, has become a clarion call about how the environment can be an implement of social injustice in our society. [*environmental justice; LULU; body burden; toxic cocktail; autism studies and environmental factors; breast milk and toxic substances; medical ecology; nanoparticles and human health; puberty and the environment, early*]

cancer villages

Villages in *China* that have well-documented high levels of cancers most likely resulting from the intense local *air* and *water pollution*. Estimates are that over 450 such villages are found throughout all of the country's provinces. [*air pollution in Beijing, China; tweets, eco-; environmental justice*]

canopy

A continuous layer of foliage in a *forest*, formed by the crowns of trees. *Tropical rainforests* can have two or three layers of canopy. Each canopy acts as a *habitat* for organisms. [*niche, ecological; biodiversity; habitat; overstorey*]

cap-and-trade

One of two types of *carbon markets* (the other being *carbon taxes*), both of which are used to reduce the amount of *carbon emissions* causing *climate change*.

Cap-and-trade legislation requires companies to keep their *carbon emissions* within certain limits considered acceptable (the cap). The cap is determined by scientists who assess acceptable levels of carbon emissions over a certain period of time. The cap is managed by the distribution of carbon permits (also called carbon credits or carbon offsets) that are auctioned off or given away to companies that produce carbon emissions. These permits are like free passes to produce a certain amount of carbon emissions, usually measured as metric tons of carbon dioxide. Companies can only produce the amount of carbon emissions allowed by their permits.

But companies are provided flexibility because these permits can be bought and sold (the trade). A company that feels it cannot meet its carbon emissions cap can buy additional permits from other companies. Companies that can figure out how to cut back on their emissions—producing less than their cap allows—can sell their permits to others; thus making a profit by being more green.

A problem with this method can occur if the value of the permits becomes too high or too low. Recently, the *Clean Development Mechanism* (CDM) established by the *Kyoto Protocol* has fallen into disarray, because the value of the carbon permits dropped to almost nothing, meaning far too many permits and not enough buyers. This particular situation occurred because countries that would have been the biggest markets for these permits did not sign the Kyoto Protocol (including the United States) or were not required by the agreement to do so (*India* and *China*). Like any market, there must be buyers as well as sellers, and at the moment, the CDM has few buyers.

Cap-and-trade plans do work, however. The CDM is not functioning well just now, but over the past seven years, it reduced carbon emissions by billions of tons, cut the cost of climate change by billions of dollars, and generated more than $200 billion in green investments. Many countries are initiating cap-and-trade programs, including China, which recently began its first carbon trading market in the city of Shenzhen.

Not all cap-and-trade plans are at the national level. Even though the United States has not passed a national cap-and-trade program for carbon emissions, California recently created the California Air Resources Board (CARB)—the first state-run program. The state's goal is, by 2020, to bring carbon emissions down to 1990 levels. Another U.S. program is the *Regional Greenhouse Gas Initiative* (RGGI).

Cap-and-trade programs have garnered more attention than the alternative carbon tax plans. [*ton of carbon dioxide look like? What does a*]

Cape Wind

This offshore *wind farm* has been approved off the coast of Hyannis, Massachusetts, on Cape Cod. It is a 24-square-acre plot in Nantucket Sound about five miles from shore. It will have about 130 *wind turbines*, each rising 440 feet from the surface. Groups concerned about the farm's impact on tourism have opposed it but lost the final legal battle. Once completed in 2015, it should provide three-quarters of all the electricity required for the Cape, including its islands, Nantucket and Martha's Vineyard. It will have a license to operate for 25 years.

Replacing the existing *fossil fuel* power plants now providing energy with *wind power* will reduce *carbon emissions* by over 700,000 tons per year. The Cape Wind project has been the subject of two documentary films. [*floating wind turbines; Global Wind Day; wind power, brief history; wind power transmission line*] {capewind.org}

captive breeding

Controlled mating and breeding of captive animals within *zoos*, aquariums, and research institutions such as universities. Unfortunately, because of the number of species *endangered* and on the verge of *extinction*, these *breeding* programs are sometimes seen as the only way to ensure a specie's survival. It is common to hear that a species is *extinct in the wild*, meaning the only remaining individuals are in these captive breeding programs. Zoos have taken this activity on as a new part of their overall mission. Because these programs are typically expensive and most zoos have limited resources, there is a debate as to which animals should be saved and which should be allowed to become extinct. [*zoos, captive breeding programs in; Zoo, Frozen; zoos, future; CITES; de-extinction; flagship species; charismatic megafauna; wildlife refuge; species recently gone extinct; Red List*]

carbamates

One of four major categories of *synthetic insecticides* commonly used to control insect pests. Carbamates are considered soft pesticides because they break down into harmless substances quickly after application—usually in only a few days or weeks.

Carbamates, along with *organophosphates*, kill by disrupting an organism's nervous system. Even though they break down quickly, they are *acutely toxic* to humans, meaning they pose a significant risk to those who apply the chemicals, are in the vicinity of the application, or ingest the harvest contaminated with the substance soon after it is applied (before it has time to break down).

Carbaryl and aldicarb are common carbamate pesticides. Aldicarb is being phased out of use because of numerous recorded incidences of people becoming sick by eating fruits or vegetables contaminated with the pesticide. Many people think it is amazing that some of these pesticides are still in use today. [*natural pesticides; biological control; Bt; pheromones; biocides; nanopesticides; pesticide dangers; pesticide residues on food; Toxic Substances Control Act*]

carbon budget

A direct relationship exists between the amount of *carbon dioxide* emitted into the *atmosphere* by human activities and the amount of *global warming* that occurs. The more carbon dioxide—a *greenhouse gas*—emitted, the higher global temperatures rise. Many scientists have set a maximum rise of 3.6 degrees F to prevent catastrophic results.

The term carbon budget refers to the maximum amount of carbon dioxide that can be emitted into the air by humans up to the year 2050 before causing global warming to go over the 3.6-degree mark. A recent report states that the carbon budget is estimated to be about 1000 *gigatons* of carbon dioxide ($GTCO_2$). [*climate change; carbon emissions; climate sensitivity; Intergovernmental Panel on Climate Change; hockeystick graph; Keeling Curve; Kyoto Protocol; ton of carbon dioxide look like?, what does a*]

carbon cycle

One of many *biogeochemical cycles*. Carbon is the primary component of all *organic* matter. The two most important parts of the carbon cycle are 1) *photosynthesis*, where carbon (from *carbon dioxide* in the *atmosphere*) and water are converted (using radiant energy from the sun) into sugar molecules that act as fuel for almost all living things and 2) cellular *respiration,* where these sugar molecules are broken back down, releasing the stored energy for the organism to use to survive. An estimated ten percent of the total amount of carbon dioxide in the air cycles back and forth between the atmosphere and organisms each year through photosynthesis and respiration.

In addition to photosynthesis and respiration—considered gaseous portions of the carbon cycle—there is a sedimentary portion as well. Carbon is also found in the solid Earth (*lithosphere*). Rocks, soil, and sediment all contain carbon. The *weathering* of rocks and the action of volcanoes return small amounts of this carbon directly into the atmosphere. However, this sedimentary portion of the cycle can take millions of years as opposed to photosynthesis and respiration, which occur quickly. [*biogeochemical cycles, human intervention in; biogeochemistry*]

carbon dioxide

Carbon dioxide makes up only about 0.039 percent of the *atmosphere* (excluding moisture) but plays a vital role in life on this planet via the *carbon cycle*. Green plants absorb carbon dioxide during *photosynthesis,* and both plants and animals produce it as an end product of *cellular respiration.*

Carbon dioxide in the atmosphere plays a major role in controlling the Earth's surface temperature and is considered the most important of the *greenhouse gases*. Because of its importance, it is studied in great detail. Instead of using a metric as vague as percent of total gases in the atmosphere (at 0.39 percent), it is measured in parts per million (ppm).

The amount of carbon dioxide in the atmosphere has been increasing—from 315 ppm in 1958 to 350 ppm in 1990. And in 2013, it reached (briefly) 400 ppm for the first time. A measurement that links the amount of carbon dioxide in the atmosphere to the amount of *climate change* is called *climate sensitivity*. It is so closely watched that there is even a famously named chart that tracks the relationship, called the *Keeling Curve.*

carbon emissions

When people burn *gasoline* to run their *automobiles*, burn *coal* to generate *electricity*, and take part in numerous other activities in our industrialized world, we produce and emit *carbon dioxide* into the *atmosphere*. When people talk about carbon dioxide emissions, they usually simply call it carbon emissions.

The most important *greenhouse gas* is carbon dioxide because it is primarily responsible for *global warming* and *climate change*. It is one of the longest lasting of these greenhouse gases, so it continues to build up in the *atmosphere*. The more anthropogenic (human-induced) carbon dioxide we produce, the more our climate warms.

About 70 percent of all carbon emissions come from energy production, with almost all of it coming from burning *fossil fuels.* This is broken down to (in 2010) about 40 percent from *electric power plants* and heating, 25 percent from *transportation* such as *automobiles* and airplanes, 20 percent from industrial processes, and 6 percent from residential emissions. Another way to look at it is by the type of fossil fuel used. Of these carbon emissions, *coal* was responsible for 40 percent, *oil* for 37 percent, and *natural gas* for 20 percent.

Most of the remaining 30 percent of carbon emissions comes from *deforestation.* As forests are cut down, the result is fewer trees to absorb carbon dioxide from the atmosphere. Forest and *peat* fires and specific industrial uses such as *cement* production also contribute to this 30 percent.

The world's five largest carbon dioxide emitters (2010) are as follows (rounded % of world total): *China*, 25 percent; United States, 19 percent; *India*, 5 percent; Russia, 5 percent; and Japan, 4 percent.

These numbers look very different when taking into account the countries' total populations. Viewed as carbon dioxide emissions per person, these same five countries are (rounded to tons of carbon dioxide) United States, 20; Russia, 12; Japan, 10; China, 6; and India, 2. [*ton of carbon dioxide look like?, What does a; United Nations Conference on Sustainable Development*]

carbon footprint

A subset of an *ecological footprint.* Where an ecological footprint includes all of the negative impacts on the planet, the carbon footprint includes only *carbon dioxide emissions* responsible for *climate change*. Many *eco-apps* let you calculate your own carbon footprint.

carbon markets

Greenhouse gas emissions, caused by burning *fossil fuels,* must be lowered to deter *climate change*. Many people agree that emitting gases such as *carbon dioxide* cannot continue to be free. Sooner or later, legislation will impose a price on burning fossil fuels and releasing these emissions. How to accomplish this is a contentious issue.

The economic solutions to this problem are called carbon markets. Two types of carbon markets are being considered in the United States and are currently used elsewhere. They are a *carbon tax* and a *cap-and-trade* program. In both cases, the end result would be fewer greenhouse gas emissions because market forces would drive them down.

carbon monoxide

Carbon monoxide is one of the five primary components of *air pollution*. It is formed from the incomplete combustion of *organic* fuels such as *oil*, *gasoline*, *wood*, and most *solid trash*. One of the largest contributors of carbon monoxide into the *atmosphere* is the *automobile*. When the car's engine isn't running efficiently, the fuel is not completely burned and carbon monoxide is produced. Inefficient *fossil fuel* power plants also emit large quantities of carbon monoxide into the air.

Carbon monoxide is produced in tobacco smoke and can affect anyone in the area as *passive smoke*. Small amounts of carbon monoxide in minute concentrations can cause headaches, drowsiness, and blurred vision. Cities full of cars and rooms filled with cigarette smoke pose a significant health risk.

carbon neutral

Carbon neutral refers to reducing or offsetting *carbon dioxide* emissions so the net result is zero. A business can minimize the amount of carbon it emits by changing how it conducts business. For example, using less material, *recycling* more *preconsumer* waste, and switching from *fossil fuels* to *renewable energy* sources can all help lower *carbon emissions*.

However, getting emissions down to zero is often impossible. To become carbon neutral, companies must often purchase or invest in *carbon offsets,* so whatever carbon emissions they do produce are negated by paying for projects that reduce carbon emissions elsewhere.

The Vatican became the first carbon neutral state in 2007. They did so not by purchasing carbon offsets but by accepting a gift from a startup carbon trading company (carbon offset provider) that planted enough trees to offset all of the Vatican's emissions for that year.

carbon offsets

Cap-and-trade and *carbon tax* programs are the two most common forms of *carbon markets*. These programs are used primarily by businesses and typically run at the state or federal levels. Carbon offsets are an important aspect of the *cap-and-trade* programs.

Here's how carbon offsets work: A business is trying to reduce its *carbon emissions* but cannot reach its target, which might be mandated by regulations or could be voluntary. If it cannot lower its emissions internally, it can purchase carbon offsets from organizations that then use the money to make purchases or investments in projects that reduce carbon emissions elsewhere. "Elsewhere" might be in less developed countries or right around the corner.

Many organizations provide this service. These carbon-offset providers invest the monies raised by selling carbon offsets in projects that reduce carbon emissions or perform *carbon sequestration storage*. For example, a company that cannot reach mandated carbon emission limits might buy carbon offsets that pay to reforest a region in the Amazon. In a cap-and-trade program, this would be considered a legal way to reduce the company's carbon emissions.

Similar programs exist for individuals, but they typically just include the "trade" component—not the "cap"—because it would be strictly a voluntary purchase of carbon offsets. Some opponents of these programs believe this is just a feel-good notion that does little, but others think it accomplishes the goal of reducing carbon emissions on a global basis.

Some of the well-respected carbon-offset providers include Gold Standard, CDM, VCS Greenhouse Friendly, Climate Friendly, and Carbon Fund. If you want to check out other carbon-offset retailers, look for those that have attained a certification from a reputable organization, such as Gold Standard or Climate, Community, and Biodiversity Standards. [*climate capitalism; Beyond GDP; CERES; corporate social responsibility; corporate sustainability rankings; decoupling; polluter pays principle*]

carbon sequestration storage (CSS)

Reducing the amount of *carbon dioxide* released into the *atmosphere* is at the forefront of the fight against *climate change*. Most of these *carbon emissions* come from *coal*- and gas-burning *electric power plants*, and some comes from the manufacture of *cement* and other industrial processes. New legislation mandates that these power plants and other industries lower their carbon emissions. For old coal-burning plants, this means finding new technologies to reduce their emissions or shutting down. One method to reduce emissions is to capture and store the carbon in a safe place so it does not get into the *atmosphere*. This process is called carbon sequestration storage (CSS).

This process has three steps. The carbon must be a) captured and processed into a form to then b) be piped to a location for c) final storage. At the moment, the technology has been shown to work but is prohibitively expensive. A typical coal-fired electric power plant must use about 25 percent of the energy it produced just to run the CSS process. New methods must be developed to make the process economically feasible.

Assuming the costs can be brought down, estimates state the world needs about 3000 such facilities by 2050. At the moment there are eight. Twenty-eight more are planned in the United States and a few other countries. The largest existing facility is in Norway, but at the moment it can only perform the first two steps because it doesn't have any place to actually store the gas on a long-term basis.

There are two options for storage—deep saline (salt) *aquifers* or depleted oil and gas fields. Most of the world's anticipated storage capacity is in aquifers, but almost all of the running CSS facilities (five out of the eight) are depleted oil or gas fields. An interesting twist to this method is that storing carbon emissions in these depleted fields means additional oil or gas can be extracted from these sites. So this process not only stores the carbon produced from burning *fossil fuel*, but also extracts more of it, so it too will have be stored somewhere—a situation that might make economic sense, but no logical sense.

Another stumbling block for this technology is the boom in cheap *natural gas*. The regulations that make CSS necessary—new laws to minimize carbon emissions—only made sense for old coal-burning plants. Today, with natural gas being extracted by *fracking*, it is more economical to convert to natural gas to reach lower emission standards than to upgrade old coal plants. So the future of CSS is in doubt at the moment.

Finally, there are environmental concerns about CSS. Some scientists believe the process can store carbon for thousands of years safely, but others believe leaks are inevitable. *[karst, carbonate; forams; geoengineering; metal organic frameworks; regenerative agriculture]*

carbon sink

A carbon *sink* is something that stores large quantities of *carbon*. *Forests* and *oceans* are both carbon sinks. In forests, the carbon is stored in the *biomass* of all the trees, as it is in all living organisms. In oceans, the carbon chemically binds with other elements and remains in the water.

In both cases, the carbon is stored within the carbon sink, so it is not released into the *atmosphere* where it would act as a *greenhouse gas* in the form of *carbon dioxide*. If it is not in the atmosphere, it is not contributing to *climate change*.

Some carbon sinks are better than others. Oceans absorb and lock up carbon for up to 1000 years, but forests and other land-based carbon sinks store carbon for far shorter periods of time. An unusual type of carbon sink currently being studied is *carbonate karst formations*.

carbon tax

One of two types of *carbon markets*. The goal of a carbon market is to reduce *greenhouse gas* emissions that cause *climate change*. A carbon tax is levied on industries based on the amount of *carbon dioxide* emissions they produce.

Many see a carbon tax as the simplest way to reduce the primary culprit of *global warming*—carbon dioxide. The amount of *carbon emissions* is quantified into units and the tax is assessed according to how many units a company produces. The tax gradually increases, forcing businesses to gradually transition to processes that reduce carbon emissions.

The goal of a carbon tax is to change the behavior of both businesses and consumers. Using *fossil fuels* produces lots of carbon emissions, so industry and businesses will try to lower this tax by switching to alternative energy sources such as *solar power* and *wind power*.

This type of tax forces companies to *internalize* the costs associated with carbon emissions. Businesses that do not adapt to a cleaner world will be in competition with those that do. For those that do not or cannot become cleaner and more efficient, the additional costs of the tax will be passed along to consumers, who will then use less of that product or service.

In addition to changing behavior, the revenue generated by the tax is used to pay for the environmental damage done by that industry or for the actions needed to prevent damage.

The level of tax assessed is critical to a carbon tax program's success. If it is too low, businesses will decide it makes more sense to just pay the tax than to change their way of business and just continue to pollute. If it is set too high, a business's costs will rise too high and too quickly for the business to survive, and even though the result might be good for the environment, it will hurt profits, jobs, and the consumer's wallet.

Carbon taxes or aspects of these taxes have been implemented in British Columbia, Quebec, *Australia*, Ireland, Sweden, Finland, *Great Britain*, and others. Boulder, Colorado, was the first city in the United States to implement a carbon tax.

carcinogen

A substance that can cause cancer.

carcinogen classification

Numerous substances can cause cancer and are therefore *carcinogens*. Many organizations categorize these substances according to the likelihood that they will cause cancer. In the United States, the most commonly used classification is created by the *Environmental Protection Agency*, and at the global level, the classifications by the International Agency for Research on Cancer is commonly used. Almost all of these groups use categories that list a substance as definitely, probably, or possibly carcinogenic. [*nanoparticles and human health; medical ecology; mutagenic*]

carnivore

Animals that only eat other animals are called carnivores, meaning meat eater. The *shark* and the dragonfly are both carnivores. [*predator/prey relationship; decapitating flies; bacteriophage; hyperparasites; virome; insectivore*]

car-pooling

Driving one *automobile* with one occupant is often cited as the single most environmentally harmful action people routinely indulge in, contributing to *air pollution* and *global warming*. More people in one car not only reduces this harm, but also is far less expensive for those in the car.

Car-pooling (also called ride-sharing) became popular during the *oil crises* of the 1970s when gas lines formed at the pump. It began informally as people got together and shared rides to work to save gas. This was often encouraged by their places of employment. At that time, more than 20 percent of commuters in the United States car-pooled to work, but its popularity has gradually faded. Today only about 10 percent do so and 75 percent drive solo every day.

Today, the Internet and *social media* have facilitated the growth of car-pooling. websites such as Zimride.com, Ridejoy.com, and others specialize in putting drivers in touch with riders. One of the largest and oldest is a German company called Carpooling.com that recently entered into the American marketplace. They have been in business (in *Germany*) for more than ten years and claim to have almost four million registered users. In the United States, Zimride.com claims 350,000 users.

There are also many *apps* focused on car-pooling such as Lyft, Zimride.com, ridejoy.com, avego.com, nuride.com, rideshare.com and eRideShare.com. [*car-sharing programs; bicycle-sharing programs; mass transit; transportation; autonomous automobiles; smart cities*]

carrying capacity

The maximum number of individuals of a particular species that a *habitat* can support and sustain is called its carrying capacity. Supporting and sustaining a *population* means all the individuals have enough natural resources such as water, food, and shelter to assure their survival and have the ability to eliminate wastes from the their environment.

The term *overpopulation* is used when the carrying capacity of an area is exceeded, resulting in a degradation of the environment, usually followed by a population decline.

Humans have an advantage over other species because they can manipulate the carrying capacity by changing how they consume resources and eliminate wastes—with technological advances. Some people believe attempts to increase the carrying capacity of our planet—with the use of technologies as well as *conservation* efforts—will be futile if the *human population* continues to increase. The current people population of Earth is 7.2 billion.

The human carrying capacity has been hotly debated for decades. Some scientists think ten billion will be the world's carrying capacity for people, beyond which there would be mass deaths because of starvation and disease. But many others believe that number can be much higher. [*human population throughout history; population explosion, human; population growth, limits of human; China's One-Child Program; safe operating space; social boundaries*]

car-sharing programs

These programs replace car ownership for people who do not need to drive to work every day or only need cars occasionally. They can result in significant reduction in local congestion, *air pollution,* and *carbon emissions.* For the consumer, the biggest advantage might be eliminating the cost of owning a car.

Where available, these programs provide 24/7 self-serve access to a network of vehicles typically strategically staged throughout a city. The cars can be reserved by the hour or day by methods such as smart-phone apps, the Internet, or a call center.

Modern-day programs began in the 1980s in Switzerland and Germany and quickly spread to Canada and then the United States. These programs are now popping up in many U.S. cities. Just in the United States, there are already more than 25 programs, boasting about 800,000 members and about 13,000 vehicles.

Car2Go is a successful program in Austin, Texas, run by Germany's Daimler Corporation, and is expanding to five more cities. Some programs are owned by major rental car companies, such as Avis/Budget, which bought ZipCar, one of the first such companies. Some nonprofit organizations are getting into the act as well. For example, City CarShare in San Francisco offers wheelchair-capable vehicles.

It is thought that more than 1.5 million people in 27 countries use these programs globally. [*car-pooling programs; bicycle-sharing programs; mass transit; transportation; automobile; autonomous automobiles*]

Carson, Rachel

(1907–1964) Rachel Carson, a *marine* biologist and writer, is best known for her 1962 book Silent Spring. In this book she described how *pesticides* cause long-term hazards to birds, fish, other wildlife, and humans but provides only short-term gains to controlling pests. As a result of her work, President John F. Kennedy formed a Science Advisory Committee to investigate her findings. They were soon confirmed and *DDT* and several other pesticides were banned in the United States six years later.

All of her books were devoted to protecting nature from an onslaught of chemicals produced by our industrial society. She coined the term *biocides* to replace the term pesticides; most pesticides do not distinguish between pests and any other organism—they just kill all life. Many believe Silent Spring helped spawn the modern *environmental movement* and consider it the most influential environmental book since *Thoreau's* Walden.

After Silent Spring was published and became popular, Carson was verbally attacked by those who disagreed with her facts—much in the same way people today attack leading *climate change* scientists.

Carson wrote many other books, including The Sea Around Us and The Edge of the Sea. Her writing style, although always based on solid fact, makes for easy, approachable reading. The following is a brief quote from Silent Spring: "Man is a part of nature, and his war against nature is inevitably a war against himself." [*philosophers and writers, environmental; body burden; toxic cocktail; breast milk and toxic substances; medical ecology; nanoparticles and human health; puberty and the environment, early*]

cartel

A group of independent enterprises, such as companies or countries, who get together with the express intent to limit competition and therefore fix the price of a product, good, or service. [*OPEC*]

Cartagena Protocol on Biosafety

See United Nations *Convention on Biological Diversity.*

catalytic converter

An emissions control device required on all (internal combustion engine) cars sold in the United States and most European countries. It reduces auto emissions by more than 75 percent compared to cars without the device.

These devices convert three of the pollutants that come out of a car's exhaust into less harmful substances. The pollutants are *carbon monoxide*, converted into *carbon dioxide* (a *greenhouse gas*); unburned *hydrocarbons*, converted into carbon dioxide and water; and *nitrogen oxides,* converted into free *nitrogen* and oxygen. [*air pollution; diesel engine fumes; greenest cars; idling your car*]

catch-and-release programs

Programs designed to maintain *populations* of certain species of fish in specific areas by requiring or requesting fisherman to throw back fish that meet certain size specifications. For example, anglers in many regions routinely release any salmon caught that measures more than 24 inches in length, because the larger fish are usually females and the larger females tend to produce more eggs than smaller females. When these programs are mandatory, they help minimize overfishing and maintain the stock. Catch-and-release programs are enhanced when the anglers use *barbless hooks.* [*fishing, commercial; fish populations; overfishing*]

Catlin, George

(1796–1872) Catlin was an American artist who painted the American West and Native Americans and became a spokesperson for natural *preservation.* He helped encourage the concept of *national parks* such as *Yellowstone.* He considered Native Americans part of our heritage and part of the natural landscape. He associated the decline of the *Native American* culture with the demise of *buffalo* on the western plains and painted these visions to instill an appreciation for both within the new country. [*Hudson River School Art; Audubon, John James*]

caulking

Caulking and *weather stripping* a home is considered the easiest, most economical way to save energy and your money. An average house (12 windows and 2 doors) requires $35 of caulking and weather-stripping materials and usually reduces heating and cooling expenses by at least ten percent. [*insulation; super insulation; R-value; indoor ecology; indoor air pollution; homes, green; zero-energy buildings*]

cellulosic ethanol

See *second-generation biofuels.*

cement

Cement production is an energy-intensive and environmentally destructive process. More than 100 million metric tons are produced annually in the United States alone. Estimates have worldwide cement production responsible for close to five percent of all *carbon emissions* and the resulting *climate change*. It has been suggested that cement is the most used manufactured product on Earth. About half of all cement is made in *China*.

A great deal of research is being done to reduce these carbon emissions and other environmental problems associated with cement production, because its use will likely only increase. The industry is looking for ways to reduce these emissions, because the possibility of some form of a *carbon tax* is looming in many nations. Some countries have already begun regulating the manufacturing process. [*hazardous air pollutants; dams, hydroelectric; carbon sequestration; air pollution*]

cemeteries and fracking

Because property owners typically have the right to lease their land, it is not uncommon for *fracking* to be allowed beneath cemeteries. This has upset some people who believe it might disturb the residents. But gas companies have stated that the most common question they receive when fracking beneath cemeteries is whether their dearly departed situated within the cemetery have the right to any of the lease royalties. [*NIMBY; YIMBY FAP*]

Census of Marine Life

This census was completed in October 2012, after 10 years of research including 540 global expeditions involving over 80 nations. The census gives a detailed look at the diversity of marine habitats and the abundance of animals and microbes found in the seas. The expeditions traveled to *coral reefs*, *hydrothermal vents*, seamounts, and, of course, the open ocean. They found organisms in areas previously thought incapable of sustaining life, including some of the hottest and coldest places on Earth and some with chemicals that would kill most forms of life instantly. The research shows microbes may make up 90 percent of ocean life mass. The census found 6000 species likely to be totally new and estimates the number of marine species to be roughly 250,000. The census was primarily funded by the United States. [*marine ecosystems; oceanic zone ecosystems; biodiversity; Map of Life; extremophiles*] {coml.org}

Center for Health, Environment and Justice

In 1981, *Lois Gibbs* created the *NGO, Citizens' Clearinghouse for Hazardous Waste* that later became the Center for Health, Environment and Justice (CHEJ). Gibbs is the

Executive Director. CHEJ offers organizational and technical assistance to grassroots groups in the areas of *environmental justice* and health. Here is a quote from their mission statement: "mentors a movement, empowering people to build healthy communities, and preventing harm to human health caused by exposure to environmental threats." [*cancer alley; cancer villages; hazardous air pollutants; Love Canal; LUSTs; Safe Chemicals Act; sunsetting; toxic cocktail; pesticide dangers; puberty and the environment, early; nanoparticles and human health; breast milk toxic substances*] {chej.org}

Center for Science in the Public Interest (CSPI)

This consumer *NGO* focuses on health and nutritional issues by identifying problems and informing the public of dangers. It regularly reports on deceptive marketing practices, dangerous food additives or contaminants, and flawed science reports, often promoted by some industries. CSPI has successfully obtained restrictions on several suspicious food additives and routinely evaluates the safety of new additives. It has a long list of accomplishments. Due in part to the organization's efforts, major fast-food chains have switched to less saturated fats, the *Food and Drug Administration* has banned many uses of sulfites (a preservative), and the beer industry has eliminated cancer-causing nitrosamines from its products.

Its <u>Nutrition Action Healthletter</u> has close to a million subscribers and is the largest-circulation health newsletter in North America. [*food dye; medical ecology; pesticide residues on food; nanoparticles in food; nonstick cookware; lipstick and toxins; nanoparticles in cosmetics*] {cspinet.org}

cephalopods

Cephalopods are *mollusks* that have tentacles; examples include the octopus, squid, and cuttlefish. Scientists consider these animals to be the most intelligent of all *invertebrates*. Most have the amazing ability to rapidly change color by using cells under their skin called chromatophores. They use color transformation both to disguise themselves and to attract mates. One species of cuttlefish has recently been discovered to use color in a more unique way. When courting, a male displays courtship coloration on one side to attract a female, while on the other side, it mimics the appearance of another female, to confuse rival males in search of females. [*camouflage, squid; giant squid in native habitat; mimicry; biodiversity*]

CERES (Coalition for Environmentally Responsible Economies)

The Coalition for Environmentally Responsible Economies (CERES) was founded in 1989 to promote corporate environmental and social responsibility. It began as a set of principles for corporations that wanted to express their concern for the environment. These principles were first called the Valdez Principles because they were initiated following the *Exxon Valdez* oil spill disaster. Signatories—companies or nations— agree to find ways to minimize *pollution*, use *renewable resources*, and reduce health risks to individuals and communities. Their mission is "to mobilize investor and business leadership to build a thriving, sustainable global economy."

They developed the *Global Reporting Initiative* (GRI) jointly with the *United Nations Environment Program* to provide guidelines about how companies and governments should report social and environmental performance along with their economic bottom line. More than 1800 companies now use the GRI. (In 2002, GRI became an independent agency.)

CERES works with 80 corporations that follow the CERES principles. About one-third of these companies are in the Fortune 500; twenty years ago, none were.

CERES also created the Investor Network on Climate Risk (INCR) in 2003 under the auspices of the United Nations and has more than $10 trillion of assets. This group of about 100 big institutional investors, such as retirement pension funds, works to help companies transition from *fossil fuels* to *renewable energy*.

In 2010, CERES published The 21st Century Corporation: The CERES Roadmap for Sustainability. This roadmap is considered an important plan for the future of the corporate world. [*corporate social responsibility; extended product responsibility; Beyond GDP; climate capitalism; corporate sustainability rankings; decoupling; integrated bottom line*] {ceres.org}

Certified Wildlife Habitats for individuals

See the *National Wildlife Federation*.

CFC (chloroflorocarbon)

Gases commonly used as refrigerants in air conditioners and refrigerators and as propellants in aerosol containers. Freon was the original CFC, developed in the late 1920s to cool refrigerators. CFCs cause *ozone depletion* and contribute to *global warming*.

Through the 1980s, the gases escaped from these products. In 1976, the concentration of chlorine (from CFCs) in the *atmosphere* was 1.25 parts per billion, but by 1989 it was twice that number. Although this might sound like a small amount, it had dramatic effects. CFCs were identified as the primary cause for a hole forming in the *stratospheric ozone layer* in our upper atmosphere. This is a serious threat because the ozone layer protects life on Earth from harmful UV light. The CFC molecules destroy *ozone* and, to make matters worse, prevent it from re-forming; a process that occurs naturally. Once in the atmosphere, CFCs linger for 50 to 100 or more years, continuing to cause harm.

This critically serious global environmental problem required a worldwide response. In 1987, many of the *more developed countries,* including the United States and Canada, *signed and ratified* the *Montreal Protocol,* which froze the production of CFCs at their existing levels with the goal of reducing the amount 50 percent by the year 2000. Today, 196 nations have signed the agreement, reducing *ozone-depleting gases.*

CFCs were replaced with *HCFCs* (hydrochlorofluorocarbons), which break down more quickly so they have less time to cause harm—HCFCs last only about 2 to 10 years. HCFCs were far better at protecting the ozone than CFCs, but technology continues to improve, and there are now refrigerants that cause no harm to the ozone at all, called *HFCs.* The Montreal Protocol has been revised many times, encouraging countries to transition to these latest substances.

Even though CFCs were banned many years ago and HCFCs have been banned since 2010 in the United States, they are still in demand to maintain old equipment. Procuring HCFCs requires a permit because it is regulated. However, a thriving black market business exists in the United States, bringing the gas in from countries such as *China.*

Today the reduction of these ozone-depleting gases on a global basis is considered one of the greatest accomplishments ever for protecting our environment. Because all of these substances linger long after their release, the hole in the ozone is taking a long time to reduce in size, but noticeable improvements were seen in 2006, and the latest studies show this continues today.

Unfortunately, there is an important relationship between *ozone-depleting gases and greenhouse gases.*

char

This is the charcoal-like organic remains of the combustion of *biomass.* [*biochar, biomass direct combustion; carbon sequestration storage*]

charcoal briquettes

Typical charcoal briquettes used for barbecues are composed of *coal,* limestone, borax, sodium nitrate, and sawdust. Burning charcoal produces the same gases as burning any *fossil fuel.* Using lighter fluid produces many compounds harmful both to the environment and to your health. Alternatives include starting your barbecue with newspaper, wax cubes, or hot (electric) irons.

charismatic megafauna

Large animals most people love and would hate to see become *extinct*. There is no formal list of them, but most conservationists would agree *gorillas*, *dolphins*, *whales*, *elephants*, tigers, and, of course, pandas are included. Notice they all are large mammals with big eyes.

Some people believe it is unfair and inappropriate to only try to save these types of organisms when many smaller species and even *insects*, nematodes, and *lichens* are *endangered* and important to *food chains* and *ecosystems*.

But others believe each of these charismatic megafauna makes for a great poster child for all endangered animals. They believe there is nothing wrong with using these species for public relations, because they draw attention to the problem of *biodiversity loss* and therefore all species benefit—called umbrella conservation. In hard economic times, when funding for such programs is limited, there remains a debate about concentrating on charismatic megafauna instead of a more balanced approach. [*conservation triage; flagship species; Ugly Animal Preservation Society; CITES; extinction, sixth mass; blobfish*]

Charity Navigator

All charities are not equal, but you can determine if a charity is truly charitable. Possibly the quickest way is to see what percentage of every dollar you give actually goes to the purpose of the charity, as opposed to overhead, salaries, and other expenses. But numerous other factors can help you determine the true worth of a charity if you know how to find the information.

The best way to determine where to donate your money for a cause is to thoroughly check the charity out first. A few organizations are dedicated to this mission, but probably the most recognized and well respected is Charity Navigator. Their goal is "to help charitable givers make intelligent giving decisions by providing information on more than five thousand charities and by evaluating the financial health of each of these charities." They clearly explain their criteria and provide excellent information. [*NGOs; news literacy; League of Conservation Voters; environmental rating systems; eco-certification; eco-labels*] {charitynavigator.org}

chemical body cleanse

Many synthetic substances found in our environment find their way into our bodies by eating, breathing, and contact. These substances include everything from *pesticides*, *heavy metals*, and *toxic substances* found in *plastics*, to *volatile organic compounds*. This load of chemicals—substances that don't belong in our bodies but make their way in—has recently become known as our *body burden*. While it is impossible to totally control what enters our bodies unwittingly, at least some of them have been found to be easily removed with a little effort—what can be called a chemical body cleanse.

The *Silent Spring* Institute, named after the famous writer and scientist *Rachel Carson*, performed experiments to see if it is possible to reduce two such substances, *BPA* and DEHP, from our bodies by simply avoiding products that contain them. Those in the experiment avoided all products containing these substances for three days, including plastics, utensils, cookware, containers, and so on.

BPA and DEHP are easily tracked in urine samples, which were taken before the experimental period and after. In all cases, the high levels initially found in those participating dropped by two-thirds for BPA and one-half for DEHP; however, they quickly returned to their high levels as soon as these people returned to their normal habits after the experimental period.

While all substances do not dissipate our bodies so quickly, it is good to know that at least for some toxic substances, it doesn't take much to cleanse our bodies of some of these burdens. [*toxic cocktail; body burden; microbiome and microbiota, human; breast milk toxic substances; geomedicine; medical ecology; nanoparticles and human health; skin bacteria*]

chemistry, green
See *Toxic Substances Control Act.*

chemoautotrophs
See *autotrophs.*

chemosynthesis
See *autotrophs.*

chemozoophobous
Plants that protect themselves from being eaten by producing noxious substances. For example, the blister beetle produces a chemical that blisters the skin on contact. [*sixth sense in animals; communications between plants; stereoscopic smell in moles; biodiversity*]

Chernobyl
One of the worst *nuclear power* plant accidents occurred in a small city north of Kiev in what used to be the Soviet Union on April 26, 1986, at 1:23 AM. (Chernobyl released six times more radioactivity then *Three Mile Island,* which was the worst such disaster

in the United States.) During a test, engineers violated regulations and turned off most of the automatic safety systems. The test resulted in two major explosions that blew the 1000-ton roof off the *nuclear reactor* and set fire to the reactor core. *Radioactive* debris flew into the air and was carried by the wind over much of Europe. Areas more than 1000 miles away became contaminated. More than 135,000 people were evacuated and the region was secured.

The Soviet Union acknowledged 36 deaths three years later, but many from within the Soviet Union say it was more like 300 dead. Medical experts estimate that between 5000 and 150,000 people in the region will die a premature death because of the accident at Chernobyl. Recent reports show the number of children with thyroid cancer has soared from one or two cases a year prior to the accident to more than 130 in 1991 in regions near the site. In 1992, Ukraine's chief epidemiologist reported a 900 percent increase in leukemia in those villages closest to the explosion.

The cleanup of the facility has cost billions of dollars and continues to this day. The reactor is now entombed in concrete but is showing signs of cracking. Chernobyl was the first nuclear disaster that caused many to fear nuclear power and question its role in our society. [*Fukushima nuclear power plant disaster; meltdown; nuclear fuel hazards; nuclear reactor safety; nuclear waste disposal; nuclear waste dilemma, the global*]

chernozem
See *soil types.*

Chief Seattle
(1786–1866) The chief's true name was Seeathl. Most of his quoted inspirational words about the environment are attributed to a speech in he gave in 1854, now called the "Fifth Gospel" speech. Many of these quotes—popularized by the media—did not come from the chief's mouth. For example, a quote about seeing the slaughter of thousands of buffalo was written by a screenwriter in 1972 and the words put into the Chief's mouth for a movie. [*Native American environmentalism; philosophers and writers, environmental; "Land Ethic," The; movies, environmental; Gaia Hypothesis*]

China, People's Republic of
The amazing rapid economic growth of China makes it an environmental focal point of the world. Choose almost any environmental problem and China has it. *Air pollution, water pollution, deforestation, desertification, hazardous waste* disposal, *biodiversity loss,* and *acid rain* are all on the list. This is important not only for the Chinese people but also for the entire planet. Depleting resources, pollution, and degradation don't

stop at a country's national boundaries. What China does over the next few decades will have enormous impact on all of us on Earth.

Air pollution created in China settles in the United States and Canada. Hydropower *dams* in China reduce water supplies in *Laos* and other countries downstream. The Chinese penchant for exotic meats and animal parts is forcing *endangered species* toward their demise—*rhinoceroses* for their horns, *elephants* for their tusks, American black bears for their gall bladders, and *sharks* for their fins, just to name a few. China's demand for exotic hardwoods reduces *old-growth forests* and increases deforestation in general. Acid rain from China settles in Japan and Korea. China recently surpassed the United States as being the largest emitter of *greenhouse gases* on the planet.

When China causes environmental harm, they do so on a grand scale, much the way the United States did years before. For example, they generate *hydropower* from the ever-expanding *Three Gorges Dam*—the largest in the world—which causes enormous environmental destruction, especially to *aquatic ecosystems*. And yet, they are planning ten more such dams up river from the Three Gorges.

The country's economic growth and accompanying multitudes of rising middle class are driving much of this global environmental degradation. Because of their economy and the speed with which it grows, China can be on both sides of an environmental problem. For example, they are one of the world's worst polluters of our air, because of their *coal*-burning electric power plants, yet they are at the forefront of *wind power* and *solar power* technology. They cause *water shortages* in other nations with megaprojects such as many huge dams, but they also are at the forefront of *desalination* projects.

More recently, they appear to be applying a green strategy to mitigate the environmental risks and probably the political risks of continuing a destructive path economically. They have programs to promote widespread use of energy-efficient lights and other electronic devices. They created "eco-areas" to promote regional *sustainable development* initiatives. More recently, they have announced water and air pollution measures along with *reforestation* and *afforestation* efforts.

The country does not lend itself to *NGOs* well, but in 1994, Friends of Nature was established and others have since found roots within China—possibly a sign of things to come. [*India*]

China's One-Child Program

Beginning in 1976, Chinese leadership decided that continued population growth was an obstacle to economic growth and adopted extreme measures to control it, mandating one child per married couple. The law is still in effect, although now more exceptions are allowed. The law has resulted in a dramatic drop in the country's population growth rate. There are concerns that infanticide is common because males are much favored in families, especially in urban regions.

Families that do have an additional child are fined. The fines, called social maintenance fees, increase with the family's income level. If the family cannot pay the fee, they can be forced to abort the baby. In 2009 there were six million such abortions.

Opposition to the one-child policy has recently been a common target for *microbloggers*. [*population explosion, human; population growth, limits of human; Population Reference Bureau; human population throughout history*]

chledophyte

A plant that can grow on rubbish heaps such as *landfills*. [*extremophiles; bioremediation; bio-ore; geoengineering*]

chlordane

A *pesticide* used on crops and lawns until 1988, when it was banned in the United States. It is on the *dirty dozen POPs* (persistent organic pollutants) list. However, it is manufactured in the United States for export elsewhere and is still commonly used. [*pesticides; body burden; Federal Insecticide, Fungicide, and Rodenticide Act; biocide; DDT*]

chlorinated hydrocarbons

One of the four major groups of synthetic *insecticides*. *DDT* was the first chlorinated hydrocarbon insecticide and probably the best known. Chlorinated hydrocarbons affect an organism's nervous system, usually resulting in death. They can be applied once and last for a long time, which is both a blessing and a curse. Because they are so persistent and remain in the environment for 2 to 15 years, they need not be continually applied to control a pest.

But these *pesticides*, like most, do not know a pest from most other forms of life, so they are harmful to nontarget (nonpest) organisms. Because they last so long in the environment, and can be absorbed and stored in an animal's body, they pass along *food chains* and accumulate in animals—including humans—in processes called *bioaccumulation* and *biological amplification*. Most chlorinated hydrocarbons have been banned in the United States because of their *chronic toxicity* and persistence. They are, however, still commonly used elsewhere in the world. [*circle of poison; pesticide dangers; Rachel Carson*]

chlorination

Chlorine has been added as a disinfectant to *drinking water* since the early 1900s. The use of chlorination to keep drinking water clean and safe has been considered a major public health success story for decades, with 7 out of 10 Americans drinking chlorinated water. But many scientists have become concerned about its safety and there is significant controversy now about its use. Many communities have prevented its use because of this debate.

Chlorine reacts with *organic* matter in the water, such as decaying leaves or grass, and produces substances called *THMs*, (trihalomethanes), as well as roughly 600 other byproducts, some of which are believed to be carcinogenic. Most of the debate waged is about whether or not chlorine should be added to a town's water. Other people are concerned about how the *Environmental Protection Agency* (EPA) regulates the chemical in the water. The EPA regulates the substance by issuing "annual" average amounts that can be present in a municipality's drinking water. This means a municipality can add high concentrations of chlorine on any given day, posing a possible health threat. Some critics don't want chlorination in their water and others want regulation reform. [*Safe Water Drinking Act; water use for human consumption; water treatment*]

chlorophyll

Green pigment found in plant cells, essential for *photosynthesis*.

chlorophyll-based solar cells

Researchers hope that someday they may create *solar-powered*, *photovoltaic cells* that use, in part, *chlorophyll* to convert light energy into electricity. It would be flexible and, of course, nontoxic. [*solar glitter; solar cookstoves; solar tube lighting; solar charging of electronic devices; photovoltaic cell film*]

christmas tree lighting

Most older holiday lights are the old *incandescent lights* that are inefficient in many respects. New *LED* holiday lights have a similar look but use 80 percent less electricity and last far longer, with some claiming 25,000 hours of use.

chronic toxicity

A harmful effect such as illness or death after long-term exposure to low dosages of a substance. [*toxic waste; toxic pollution*]

chytrid fungus

Most members of this group of *fungi* feed on decomposing organic matter; others are *parasites*. A new chytrid fungus was discovered in 1999 that is parasitic on *amphibians*, called Bp for short or *Batrachochytrium dendrobatidis*.

From an environmental perspective, the importance of this fungus cannot be overstated. In scientific papers, it has been called "the worst infectious disease ever recorded among *vertebrates* in terms of the number of species impacted, and its propensity to drive them to *extinction*." The species referred to are amphibians–most specifically frogs. However, it appears it could infect almost all of the world's roughly 6000 amphibian species. Scientists are advocating for a large-scale research effort to stave off a possible oncoming disaster for this important component of most *ecosystems*. At the moment, 100 species are known to have contracted the disease.

The fungus attacks the amphibian's skin. Most amphibians "drink" water by absorption through their skin (not with their mouths). The fungus appears to interfere with the proper absorption of water and important chemicals through the skin. There are known remedies, but nothing that can be used on the scale needed to curb the spread of this disease. [*biodiversity, loss of; amphibian decline; Amphibian Ark project*]

CIA and climate change

The U.S. Central Intelligence Agency feels that *climate change* is a national security threat and began funding studies to consider the risks and benefits of *geoengineering* solutions. Possible methods being analyzed include *carbon sequestration storage* and reflecting solar radiation back into space with sulfate *aerosols* high in the atmosphere.

cigarette butt litter

Cigarette butts are the most common form of *litter*. Estimates are that 65 percent of all smokers routinely throw butts out of car windows. *Keep America Beautiful* states butts make up 38 percent of all litter along roadsides. In 2011, during an annual beach cleanup event in California, over one million butts were collected. *Cadmium*, arsenic, *lead*, and nicotine are known to *leach* out of them into the *environment*. (Not to mention in your lungs.) One butt in a liter of water with minnows swimming in it will become toxic enough to kill all of the fish within 96 hours.

For some reason, many people think these butts are *biodegradable*, but they are not, because most contain a *plastic* component. RethinkButts.org is a campaign to get people to stop this type of littering. A company called TerraCycle is working with a tobacco company to recycle cigarette butts into shipping pellets. [*plastic pollution; garbage*]

Cinderella species

See *flagship species*.

circle of poison

Some *pesticides* are banned in the United States but still exported to other countries. There they are used on crops, which are then imported back into the United States and sold in American supermarkets. Some of this produce contains *pesticide residues* that would be illegal if grown and sold in the United States. Some people believe this occurs only occasionally and is not a major concern. Others believe it to be a serious problem. [*pesticide dangers; pesticide regulations; Toxic Substances Control Act; Federal Insecticide, Fungicide, and Rodenticide Act*]

circular economy

See *recycling metals*.

circus elephant ban

Many animal rights advocates have fought long and hard to ban the use of circus *elephants*, stating they are treated inhumanely. Many cities have such bans; California has at least six cities with these bans. Los Angeles is considering implementing one, which would make it the largest city in the country with such a ban. [*animal rights movement and animal welfare; farm animal cruelty legislation; primate research; greyhound dogs; sealing; zoo*]

CITES (Convention on the International Trade of Endangered Species of Wild Flora and Fauna)

CITES was established in 1973 to monitor and regulate the international *wildlife trade*—a multibillion-dollar-a-year business. As of 2013, 178 countries participate. It meets every other year to assess which organisms require protection and then implement restrictions based on this assessment.

The organization has three designations: The first prohibits all international trade of an *endangered species*. The second permits some trade of species and the third designations leaves the regulations up to the countries involved. Recently CITES agreed to regulate the trade of many *threatened* species, including five shark species and manta rays. This was especially important for many shark populations being decimated because of *shark finning*.

To be assigned to one of these three designations, a species does not need to be on the *Red List of threatened species*. This is important, because many nations only provide protection for species on this list. But countries abiding by CITES are required to protect species even if they are not on that list. CITES currently provides some level of protection for about 35,000 species of plants and animals.

Although CITES passes laws that member countries are required to abide by, there are concerns. Adhering to CERES regulations by participating members remains voluntary, loopholes can be found, and enforcement is often difficult.

In the United States, CITES is implemented by the *Endangered Species Act* and enforced by a patchwork of agencies, including the *Fish and Wildlife Service*, the Animal and Plant Health Inspection Service (part of the *Department of Agriculture*), the *National Marine Fisheries Service* (part of the Department of Commerce), and the Customs Service (part of the Department of Treasury). [*poaching; ivory trade; polar bear trade; biodiversity, loss of; extinct in the wild; rhinoceroses and their horns; species recently gone extinct; wildlife trade and trafficking; conservation triage*]

cities, ten best American green

Many organizations rank cities according to how environmentally conscious they are. The Mother Nature Network used criteria such as air and water quality, *recycling* programs, *waste* management, the percentage of *LEED*-certified buildings, the amount of green space, and many others. They ranked the top ten best as 1) Portland, OR; 2) San Francisco, CA; 3) Boston, MA; 4) Oakland, CA; 5) Eugene, OR; 6) Cambridge, MA; 7) Berkeley, CA; 8) Seattle, WA; 9) Chicago, IL; and 10) Austin, TX. [*universities, top ten green; countries, ten greenest; countries, ten most energy-efficient; NFL stadiums, green*]

Citizens Clearinghouse for Hazardous Waste

Originally founded in 1981 by activist *Lois Gibbs* following the *Love Canal* disaster, this *NGO* provided information about *hazardous wastes* so it was readily available to environmental groups and decision-makers. It is no longer in operation but has evolved into the *Center for Health, Environment and Justice.*

civil disobedience and environmental movement

See *environmental movement.*

Clean Air Act

This is a cornerstone piece of legislation, passed in 1970, designed to improve the quality of America's air by controlling *air pollution*. Its goal is to assure Americans that the air they breathe poses no health risk. The law is enforced by the *Environmental Protection Agency* (EPA).

In 1990 the Act was renewed and amended to include many new important mandates: Most major cities were given timetables to reduce emissions of primary air pollutants that cause *air pollution* and *smog*. (Many cities missed these original deadlines.) The Act required reductions in *automobile* tailpipe emissions and the use of special nozzles on gasoline pumps to reduce *volatile organic compound* (VOCs) fumes. It required industries emitting any of 189 identified *toxic* chemicals to use the *best available technology* to reduce these emissions. Limits were also placed on the amount of *sulfur dioxide*, partly responsible for *acid rain*, released from *fossil fuel*-burning electric power plants.

In 2013, the EPA proposed amendments to the Act, including new, more stringent *carbon emissions* standards for most *electric power plants*. The proposal includes the following regulations: Any new large *natural gas* electric power plants must meet a limit of 1000 pounds of *carbon dioxide per megawatt-hour*. New, small natural-gas electric power plants must meet a limit of 1100 pounds of carbon dioxide per megawatt-hour. New *coal*-burning electric power plants must meet a limit of 1100 pounds of carbon dioxide per megawatt-hour.

The EPA has the authority to create these regulations for new power plants. However, it has less flexibility in controlling the emissions of existing, older facilities responsible for much of the pollution. Older coal-fired electric power plants spew out the most *greenhouse gases*, contributing the most to *climate change*. The Obama administration had promised to change these regulations by mid-2014, but there are doubts whether this can be accomplished.

Clean Development Mechanism

Carbon markets were first created at the international level in the *Kyoto Protocol*. This protocol created a *cap-and-trade program* designed to meet the *greenhouse gas* emission limits previously established at the *United Nations Conference on Environment and Development*. This cap-and-trade program included a special plan that allowed carbon emissions to be traded by using the Clean Development Mechanism. The idea was to create projects that reduce greenhouse gas emissions in *developing countries*. Then, to let these countries sell credits called *carbon offsets* to *developed nations*.

The money collected by developing nations would pay for the emission-reduction projects, and the money paid by the developed nations for these offsets could be used as a credit against their own emission-reduction requirements. More than 2400 projects have been registered and they have been responsible for a reduction of two billion tons of carbon dioxide.

Clean Fifteen

The *Environmental Working Group* (EWG) creates an annual list of the 15 fruits and vegetables least contaminated with *pesticides*, called the Shopper's Guide to Pesticides in Produce. The 2013 list is (from best): 1) asparagus, 2) avocados, 3) cabbage, 4) cantaloupe, 5) sweet corn, 6) eggplant, 7) grapefruit, 8) kiwi, 9) mangoes, 10) mushrooms, 11) onions, 12) papayas, 13) pineapples, 14) sweet peas (frozen), and 15) sweet potatoes.

The EWG estimates that if you eat five servings a day of fruits and veggies from the Clean Fifteen, as opposed to from the *Dirty Dozen Fruits and Vegetables*, you would lower your intake of pesticide residues by 92 percent.

Clean Water Act

This is the cornerstone piece of legislation designed to protect U.S. waters from pollution. Its stated purpose is "to restore and maintain the chemical, physical, and biological integrity of the nation's waters." It began as the Federal *Water Pollution* Control Act of 1972, but when combined with amendments in 1977, 1981, 1987, and 1993, it has become known collectively as the Clean Water Act. It has resulted in a dramatic improvement in the quality of U.S. waters. The law mandates each state to adopt water quality standards for all *surface waters* and restricts industry and municipal waste discharges into the water. It is also meant to protect *wetlands*. It is enforced by the *Environmental Protection Agency*.

The portion of the act responsible for wetlands protection has been a political battleground for many years. The battle is over private property owners' rights regarding their wetlands and the federal or state rights that protect these wetlands from destruction. The lines are usually drawn in courts and Congress over the legal definition of what exactly is a wetland. Some bills proposed in Congress that favor redefining wetlands would result in the destruction and development of most remaining wetlands. Other bills would leave the definition of a wetland up to scientists instead of politicians. This debate is not yet settled. [*water pollution; water use for human consumption; Safe Drinking Water Act*]

Clean Water Act and fracking

The *Clean Water Act* has become news lately for something it does "not" do. Many people are concerned about the environmental harm caused by *fracking* for *natural gas*. They are often under the false impression that the fracking process must adhere to Clean Water Act laws. That would mean fracking would not be allowed to proceed if it polluted bodies of water near these fracking sites and would lessen the damage caused by fracking.

However, that is not the case. *Natural gas* development, including fracking, has an exemption from any regulations put forth by the Clean Water Act. This is a highly contentious exemption because few people see any logical reason for exempting this one form of resource extraction from this important law. The bottom line is—if you read that fracking must adhere to the Clean Water Act, it is not factual. [*natural gas loophole*]

clean technology

Clean technology (also called "clean tech") is an informal term used to describe products or processes that use *waste minimization* techniques and as few *nonrenewable resources* as possible. [*best available technology; green jobs; green buildings*]

clear-cutting

Clear-cutting is a *forest* logging method in which every tree in a region is cut. This is one of the oldest methods of harvesting forests and is used globally. Clear-cutting is economical for the logging industry but often devastating to the forest. When large tracts of forest are cleared, the *habitat* of most animals is lost and the original *ecosystem* for the entire region decimated. It is important, though, to distinguish between clear-cutting as a *forest management* practice and clear-cutting resulting in *deforestation*.

Clear-cutting, when properly managed and used as part of a sustainable program, is considered by many an acceptable form of forest management. But this is a big "if." Clear-cutting large tracts of the *Amazon* for development with no plans for sustainability is just as defined—devastating to the forest and the environment.

Clear-cutting, especially on sloped terrain and where regrowth is slow, causes *soil erosion*, making recovery of the ecosystem impossible. Global clear-cutting as part of deforestation is believed to be affecting the overall balance of *carbon dioxide* throughout the *biosphere*. For decades, clear-cutting in U.S. national forests has angered environmentalists, who try to curtail further destruction.

Clear-cutting relatively small tracts of forests with little slope and rapid regrowth has often been done successfully. This *patch clear-cutting* removes only small areas of forest, but leaves the surrounding growth untouched, causing less harm to animal habitats and being less likely to result in soil erosion. [*forest health, global; forest management, integrated; forest, pop-up; forest, sustainable; logging, illegal; rainforest destruction; redwood forest destruction; timber, DNA sequencing of; tree farms*]

climate capitalism

See *natural capitalism*.

climate change

Greenhouse gases produce the *greenhouse effect,* which traps heat at the Earth's surface, maintaining a relatively constant temperature. Human activities increase the amount of greenhouse gases in the *atmosphere*, resulting in an increase in the Earth's surface temperature. This process is often called *global warming* and results in what is commonly called *climate change*.

Carbon dioxide is responsible for the majority of global warming. It occurs naturally, but large volumes of it are released when we burn *fossil fuels* (*coal, oil, natural gas*). Other greenhouse gases include *methane, nitrogen compounds*, and *ozone*, all occurring in nature but increased by human activities. About 80 percent of global warming is caused by these gases. Other causes of global warming are not from the addition of these gases but from humans changing natural cycles.

For example, *deforestation* is believed to account for much of the remaining 20 percent of global warming. Plants incorporate *carbon dioxide* into their tissues during *photosynthesis*. Fewer trees means less intake of carbon dioxide; so if it's not in their tissues, it's in the atmosphere. In addition, burning this wood sends carbon dioxide back into the *atmosphere* at an accelerated rate.

In 1988, the *United Nations Environment Programme* (UNEP) established the *Intergovernmental Panel on Climate Change* (IPCC) to study this phenomenon on a global scale and report to the world its findings. They have published five comprehensive *IPCC reports* (1990, 1995, 2001, 2007, and 2013) that provide scientific data documenting this complex problem.

Scientists have debated how much warming the planet could withstand before catastrophic results. The consensus is that we must stay below 2 degrees C (3.6 degrees F) to prevent profound negative impact on our planet, such as a *sea-level rise* that would displace millions of people in low-lying countries such as Bangladesh and even many of the U.S. coastline cities. Many scientists believe we might be reaching a *tipping point* that—once passed—would become impossible to reverse.

The most important international agreement to slow greenhouse gas emissions is the *Kyoto Protocol*. The *developing nations* and *developed nations* are divided as to who should be responsible for "fixing" this problem—the developed nations who are primarily responsible for causing these emissions or the developing nations who often bear the brunt of the negative environmental impact.

It appears that science and logic have not convinced most of the nations of the world that we must initiate drastic changes to prevent climate change. Many people believe the only things that get nations to finally act are *economics* and profit/loss statements. They point to the $100 billion dollars per year it will take for the world to adapt to climate change when it does occur between now and 2050, as projected by the World Bank. Once the world comes to grip with these numbers, possibly something might be done.

Some people—usually called *climate change deniers*—consider climate change to be overstated and a natural problem, not human-made. Others believe the entire scientific body of evidence is a hoax. (Others still believe the world is flat, as well.) [*anthropogenic stress; Agenda 21; archeologists, ice-patch; assisted migration; Bangladesh and sea rise; climate change and human health; climate change and biodiversity; climate change and marine life; climate change legislation; Climategate; vineyards and assisted migration; oysters and climate change*]

climate change and biodiversity

Scientists have predicted major changes to *ecosystems* as global temperatures warm due to *climate change*. The pine *beetle* in the Rocky Mountains demonstrates how small changes in temperature can have a big impact on an entire *ecosystem*.

Many types of trees have evolved chemical defenses again *insect* attacks. Lodgepole pine, common at lower elevations in the mountains, produce toxic chemicals that prevent the pine *beetle* from attacking the trees. As temperatures have warmed, the pine beetles have expanded their range and moved up to higher elevations, where whitebark pine is typically found. But these trees did not coevolve with pine beetles and therefore have no toxic chemical defense mechanism. Nor do the beetles have any natural *predators* in this new range.

Whitebark pine with no natural defenses against the beetle and the beetle with no predators gives significant potential for major outbreaks of this pest. Whitebark pine is a critical *habitat* for grizzly bears and plays a major role in the *water cycle* of the region by shading snow, which slows the flow of meltwater.

The exact impact of climate change on biodiversity is unknown, but it is assumed that it will be far reaching.

climate change and human health

Studies are being conducted to see how *climate change* may affect or already is affecting humans. The most obvious impact has started with the lengthening of the ragweed pollen-producing season and the production of more pollen. The ragweed season in the Midwest increased from 13 to 27 days over the span of 1995 to 2009. But these types of studies are in their early stages and much more needs to be done.

climate change and marine life

As the *atmosphere* warms—because of *climate change*—so too will the *oceans*, and this has been the case. Moreover, recent studies show this warming trend affecting far deeper into the waters than first expected. Most early projections stated the warming effect would be down to about ten feet beneath the surface, but recent research shows warming changes down to almost 3000 feet. Most marine organisms live within the top 1300 feet, so this is big news, because climate change is now believed to impact almost all *marine ecosystems* instead of just a few as first thought. [*jellyfish*]

climate change deniers

A substantial contingency of people do not believe *climate change* is occurring—or if it is, it is not a result of human actions. The group is well funded and far more vocal than knowledgeable.

About 99 percent of the scientific community states that research clearly shows climate change is occurring and does result from human actions. The *Intergovernmental Panel on Climate Change* is a collection of the world's finest scientists, who spend all of their time studying this field—about 800 scientists from about 120 countries. Their *IPCC reports* explain why they believe climate change is human-made. Many newspapers, websites, magazines, and blogs translate these reports into plain English (and every other language), so anyone can understand them.

To deny the latest scientific evidence about climate change is similar to going to a doctor and being told you have a terrible disease and need surgery. You go to another doctor for a second opinion and are again told you need surgery. You go to 99 doctors and they all say the same thing. Then you go to one more doctor, who says you are fine and the other specialists don't know what they are talking about. So you don't have surgery. That hundredth doctor is probably a climate change denier. [*Agenda 21 conspiracy; argumentum ad ignorantium; pseudoscience; science; opinions and science*]

climate change legislation (U.S.)

As of 2013, the U.S. Congress did not appear to be capable of passing any meaningful legislation about *climate change*. If things did not change, President Obama stated his plans to use an executive order (which does not require Congressional approval) to get something accomplished. He stated (about climate change) in February of 2013, "If Congress won't act soon to protect future generations, I will." Whether he does or not remains to be seen.

Climategate

In 2009, the e-mail of many of the world's most highly respected climate scientists was hacked and exposed to the public. As with anyone, having an unintended audience read your e-mail was embarrassing to say the least. Their informal talk about *climate change* and convincing people of its importance looked to some as if they were not serious about their science. However, to those who investigated the episode, this was not the case. After investigations in Britain, the scientists were absolved of having done anything wrong. But to *climate change deniers*, who must find any scrap of evidence to help their cause, it was a scandal of the highest proportion and they made the most of it, including attacking the scientists personally. The biggest crime was probably the damage done, for a short period, to convincing people about the urgency of climate change. [*Intergovernmental Panel on Climate Change; Agenda 21 conspiracy*]

climate sensitivity

Carbon dioxide is a *greenhouse gas.* The more carbon dioxide we put into the atmosphere, the more global temperatures rise. But how much carbon dioxide causes how much temperature rise? Climate sensitivity is a measurement that combines these two factors: How much carbon dioxide does it take to increase global temperatures how many degrees? Climate sensitivity gives us a better understanding of *climate change*—what is going to happen and when.

The standard method of measuring climate sensitivity is as follows. If we double the amount of carbon dioxide found in the atmosphere prior to the *Industrial Revolution* (when we first started spewing carbon into the air), how much will temperatures rise? Prior to that time, carbon dioxide in the atmosphere measured 280 parts per million (ppm)—doubling that is 560 ppm. So to determine climate sensitivity, the question becomes: How much will global temperatures rise when our activities raise carbon dioxide levels to 560 ppm? The <u>*Intergovernmental Panel on Climate Change Fifth Assessment*</u> report stated it would be 2.7 F at the low end and 8.1 F at the high end. (In 2013, carbon dioxide levels were recorded for the first time to hit 400 ppm.)

Climate sensitivity only tells us what the temperature might be— but not definitively what it will be. That is what climate experts are trying to determine and spend a great deal of time debating. [*carbon budget; Keeling Curve*]

climax community

The final, relatively stable, and self-perpetuating *community* of organisms in an *ecosystem*. The climax community is the final stage of *succession*. The climax community of a terrestrial ecosystem is often called a *biome*. [*anthrome*]

cloud computing and the environment

See *data centers*.

Club of Rome

A group of influential businessmen and scientists formed this think tank in 1972 (the same year as the *Stockholm Summit*), with the intent to study the limits of economic growth. The group published the well-known, well-respected, and still read *Limits to Growth* in the same year. The book provides models that correlate our use of natural resources to our population growth and consumption and projects the resulting inability of our planet to sustain *human populations*. This book was one of the first to sound this alarm.

In 2012, the Club of Rome released a 40-year follow-up report titled <u>2052 – A Global Forecast for the Next 40 Years</u> that revisits many of the issues introduced in the original book. [*human population throughout history; overpopulation; population explosion, human; Beyond GDP*] {clubofrome.org}

coal

A *fossil fuel* that supplies about 28 percent of the world's energy. Most of this coal is burned in boilers to produce steam that generates electricity. In the United States, almost 40 percent of our electricity is produced from coal in this manner, although new regulations and the boom in *natural gas* are reducing this amount. Most coal is extracted from the Earth by either *strip mining* or *subsurface mining*.

China is the biggest coal consumer, accounting for almost half of world use, and also the biggest coal producer, accounting for almost half of all global production. But the United States has the largest amount of coal by far, with almost 30 percent of global *reserves*.

Coal reserves (the amount that can be extracted today) are projected to last for more than 200 years if existing demand continues. The projected amount of coal *resources* (unidentified deposits), however, might last up to 900 years with the current demand.

Coal contains high heat content at economical costs but is a major cause of many of our most pressing environmental problems. Coal is the dirtiest fossil fuel to burn, so sophisticated *air pollution* control devices are necessary. Burning coal releases pollutants that contribute to *acid rain*. But the biggest problem is it produces more *carbon dioxide* than any of the other fossil fuels, contributing to *climate change*.

Strip mining for coal devastates an area and usually results in serious *erosion*. Even when attempts are made to reclaim the land with grading, return of *topsoil*, and replanting, the *ecosystem* never fully recovers its *biodiversity*.

Newer technologies burn coal more efficiently and more cleanly. This includes a *fluidized-bed combustion method* that converts solid coal into gas or liquid fuels called *synfuels*. However, "clean coal" is a misnomer because coal, even burned with the newest technologies, is not anywhere close to being as clean as *natural gas* or any of the *alternative energy sources*.

coalbed methane gas production

Hydraulic fracturing, or *fracking* for short, is a technology used to extract *natural gas* and *oil* from geological formations where low permeability made it impossible or too expensive to retrieve in the past. There are three such types of formations where this is a problem: *shale beds*, tight sands, and coalbed methane beds.

Fracking has been used to extract natural gas from coalbed methane beds far longer than the other two types of formations. Coalbed methane beds consist of *coal* that has absorbed large amounts of *methane gas* and water.

Fracking enlarges existing cracks or fissures within the coal. When the size of the fissures is increased and more are made, the water is released and removed through the well. As the water is removed from the coal, the pressure within the coal bed is lowered, allowing the methane gas to escape as well. The escaping gas can then be extracted along with the water.

Many environmentalists are concerned about a serious problem associated with this form of extraction. As the process proceeds, the *water table* is lowered to extract the gas. In addition, the fracking process injects large amounts of *toxic* chemicals into the wells. So, not only are these wells lowering the water table, but also chemicals are added that can contaminate the water that remains.

Many of these coalbeds exist in formations considered underground sources of *drinking water*, so there are serious concerns that this extraction method is destroying *aquifers* that supply much of our water supplies.

coal formation

Coal began to form about 300 million years ago (during the Carboniferous Period) when large regions of the Earth were covered with tropical *swamps* containing dense vegetation. After the vegetation died and accumulated, it was submerged under the water and formed a material called *peat*—the first step in the formation of coal. Peat is composed of about 90 percent water, 5 percent carbon, and 5 percent other substances. The peat was then gradually covered by *sediment*.

Over time, pressure squeezed out much of the water and compressed the peat into *lignite* coal (also called brown coal), which contains about 40 percent water. With heat from the Earth and continued pressure, lignite was transformed into a soft type of coal called *bituminous* coal, which only has about 3 percent moisture. With continued heat and pressure, hard coal called *anthracite* could finally form.

The entire process took hundreds of millions of years. The products produced during each of these stages are found today. In some *less developed countries*, peat is dried and burned, as a form of *biomass energy*, but it has low heat content and makes for a poor energy source. Lignite also has low heat content and is not a good source of energy. Bituminous coal has high heat content and is the most common type of coal. Unfortunately, it has high *sulfur* content, making it more harmful to the environment. *Air pollution* control devices are used to reduce these emissions. Anthracite has high heat content and low sulfur content, making it the most desirable energy source. This coal, however, is in limited supply and expensive. [*geological time scale*]

coal gasification and liquefaction

See *synfuels*.

coal gasifier

Almost any fuel will burn more cleanly and efficiently if it is first converted to a gas. When this principle is applied to organic matter (*biomass* or *fossil fuels*), it is called *gasification*. A promising form of gasification involves using coal gasifiers that burn *coal* far cleaner than conventional coal combustion, producing less *air pollution* and fewer *greenhouse gases*. It also burns more efficiently, getting more energy out of the same amount of coal when compared to conventional coal burning. However, even the most efficient coal-burning technologies cause more environmental harm than either *oil* or *natural gas*.

coastal flooding

Sea levels have been rising and major storm frequency increasing, most likely because of *climate change*. In spite of this, more people continue to move closer—not further away from—the coasts. The results have been catastrophic and are expected to get worse. Because we keep building near the shore, damages caused by coastal flooding will increase. For example, Hurricane Andrew, a Category 5 storm, hit the southeastern coast of Florida in 1992, causing $23 billion in damages. If this same storm were to happen today, it would cause about $50 billion in damages, because more people live in these same coastal areas.

Regions in low areas are most prone to destruction. South Florida, areas surrounding San Francisco and Seattle, parts of Houston, Norfolk, North Carolina's Outer Banks, and the Eastern Shore of Maryland are all in jeopardy of coastal flooding. Some areas, such as the central valley of California, are protected by levees and are likely candidates for flooding. New York City, which showed its vulnerability during the recent Superstorm Sandy, illustrates how we have exacerbated the problem by building up and building on areas that were once inundated with water. Virtually the entire part of Manhattan that was originally built on human-made land became flooded during Sandy.

Most of this damage is because of building on these lands that should not have been built on in the first place. And the problem has been exacerbated by the destruction of coastal trees and vegetation that naturally absorb water and the *coastal wetlands* that act like natural sponges, soaking up water and storing it to be released slowly. With these natural buffers gone, coastal flooding has become far worse. Some cities are trying to reestablish these natural barriers—but re-creating wetlands and natural buffers is a much greater task than not destroying them in the first place. Other coastal cities are trying to defend themselves with massive high-tech storm barriers. [*urbanization and urban growth; smart growth; green infrastructure; no net loss; floodplain; Bangladesh and sea rise*]

Coastal Society, The (TCS)

The Coastal Society, an *NGO* chartered in 1975, serves as a forum for *coastal* resource professionals and others interested in promoting a better understanding and sustainable use of coastal resources. Its goals are to: encourage cooperation and communication, promote *conservation* and wise use of coastal resources, help government and industry balance development and protection along the world's coastlines, and advance public education and appreciation of coastal resources.

Their mission statement says they are "an organization of private sector, academic, and government professionals and students. The Society is dedicated to actively addressing emerging coastal issues by fostering dialog, forging partnerships, and promoting communications and education." [*coastal flooding; coastal wetlands; marine ecosystems*] {thecoastalsociety.org}

coastal wetlands

Land that is flooded for all or part of the year is called a *wetland*. If it contains saltwater, it is called a coastal wetland (as opposed to *inland wetland*). Coastal wetlands in temperate regions include bays, lagoons, and salt marshes, and all have grasses as the *dominant* vegetation. In tropical regions, coastal wetlands are primarily *swamps* inhabited by *mangrove* trees. [*marine ecosystems; coastal flooding; eco-services*]

coastal zone

See *neritic zone, ocean.*

Coastal Zone Management Act

This act created the Office of Coastal Zone Management as well as the *National Oceanographic and Atmospheric Administration (NOAA)*. The office is responsible for protecting the U.S. coastlines from development by requiring states to provide details about activities impacting coasts. The term "coasts" was expanded in recent years to include *wetlands, floodplains, estuaries, beaches, coral reefs,* and fish and wildlife *habitats*.

coast redwood

The coast redwood is the tallest species of tree in the world and can be one of the oldest. An average mature tree stands 220 feet tall, but some have measured in at 360 feet with a trunk diameter of 10 to 15 feet. The average age of the *old-growth* redwoods is 500 to 700 years, but a few of these trees have lived for over 2000 years. They are only found in a narrow band along the Pacific Ocean from the southern tip of Oregon to central California. They have been under constant threat of *deforestation* and there have been large-scale acts of civil disobedience to save them. However, 96 percent of all the old-growth coast redwoods have been cut down, which many conservationists and environmentalists in general consider a travesty that should never have been allowed. The *Save-the-Redwoods League* has been helping to preserve these magnificent last stands of trees since 1918. [*redwood forest destruction, Luna Redwood; ancient forests; clear-cutting; dual-use policy; forest health, global; forest management; forest, sustainable; logging, illegal*]

cod

Probably the most dramatic example of the ecological and economic impact of *overfishing* can be seen in Newfoundland, Canada. In the early 1990s, a thriving cod fishing industry ended suddenly when the fish basically disappeared. Decades of fisheries mismanagement resulted in a 97-percent drop of the cod population, 40,000 people losing their livelihoods, and a destroyed *ecosystem*.

Twenty years after the collapse, many fishermen in this region are still waiting for cod to return to sustainable levels, although no one is sure it will ever happen. Changes made to the fisheries management has led to a comeback—but only minimally.

The problems with cod overfishing are global—occurring almost everywhere fish are commercially caught. The U.S. Commerce Department issued a disaster declaration in September of 2012 for the entire northeastern United States commercial fishing grounds, where much of the U.S. cod catch is found. In 2012 the cod quotas in this region were planned to be cut from 32,000 to 26,500 tons; many believe that is still far too high, forcing the depleted cod populations still closer to total collapse. [*fishing, commercial; bycatch; factory ship; fish populations and food*]

coffee, environmental impact of

Every day, Americans alone drink more than 400 million cups of coffee. Coffee might not seem like something that harms the planet, but it, like most crops, has a large *ecological footprint*. Coffee farms are *monocultures* that require vast amounts of synthetic *pesticides* and *fertilizers*.

Eco-certification programs identify coffee companies that use *sustainable farming* and *fair trade* practices. These certifications place emphasis on protecting regional *biodiversity*, reducing waste, and protecting local *drinking water*. Unfortunately, less than two percent of the world's total coffee crop is properly certified. However, with *social media* campaigns getting the word out about the environmental harm caused by your morning cup of coffee, that percentage is on the rise.

cogeneration

About one-third of the energy produced by a conventional *fossil fuel* electric power plant is converted into electricity; the rest is lost as heat. Cogeneration refers to capturing and using that lost heat to be used to heat the facility or recaptured to generate more electricity. [*black liquor, waste-to-energy power plants; alternative energy sources; alternative fuels*]

cold resistant

The ability of an organism to survive freezing temperatures. Some *insects* produce an antifreeze substance to allow them to survive the winter. [*adaptation; estivation; hibernation; extremophiles; quiescence; torpor; xenobiotic*]

Cole, Thomas

(1801–1848) Cole is one of America's great artists and a leading force behind the *Hudson River School art movement*—a group of artists primarily working in New York's Hudson River Valley region. They painted beautiful landscapes, using illumination to highlight and intensify the views. Cole and fellow artists of this movement helped people in their new country appreciate its magnificent natural wonders.

college divestment campaigns

During the 1980s a successful campaign was forged by students to force colleges to divest in companies associated with South Africa's apartheid policies. The 1990s saw another successful national effort, this time to divest in companies associated with the tobacco industry. Today, there are the rumblings of another such divestment movement on college campuses: divestiture of companies associated with *climate change.* [*payroll deduction, environmental; environmental movement; environmentalist*]

colony collapse disorder

About one-third of all the food you eat would not exist if honeybees didn't exist. Honeybees pollinate much of our plant life, including a great deal of our food. In 2007 something strange began to happen—honeybee populations started crashing mysteriously. In 2012 almost half of all hives were lost to some unknown disease. The declines continue today and fears are increasing that something must be done to prevent a significant impact on our food supply, not to mention damage to *ecosystems* in general.

Research shows that honeybees, as well as other bees such as bumblebees, are sensitive to a relatively new type of *pesticide* called *neonicotinoids,* which is related to a substance found in tobacco plants. These pesticides are applied to the crop seeds so they travel systemically through the entire plant. The toxic substance in the neonicotinoids attacks the insect's nervous system, paralyzing the insect. As is often the case, pesticides usually do not distinguish between insects we don't like (pests) and insects we need (beneficial insects).

The first research—performed by the pesticide manufacturers—showed that the bees are not harmed by this substance. However, more recent field studies have indicated that the substance changes the bee's behavior patterns; the bees collect less food, contributing to the colony's collapse. At the moment, this pesticide remains the primary culprit, but conclusive evidence is not yet in. According to other research, some of the damage might be attributed to a virus.

A few countries have banned the pesticide and others are considering a ban until more research is complete. In the United States, the *Environmental Protection Agency* says it won't make any decisions until they have more research, which won't be completed until 2015. [*honeybees, pollination, and agriculture; honeybee sperm bank; white-nose syndrome; zombie bees*]

Colorado River

The Colorado River is the longest river in southwestern United States, traveling almost 1500 miles from the central Rocky Mountains to the Gulf of California in Mexico. It is impeded by 29 dams for *hydropower* production and tapped for *irrigation* and municipal *water* supplies all along its route. About 40 million people use water from the

Colorado. And this is why it no longer runs all the way to the sea—drying up a few hundred miles before the mouth of the river.

When this many people use a single natural resource, problems result, as seen by the many court cases where states fight over their share. Estimates show that by 2060, the river will come up short by about 3.2 million acre-feet of water supply—or about five times what Los Angeles uses in a year. The problems seen with the Colorado River today are viewed by many as a sign of things to come elsewhere.

Scientists are looking for solutions. They have ranged from the bizarre, such as dragging huge icebergs from the *Arctic*, to the sensible, such as more *conservation* measures. (Both Los Angeles and Phoenix use less water now than they did a few decades ago, thanks to conservation efforts.) Although conservation does work, it cannot come close to resolving this shortage.

The Colorado River cannot meet the demands. There is no current solution—only stop-gap measures such as California municipalities paying farmers to leave land fallow because the water must bypass their farms to supply San Diego. Only time will tell if viable solutions can be found. [*rivers in decline; water shortages, global; water grabs; running water habitats, human impact on; water diversion*]

combined sewer systems (CSS)

Many *sewer* systems collect both storm-water runoff collected on streets and sewage waste from buildings and send them both to *sewage treatment facilities*. There, all of this water is treated before being released back into the environment (usually bodies of water, such as off the coast or in large lakes). Because they collect both types of water, they are also called combined sewer overflow systems.

During downpours or floods, the storm-water *runoff* can overpower the system and the water cannot be treated. When this happens, all of the water, including the sewage wastewater, is released directly into the environment, often causing serious *water pollution* problems.

Almost 800 U.S. cities use CSS. In New York City, for example, more than 30 billion gallons per year of sewage mixed with storm-water runoff are released untreated, directly into bodies of water after heavy rainfalls.

These systems are no longer being built, but cities are stuck with them and must work to control the damage when these events occur. Anything that reduces rainwater runoff helps minimize the potential damage caused by CSS. Overflow events can be reduced by managing city rainwater to slow the collection of excessive water during storms. [*beach water quality and closings; Clean Water Act; coastal flooding; effluent; point pollution; gardens, rain*]

comic books, environmental

Few comic books have an environmental theme but "The Massive" is a comic book with an environmental hero and villains who are a mysterious collection of environmental disasters known as "the Crash." [*Ranger Rick; movies, environmental; hip-hop, green; jazz, Earth; music, Earth; magazines, eco-*]

command-and-control regulation

See *industry self-regulation*.

commensalism

Commensalism is a *symbiotic relationship* in which one organism benefits from the relationship and the other is unaffected by it. Many mosses, *lichens*, and vines grow on trees in this type of relationship. The classic textbook example is that of a fish, called the remora, that attaches itself to *sharks* and feeds on the food scraps left by the shark's feeding. The shark appears indifferent to the relationship. [*adaptation; natural selection; mutualism; parasitoidism; predatory; prey*]

Commoner, Barry

(1917–2012) Barry Commoner was one of the early group of scientists who turned to writing in the 1960s to make the public aware of the environmental degradation occurring because of our technological development. His first popular book, The Closing Circle, made him a household name in environmental circles. It focused on the problems caused by our use of *detergents*, *pesticides* and synthetic *fertilizers* and their impact on natural *ecosystems* such as Lake Erie. Commoner made a good case in his book Making Peace with the Planet about how science, politics, the private sector, and public policy must be examined and studied as one if we are to preserve our natural resources. [*philosophers and writers, environmental; environmental movement; environmentalist; Our Common Future*]

communications between plants

Plant-fungus *symbiosis* is common. For example, a tomato plant's roots live in a symbiotic relationship with *fungi*; the tomato plant provides the fungi with food (created during *photosynthesis*) and the fungi provide the plant with minerals it can absorb but the plant cannot.

New research shows that fungi hyphae (root-like structures) that intermingle in the roots of many different tomato plants can send chemical messages from one plant to another. The fungi hyphae appear to recognize that a particular tomato plant is being attacked by aphids, and it chemically sends a message to other tomato plants warning them of this attack. This gives the other plants time to produce substances to ward off the attack of the aphids when they approach. [*sixth sense in animals; pheromones; smell sensitivity in albatross; animals experiencing grief; beehive sound research; bumblebee senses*]

community

A collection of all the *populations* inhabiting an area at a given time. [*ecosystem; pioneer community; climax community*]

community ecology

See *synecology*.

community gardens

Plots of city land offered to local residents as a space to grow gardens. Often, if funding is available, the seeds, supplies, and tools are also provided. The produce can then be harvested by those that grow and care for the gardens. These gardens not only provide food—typically for low-income individuals—but they also teach sustainability, encourage community relations, and even help reduce urban *rainwater runoff*. They are historically rooted in World War II Victory Gardens. [*gardens, rain; straw-bale gardening; urban gardens; microgreens and urban gardening*]

Community Right to Know Act, Emergency Planning and

In 1986, following the *Bhopal* disaster, Congress enacted the Emergency Planning and Community Right-To-Know Act (commonly called the community right-to-know act). The goal was to ensure Americans that their communities would be capable of preventing, preparing for, and knowing how to handle a chemical spill disaster. It mandates that communities must prepare emergency procedures to protect the public from chemical accidents, prepare procedures to warn and evacuate the public in case of such an emergency, provide citizens with information about hazardous substances and their accidental releases when they happen in their communities, and assist in the preparation of public reports on the annual release of *toxic chemicals*. [*hazardous waste; Comprehensive Environmental Response Compensation and Liability Act; Toxic Substances Control Act; cancer villages; cancer alley*]

community-supported agriculture (CSA)

The *farm-to-table movement*, *locavores*, the resurgence of *farmers' markets*, and many other social phenomena are affecting our relationship with food. Possibly the most emphatic statement about how people want to know and—in part—control their food supply, is community-supported agriculture—also called subscription farming. These are typically *organic farms* where individuals pay a subscription each season to get a share of the harvest. Their share might be a weekly delivery (or pickup) of whatever has been produced that week. The costs vary greatly, from possibly a hundred dollars per season to more than a $1000 for the season. Depending on your perspective, this might be a good deal or maybe not. They differ widely on how they work. Some are just produce and others have eggs, dairy, or even meat. Some have different types of shares, with a wide range of offerings, and others are locked into whatever becomes available, for better or worse. The only way to have more control over your food is to grow it yourself. [*farm-to-school movement; food miles; Non-GMO Project Verified label; seasonal foods; tomatoes—tasteless but red; vegan and vegetarian diets and the environment; vegetable seeds, heirloom; WWOOF*]

compact fluorescent light bulbs (CFLs)

CFLs are a relatively new *light bulb technology* that dramatically improves the efficiency of light bulbs when compared with *incandescent bulbs* but has met with poor acceptance from consumers.

CFLs emit four times as much light per watt as an incandescent bulb, and they convert far more energy into light instead of heat when compared to incandescent bulbs. However, Americans were not happy when they were forced to switch to CFLs because of *light bulb legislation* and then found the CFLs did live up to their expectations. They did not last as long as they were supposed to and most don't start up at full brightness immediately, increasing in brightness over a long time. Plus, they contain mercury—a *toxic* substance. They are supposed to be disposed of as a *hazardous waste*, but most people simply throw them out—meaning they end up in *landfills*, causing a great deal of potential harm. New technologies, such as *LED bulbs*, are dramatically better and will probably result in the phase out of CFLs. [*fluorescent lighting; light bulb - recent history; smart meters; vampire energy; hotel load; phantom load*]

companies, top ten green global

<u>Newsweek</u> conducts an annual green companies list. It is based on criteria such as *ecological footprint* (*greenhouse gas* emissions and *water use*), management (environmental policies and programs), and environmental disclosure (reporting). So, at least according to <u>Newsweek</u> magazine, the greenest companies globally are 1) IBM, 2) Hewlett-Packard, 3) Sprint Nextel, 4) Dell, 5) CA Technologies, 6) Nvidia, 7) Intel, 8) Accenture, 9) Office Depot, and 10) Staples.

The top 10 companies on the global list are 1) Santander (*Brazil*), 2) Wipro (*India*), 3) Bradesco (Brazil), 4) IBM (U.S.), 5) National Australia Bank (*Australia*), 6)

BT Group (U.K.), 7) Munich Re (Germany), 8) SAP (Germany), 9) KPN (Netherlands), and 10) Marks & Spencer (U.K.). [*CERES; corporate social responsibility; corporate sustainability rankings; integrated bottom line; economy vs. environment; internalizing costs; steady-state economics*]

competition

When two individuals of the same species compete for the same resource, it is called intraspecific competition. The resource might be food, such as a plant or *prey*, or it might be an abiotic (nonliving) factor such as sunlight or water. If the two individuals competing for a resource are of different species, it is called interspecific competition. [*adaptation; apex consumers and ecosystems; synecology; competitive exclusion, principle of; computer modeling, ecological; convergent evolution; symbiotic relationships*]

competitive exclusion, principle of

Within an *ecosystem*, only one species can fill an *ecological niche*. If two or more species inhabit the same or overlapping niches, the one that is better adapted or has a better fit will succeed, while the others will either have to adapt to another niche, migrate elsewhere, or die. This is called the principle of competitive exclusion.

compost

One of three types of *organic fertilizer*. It is a rich *soil* containing a large amount of decomposed *organic* matter. Compost is created from wastes such as *lawn clippings*, *leaf litter*, *food waste*, and animal droppings that are mixed with *topsoil* and then decomposed by populations of microbes and other organisms such as *insects* and *earthworms*. These organisms decompose the organic matter, making it available for future plant growth. About 29 percent of the U.S. *municipal solid waste* consists of food wastes and yard clippings that are compostable. Although the amount of waste composted has increased significantly, much more can be done.

In addition to the traditional composting of food wastes and yard clippings, new methods have developed. Some facilities now use sewage *sludge* from *sewage treatment facilities* as the organic material to create compost. [*composting, large-scale; sludge disposal; San Francisco and compostable waste; biosolids and wastewater*]

composting

Composting is the process of converting *organic* waste into *compost*. *Food wastes*, *grass clippings*, leaves, animal wastes, and *sewage sludge* are all candidates for composting. This organic material is broken down (decomposed) into the essential *nutrients*. When finished, the material is no longer identifiable and can be used as a rich *fertilizer* for another generation of plants. Composting can be done on a small scale such as in backyard gardens with composting bins, or can be processed in *large-scale composting* facilities that use either *municipal solid wastes* or sewage sludge.

The biggest advantage of composting is that these wastes do not end up in *landfills*. When organic matter does go in landfills, it decomposes *anaerobically* and produces *methane gas,* the most potent of all the *greenhouse gases*. The composting process redirects this waste from potentially contributing to *climate change*, converting it into rich, *organic fertilizer*. [*topsoil; humus*]

composting, large-scale

Some European countries and a few U.S. municipalities have compost plants that *recycle organic* wastes from *municipal solid wastes* or *sewage sludge* and sell or give away the *compost* as *fertilizer*. Even though the compost might be sold to generate revenue, most of the cost savings comes from reduced *landfill* fees, because there is less waste to dump. These large-scale compost plants agitate and aerate the organic mix to speed up the natural composting process. The *bacteria* that *decompose* the organic matter thrive with the added air and moisture. What takes many months to decompose in nature or in small garden compost bins takes only a few weeks in these compost plants.

Large-scale composting facilities are beginning to be used on the vast amounts of animal wastes produced by dairy farms. One company, WormPower, converts about ten million pounds of cow manure into 2.5 million pounds of compost, which is then sold. Different companies are trying to fine-tune their compost to make it more valuable than just plain compost. [*San Francisco and compostable waste*]

Comprehensive Environmental Response, Compensation and Liability Act (CERCLA)

See *Superfund*.

comprehensive water management planning

The process of developing but still protecting a region's water resources is called comprehensive water management planning. This process often requires the integration of many agencies, organizations, companies, and other groups that use the resource and/or have jurisdiction over it. The water management plan involves data collection and analysis, problem determination, and recommendations for resolutions. It often includes studies that require the continued monitoring of the resource over long periods. [*water use for human consumption; domestic water use; water conservation; Clean*

Water Act; cultural eutrophication; groundwater mining; water grabs; water shortages, global]

computer modeling, ecological

The use of high-speed supercomputers that simulate changes in biological systems so predictions can be made about their impact. The biological systems studied may be a single *population*, a *community*, or an entire *ecosystem*. These models assist in all forms of *ecological studies*, predict what might happen to an ecosystem in the future, and help develop new ecological theories. [*apps, eco-; biotechnology; cloud computing and the environment; crowd research, scientific; data centers, energy consumption of; geoengineering; nanotechnology; robotic bees; satellite mapping; synthetic biology; terraforming*]

concentrated animal feeding operations (CAFO)

See *CAFO*.

concentrated solar technology

See *solar thermal power plants*.

coniferous forest

See *taiga*.

conservation

In biology, conservation is the wise use and proper management of *natural resources* to provide the maximum, continuous benefit of those resources for people over long periods of time—meaning they are still there for future generations to enjoy. By natural resources, we mean all natural *ecosystems* and all the *biodiversity* found within.

In environmental studies, conservation is also an attitude or a belief or a school of thought. The *preservation* versus conservation debate has been waged ever since *John Muir* wrote to preserve nature and *Gifford Pinchot* debated to conserve it in the late 1800s and early 1900s. Preservation is saving nature exclusively for nature's sake with no extraction or degradation at all. Conservation is sustaining nature while it is used for economic benefit, including extraction such as *mining* and the degradation that comes with it.

When people talk about *sustainability*, they are talking about the conservation of nature—using it without losing it for future generations. It is striking a balance between the two. However, finding that balance is much easier said than done, and can easily sway to one side or another depending on many factors. One major factor is who is in the White House and the Congress and what laws are passed or fail, enforced or ignored, and where the emphasis is placed. On an international scale, it is the same—the people put into office determine the direction a government pursues.

conservation agriculture

A more environmentally friendly form of *agriculture,* where fewer synthetic *pesticides* and *fertilizers* are needed, when compared with *conventional agriculture.* Conservation techniques include *conservation tillage,* using *cover crops, crop rotation,* special watering techniques, and *precision farming.*

It also includes various techniques to minimize the environmental harm caused by *high-impact agriculture.* Planting trees in rows between crops or on bordering fields in parts of Africa have improved soil and water quality. *Windbreaks* can keep dry *topsoil* from eroding away with the wind. Trees can cool local temperatures and are natural filters between the fields and nearby waterways, keeping the soil moist. *Integrated pest management*—everything from *biological control methods* to synthetic *pesticides*—is used to control pests.

Conservation agriculture is similar in some ways to *organic farming,* but conservation agriculture allows for pesticides and fertilizers when needed—only after other techniques are tried first.

Many aspects of conservation agriculture have been used for hundreds of years. (Some people have called some of these methods Early American agriculture.) In many aspects it is also similar to *subsidence farming.* During tough times, such as drought, conservation farming is considered superior to conventional farming because the crops are more resistant, probably because of increased *biodiversity* within the *habitat.*

However, regardless what it is called, conservation agriculture remains a small fraction of total farming because it requires additional expense and can often result in lower yields, at least in the beginning—things farmers won't embrace without incentives. High-impact agriculture remains the standard.

conservation districts

Throughout the United States, 3000 local conservation districts help decide the ultimate environmental quality of a town or city. These conservation districts are authorized under state laws to assess environmental problems, set priorities, and coordinate and carry out efforts at the local level to address these problems. Each district is governed by either appointed or elected officials. [*smart cities; zoning, land use; urban open space; handshake buildings; eco-cities; infrastructure, green; interim use; urbanization and urban growth*]

conservation easement

A legal agreement between a landowner and a conservancy organization—typically a *NGO*. The easement allows the landowner to retain ownership of the land, but puts the land under the stewardship of the conservancy for protection, typically in perpetuity. It is a voluntary but legally binding agreement that restricts the development of a property and preserves its natural value. This is one of the most common ways for NGOs, municipalities, and governments to conserve land so it remains in its natural state. The IRS allows these agreements if the land has significant *conservation* value. [*Nature Conservancy; habitat loss; habitat management; inholdings; no net loss; virgin habitat; wilderness; land grabs; land trusts; zoning, land use*]

Conservation International

Based in the United States, this *NGO* works to conserve global *biodiversity* by using science, economics, and policy along with community involvement. [*CITES; Convention on Biological Diversity; debt-for-nature swaps; Wildlife Conservation Society; wildlife trade and trafficking*] {conservation.org}

Conservation Reserve Program

This program was created in 1985 and is administered by the *United States Department of Agriculture* and a few other federal agencies. It offers financial incentives to agricultural landowners who set aside portions of their cropland that are sensitive to erosion. For example, it might pay a landowner to take land currently used to grow crops and convert it to natural grasses or trees to prevent *soil erosion*. Today, over 700,000 agreements are protecting over 30 million acres.

When these lands are protected, soil does not erode and far fewer *pesticides* and *fertilizers* make their way downriver, resulting in less *water pollution*. The results also encourage more *biodiversity* on these lands.

In spite of these advantages, there are those in government trying to cut back the program under the premise that this farmland is too valuable to be removed from economic production just be set aside to protect the environment. [*conservation districts; conservation easement; high market value farming counties; farmland lost; Natural Resource Conservation Service*]

conservation tillage

Conventional tillage, used on *high-impact agriculture*, invites *soil erosion* because it involves turning over the soil in the fall, leaving it bare through the winter, and planting crops in the spring. Conservation tillage methods, on the other hand, use special tillers that break up the subsurface but don't disturb the surface layer, leaving a protective layer of *organic* matter. Seeds, *fertilizers*, and *pesticides* are injected through the surface layer to the layer below. Leaving the top layer intact greatly reduces the amount of erosion.

Conservation tillage is used on about 40 percent of U.S. crops. This is up from about 33 percent 20 years ago. About 60 percent of farmers use reduced tillage, which is a less intense version of conservation tillage and at least a move in the right direction. It is estimated that if 80 percent of the U.S. croplands used conservation tillage, about half of all the soil lost to erosion would be saved. [*conservation agriculture; low-input sustainable agriculture; organic farming and food; permaculture; precision farming; sustainable agriculture*]

conservation triage

A new and controversial idea that conservation efforts should be spent on species with the best chances for survival, instead of those that are often called *flagship species*. As an example, the program to save the California condor from extinction costs $4 million annually. Some scientists believe the limited funds available for conservation could be better spent elsewhere—many species could probably be saved from extinction for the same cost now spent on this one bird. [*charismatic megafauna; biodiversity, loss of; de-extinction; dodo bird; endangered and threatened species lists; extinct in the wild; extinction, mistaken; extinction, newly identified species; god squad; Lazarous taxon; Lonesome George; Ugly Animal Preservation Society; Wildlife Conservation Society*]

conspicuous consumption

Purchasing goods or services for the specific purpose of displaying one's wealth or to create the appearance of wealth. The term was coined by the American economist and sociologist Thorstein Veblen in his 1889 book <u>The Theory of the Leisure Class</u>. While it is typically associated with the wealthy today, it originally described the consumption of the developing middle class during the 19th and early 20th centuries. The concept of "consumerism" probably stems from conspicuous consumption.

To many people, conspicuous consumption is one lifestyle, of many, that encourages the production of excessive, unwanted, and unneeded goods—all of which must be consumed to justify their continued production. And their continued production is necessary to ensure a growing economy as measured by *gross domestic product* (*GDP*). There is an ongoing discussion that economic growth, of which consumerism is a vital component, must be *decoupled* from environmental degradation. This is easy to talk about but will be very hard to accomplish. [*Beyond GDP; economy vs. environment; sustainable development; credit cards, green; planned obsolescence; Small is Beautiful; stonewashed jeans; municipal solid waste disposal*]

constancy

See *population stability*.

constant species

A species that is likely to be found in a specific *community*. A constant species is found in at least 50 percent of the samples taken from that community. [*ecosystem; indicator species; indigenous; native species; invasive species; biodiversity*]

consumer

See *heterotrophs*.

consumers, green

You can show your concern about environmental issues in many ways. Becoming a green consumer is one of three ways that might be considered the most important. The first of the three—and probably the most important because the other two are based on this one—is to be *environmentally literate* so you understand the science behind all the rhetoric. Second is to use what you know to vote politicians with similar views as yours into office. It is your elected officials at all levels who create or dismantle environmental policies and regulations.

The third is to become a green consumer, which means making purchasing decisions for anything and everything based on your knowledge of environmental issues. Voting for elected officials can only be expressed every few years during elections, but green consumers can express their concerns every time they take out their wallet.

Numerous companies have become environmentally and socially conscious because that's what their consumers want. Some of these companies have worked hard to lessen their *ecological footprint*, to advance their *corporate social responsibility,* and to accept *extended product responsibility,* to mention a few improvements. Green consumers can accelerate and increase this encouraging movement by patronizing those companies that accept these responsibilities.

But an environmentally literate consumer knows that just because a company calls itself green doesn't mean it is. That's why there are terms such as *greenwash, green scam,* and *greenspeak.* Consumers must use *environmental rating systems* and know which *eco-labels* are useful and which are not. As consumers make more and more knowledgeable purchasing decisions, more and more companies will join the move toward sustainability—the endgame of environmentalism. Throughout history, changing the world for the better always comes back to the individual. [*ecological transparency; credit cards, green; green marketing; green products; holiday waste;*

hotels, green; ecotravel; ecotourism; lodging, green; sneakers, green; tropical fish tanks; wine-bottle cork debate]

continental shelf

That portion of the ocean floor between the shore and the continental slope (which is where the floor drops steeply into the deep sea). The continental shelf extends out about 100 miles from shore and reaches depths of about 600 feet before reaching the slope. The shelf is the most productive region of the ocean for commercial fishing and marine life in general. Unfortunately, it is also one of the most productive regions for offshore *oil drilling*, bringing *environmentalists* and the petroleum industry in a constant battle between protection and extraction. [*fishing, commercial; oceanic zone ecosystems; marine ecosystems*]

contour farming

Crops grown on gradually sloping land are prone to *soil erosion*. Contour farming is a *soil conservation* method in which crops are planted at right angles to the slope instead of up and down the slope. Each row creates a sort of mini dam that holds the water (and soil) in place. This dramatically reduces the amount of *runoff* and thereby reduces *soil erosion*. [*conservation agriculture; organic farming and food, precision farming; sustainable agriculture*]

Convention on Climate Change, The United Nations Framework

One of five documents discussed during the *United Nations Conference on the Environment and Development,* held in June 1992. The issues discussed in this *convention* included reducing *carbon emissions*, the primary cause of *global warming*; controlling emissions of other *greenhouse gases*; and the need for financial and technical aid to developing countries now dependent on *fossil fuels*. The main objective was "to stabilize the greenhouse gases in the atmosphere to prevent dangerous interference with the climate system, to ensure that food production is not threatened and to enable economic development to proceed in a sustainable manner." [*climate change*]

Convention on Biological Diversity, The United Nations

One of the five documents discussed during the *United Nations Conference on the Environment and Development,* held in June 1992. The primary goal of this *convention* was to protect *biodiversity* and promote *sustainability* of biological resources, based on the *precautionary principle*. It includes protecting not only species but also the genetic variation with species as well as entire *ecosystems*.

This document was appended in 2000 with an agreement called the Cartagena Protocol on Biosafety, aimed to protect biodiversity by using *biotechnology*. A Biosafety Clearing-House was created to exchange information between countries regarding *genetically modified organisms*.

Many environmentalists believe this agreement has done little to protect *marine ecosystems*, as originally promised. Only 1.6 percent of oceans have been designated as *marine protected areas*—far below what was hoped.

Convention on the Conservation of Antarctic Marine Living Resources (CCAMLR)

One of three agreements produced by the *Antarctic Treaty*. This *convention*, signed in 1980, regulates relations among nations vying for land rights in *Antarctica*. It was originally created to manage krill harvesting but now prohibits military activity, *nuclear* tests, and nuclear *waste disposal*. It adopts the *precautionary principle*. [*arctic; Arctic Council; arctic drilling*]

Convention on the International Trade of Endangered Species

See *CITES*.

Convention on the Prevention of Marine Pollution by Dumping of Wastes and other Matter

Commonly called the London *Convention*, it was signed in 1972, put in force in 1975, and amended six times, with the most recent in 2006. In 2013 there were 82 signatories. It uses the *precautionary principle* and the *polluter pays principle*. Its primary goal is to provide codes on the prohibition of deliberate dumping of wastes and other matter at sea. [*MARPOL; marine ecosystems; Ocean Conservancy; Ocean Dumping Ban Act; ocean pollution; oceanic zone ecosystems; sea gliders*]

Convention on Wetlands

More commonly known as the Ramsar *Convention*, it was signed in 1971 and, as of 2013, has 168 signatories. The primary purpose is to protect flora and fauna that depend on *wetland habitats*. It encourages countries to employ wise-use practices. They produced a List of Wetlands of International Importance that has more than 1600 entries. [*coastal wetlands; inland wetlands; eco-services; biodiversity, loss of; estuary and coastal wetlands destruction*]

Convention to Combat Desertification, The United Nations

One of three *conventions* created at the *United Nations Conference on the Environment and Development,* held in June 1992. Its purpose is to prevent *desertification* and drought in Africa. [*drought, the latest; Great Green Wall of Africa, the*]

convention vs. treaty

Treaties and conventions are very similar. They both are agreements between nations. Once *signed and ratified,* it typically becomes an international law the nations are supposed to abide by. A treaty usually pertains to only two nations and is often an effort to stop hostilities. A convention usually pertains to many nations; it can be about almost anything but usually is about stopping something before it happens. For example, most of the agreements created by the *United Nations* to reduce or prevent various types of environmental harm are called conventions, such as the *Convention on Biological Diversity.*

The nations all come together at some type of international conference, such as the *United Nations Conference on the Environment and Development* and create a convention. Then representatives from each participating nation can sign, ratify, or simply ignore it. But with this all said, there are no carved-in-stone rules about which term can be used and when.

conventional agriculture

See *high-impact agriculture.*

convergent evolution

The *evolution* of structures that have similar function but are found in different organisms. For example, flight evolved in birds, *bats*, and *insects*. They are totally different groups of organisms, but all evolved structures to fly. [*ecological studies; adaptation; natural selection*]

cooling ponds and towers

Large volumes of water are used for *nuclear power*. This water becomes heated during the processes and must then be cooled down before it can be released into the *environment*. Cooling ponds and towers hold the water while it cools down. Towers are of particular concern to environmentalists and health professionals because they have been known to act as breeding grounds for pathogens and to collect and release pollutants. [*water use; water pollution; industrial water use; industrial water pollution*]

Copenhagen Accord

In 2009, at the *Intergovernmental Panel on Climate Change (IPCC) conference*, promises were made to reduce *carbon emissions* by investing in clean energy sources and developed countries agreed to assist developing nations financially in doing so. It also called for lowering the previous target for carbon dioxide emissions. These promises (as well as others) were entered into an agreement—called the Copenhagen Accord because it occurred in Copenhagen. Since then the IPCC and *IPCC Climate Change reports* have been revised and updated. [*climate change*]

coprophagous

Organisms that feed on animal dung. Many species of dung *beetles* live most of their lives in a pile of dung. Some prefer a certain type of dung—cow or horse, for example. Some, such as the tumblebug (which is really a *beetle*) chew off little pieces of dung and roll them into balls. They then roll the balls to a safe location, lay their eggs in the ball, and bury them. When the eggs hatch, the young are protected in the soil and fully stocked with food. [*decomposer; niche, ecological; saprophyte; scavenger; eco-services; night soil; chledophyte*]

coral bleaching, causes of

The death of coral and the entire *coral reef* ecosystem in a region can occur for many reasons, such as *sewage* or other waste flowing downstream and into the oceans, silt accumulation from *soil erosion* caused by improper *logging* and *agriculture*, or contaminants such as *pesticides*. More recently, *ocean acidification* and *climate change* have become the biggest threats to coral. [*biodiversity, loss of; marine ecosystems; climate change*]

coral reef

Marine ecosystems along the shore contain many unique *habitats* such as coral reefs. Coral reefs are typically found in warm tropical and subtropical regions. The reef is built by tiny *coelenterates,* which are cylindrical in shape with a pouch-like mouth surrounded by tentacles to capture *prey* and bring it into the mouth. (Sea anemones and jellyfish are other types of coelenterates—but much larger.)

These organisms attach themselves to the floor of coastal waters and secrete calcium that acts like a substrate for more individuals. As these calcium deposits increase, large coral reefs form. Only the surface of the reef contains living coral organisms—called the live cover; the rest is just the remaining calcium deposits.

Some coral have a *symbiotic relationship* with *algae* that lives within the coral's cells. The coral captures food at night, but during the day the coral are nourished directly from the *photosynthesizing* algae living within their bodies.

Coral reefs, along with other *coastal zone* habitats, are among the most productive on Earth—as productive (*net primary production*) as *tropical rainforests*. More than 4000 species of fish as well as numerous other organisms, such as sea urchins, feed on the algae, *bacteria*, and other microorganisms that abound in coral reef *ecosystems*.

Even though coral reefs only cover about one percent of our planet's surface, the resources they provide are estimated to be worth more than $300 billion dollars annually in *eco-services* and about $100 million in commercial value each year. Most of this value is because these reefs are home to thousands of commercially important species of fish—at least during some portion of their life span. In addition to providing this seafood, the reefs buffer shorelines against storms, reducing *erosion*.

Coral reefs are considered one of the most fragile ecosystems. They are under constant threat from recreational diving, destructive fishing practices, *water pollution*, and *sedimentation*. However, the biggest threat is now thought to be *climate change*. As the oceans warm, the photosynthesizing algae die. With the loss of the coral's symbiotic partner, they too die. But even if the algae do not die, the organisms are stressed and vulnerable to infections that can kill the coral. In addition, *climate change* contributes to *ocean acidification*—another major factor in the decline of coral.

The loss of coral reefs in recent years is staggering. In the Caribbean, the live coral coverage in the 1970s was about 50 percent; today only about 7 percent of the reef is covered with live coral. In the *Great Barrier Reef*, even with all of the legal protection it enjoys, the live coral cover is down from about 40 percent in the 1960s to about 20 percent today.

When these reefs die, they become a bleached white color—a process known as *coral bleaching*. The loss of one of our planet's richest and most beautiful *keystone species* means an entire ecosystem is threatened. Roughly 100 countries have coral reefs offshore, with the most famous being *Australia's* Great Barrier Reef. [*coral reef, cold water*]

coral reef, cold water

When most people think of *coral reefs*, they think of warm water reefs found in lush, tropical waters with scuba divers investigating their beauty. However, one group of coral reefs, called cold water reefs, is found in higher latitudes and in much deeper waters, so they are not typically seen or photographed. Unfortunately, they appear to be just as susceptible to *global warming* and *ocean acidification* as their warm-water relatives. [*coral reef sperm bank; biodiversity, loss of*]

coral reef sperm bank

Coral reproduces both asexually and sexually. *Asexual reproduction* occurs when fragments of living coral break off and grow into clones of their parents. But sexual reproduction is more important to the future survival of coral because it maintains genetic diversity and keeps populations robust and capable of long-term species survival. *Coral bleaching* and other threats have scientists estimating that coral reefs might be completely gone by 2050, if current trends continue.

Hopefully, that won't be the case, and scientists are working on many ways to protect these organisms. One of the more unusual solutions is the creation of a coral sperm bank. A research scientist at the *Smithsonian Institution* has created the only such bank, storing about one trillion frozen coral sperm, enough to fertilize a half-billion to 1 billion eggs. The idea is they can be used to restore or rebuild damaged reefs if needed. [*biodiversity, loss of; seed banks and vaults; doomsday seed vault; endangered and threatened species lists*]

corn to biofuel

In the United States, some of the corn grown is converted into *bioethanol*, a *biofuel*. This has become a contentious issue because it is competing with corn for food. Many believe that in spite of environmental advantages to this process, the overall impact on the environment is negative. Corn has been used for biofuel production for many years and is called a first-generation biofuel. Scientists are now developing *second-generation biofuels* that appear to be superior in many ways.

corporate acquisitions of small green companies

Many companies made a name for themselves by being *green*, truly green. And in doing so, they created a loyal customer base wanting a *green product*. Companies like Ben & Jerry's, Honest Tea, Burt's Bees, Cascadian Farms, Kashi, Stonyfield Farm, Tom's of Maine, Naked Juice, The Body Shop, and Lands' End are some examples.

The problem is that all of the green companies mentioned have been acquired by large corporations. For example, Kashi is part of Kellogg, Ben and Jerry's is part of Unilever, and Toms' of Maine is part of Colgate-Palmolive. The question is, what happens to the greenness of these companies when taken over by major corporations?

Of course, the results differ from company to company, but the consensus is that the products no longer maintain their original intent and gradually—in some cases—become nothing more than a *greenwash*. Some companies start to slip in ingredients that were not there, or replace organic and sustainable ingredients with others. In time, consumers learn about these changes and move on to truly green products.

There are many good *environmental rating* groups to determine the true greenness of products and services. [*consumers, green; green fatigue; green marketing; green scam; CERES; corporate social responsibility; companies, top ten green; waste-conscious product development; triple bottom line; sustainable marketing; shareholders, corporate*]

corporate average fuel economy (CAFÉ)

The Energy Policy and Conservation Act of 1975 resulted in the creation of the Corporate Average Fuel Economy (CAFÉ) standards. This was initiated as a response to the 1973 oil embargo. CAFÉ set the average new vehicle fuel economy—measured as miles per gallon (mpg)—that a manufacturer's fleet must attain beginning in 1978. The initial goal was to double the average mpg to 27.5 within 10 years, 1985.

In 2007 CAFÉ was updated for the first time since its conception, to achieve 35 mpg by 2020. There were also changes that grouped many light trucks and large SUVs into the same category so they cannot bypass these requirements.

In 2009, the *Environmental Protection Agency* (EPA) accelerated this CAFÉ goal to 35 mpg by 2016. President Obama is proposing an agreement with automakers to raise the mpg to more than 50 by 2025. However, this is just a discussion at the moment and not a mandate and many believe it will never come to be.

In 2013, the EPA proposed new improved targets for auto manufacturers that increase mpg as well as requiring cleaner-burning fuels. These new regulations would reduce *VOC* and *nitrogen oxide* emissions by 80 percent and *particulate matter* by 70 percent. The *cost/benefit analysis* shows that for every one dollar spent implementing these new regulations, seven dollars in health care costs will be saved. [*automobile; gasoline, saving on auto; greenest cars, top ten; automobile propulsion system, alternative*]

corporate social responsibility (CSR)

This business approach first appeared in the 1970s but only recently is becoming mainstream. It implies that companies have a social and an environmental responsibility. Most global companies now have a CSR statement. Advocates believe this is a positive effort to shift the corporate world from a foundation based solely on economics toward one of sustainability. Critics believe it is nothing more than a simple marketing ploy to appease customers—calling it *greenwash*.

However, it has become clear that the public wants environmentally friendly products and services and companies that work to that end are finding it good for business and their bottom line, regardless of what their true environmental beliefs might be. [*CERES; climate capitalism; corporate sustainability rankings; triple bottom line; economy vs. environment; extended product responsibility; industry self-regulation; internalizing costs; natural capital; Sustainability Imperative*]

corporate sustainability rankings

Not too many years ago, you would be hard pressed to find rankings of how well companies are doing in relation to their environmental impact. Today, we have the opposite problem. There are so many rankings—and many of them contradicting each other—it is difficult to know what to believe. Well over 100 such ranking systems can be found, including the Dow Jones Sustainability Index, the Newsweek Green Rankings, the Global 100 List, and the Carbon Disclosure Project.

Each year, some rankings will have certain companies in the top ten percent, while other rankings have the same companies in their bottom ten percent. Who to believe? Help will arrive soon. *CERES* is working with the Tellus Institute to develop a standard for these rankings, called the Global Institute for Sustainability Ratings (GISR). It will rank the rankings. GISR will provide principles and methods to standardize the rankings made by others.[*corporate social responsibility; companies, top ten green global*]

cosmetics, labeling of

The labeling of cosmetics is regulated by the *Food and Drug Administration* (FDA)—but only the label. Contrary to what many people think, no regulatory process exists for the ingredients of these cosmetics (unlike drugs, which are regulated). Cosmetics are defined by the FDA as "articles intended to be rubbed, poured, sprinkled or sprayed, introduced into, or otherwise applied to the human body. . .for cleansing, beautifying, promoting attractiveness, or altering the appearance." Toothpaste and deodorant are considered cosmetics. Some soaps are classified as drugs if they are used for a therapeutic purpose.

Since cosmetics are not regulated, consumers do not really know what they are putting on. Let the buyer beware. [*lipstick and toxic substances; nanoparticles in cosmetics and personal care products; essential oils; Skin Deep, EWG's; green products; consumers, green; wine, organic and biodynamic*]

cosmetics, cruelty-free

In 2013, the European Union passed legislation banning the use of animals to test cosmetics and personal care products. This is a far cry from the United States, where there are no laws—only voluntary programs established by some manufacturers, such as Paul Mitchell products or those carrying the Leaping Bunny *eco-label*. Contrary to most other nations, *China* actually requires animal testing before a cosmetic can be sold. [*animal rights movement and animal welfare; vivisection; farm animal cruelty legislation; dissecting frogs in school*] {leapingbunny.org}

cost/benefit analysis

The economic analysis of what a product, service, or law will cost versus the benefit that might be gained by it. For example, studies of the *Clean Air Act* from 1990 to 2020 show direct benefits to Americans of almost two trillion dollars. This includes the number of sick days not taken, lives saved from health-related issues (estimated at 230,000), and many other factors. Those are the benefits. The costs to implement this act over the same time period? Roughly 65 million dollars. So, in this cost/benefit analysis, the benefits outweigh the costs by about 30 to one.

Cost/benefit analyses can be performed on almost anything and by almost anyone. You could perform a cost/benefit analysis about buying a *hybrid car*, for example. The government requires these studies be performed on most major projects. For example, the *National Environmental Policy Act* requires agencies to create *environmental impact statements* for every project that might have environmental consequences. These statements are basically a cost/benefit analysis. [*risk assessment, environmental; risk communications; risk management, environmental; risks, true vs. perceived*]

couches and indoor pollutants

We spend far more time indoors than out, and *indoor air pollution* has become a hot topic and one fraught with issues. Couches are part of most of our indoor environments. In recent studies, 85 percent of the couches tested in the United States contained *toxic* substances or untested *flame retardants*, which unfortunately have been shown to make people sick but do little to reduce the risk of fire. The most common substance found was a flame retardant belonging to a group of compounds called PBDEs, or polybrominated diphenyl ethers. These substances break down and are commonly found in household dust. They are long lasting and associated with many health concerns, including interfering with brain development in infants and with memory, and they are considered endocrine *hormone disruptors*. [*heating, ventilation, and A/C system; air conditioning; humidifier fever; indoor ecology; mattresses, recycling*]

countries, ten greenest

Many organizations have ranked countries according to their environmental policies, each with its own selection of criteria. Yale University ranked 149 countries according to an environmental performance index (EPI) that takes into account their carbon and sulfur emissions, water purity, and overall conservation practices. The top ten are 1) Switzerland, 2) Sweden, 3) Norway, 4) Finland, 5) Costa Rica, 6) Austria, 7) New Zealand, 8) Latvia, 9) Colombia, and 10) France. The United States ranked 39. [*cities, ten best American green; universities, top ten green; countries, ten most energy-efficient; companies, top ten green global; greenest cars, top ten; ecotourism, ten best countries for; energy efficient states, top ten U.S.; bicycling cities, ten best; GMOs in your food, the five most common*]

countries, ten most energy-efficient

The ACEEE (American Council for an Energy-Efficient Economy) publishes an International Energy Efficiency Scorecard that ranks the 12 largest economies in order of energy efficiency. They use a complex scorecard with numerous criteria. The ten best are 1) United Kingdom, 2) Germany, 3) Italy, 4) Japan, 5) France, 6) European Union, 7) *Australia*, 8) *China*, 9) United States, and 10) *Brazil*.

Forbes also publishes a list based on energy-intensity figures for the 75 largest countries in terms of *GDP*. The ten best are 1) Japan, 2) Denmark, 3) Switzerland, 4) Hong Kong, 5) Ireland, 6) United Kingdom, 7) Israel, 8) Italy, 9) *Germany*, and 10) Austria. [*cities, ten best American green; countries, ten greenest; countries, ten most energy-efficient; companies, top ten green global; greenest cars, top ten; ecotourism, ten best countries for; energy efficient states, top ten U.S.; bicycling cities, ten best; GMOs in your food, the five most common*]

country of origin label (cool)

The United States *Department of Agriculture* requires that fresh and frozen fruits and vegetables, frozen fish, and shellfish have a label identifying the country of origin. This could be useful when considering *imported food dangers*. These labels are not required on any mixed or blended foods. [*Dirty Dozen - fruits and vegetables; seasonal foods; circle of poison; slow food movement; Clean Fifteen*]

Cousteau, Jacques Yves

(1910–1997) Jacques Cousteau was an inventor and scientist, but most of all he was an explorer. He explored every major body of water on our planet. So they could explore the world underwater for long periods of time, Cousteau and Emile Gagnan invented the Aqualung. He is also credited with developing the first underwater cameras with television capabilities. Through this medium, he was not only able to study the underwater world but to teach everyone about it as well. In 1973, he founded the Cousteau Society, which remains a leader in marine research and education. Currently it has 50,000 members worldwide. [*marine ecosystems; dead zones; garbage patch, marine; Ocean Conservancy*]

cover crops
Crops planted between the rows of the main crop to reduce *soil erosion* and act as a *organic fertilizer*, improving *soil texture*. This method is used in *conservation agriculture*. The cover crop can then be sold separately as forage or as a feed crop for livestock.

crab pots, biodegradable
Crab pots are used to catch many species of crab, such as those used by crabbers in search of blue crab in the Chesapeake Bay in Virginia. A large number of these pots are lost every year and sink to the bottom, where they continue to catch crabs and other organisms such as *sea turtles*. The number of these lost pots is vast so the damage is serious. Scientists have devised a solution to prevent continuing damage of lost pots. A small biodegradable panel gradually breaks down as the pot remains in the water, so in time the animals will no longer be trapped. However, crabbers may not use them, because the pots are more expensive. [*turtle excluder device; ghost fishing; fisheries and seabirds; dolphin-safe tuna*]

cradle-to-cradle
This takes the *cradle-to-grave* concept to the next logical step, that a product should be designed so it readily decomposes to be ready for reuse by nature or in another product. In other words, man-made products should work the way natural products have always worked. They should become part of natural *biogeochemical cycles*, and if they cannot, they should at least be readily available for use in another man-made product.

cradle-to-grave
Originally, this was a phrase used by the *Resource Conservation and Recovery Act* to describe how *hazardous wastes* must be tracked and documented throughout the entire lifetime of the product. However, it is now also used to track the overall negative

environmental impact of any product or process throughout its entire life-cycle. For example, instead of just considering the harm an *automobile* causes when it is driven, the impact should be considered from the time it is manufactured until it is discarded and how it is dealt with once junked. All products have these cradle-to-grave environmental costs. The only way to truly understand a product's environmental impact is to know this indicator. [*extended product responsibility; CERES, economics and sustainable development; internalizing costs; Beyond GNP; life-cycle assessment; natural capital; slow money movement; triple bottom line*]

credit cards, green

Credit cards are often co-branded with different types of organizations, many of which are environmental *NGOs*. The idea is that the bank issuing the card provides some sort of benefit to the environmental group it is co-branded with each time a purchase is made. For example, they might make a direct donation to the NGO or *purchase carbon emission offsets* every time you use the card. This can be a great way to help environmental causes on a daily basis, if you are careful to choose a card that truly does what it says, versus those using this as nothing more than *greenwashing*. By using a green credit card, you become a *green consumer*. You can ensure you choose wisely by checking some of the *environmental rating systems*. [*green marketing; green products; green scam; college divestment campaigns; payroll deduction, environmental*]

cremation, human

In some countries, the rituals associated with cremation are considered environmental issues. *India* has 1.25 billion people, with 80 percent of them being cremated—that's about 8 million cremations per year. Most of these cremations occur on a wooden pyre consisting of from 400 to 1000 pounds of wood. About 750 square miles of forested areas are needed each year to supply the wood. In addition to the *deforestation*, the burning of all this wood releases tons of *carbon dioxide* and *pollutants*, causing *air pollution*, and the remaining *ash* is typically emptied into the Ganges or other sacred rivers, causing *water pollution*.

Mokshda, a company in Delhi, and the Oil and Natural Gas Corporation of India are looking for ways to reduce this environmental harm while retaining respect for the Hindu religion. [*human population throughout history; population growth, limits of human; United Nations Population Fund*]

criteria pollutants

See *air pollution*.

crop rotation

Crop rotation is a *conservation agriculture* practice that prevents *soil* nutrients from becoming depleted. Some crops rapidly deplete the soil of nutrients, while others replenish them. Crop rotation involves alternating both kinds of crops so the nutrient level in the soil remains stable. Corn, tobacco, and cotton deplete nutrients (especially *nitrogen*), but *legumes* such as oats, barley, and rye add nitrogen by a process called *nitrogen fixation*. Crop rotation also reduces pest infestations and plant diseases because the pest populations cannot build up over many years. [*low-input sustainable agriculture; sustainable agriculture; organic farming and food*]

crowd funding, green

Appealing to the online community to fund charities, projects, research, businesses, or just about anything someone feels is worthy of an online request. Many websites exist to facilitate these efforts. Some of these projects are scientific and some of those, environmentally based.

Those initiating the requests may include graduate students, desperate researchers in search of funds, or a person or group in search of charitable donations to support their specific green agenda. Some sites host contests to advertise those looking for funding.

The site ioby.org—stands for "in our backyard"—specializes in helping fund environmental projects. Other sites, such as RockthePost.com, do not specialize in ecoprojects but have a significant portion that do so.

Some of these projects were posted more for fun than reality. For example, requests have been made for $24,000 so the Gotcha Collective could buy a Prius and place it in Dick Cheney's garage and $40,000 so the Green Project could reinstall the solar panels originally put on the White House roof by Jimmy Carter, but removed by Ronald Reagan. [*crowd research, scientific; crowdsourcing, eco-; microblogs, green; blogs, green; online petition sites and environmentalism; social media, environmental, tweets, eco-; tweetstorm, Rio+20*]

crowd research, scientific

Some researchers, especially those short on funding, are getting their research done by relying on everyone and anyone. A lot of research depends on data collection, a perfect type of research for anyone to help. It can be measuring plant growth or water depth, counting tadpoles in a stream, or almost anything else that can be quantitatively measured—so anyone can do it.

For example, a research scientist in upstate New York stuck a large ruler in a stream, because his research required measurements of depth taken at intervals. He did not have funding for research assistants to do the work, so he placed a sign on the ruler, with his cell phone number, and asked anyone passing by to assist him with his research by texting to him the height of the stream at that moment. He got an excellent response and continues to collect data via crowd research. The possibilities are endless. [*crowd funding, green; crowdsourcing, eco-; microblogs, green; blogs, green; online petition sites and environmentalism; social media, environmental, tweets, eco-; volunteer research, scientific; voluntourism, environmental; WWOOF; tweetstorm, red panda on the loose*]

crowdsourcing, eco-

Obtaining services, ideas, or content by soliciting contributions from a large group of people, especially from an online community. It can also be thought of as distributive problem-solving. Crowdsourcing has been used for many environmentally pertinent issues. For example, the Governor of Montana used crowdsourcing to assist in tracking a *pipeline oil spill* as it flowed downstream. The possibilities are enormous for this type of activity in pursuit of helping with environmental issues and protection. [*crowd funding, green; crowd research, scientific; microblogs, green; blogs, green; online petition sites and environmentalism; social media, environmental, tweets, eco-*]

cruelty-free movement

See *animal rights movement and animal welfare.*

cruise-ship pollution

Cruise ships are allowed to dump huge amounts of waste at sea. As much as 25,000 gallons of raw sewage from *toilets* and 143,000 gallons of *gray water* from sinks, galleys, and showers every day are disposed of from cruise ships each day. Most state and federal laws allow these ships to dump this waste as soon as they are three miles from shore; however, they don't even have to wait to get outside the three-mile limit. Closer to shore, they can dump toilet *sewage* if it has been treated by marine sanitation devices. These wastes can increase *bacteria*, pathogens, and *heavy metals* in *coastal ecosystems*. Some cruise lines have agreed to voluntarily improve their policies by installing advanced devices to sanitize the waste before it is dumped into the sea. [*ship pollution; ocean pollution; marine ecosystems; marine reserves; MARPOL; Marine Stewardship Council; sea gliders*]

crustaceans

A class of arthropods (*invertebrates* with jointed appendages) containing about 35,000 species, many of which are important components of *food chains*. The group includes shrimp, crabs, and lobster, which are an important human food source and all of which live in *freshwater* or *marine* environments. Wood lice are an example of a terrestrial species. One of the smallest crustaceans is one of the most important and found in the largest numbers; krill makes up much of the *plankton* that floats at sea and is a primary consumer in many *marine ecosystems*. Krill is the primary food source for *whales*. [*shrimp farms, aquaculture; aquaculture; aquatic ecosystems*]

cryptic species

Species that biologists had previously thought to be the same, but with the advent of *DNA sequencing technology*, have been found to be different species. For example, the African *elephant* was thought to consist of one species but was found, with recent DNA analysis, to be two separate species, incapable of breeding and now called the African elephant and the African Bush elephant. [*weird life; convergent evolution*]

cryptozoology

The study of small terrestrial animals that live in crevices, under stones, and in *leaf litter*; includes many small *insects* and *crustaceans*.

cullet

Crushed glass used during *recycling*. [*glass recycling; recycling metals; downcycling*]

culling

The selective removal or killing of individuals to reduce the size of the population. [*Animal Damage Control program; carrying capacity; community ecology*]

Cuyahoga River Fire

During the 1960s, when the *environmental movement* was in its formative years, a few events took center stage and helped build public opinion to further the cause. One of these events was when this river, in Cleveland, Ohio, actually caught fire. *Oil* pollution in the river had become so bad that the river caught fire in 1960 and burned for eight days, burning down two bridges. The event made national news headlines. The outrage resulted in a demand for better water quality and protection and led to the monumental passage of the Federal Water Pollution Control Act (later changed to the *Clean Water Act*) of 1972. [*water pollution; effluent; industrial water pollution; rivers in decline; water shortages, global*]

cyanobacteria

Also called blue-green bacteria (previously called blue-green algae), these *bacteria* are *autotrophs* since they are capable of *photosynthesis*. They are one of, if not the oldest of all organisms, dating back 3.5 million years. Some are capable of *nitrogen fixation* while others live *symbiotically* with *fungus*, forming *lichens*.

These bacteria were the first to photosynthesize and were responsible for changing the Earth's *atmosphere* from being very inhospitable to life into what it is today—rich in oxygen. They are also the ancestors of today's plants. [*archaea; symbiotic relationships*]

dairy cow output

In 1961, a dairy cow in the United States produced a little over 7000 pounds of milk a year. In 2011, a dairy cow living on a *factory farm* produced well over 21,000 pounds of milk a year. [*egg farms; selective breeding; genetically modified organisms—animals; Franken-salmon*]

Daly, Herman E.

(1938–) Daly is an *ecological economist* who first promoted the phrase sustainable economy, commonly called a *steady-state economy*—an economy that maintains a constant state of the number of people and of the physical wealth of these people. He also introduced the phrase *index of sustainable economic welfare (ISEW)* in his book <u>For the Common Good</u>. The ISEW is meant to improve the concept of *Gross National Product* (GNP). Daly argues that increased GNP—although looked on favorably by most—actually results in a decrease in the quality of life. [*Beyond GNP; economy vs. environment; economics, environmental; triple bottom line*]

dam, hydroelectric

Large dams are technological marvels but often environmental disasters. They are constructed on rivers or streams where they slow the flow of water or create an artificial body of water—the reservoir. The backed-up water is then allowed to flow rapidly, turning turbines that generate electricity. *Hydropower* is a clean, *renewable energy* source and they do generate large amounts of power, but most environmentalists agree that huge dams cause more environmental harm than good.

A dam displaces people and also damages or destroys an entire *aquatic ecosystem* within which it resides, extending far downstream and even into the surrounding terrestrial ecosystems. Dams interrupt migratory fish, often resulting in the complete loss of a river's species. They accumulate sediment that changes water depth, change water speed, and cause erosion along portions of the river and many other alterations that change the ecosystem drastically.

The magnitude of some of these huge construction projects—megaprojects—leads to environmental degradation. For example, just the *cement* production for some of these projects contributes to *global warming*.

The three largest dams in the world are the *Three Gorges* in China (opened in 2006 but in 2012 put online its final turbine, with a total 22.5 GW); Itaipu on the border of *Brazil* and Paraguay, with 14 GW; and the *Belo Monte* in the *Amazon* Basin in Brazil (construction began 2013, with 11.2 GW).

Large dams such as these can also cause international tensions when a dam built in one nation is detrimental to the water supply of nations downstream. This is the case between *Laos* and its neighbors to the south, *Thailand*, Cambodia, and *Vietnam*. Laos is building the Xayaburi dam and has proposed 11 other dams along the Mekong delta. These dams will generate a great deal of electricity but will result in shortages of water downstream, seriously impacting *agriculture* of the other nations. [*hydropower, nontraditional; run-of-the-river hydropower*]

dam removal

To generate *hydroelectric power*, the norm had been to build large dams, but these cause a great deal of environmental harm, including *soil erosion*, a collapse of salmon populations that rely on unencumbered access up rivers to spawn, and the destruction of *aquatic ecosystems*. In the United States today, the move has been to not only build smaller, *run-of-the-river dams* that are less harmful but also to remove existing dams—to undo the damage previously done. Over the past 30 years, dams have been coming down in the United States—about 1000 so far and more each year. This occurs in part because they are aging and maintaining and upgrading them to remain productive is cost prohibitive.

But the driving reason for these dam removals is a concern about the loss of commercial fish stocks of salmon and other wildlife. On many rivers, dams prevent these fish from migrating back to their spawning grounds each year. Even with retrofitted, high-tech salmon ladders that help fish get past the dams, wild salmon *migratory* runs are a fraction of what they were before the dams were built. Thirteen species of salmon and steelhead are on the *endangered list* and will be commercially nonviable if actions are not taken to save them. In addition, the dams have reduced populations of sturgeon, alewives, *eels*, and smelt.

Some of the U.S. dams coming down include those on the Elwha River in the Northwest, where salmon runs are ten percent of what they once were, and on the

Penobscot River in the Northeast, where their numbers have dwindled from 75,000 before the dam to 3000 today.

Even though large dams are coming down in the United States, they continue to be built in many other countries, including many megaproject *hydroelectric dams.* [*habitat loss*]

dams, ten largest

You can find about a dozen such lists, depending on which factor you are using for "the largest." It can be the largest storage capacity of water, the largest amount of energy generated, the tallest, the widest, and so on. It also is complicated by the fact that some of these dams are mega-engineering projects that take decades to complete, and they grow as each phase of construction is completed.

If you are mainly interested in the impact a dam has on the local *habitat*, volume is the most important. So here are the ten largest dams (by storage capacity of water in billions of cubic meters): 1) Akosombo Dam, Ghana (144); 2) Guri Dam, Venezuela (135); 3) W.A.C. Bennett Dam, Canada (74); 4) Ataturk Dam, Anatolia, Turkey (48); 5) *Three Gorges Dam, China* (39); 6) Hoover Dam, United States (36); 7) Garrison Dam, United States (30); 8) Itaipu Dam, *Brazil*/Paraguay (29); 9) Oahe Dam, United States (29); and 10) Fort Peck Dam, United States (23).

If you are interested in the most energy produced by a dam, the two largest are the Three Gorges Dam, generating 22,500 *MW* of power, and the Itaipu Dam, generating 14,000 MW. [*habitat loss; dam, hydroelectric; hydropower, traditional*]

Darling, J. N. "Ding"

"Ding" Darling was an eminent political cartoonist during the first half of this century. With his cartoons, he brought the idea of *conservation* to the forefront before most people knew what the term *ecology* even meant. His interest in the *environment* did not stop with the pen. He became the chief of the organization that later became the *U.S. Fish and Wildlife Service;* originated the *Federal Duck Stamp Program*, which today generates millions of dollars in revenue for *wildlife refuges*; and then founded the *National Wildlife Federation.*

The *J. N. "Ding" Darling Foundation* was created in 1962, after his death, by colleagues and friends, to continue his efforts in conservation education. The foundation funds educational programs at the elementary school level and provides grants at the university level, all focused on *conservation* communications and *environmental education.* The foundation has no paid staff, so every dollar contributed is used for its programs.

Darwin, Charles

(1809–1882) The British naturalist who developed the accepted theory of evolution as based on the process of *natural selection*. Darwin formed his ideas while on a five-year trip onboard a research ship, the *HMS Beagle*, and its famous stopover in the Galapagos Islands as well as many other locations. Much of his research involved many species of finches he observed on these islands. His theory was published in journal papers but popularized in his book, The Origin of the Species by Means of Natural Selection. He told about his travels onboard the HMS Beagle in The Voyage of the Beagle.

data centers, energy consumption of

Have you ever thought about all that data you store "in the cloud," such as in *Facebook*, Google, Amazon, Yahoo, or any other online service? Where are all of your pictures and reports and videos and music? They reside in data centers owned by all of the above. More than three million data centers occupy more than 600 million square feet, and the number is growing at lightning speed because of the upswing in the number of smartphones. These vast data centers need lots of power to run and to cool them. One large data center uses the same amount of energy as a medium-size city. More than two percent of all U.S. energy consumption comes from these data centers, and it continues to grow rapidly.

Some of these companies are better than others at reducing their energy consumption. Some cloud computer centers are even classified as "brown clouds" for inefficient centers with a large *carbon footprint* and "green" for centers saving energy with a smaller footprint. All of these companies will be working to reduce energy, because it is good not only for the environment but for their bottom lines as well.

Chip makers are working to build more efficient chips for servers used in these data centers. Most of the companies mentioned, plus giants such as Apple and Microsoft, are working on using *renewable energy* sources for their data centers or manufacturing plants. To handle its transactions, eBay has stated it is building a data center that will run on 100 percent renewable power, using nothing from the *power grid*. [*e-waste; energy efficient states, top ten; baseball stadiums, green*]

DDT

Dichlorodiphenyltrichloroethane—or DDT—is a *chlorinated hydrocarbon* type of *insecticide*. First synthesized in the early 1940s, it was considered a panacea for pest control. DDT was deadly to insects and appeared to cause little harm to humans. It was used to control mosquitoes that carried malaria and sleeping sickness, and is attributed with saving more than five million lives during that time. Its benefits cannot be argued. (The entomologist who discovered it, Paul Mueller, won the Nobel Prize in Physiology or Medicine in 1948.) However, by the early 1950s, mosquitoes had become *resistant* to DDT so it was no longer effective.

DDT, along with other chlorinated hydrocarbons, was later found to harm more than just the targeted insect pest. These *pesticides* are *persistent*, meaning they remain in an *ecosystem* for long periods of time. Many different types of animals ingest the substance and it accumulates within their bodies. DDT became a symbol of pesticide dangers when traces of it (in the form of DDE) were found in many kinds of animals, including humans, as a result of *bioaccumulation* and *biological amplification*. Today, it is found in almost all of us and is part of our *body burden*.

During the 1960s, DDT was responsible for the dramatic decline in the number of *predatory* birds such as ospreys, cormorants, and bald eagles. These populations declined because the insecticide reduced the amount of calcium available for egg production, resulting in thin egg shells that broke.

DDT was banned in the United States in 1972, thanks mostly to *Rachel Carson's* famous book <u>*Silent Spring*</u>. The ban was followed by an increase in the populations of affected birds. Recently, declines have begun once again, however. This is thought to be caused by the DDT used in Latin America, where it remains legal, or possibly as a result of weak regulations that allow substantial quantities of DDT to remain present as an impurity in other pesticides.

Even though DDT is banned for use in the United States, it is still manufactured in the United States and then distributed to other countries, where it remains legal. [*circle of poison*]

decapitating flies

The smallest species of fly known to exist is a *parasite* with an interesting and unusual life cycle. The female lays her eggs in the heads of a certain species of ant. The eggs hatch, the larvae feed on the tissue in the ant's head until the ant dies, the head falls off, and the young flies emerge to begin anew. [*hyperparasites; zombie bees; parasitoid*]

deadtime

This is the amount of time it takes an item to *decompose* (biodegrade) into usable *organic* materials future generations of plants can use for *nutrients*. The shorter the deadtime, the more biodegradable an item is said to be. The deadtime of an item differs greatly depending on whether it is in a *landfill*, where water and air are scarce, or exposed to the elements.

For example, most paper products break down quickly under normal conditions but have a deadtime of many decades in a landfill. Most organic yard wastes, such as *grass clippings*, become usable *organic fertilizer* within a few days if left on the *lawn* but have a deadtime of years in a landfill. Most *plastics* have a deadtime of hundreds or thousands of years, whether they are in or out of a landfill. [*recycling; grass cycling; paper recycling; plastic recycling; plastic pollution; biogeochemical cycles; biogeochemistry*]

dead zones

Huge swaths of *ocean* have become dead zones, meaning they contain very little life compared to the normal *oceanic zone ecosystem* that would otherwise exist. This is primarily because of *nutrient enrichment* that results in little to no oxygen available for life to exist. For example, the Mississippi river—where the nutrient enrichment occurs from the *high-impact agriculture* found along its banks—releases its waters into the Gulf of Mexico, resulting in a dead zone that begins about 60 miles from shore and encompasses more than 9500 square miles.

There are many other dead zones. The largest in the world is found in most of the Black Sea, where almost the entire bottom 500 feet is considered a dead zone.

Dead zones cause many aquatic species to move elsewhere. Populations of those that remain in the area usually plummet. Some species have been found to still exist but many individuals are deformed, probably from exposure to limited oxygen, called *hypoxia*. [*marine ecosystems; ocean pollution; eutrophication, cultural; Ocean Conservancy; marine reserves; ship pollution; red tides; whale earwax*]

debt-for-nature swaps

In 1987, countries experiencing *rainforest destruction* and a large national debt received interesting proposals, called debt-for-nature swaps. Some *NGOs* made arrangements to have a portion of the country's debt paid if these countries began serious rainforest *conservation* efforts. About $60 million dollars changed hands in these nature swaps over many years. The countries involved included Costa Rica, Madagascar, Philippines, Bolivia, and Ecuador. The NGOs most often involved in these swaps were the *World Wildlife Fund*, *The Nature Conservancy*, and Conservation International.

Today, debt-for-nature swaps have morphed into far bigger international agreements, usually under the auspices of the United Nations, beginning with agreements created at the *United Nations Conference on Environment and Development* in 1992 and *REDD*. [*deforestation; tropical rainforests*]

deciduous

Plants that lose their leaves during cold weather or other environmental change such as drought. [*temperate deciduous forest; cold resistant; estivation; hibernation; quiescence; torpor; natural selection*]

decomposer

Organisms can be categorized according to how they obtain food to survive. Producers, or *autotrophs* (green plants), make their own food, while consumers, or *heterotrophs*, eat the plants or other animals. Decomposers, however, obtain nourishment by eating dead and decaying organisms, called *detritus*. Decomposers include many *insects*, worms, *bacteria,* and *fungi* (molds and mushrooms). They reduce the detritus to the basic *organic nutrients* so they can be reused by a new generation of plants. Without decomposers, many *biogeochemical cycles* would stop cycling. Decomposers keep the *environment* free from the remains of all preceding generations of life on Earth. [*food webs; ecosystems; biogeochemical cycles; biogeochemical cycles, human intervention in; scavenger; coprophagous; scavengers*]

decoupling

Today's economic model is based on constant growth, typically associated with constant environmental degradation. Decoupling is a new, ongoing topic of discussion that attempts to break the link between the prosperity and growth of successful economies, on the one hand, and the resulting environmental degradation it causes, on the other hand.

Existing economic models are based on constant growth that is considered by most scientists and environmentalists as unsustainable. *Sustainable growth* cannot be achieved, because we live on a planet with finite resources to extract and finite space to put our waste. Sooner or later, growth cannot continue without serious environmental consequences. Decoupling is a simple word describing a monumental sea change that no one currently knows how to accomplish. [*Beyond GNP; triple bottom line; steady-state economics; natural capital; Genuine Progress Indicator; economics and sustainable development; CERES*]

deep ecology

Deep ecologists believe humankind to be the primary threat to our planet's survival. The term was first coined by Arne Naess in 1973. It questions the fundamental purpose and value of life. It is at odds with materialism, consumerism, and capitalism and attempts to integrate religion and philosophy into ecological ethics. It urges digging deeper into the meaning of life.

In the book <u>Deep Ecology: Living as if Nature Mattered</u> by Devall and Sessions, the following deep ecology principles are proposed. 1) Humans have no right to reduce the richness of life except to satisfy basic needs. 2) The quality of human life and culture is linked to a substantial decrease in the *human population*. 3) A decrease in the human population is required, if nonhuman life is to flourish. [*ethics, environmental; shallow ecology; philosophers and writers, environmental*]

deep sea trenches

The deepest parts of oceans reside in what are commonly called trenches, where the ocean floor rests about 32,800 feet down. These regions—properly called Hadal Zones—are found between 20,000 and 36,000 feet deep. Not long ago, it was believed that no life could survive at these depths. However, recent expeditions have shown these areas to be biologically (as well as geologically) active.

James Cameron, the movie producer, took his one-man submersible down into the Pacific Ocean's Mariana Trench in 2012. A team from the Woods Hole Oceanographic Institution is in the midst of a multi-year program to study the Kermadec Trench off the coast of New Zealand. These studies hope to identify and explain how life survives in these Hadal *ecosystems*. [*extremophiles; marine ecosystems; hydrothermal ocean vent ecosystems*]

Deep Water Horizon oil spill

See *BP Deepwater Horizon oil spill*.

deep-well injection sites

Hazardous wastes have been disposed of for decades by being injected into deep wells in the Earth's crust. These wells can be anywhere from 20 feet to several thousand feet deep. Many of these wells have polluted *groundwater aquifers* that supply *domestic water* to millions of homes.

Today, these sites are typically used to dispose of the contaminated wastewater produced during *fracking*. [*groundwater pollution*]

de-extinction

Far-fetched as it might sound, there are scientists who want to resurrect species that have gone extinct. They propose using cloning, *DNA sequencing*, and additional technologies not yet invented, to restore what humans have destroyed. Whether this becomes possible or not, other scientists and *environmentalists* are strongly opposed to this idea. They believe we should spend our efforts saving what we still have instead of trying to bring back what we have lost. They also believe, if it ever came to be, de-extinction would lessen the concerns about *extinction* in the first place. [*biodiversity; Lazarus taxon; captive breeding; debt-for-nature swaps; dodo bird; doomsday seed vault; extinct in the wild; species recently gone extinct*]

deforestation

In 1988, satellite images began to reveal to the public a new threat to our planet. *Tropical rainforests* in many parts of the world were being cut down (with much of it burned) at a rate that could change the planet as we knew it. This permanent destruction of forests was coined deforestation.

Forests are destroyed to make room for agricultural crops and for grazing animals such as cattle, as well as for a timber harvest. Deforestation has a significant impact on global environmental problems such as *soil erosion, air pollution,* and *climate change*. In addition, the loss of these *habitats* is forcing many species to the brink of *extinction* and dramatically reducing our planet's *biodiversity*. Tropical rainforests are home to half the species of the world, but more than 30 million acres were destroyed annually from 2000 to 2009. In 1950, 30 percent of the Earth's land was covered in these forests, but today only 7 percent remains.

These forests are often burned to add the *nutrients* locked up in the trees back into the soil as *fertilizer* for crops (called *slash-and-burn cultivation*). This large-scale burning releases vast quantities of *carbon dioxide*, a *greenhouse gas*. About one-quarter of all human-created *carbon dioxide emissions* comes from deforestation.

Most deforestation now occurs in Asia, Africa, and South America. *Indonesia* and *Brazil* are especially hard hit by deforestation because of destructive *logging* practices, where slash-and-burn methods are still routinely used.

In some developed countries, especially those with temperate forests, *reforestation* is helping to mitigate deforestation. However, the tree plantations planted as part of reforestation do not possess the same biodiversity as the native forests they replace.

No international agreement is in place to prevent deforestation, but we sorely need one. Many *NGOs* work to protect forests, including the *Rainforest Action Network, Rainforest Alliance, Greenpeace,* and the *World Wildlife Federation*. Other organizations, such as the *Forest Stewardship Council*, work with timber companies to create *eco-certification programs* that establish sustainable harvests. [*United Nations Conference on Sustainable Development*]

deforestation and climate change

Forests are *carbon sinks. Deforestation* frees vast amounts of carbon that was stored in the trees and releases it into the *atmosphere*, contributing to *climate change*—so much so that deforestation is believed to cause more climate change (15 percent of the total) than all of the *automobiles* and trucks on the roads worldwide (14 percent of the total).

International agreements such as *REDD* try to get developing nations to slow deforestation with financial incentives from developed nations.

deicing roads

Spreading road salt is the standard method of melting ice on frozen highways. The salt often damages or kills trees and vegetation and can contaminate *groundwater*. It has been suggested that one ton of road salt causes about $600 of environmental damage.

There are environmentally friendly alternatives to road salt, but they are all far more expensive and rarely used. With competition, more acceptance, and—especially—environmental awareness, the hope is these alternatives will become more competitive and used routinely. [*POPs, dirty dozen; PERC; POPs, dirty dozen; Safe Chemicals Act; sunsetting; Superfund; tire fires*]

demographic transition, theory of

Demographers who studied human populations of western European countries during the 19th century were the first to identify what appears to be a link between a nation's *standard of living* and *population growth*. They proposed the theory of demographic transition. The transition refers to stages these countries went through—and other countries might go through—as they develop.

The first stage (preindustrial stage) has high death rates because of harsh living conditions and high birth rates to compensate, hence a stable population. As the country becomes more industrialized (the transitional stage), the death rate lowers because of improved conditions such as better health care, but the birth rate remains high, resulting in an increase in population. Most of today's *less developed countries* are at this stage now.

In the next stage (industrial stage), further technological advances result in lower birth rates because of (for example) increased education and income, each individual's desire to advance with the society, and the belief that children will hinder them in doing so. The birth rate begins to approach death rates. The majority of *more developed countries* currently fall in this stage. In the last stage (postindustrial), the birth rate reaches the same level as the death rate, resulting in *zero population growth* and finally a reduction in population. This has already occurred in a few countries, such as *Germany*, and looks like it will begin to occur in many others, as well.

Many people do not believe what has happened in the past in some countries will necessarily hold true in the future for other countries. Rapid and long-term economic growth is necessary if less developed countries are to make the demographic transition. Even if the entire theory is not accurate, two components of this theory remain consistent throughout recent history: increased poverty results in increased population growth, and increased education (particularly of women) results in decreased population growth.

An interesting theory on why humans pass through this transition—from an evolutionary perspective—states that people within nations are at first *r-strategists* and then transition into *K-strategists*. [*human population throughout history; doubling time in human populations*]

demography
The study of populations and the factors that make populations increase or decrease in size. [*age distribution in human populations; overpopulation; baby boom*]

dendrochore
That part of the Earth's surface covered by trees. [*forests; deforestation; Save the Redwoods League; logging, illegal*]

denitrification
An essential part of the *nitrogen cycle* that occurs in soil and water when *organic* forms of nitrogen (ammonia and nitrates) are converted back into free nitrogen by certain types of *bacteria* and then released into the *atmosphere*. This is an *anaerobic process*.

Denitrification is also artificially induced during the secondary water treatment process used for *sewage waste treatment*. [*biogeochemical cycles*]

Department of Agriculture, U.S.
The executive branch of the U.S. government responsible for creating, executing, and managing policies on *agriculture*, farming, and forestry, as well as aspects of our food. It supports agricultural research and technological development in areas such as the care of agricultural lands, *forests*, and rangeland; plant disease and its control; food safety and nutrition; animal reproduction; ways to support rural communities; and ways to reduce hunger. The department has more than 30 agencies, including the *U.S. Forest Service*, the *Natural Resources Conservation Service*, and the Animal Health Inspection Service.

Department of Energy, U.S. (DOE)
The executive branch of the U.S. government responsible for creating, executing, and managing policies on all forms of *energy* production and use, including energy conservation, as well as *nuclear power* and *radioactive waste*. It is also responsible for the management of *nuclear weapons*. Most of the U.S. research in the physical sciences is

conducted with the DOE through its many national laboratories, such as Los Alamos and Brookhaven National Laboratories.

Department of the Interior, U.S.

The executive branch of the U.S. government responsible for managing the country's cultural and *natural resources*. It includes many agencies and offices responsible for protecting (as well as using) our public lands. This includes: the *National Park Service*, *U.S. Fish and Wildlife Service*, *Bureau of Land Management*, U.S. Geological Survey, Bureau of Ocean Energy Management, Bureau of Safety and Environmental Enforcement, Office of Surface Mining, Reclamation and Enforcement, Bureau of Reclamation, and the Bureau of Indian Affairs. [*dual-use policy*]

deposition

The natural laying down and accumulation of sediments that have been transported from other areas. For example, river water washes away and carries sediment downstream to the mouth of a river (for example, in an *estuary*) where it accumulates. Also, wind sometimes carries away *topsoil* and transports it elsewhere, and *glaciers* carry soil with them as they slowly move, leaving moraines where the soil and rock has built up in front of the glacier.

Deposition can be a positive force when it occurs naturally, which usually means slowly. However, it can also be a dangerous force if it occurs quickly, usually caused by human activity. For example, mudslides usually result from human-induced *soil erosion* or topsoil loss from poor farming practices, as during the *Dust Bowl*. [*sedimentation*]

desalination

The process of removing salt from salt water located in the *oceans* or in *groundwater* saline *aquifers* to produce *drinking water*. It appears odd that we live on the water planet, consisting mostly of water, but millions of people live without access to clean drinking water. Of course, the problem is most of this water contains salt. But you would think with technological advances, we should be able to desalinate this water. And we can, but the process of desalination is very energy intensive and therefore very expensive.

The process typically uses reverse osmosis, where the salt water is put under pressure and forced through a permeable polymer membrane that allows water molecules to pass through, but not the sodium and chloride ions of salt. New technologies such as using a membrane made of graphene, a natural substance, instead of man-made polymer membranes might improve the process and lower the cost.

Some important advances have been made, though. Technological improvements have reduced the amount of energy needed for the process by 90 percent over the past 40 years (even though it still is expensive).

In Texas, where water shortages are common, more than 40 such desalination facilities have been built that tap into underground saltwater aquifers. They are small but productive, yielding around three million gallons per day—a quantity helpful for rural small towns. Some larger ones can produce up to 30 million gallons per day, but because of the high costs, do not run at capacity. Texas hopes to produce about 3.5 percent of their fresh water with desalination by 2060.

But some regions of the world are desperate for clean water and can afford to pay the price. For example, wealthy countries such as the United Arab Emirates get almost all of their *freshwater* from these facilities. In *China*, the Beijing Power and Desalination Plant has opened recently and draws water from the mouth of the Yangtze River. It cost more than four billion dollars to build. [*water grabs; water shortages, global; saltwater intrusion; water shortages, global*]

descriptive ecology

Descriptive *ecology* concentrates on describing the variety of *environments* found on our planet and the components of each. This was the first approach used to study ecology and was very popular in the first half of the 1900s. It is still an integral part of modern-day ecology but has been eclipsed by newer approaches, including experimental and *theoretical ecology*.

deserts

One of several kinds of *biomes*. The primary factors that differentiate biomes are temperature and precipitation. Deserts receive less than 25 centimeters (10 inches) of precipitation per year and therefore are inhospitable to most forms of life. Rain is erratic but when it comes, it is intense, so most of the water runs off. Bursts of life follow these heavy but rare rains. Many species of plants are adapted to complete their entire life cycle and produce seeds within a few days, before the water dries up.

Although all deserts have low precipitation, different temperature ranges delineate three types. Tropical deserts (Sahara) are hot all year; temperate deserts (Mojave) are hot in summer and cool in winter; cold deserts (Gobi) are warm or hot in the summer and cold in winter. All deserts experience wide temperature differences between day and night.

The relative number of species in a desert is low compared with other biomes, but there is still considerable *biodiversity*. Organisms living in the desert possess specialized adaptations to survive. Plants have small or no leaves, to reduce surface area, which in turn reduces water loss by evaporation. Some plants store water in their fleshy tissues, and others go through rapid life cycles that coincide with the erratic rains.

Many kinds of animals live in the desert, but they are few in number. They include *insects*, snakes, lizards, some grazing *mammals,* and a few *carnivores*. Birds are also common in many deserts. Most desert animals obtain water via the food they eat. They have a waterproof skin or waxy cuticle (insects), and many live underground to avoid the day's heat.

Due to the severe conditions, the majority of plants grow slowly, resulting in fragile *ecosystems*. Damage by human activities, such as off-road vehicles, takes decades to recover. [*anthromes; natural selection; succession; convergent evolution; Desert Protective Council, Inc.; cacti*]

desertification

When rangeland is *overgrazed* or cropland overcultivated, *soil nutrients* are lost and *soil erosion* is likely. As a result, the land becomes more like a *desert ecosystem* than it once was. If the land becomes at least 10 percent less agriculturally productive than it once was, the process is called desertification. "Severe desertification" means the productivity of the land has been reduced more than 50 percent. *Logging* for timber and collecting firewood also contribute to the process. Many believe c*limate change* will become the primary factor driving desertification in the future.

Desertification can be controlled, in part, with proper *land management* practices that prevent *soil erosion*. Many s*ustainable agriculture* practices are also designed to prevent desertification.

Drought can cause desertification to occur naturally along the fringes of existing deserts.

The term was first coined in the 1920s. It was thrust into the environmental spotlight during a terrible drought in western Africa from 1968 to 1973, when 200,000 people and millions of livestock died. In 1996, the *United Nations Conference on Environment and Development* created the *United Nations Convention to Combat Desertification*. The United Nations has designated June 17 as World Day to Combat Desertification.

Desert Protective Council, Inc.

The Desert Protective Council was established in 1954. Its purpose is to protect *desert* plants and animals, cliffs and canyons, dry lakes and *sand* dunes, and historic and cultural sites. They focus primarily on deserts of southern California but assist in desert conservation throughout the southwestern United States. [*desertification; anthromes; biomes; NGOs*] {dpcinc.org}

detergents, phosphate in

Most detergents are synthetic compounds used in household cleaning products to remove oil, soil, and dirt. In the past, *phosphates* were commonly used in these laundry detergents and are still used in dish detergents. Phosphates are added to detergents because they help break down and remove dirt. Because phosphates were common in household detergents—and were also used in *sewage treatment facilities*—large volumes of the substance began entering the *waste stream*. This wreaked havoc in *aquatic ecosystems*.

Phosphates are important *nutrients* in the *soil*, but when found in abnormal amounts, as began to happen, a process called *nutrient enrichment* occurs. This in turn causes explosions of algae growth, called *algal blooms*. And finally, these blooms cause *eutrophication*, resulting in damaged *aquatic ecosystems* in lakes and ponds where the wastewater found its way in.

Phosphates were banned in the United States in laundry detergents in 1993 and most sewage treatment plants upgraded to no longer use them. Using phosphates in dish detergents is still legal, but beginning in 2010, many states started banning phosphates in dish detergents, as well. Those that don't contain phosphates are usually clearly marked so environmentally concerned consumers can find them. [*green consumers; dry cleaning; green marketing; green products; greenwash*]

detritus

Partially *decomposed organic* matter. [*scavengers; trophic levels; anaerobic*]

developing and developed countries

Developing countries and *less developed countries* (LDC) are terms often used interchangeably. Developed and *more developed countries* (MDC) are also used interchangeably, as well as the term industrialized countries.

All of these definitions come from The World Bank and the *United Nations*—which use different criteria to specify which countries fall into which category. The most important criterion used is the gross national income (GNI) per capita (previously referred to as gross national product or *GNP*). High GNI, a significant amount of economic development, and advanced technologies are indicative of MDCs, while low GNI, little economic development, and few technologies are indicative of LDCs.

The number of LDCs ranges from about 100 to 150, depending on which criteria are used, and the number of MDCs range from about 45 to 66. [*water shortages, global; population explosion, human; hunger, world; Toilet Day, World*]

developed countries

See *developing and developed countries.*

diatoms

A type of *algae* with oddly shaped cell walls containing silica. Diatoms make up a large portion of most *phytoplankton.* [*plankton; photosynthesis; tropic levels*]

dientomophilous

A species of plant that has two different types of flowers for attracting two different kinds of *insects* to pollinate the plant. [*colony collapse disorder; bumblebee decline; bumblebee senses; communications between plants; honeybees, pollination, and agriculture; neonicotinoid pesticides and honeybees; sixth sense in animals*]

diesel engine fumes

Modern diesel engines, in *developed nations*, burn cleanly and industries such as *mining* have strict limits about exposure to their fumes. They also burn low-*sulfur* fuel, so they emit a fraction of the amount of *particulate matter* and even less *nitrogen oxides* than older diesel engines. However, in developing nations, with few if any regulations and old diesel engines still in regular use, the engine fumes released are dangerous to health and the environment. The World Health Organization in 2013 relabeled diesel fumes as a known *carcinogen.* [*biodiesel fuel; catalytic converter; hotel load; sunflower oil as a biodiesel fuel; soot*]

digester

See *methane digester.*

Dillard, Annie

(1945–) Annie Dillard is an American author of fiction and nonfiction. She has published poetry, essays, prose, novels, and literary criticism. In environmental circles, she is best known for Pilgrim at Tinker Creek, written in 1974. It won the Pulitzer Prize for General Nonfiction and was listed in Random House's survey of the century's 100 best nonfiction books. Dillard taught for many years at Wesleyan University. [*philosophers and writers, environmental*]

dioxin

Dioxin is one of the *dirty dozen persistent organic pollutants*. It is a toxic chemical byproduct of many industrial processes and considered an unintentional pollutant because it is not manufactured for a particular use. Instead, dioxins form when chlorine is exposed to extreme heat. Because chlorine is found in many synthetic substances, dioxins form when these substances are either manufactured (with heat) or disposed of by *incineration*.

Dioxins are believed to cause cancer and birth defects and to affect the human nervous system. It is found in *industrial waste effluent* entering the *water* and in *soils*, and traces have even been found in mother's *breast milk*. It is part of the human *body burden*. It is especially prevalent in effluent from pulp and paper mills. Dioxins are heavily regulated but difficult to manage, because they are a byproduct instead of an actual product. [*biological amplification; chemical body cleanse; Rachel Carson; Toxic Substances Control Act*]

directive coloration
Markings on an organism that diverts the attack of a *predator* to nonvital body parts. For example, eye spots on a butterfly's wings draw the attention of a bird away from the insect's body. A butterfly can survive a bite to the wing far better than one to its body. [*aggressive mimicry; aposematic coloration; Batesian Mimicry; bioluminescence; mimicry; predator/prey relationship*]

Dirty Dozen – fruits and vegetables
The *Environmental Working Group* creates an annual list of the top 12 fruits and vegetables contaminated with *pesticides*. It is included in the Shopper's Guide to Pesticides in Produce. The 2013 list includes (from worst): 1) apples, 2) celery, 3) cherry tomatoes, 4) cucumbers, 5) grapes, 6) hot peppers, 7) nectarines (imported), 8) peaches, 9)potatoes, 10) spinach, 11) strawberries, and 12) sweet bell peppers—and for those who prefer a baker's dozen: 13) kale. [*rice, contaminated; Clean Fifteen; apples, antibiotics in organic; community gardens; community-supported agriculture; fair trade certification; farmers' market; guerrilla gardening; heirloom plants; organic farming and food; food miles; Non-GMO Project Verified; tomatoes, tasteless but red; WWOOF; microgreens and urban gardening; wine, organic and biodynamic*]

dishwasher rinse-hold cycle
You can conserve about five gallons of water each wash by not using the rinse-hold cycle. [*water conservation; bathroom water-saving techniques; Christmas tree lighting; Energy Star rating; energy, ways to save residential; energy-saving apps; smart energy*]

disposable diapers
More than 25 billion disposable diapers are used in the United States annually. It is estimated that roughly four percent of U.S. *municipal solid waste* consists of disposable

diapers. Almost all of it ends up in *landfills*. The diapers are made of *plastic* (*PET* and *PP*), so they decompose very slowly. Estimates state they probably will take a few hundred years, at least, to decompose in a landfill.

Cloth diapers, when discarded, decompose within months. About 5000 diapers would be used before a baby becomes potty-trained, but about 35 cloth diapers would do the same job. Some studies state that washing cloth diapers causes as much environmental harm (in the form of *water use* and the *detergents* used with it) as do disposable diapers (a waste problem). However, many environmentalists dispute that assessment.

Biodegradable disposable diapers are claimed to break down faster, with more biodegradable components. Studies show they do break down more quickly in the tests but probably won't in a landfill. [*dog waste; municipal solid waste disposal; toilets and biosolids; ecological footprint; plastic pollution; deadtime*]

disposal fee

These fees help pay for the disposal costs of products that cause environmental harm and require a special method of disposal. These fees are usually charged to the consumer but can be charged to the manufacturer. They can be front-end fees, charged when the product is purchased or back-end fees, charged when the product is disposed of.

Back-end fees include entrance or dumping charges at *landfills* or *incineration* facilities. Front-end fees include taxes or deposits on products such as car *batteries* and *tires* or beverage bottles. For example, many states charge a $5.00 deposit on purchase of a car battery because they must be disposed of as a *hazardous waste*.

Disposal fees paid by the manufacturer are one aspect of *extended product responsibility* and is a method to *internalize* the cost. When consumers pay for these types of fees, the manufacturer *externalizes* the cost.

dissecting frogs in school

Dissecting frogs has been part of the curriculum in most schools for decades. Many people believe it to be inhumane and unnecessary, considering so many high-tech alternatives are now available. In 2012 alone, more than 300,000 wild-caught frogs were captured in Mexico for just one biological supply company and transported to the United States for use in science education labs in schools. The many excellent computer virtual dissection programs available are typically less expensive than buying the real thing and mean less contact with possible harmful chemicals. (Real frogs are available in nontoxic preservatives, at higher costs.) Many schools might find this change a way to save money, cause less environmental harm, avoid (or make a point of) ethical issues, and make a lot of students happy. [*biodiversity, loss of; vivisection; primate research; animal rights movement and animal welfare*] {savethefrogs.com}

DNA sequencing technology

DNA looks like a twisted, spiral ladder with steps consisting of components—called nucleotides—of four types: adenine, guanine, cytosine, and thymine. Cells become specialized into different types based on this sequence. DNA sequencing is used to determine the exact order of these nucleotides. One of the most important aspects of this technology is that only a small piece of genetic material is needed for study instead of the actual organism. DNA sequencing can be thought of as the ultimate fingerprint.

This technology was used in the Human Genome Project, which mapped the entire genome of humans (their entire genetic map). Today it is being used for a wide range of applications from the *Human Microbiome Project*, where it is discovering which *bacteria* live within our bodies and what they do there, to determining if timber was illegally logged. [*timber, DNA sequencing of; cryptic species; de-extinction; genetically modified organism (GMOs)—basics about; recombinant DNA; transgenic crops; Franken-salmon*]

dodo bird

This bird is often used as a symbol of *extinction*, probably because of its name and odd features. It became extinct in the 1680s. It is one of 34 bird species that have become extinct on the island of Mauritius in the Indian Ocean. Considering only 45 species were found on the island originally, this is a staggering number. [*biodiversity, loss of; CITES; de-extinction; endangered and threatened species lists; Endangered Species Act; extinct in the wild; extinction and extraterrestrial impact; extinction, mistaken; extinction, newly discovered species and; species recently gone extinct*]

doggie bags

Food waste is a worldwide problem. It begins on the farm and ends on our tables. If the table is in a restaurant, Americans often ask for a doggie bag, reducing the amount of food that goes to waste (no matter who actually eats the food, you or your dog). But in many other countries, especially in Europe, doggie bags are rarely used and the food simply is tossed into the garbage. This occurs because people don't want to appear cheap by asking for a doggie bag or they worry that the food will become contaminated. If you are travelling to Europe, you might be forewarned not to ask for one so as not to be embarrassed.

Environmentalists in Europe are trying to remedy this small portion of the food waste problem. The Sustainable Restaurant Association in Britain is running a Too Good to Waste program promoting the use of doggie bags in restaurants. Americans lag behind Europeans in many environmental aspects, but we appear to have the upper hand on this issue. [*composting; slow food; organic fertilizer; biosolids and wastewater*]

dog waste

Pet dogs produce lots of solid waste. The 78 million dogs in the United States produce over 10 million tons of solid waste each year. Parks and beaches are nice places to walk dogs but often collect large amounts of their dung, which pollutes the nearby waters.

Many people scoop up the waste, but the plastic bags they use cause their own problems. Environmentalists and entrepreneurs often find a way to turn a problem into a business. For example, a state park in New York provides corn-based, compostable dog-poop bags. The bags are collected regularly and, along with yard waste and wood chips collected from other sites, is delivered to a *large-scale composting* site. The *compost* makes great *fertilizer* and can then be provided free to local residents or sold. [*organic fertilizer; night soil; San Francisco and compostable waste curbside pickup*]

dolphin-safe tuna

See *driftnets* and *Flipper Seal of Approval*.

dolphins and porpoises, true

These aquatic *marine mammals* belong to the order of toothed whales. They are *carnivores* and *prey* on *fish* and squid. Studies have shown these animals to be intelligent and capable of extensive communications between members of their species. Their intelligence is why they have been captured and raised in captivity to perform in aquariums and *zoos*. The U.S. Navy also uses them for search and rescue as well as surveillance missions. Many environmentalists protest these uses, establishing campaigns to limit or stop these activities.

In addition, the survival of dolphins and porpoises has been stressed by *pollution, sonar testing* and their unintentional capture during *commercial fishing*, called *bycatch*. [*seismic air guns, whales and; whistle names, bottlenose dolphin; whales; whale and commercial ship collisions; biodiversity; giant squid in their native habitat*]

domestic water conservation

Any steps taken to reduce *domestic water use* is considered domestic water conservation. This includes stopping leaky faucets and valves and using low or ultra-low flush toilets, front-loader washing machines, high-efficiency shower heads, and faucet aerators. These devices can reduce the amount of water used in the home by more than 30 percent. [*water conservation; bathroom water-saving techniques; Christmas tree lighting; Energy Star rating; energy, ways to save residential; energy-saving apps; smart energy; nutrient enrichment*]

domestic water pollution

After domestic *water* has been used, it returns to the environment. It might be released into a *septic system* or a *sewage treatment facility*. A septic system releases the water into the *soil,* and it might end up in the *groundwater*. The treatment system returns water to streams, rivers, or the *ocean.*

Domestic wastewater includes large amounts of *organic* matter (leftover food and human waste) that act as food for microbes. High concentrations of these microbes can deplete oxygen from the water, often destroying the *aquatic ecosystem.*

Domestic wastewater also contains soaps and detergents, some of which include chemicals that can affect aquatic ecosystems. For example, many detergents contain *phosphates*—a *nutrient*. When wastewater enters a pond or lake, increased concentrations of phosphates produces a population explosion of *algae* that fouls the water, resulting in *eutrophication* and the possible demise of the ecosystem. For this reason, many states and countries in the European Union (EU) have banned the use of phosphates in detergent and improved their *water treatment* procedures to remove these substances. [*domestic water use; water conservation; water use for human consumption*]

domestic water use

One of four types of *water use for human consumption*. A typical family of four in the United States uses more than 100 gallons of water each day for flushing *toilets* and another 100 for watering the *lawn*. About 75 gallons are used each day for bathing and 30 for *laundry*. Another 18 gallons are used each day for dishes and about 6 for *drinking* and cooking. More than 25 billion gallons of water are used each and every day in the United States for domestic water. Most domestic water comes from underground *aquifers.*

dominants

Some *ecosystems* have a single species that is so abundant it dictates the overall characteristics of the area. For example, the sugar maple is the dominant species in eastern U.S. *forests* and exerts considerable control over what other types of organisms can survive in the region. [*keystone species; apex consumers and ecosystems; niche, ecological; acoustic niche; organism, most abundant*]

doomsday seed vault

The Svalbard Global Seed Vault is commonly called the doomsday seed vault. Its primary function is to act as a backup depository of the world's crop seeds in case some catastrophe occurs, resulting in a complete loss of important crop seeds. It is located deep in a mountain on a remote island halfway between mainland Norway and the North Pole. Countries from all around the world send seeds to be stored in this five-year-old, refrigerated safe haven. [*seed banks and vaults; seeds and climate change, Project Baseline; biodiversity, increased disease with loss in; extinction, sixth mass; suburbia and biodiversity*]

Dot Earth Blog

An excellent <u>New York Times</u> *environmental blog* worthy of reading. Its goal is to "examine efforts to balance human affairs with the planet's limits." Written by environmental science author Andrew Revkin, the blog always contains up-to-date information about environmental dilemmas. His blogs are full of fact but always followed with his personal thoughts—probably the reason the blog was recently moved from the news to the opinion section of the paper. Many consider this, along with another <u>New York Times</u> environmental blog called *GREEN: A Blog about Energy and the Environment,* the two best blogs available. (At least until GREEN was discontinued in 2013.) [*Yale Environment 360 blog; blogs, green; microblogs, green; magazines, eco-; E Magazine; OnEarth magazine*] {dotearth.blogs.nytimes.com}

double genesis

See *weird life*.

double value coupon program (DVCP)

This program, now in about half of all U.S. states, allows food stamps to be used at local *farmers' markets*. This means local farms get an economic benefit from food stamps and the recipients can purchase locally grown, higher quality, often *organic foods*. The initial funding for the program came from Newman's Own Foundation. [*farm-to-table movement; farm-to-school movement; food desert; food miles; food's carbon footprint-the best and worst; fregan; hunger, world; locavore; Meatout Mondays*] {wholesome-wave.org/dvcp}

doubling time in human populations

The world's *human population* is growing at around 1.2 percent per year, which seems deceivingly small. But one of the easiest methods of understanding the magnitude of this seemingly small increase is to look at how long it takes for our existing population to double in size based on the growth rate. (This calculation does not take *emigration* or *immigration* into account.) Countries with high growth rates (4.0 percent or higher), such as Afghanistan, have a doubling time of only about 14 years. At the opposite end of the range, growth rates between 0 percent and 0.99 percent, such as the United States, Canada, and most of Europe, double roughly every 71 years. (Some countries, such as Sweden and *Germany*, have declining populations.) [*carrying capacity; overpopulation; China's One-Child Program; demographic transition, theory of*]

douglas fir

The Douglas fir is a classic *old-growth forest* tree, cherished by environmentalists and prized by foresters. A typical tree lives 400 to 1000 years and grows 300 feet tall and 50 feet around. These trees are found throughout the Rocky Mountains, from the Pacific Northwest, and down to western Texas. A living tree removes over 400 tons of carbon from the air during its lifetime and provides a habitat for hundreds of animals, small and large. Foresters are more interested with the fact that a single tree provides enough wood to build a single family home. [*ancient forests; deforestation; logging, illegal; redwood forest destruction; Save the Redwoods League; Luna Redwood; timber, DNA sequencing of*]

Douglas, Marjory Stoneman

(1890–1998) Marjory Stoneman Douglas was a writer and reporter and the founder and first president of Friends of the Everglades, a support group established to prevent the development of the *everglades*. Many people believed the Everglades (along with other *wetland* habitats) were nothing more than a snake-infested *swamp*. Ms. Douglas had other ideas, which she researched and wrote about in a book published in 1947, The Everglades: River of Grass. [*coastal wetlands; inland wetlands; estuary and wetlands destruction*]

Douglas, William O.

(1898–1980) Douglas was a U.S. Supreme Court Justice from 1935 to 1975, the longest serving in history. He is known as the justice who stood up for the defense of the environment. In the 1900s he declared equal rights for nature, which was a radical belief at that time. His statements about protecting public lands remain legendary, such as "The public domain was up for grabs and its riches were being dispersed by the federal bureaucracy to a favored few." He wrote numerous books that praised nature and considered its destruction by man an unpardonable sin. His most noteworthy books include A Wilderness Bill of Rights and Farewell to Texas: A Vanishing Wilderness.

His most important and lasting action was his position in the case of the *Sierra Club* v. Morton. He stated that the land needed to be represented in the court of law and deserved the same extended protections as afforded women and minorities. He agreed that the Sierra Club had the right to represent and speak for the land, a legal reprise of *Aldo Leopold's Land Ethic*. [*ethics, environmental; environmental movement; environmental justice; eco-services; safe and just operating space; tragedy of the commons*]

downcycling
See *glass recycling*.

dredging
Dredging is the mechanical removal of sediment that builds up at the bottom of a body of water. Once dredged, the sediment is disposed of. In industrial regions, this sediment often contains contaminated wastes. Dredging these bodies of water poses environmental problems in two forms. First, during the dredging process, much of the contaminated sediment becomes suspended and re-contaminates the *aquatic ecosystem*. Second, the contaminated sediment must be disposed of somewhere and becomes part of the *NIMBY* (not in my backyard) syndrome. [*running water habitats, human impact on; rivers in decline; eutrophication, cultural; hazardous waste; aquatic succession; deposition*]

driftnets
Driftnets are sophisticated fishing devices used to catch large volumes of tuna, salmon, and squid in a relatively short period of time. In addition to this intended catch, the nets entangle large quantities of fish and other *marine mammals* not sought after. This unintentional catch is called *bycatch*.

Driftnets are finely woven nylon mesh nets up to 50 miles in length that drop into the deep waters about 40 feet. As unsuspecting fish swim into these huge walls of mesh, their gills become entangled and they die from asphyxiation. It is estimated that 2500 ships put out more than 50,000 miles of driftnets and their coastal water equivalent, *gillnets,* every night. Hundreds of miles of these nets become entangled or lost at sea each year, where they continue to trap countless fish and marine mammals, called *ghost fishing*.

These nets not only decimate entire populations of targeted fish but also wipe out tens of thousands of bycatch species, including *dolphins, sharks, whales, sea turtles*, and sea birds. Tuna and dolphins often swim together in close proximity, and by the 1980s, the tuna industry was killing more than 100,000 dolphins per year in these nets. The killing of these nontargeted species in these numbers severely damaged many *marine ecosystems.*

In 1987, the United States enacted the Driftnet Impact, Monitoring, Assessment and Control Act, which limited the length of nets used in U.S. waters. In the early 1990s, public pressure in the United States resulted in many tuna canners selling *dolphin-safe tuna*, which affords some level of protection to dolphin, but it does not mean that no dolphin were caught in driftnets or *purse seine nets*. More stringent *eco-labeling* exists, such as the *Flipper Seal of Approval*, managed by an environmental *NGO*.

On the global stage, in 1992 the *United Nations* banned the use of drift nets in international waters. Since then, many nations have agreed to ban these nets in their coastal waters (called their *exclusive economic zone*), but other nations still use the device.

Unfortunately, even though there is now a ban on these nets in international waters—it's a big place and enforcement is rare, so their use continues albeit in lower numbers. [*tragedy of the commons*]

drinking water

Drinking water is a *domestic water use* and certainly the most critical to life, and it is the only use required for survival. (Other uses are flushing *toilets*, watering *lawns*, bathing, doing *laundry* and dishes, and cooking.) In *developing countries* throughout the world, safe drinking water is rare and waterborne diseases common.

In the United States, about half of the drinking water comes from *groundwater aquifers* (wells) and the other half from *surface waters* (lakes, rivers, etc.). Almost 90 percent of all this drinking water is supplied by municipalities that treat the water before it is made available. This almost always includes the use of *chlorination* as a disinfectant. People who have their own wells do not have treated water unless they do it themselves. In more developed countries with little fresh water—for example, many middle-eastern nations such as the United Arab Emirates—drinking water is supplied by *desalination facilities*.

Providing more clean, safe drinking water in developing countries is one of the stated goals of the *Millennium Development Goals*, one of the few that have met with considerable success.

Studies have found that most U.S. municipal drinking-water supplies contain more than 200 contaminants—including cleaning solvents, *pesticides, heavy metals, gasoline* additives, *refrigerants*, and *PERC*—that come from *agriculture*, factory wastes, consumer products, *runoff*, and *sewage treatment facilities*. These substances contribute to the *body burden* we all carry with us.

drip irrigation

Most types of *irrigation* methods lose large volumes of water by evaporation and seepage. In the 1950s, Israel developed the drip irrigation method, where perforated pipes are installed just below the soil surface in the root zone of the plants. As the water drips out, it is delivered directly to the plant roots, reducing the amount of water lost from the usual 50 percent to less than 25 percent.

Drip irrigation is one specific type of irrigation from a new category called micro-irrigation. These types of irrigation are gaining popularity and have increased from about 6000 to more than 38,000 square miles in recent years, although it remains a small portion of the total irrigated land. *Water shortages* are projected to significantly increase over the next decade, which will probably result in greater demand for these more efficient irrigation techniques. [*conservation agriculture; low-input agriculture; permaculture; precision farming; regenerative agriculture*]

drones, conservation

Drones—small pilotless aircraft—became well known in wartime regions but are now finding many peaceful applications as well. Imagine large nature preserves where rare, *endangered species* live in what is supposed to be a protected area; except for poachers who can make the equivalent of 30 years of income in one night by killing a *rhino* and selling its horn. The *ivory trade* and *poaching,* in general, are big business. Even if the animals are protected, there are never enough guards to cover these large areas where the endangered species roam.

Enter, new high-tech drones that can cover 100 times as much territory as one armed guard. Because the areas can be so vast, animals can be fitted with radio-frequency ID tags the drones can quickly locate. When someone monitoring the drones sees poachers entering the preserves, the armed guards can be sent to the proper location immediately. The amount of protection that can be provided in this manner can be a game changer to protecting rare species.

As an example, the 140-square-mile Ol Pejeta Conservancy in central *Kenya* is home to some of the rarest animals on Earth, including the only seven remaining northern white rhinos on the planet. The conservancy raised $35,000 through *crowd funding* to purchase their first drone. Just the mere fact that poachers hear a drone flying above acts as a deterrent. In the *Arctic*, drones are planned to watch for *oil spills* and to track ice floes and migrating *whales*. The possibilities for environmental protection are enormous. [*sea gliders; beetlecam; robotic bees; apps, eco-; biosensors; biotechnology; cloud computing and the environment; Landsat; nanotechnology; satellite mapping; Wide-Field Infrared Survey Explorer*]

drought, the latest

Abnormal dry weather over long periods, because of a lack of rain, result in drought. Long lasting drought has severe impact on natural *ecosystems* and on *agriculture*. The 2012 to 2013 North American Drought, as some are calling it, has much of the western half of the continental United States in drought conditions. It began in the south and has gradually moved northward, damaging crops and reducing water supplies. It is considered worse than the last serious drought, 1988 to 1989. The only two U.S. droughts considered worse where in the 1950s and in the infamous 1930s, known as the *Dust Bowl.* [*soil; soil conservation; soil erosion; topsoil; desertification*]

dry cleaning

The most common cleaning agent used for dry cleaning—which is not a dry process because it uses water—is perchloroethylene, or perc for short. Perc causes dizziness, nausea, and headaches and is listed as a possible human *carcinogen.* It slowly evaporates from the clothes into the air, where it creates a sharp, sweet smell and can collect and become a danger to those inhaling it. Perc is dangerous to anyone working in these facilities, but it remains on clothes and can contaminate indoor air, so it is a threat to others as well.

A so-called safe alternative method of dry cleaning, called "hydrocarbon" cleaning, is no safer than the original method. However, newer, greener chemical agents are used as alternatives to perc. An alternative to "dry cleaning," called "wet-cleaning," and another called "carbon dioxide cleaning" are both considered safer.

Unfortunately, at this time, there are no *eco-certification programs* or *eco-labeling* efforts to indicate the safety of such a process. Some states—California, for example—have passed a ban on perc for dry cleaning, requiring facilities to begin phasing in safer alternatives over the next few years. All municipalities are supposed to test for concentrations of perc in their *drinking water*, but many do not. [*hazardous waste; outgassing; sunsetting; toxic pollution*]

dry deposition

See *acid rain.*

dual-use policy

First described by *Gifford Pinchot*, this is the overriding policy of the *U.S. Forest Service:* the use of national forests, as well as other *public lands*, for two opposing purposes. The two purposes are *preservation* of these natural resources and exploitation of them for economic benefit. This dual use has been debated for more than 100 years and continues to this day. [*forest management, integrated; forest health, global; John Muir; wise-use movement; multiple use policy; conservation; utilitarian environmental ethic*]

Dubos, René

(1901–1982) Dubos was a French-American microbiologist, ecologist, educator, and Pulitzer Prize author. He changed the study of microbiology by studying *bacteria* in the *soil* instead of in a laboratory, where everyone else did all their research at that time. He performed much of the ground-breaking work on *antibiotics*—produced by microorganisms—to treat wounds. He is credited with coining the now popular phrase "Think locally. Act globally." He founded the Dubos Center for Human Environments. Some of his work is considered controversial because he believed man-made environments could be as beneficial as natural environments. However, he is considered an important figure in the *environmental movement*. [*medical ecology; philosophers and writers, environmental*]

Ducks Unlimited

Founded in 1937, this *NGO* is a leading protector of *wetlands* and waterfowl resources. Although hunting is their primary focus, they have done a great deal to protect these important *habitats*. {ducks.org}

duff

When organic litter (dead leaves, bark, twigs, for example) decomposes to the point that the source of the litter is no longer discernible, it is called duff. [*soil horizons; soil profile; topsoil*]

dulosis

A relationship in which worker ants of one species capture the brood of another and rear them as slaves. [*ant slaves; hyperparasites; zombie bees*]

Dust Bowl

Between 1933 and 1939, about 150,000 square miles of farmland in the Great Plains of the United States was denuded of *topsoil*. Huge dust storms turned day into night. The land had been overplowed and overplanted for decades, turning the once fertile soil to powder. On May 11, 1934, one storm is believed to have blown away 300 million tons of topsoil. In many of the affected states, 60 percent of the entire population was forced to move to other locations in search of a livelihood.

The causes of *soil erosion* during the Dust Bowl are understood and can be significantly controlled by *soil conservation* measures, including *conservation tillage*, *contour farming*, *terracing*, *strip-cropping*, and *windbreaks*, among others. Poor farming practices, however, still exist on many U.S. farms and especially in many *less developed countries*, resulting in devastating losses of topsoil on a global basis. [*desertification; precision agriculture; drip-irrigation; weathering; Natural Resources Conservation Service*]

dust dome

Heat produced and absorbed by cities forms an *urban heat island*. Under certain weather conditions this heat island generates its own air currents, acting like a microclimate above the city. The circulating air in this heat island traps *pollutants* and dust particles, forming a dust dome or bubble over the city and significantly increasing levels of *air pollution*. The dome is usually blown away when a strong cold front moves through the area. [*urban sprawl; Alliance for a Paving Moratorium; eco-cities; urbanization and urban growth*]

Dutch elm disease

A fungal infection that kills all species of elm trees and has wiped out entire populations in some regions. The *fungus* is carried by the elm bark *beetle*, which lays its eggs in the tree's bark. The fungus gets into the tree and spreads throughout; leaves soon die, then entire branches, and finally the entire tree.

The first outbreaks started in the 1930s in the Netherlands and spread to Europe and North America, in most cases destroying almost the entire populations of the English, Dutch, and American elm trees. Although there is no cure, preventive measures can be taken to lessen the likelihood of the tree's demise. [*ash tree demise; fire blight on apples and pears; biological control methods; Bt*]

E10 and E15 gas

E10 is a mixture that contains 90 percent gasoline and 10 percent *bioethanol*, a *biofuel*. E15 contains 15 percent of the bioethanol. E10 is now the standard *gasoline* formulation sold in the United States, but a push has been to switch to E15. However, some people are concerned that the E15 mix will harm older, pre-2001 cars. [*alternative fuels; biodiesel fuel; biofuels, second generation; biomass energy; corn to biofuel; flex-fuel engines; plant-oil fuel*]

E85 gas

E85 is a mixture of 15 percent regular *gasoline* and 85 percent *bioethanol*, a *biofuel*. *Flex-fuel vehicles* can run on E85 fuel.

Earth Communications Office (ECO)

ECO is an *NGO* of actors, movie directors, producers, and writers dedicated to making the public aware of environmental concerns. Because celebrities readily receive attention, they use TV, movies, online venues, and magazines to heighten the public's awareness on environmental issues. They believe the entertainment industry can greatly influence the public by conveying these messages through public service announcements as well as incorporating environmental issues into their story lines. [*movies, environmental; Environmental Media Association; grant-making foundations; Lighthawk, the Environmental Air force; News Service, Environment; Student Environmental Action Coalition; Student Conservation Association*] {oneearth.org}

Earth Day

The first Earth Day was the brainchild of Gaylord Nelson, a senator and governor from Wisconsin, and was held on April 22, 1970, across the United States. Many people credit the first Earth Day as the unofficial beginning of the modern-day *environmental movement*. Millions of individuals, businesses, and government agencies participated in *environmental education* and activism. Polls indicated a dramatic increase in environmental awareness following the event, which has been held every year since.

The first Earth Day, along with the release of books such as _Silent Spring_ and The Population Bomb, was followed by the creation of the _U.S. Environmental Protection Agency_ and a surge in environmental legislation. By 1990, the 20th anniversary of the first Earth Day, it had evolved into a worldwide event for environmental awareness and education.

In recent years the enthusiasm is not what it used to be, but the day is still celebrated and will probably surge once again, because interest appears to be cyclical. In recent years, more emphasis has been placed on global _climate change_ along with the continuing emphasis on educational activities at the local level. Some large-scale initiatives have been successful, such as the Canopy Project with its goal to plant ten million trees, kicked off with the release of the movie Avatar.

Earth Day activities are organized by the _NGO Earth Day Network_. In 2013, the theme was "The Face of Climate Change." They state that they worked in 92 countries with more than 22,000 partners and that a billion people took some part in that year's event. They also run a program called A Billion Acts of Green, which is the largest green service campaign in the world. It works to inspire individuals, businesses, and governments to perform environmentally worthwhile acts. They have already reached their one billion mark and now are working toward two billion. {earthday.org}

Earth First!

Originally founded as a small group of environmental extremists in 1980, this group promoted _biocentrism_ through education, litigation, and civil disobedience. They became front page news in the 1980s by _tree-spiking_ and other acts they called _ecotage_ or acts of environmental sabotage, also called _monkeywrenching_. They were influenced by the _deep ecology_ philosophy. Even though no longer front page news, the group still exists and has chapters in many countries within and beyond the United States. Currently, the group continues to operate, with newer missions such as the _Marcellus Shale_ Earth First! group that temporarily shut down a _fracking_ facility in Pennsylvania. [_Hill, Julia, "Butterfly"; Abby, Edward_] {earthday.org}

Earth Island Institute

This _NGO_ was founded by _David Brower_ in 1982. As stated on their website, the group's "Project Support program acts as an incubator for startup environmental projects, giving crucial assistance to groups and individuals with new ideas for promoting ecological sustainability." They have created and managed well over 100 innovative projects, each focusing on a major environmental concern. All the projects are education oriented. Since its inception in 1982, the Institute has established a respected worldwide network of environmental leaders and has more than 30,000 members. [*environmental movement*] {earthisland.org}

Earthrise photograph

The picture was taken on Christmas Eve 1968, during the Apollo 8 mission to the moon. It shows a small but vivid blue Earth rising above the stark landscape of the moon—both against the jet black void of space. Many consider this to be one of the most influential photos ever made. It gave birth to analogies such as *Spaceship Earth* and impressed anyone who bothered to think about just how amazing Earth is and how vulnerable an existence we all live in. [*Blue Marble photographs, original and update; environmental education; environmental movement; outer space; near earth objects; space debris; Ansel Adams*] {nasa.gov/multimedia/imagegallery/image_feature_102.html}

Earth Summit

A name commonly used for two United Nations conferences that brought together world leaders to discuss and try to agree on how to deal with global environmental issues. They have had great success "discussing" but little success agreeing. The first Earth Summit, held in 1992, was properly called the *United Nations Conference on Environment and Development,* and the second, in 2012, was properly called The *United Nations Conference on Sustainable Development.* You also hear people call these conferences the Rio and the Rio+20 summits because both were held in Rio de Janeiro, 20 years apart.

EarthWorks

Mineral exploitation causes serious environmental damage and most environmentalists believe it to be poorly regulated. EarthWorks (formerly the Mineral Policy Center) is an organization dedicated to cleaning up the impact of *mining* in the United States and assuring that the public has a say in the process. They are concerned with the destruction of land by mining and the *toxic wastes* that are often a result of these operations. One major goal of the Center is to replace the *Mining Law of 1872.* { worksaction.org}

earthworms

In their quest for food, earthworms eat through *soil*, mixing *inorganic* with *organic* material. Their tunneling improves the soil's ability to absorb air and drain water. Two-and-one-half acres of rich soil can contain 500,000 earthworms that will eat through 10 tons of soil a year.

As the soil passes through their guts, they add microbes that help decompose organic matter such as animal wastes and decaying plant matter, releasing the *nutrients*. Then it is excreted as castings, which contain the decomposed organic matter and microbes as a crumbly textured, well-conditioned soil for plant growth. These castings are indicative of healthy soil—the more the better.

Earthworms are essential to *composting* organic matter. Most composting is done on a small scale, but some cities and *factory farms* are planning *large-scale composting* facilities where earthworms play a key role.

e-bikes

Bicycling is hugely popular in many countries and some U.S. cities. A more recent trend has been to motorize these bikes with small electric motors, called e-bikes or motorized bikes. The battery power contributes to pedal power only when needed. Estimations are that more than 430,000 of them were sold (mostly in Europe) in 2013. In France, regular bike sales dropped by 9 percent, while e-bike sales rose by 15 percent. Sales are expected to climb as people who wouldn't typically ride a strictly pedal-powered bike—as an alternative to a car—might drive an e-bike. The next step is for e-*bike-sharing programs* to sprout up just as they have for regular bikes.

eclosion

An adult *insect* emerging from a pupal case. For example, a butterfly emerging from its cocoon. [*metamorphosis; migration*]

eco-

"Eco" is derived from the Greek word for house. The prefix has become the buzzword—or "buzzprefix"—for anything pertaining to our *environment*. There are *eco-preneurs*, *eco-magazines*, *eco-warriors*, and *ecotourism*, just to name a few. Eco- is also often used in many *green marketing* campaigns to make consumers believe the manufacturer is environmentally concerned. But just using the prefix eco- does not make something truly good for the environment— it might be just a form of *greenwash*.

eco-conservative

A person who is cautious about making changes that might harm our *environment*. An eco-conservative only allows change to occur if there is a scientific basis that the change will not have a negative impact on the planet. Eco-conservatives allow or disallow changes to occur about the environment by voting for or against politicians and referendums at the local, state, and federal levels. [*environmental literacy; precautionary principle; consumer, green*]

eco-certification

Consumers want companies to be environmentally responsible, both in how they operate their businesses and how they make their products or provide their services. Companies are learning that consumers award this *corporate social responsibility* with their wallets. *Green consumers* patronize companies that help protect our environment.

Eco-certification is a way for companies to prove they are environmentally and socially responsible. Certification organizations provide specific criteria, describe what a company must do to meet these criteria, and assess whether they do so. Meeting these requirements results in some form of *eco-label* that companies proudly advertise. Some eco-labels are more reliable than others. It is not uncommon for some certification programs and the eco-labels they dispense to be a form of *greenwashing*, and should be considered nothing more than paid advertising.

So the question becomes—how do you know if an eco-certification organization and its eco-label are meaningful or not? A number of them are listed below and throughout this book. However, if you see one you are not sure about, you can find out if it is real or greenwash.

Most eco-certification programs are scrutinized by *NGOs* or nonprofit organizations. For example, the NGO ForestEthics does a lot of great work, one of which is analyzing eco-certification programs that provide eco-labels pertaining to *forest sustainability*. To be sure the certification you find on a product you plan to purchase is not causing deforestation, you can check with them. In this particular case, you will find that of the two biggest forest certifications, one is considered far more reliable than the other.

An easier method, at least for some types of products and services, is to look at *environmental rating systems* that directly rate products, foods, and services as they pertain to environmental harm and often health.

The following eco-certifications are respected in their fields: Totally Chlorine Free (paper), Recycled Content Certification (paper), Processed Chlorine Free (paint), Global Organic Textile Standard (clothing), SCS Certified Biodegradable (cleaning products and wood), Green Seal (many products and services), Basel Action Network (electronics recycling and disposal), *Forest Stewardship Council* (forestry, wood, and paper products), *Rainforest Alliance* Verified (tourism and forest products), Rainforest Alliance Certified (paper and wood), *Non-GMO Verified* (personal care products), USDA National Organic Program (clothing), *Fair Trade Certified* (clothing), Fair Labor Practices and Community Benefits (clothing), EPA-*Energy Star* (energy efficiency), *EPA* Design for the Environment (cleaning products), EPA Smartway Vehicle Certification (auto fuel efficiency), EPA Watersense (water use), EPA Burnwise (wood-burning energy efficiency), and EPA Environmentally Preferable Purchasing (many products and services).

eco-cities

About 75 percent of all people living in industrialized countries and 50 percent of those in the *less developed nations* live in cities. Proposed cities that work with the environment, instead of against it, are referred to as eco-cities and will require what is called *smart growth*.

Most ecologists that study *urbanization* don't think anything is inherently wrong with cities, but something is wrong with the way they are currently designed and built. Eco-cities emphasize sustainability by conserving land and other resources and polluting less. One of the primary offenders in cities is the need to commute and the *automobile* that fulfills this need. Eco-cities would be designed to contain relatively dense populations, reducing the need to get from here to there. *Walking* and *biking* should be the normal forms of *transportation* to and from work or for shopping. *Mass transit* would be the only acceptable alternative. This means homes, businesses, and shopping areas must be close to one another. They would also be near resources such as water and farms so they would not have to be transported in.

Most eco-cities are still on the drawing board, but some do exist. Songdo in Incheon, South Korea, has 30,000 residents and plans to grow to 65,000 upon completion in 2017. One of the more unique aspects of this city is that waste is collected by a tube rather than a truck; something like a large version of the pneumatic tubes used at drive-in teller windows.

Fujisawa Sustainable Smart Town in Wanagawa Prefecture, Japan, is scheduled to open soon but with only 3000 residents. It will include features such as *solar panels* on every house that are all connected to and work with the *power grid*.

Probably the largest of all planned eco-cities is Masdar City in Abu Dhabi, United Arab Emirates. If the money being invested is any indication of the effort, it is huge. It seems odd that one of the richest oil countries on Earth (it ranks 6th in *oil reserves*), would be interested in creating a city using exclusively *renewable energy* sources. But they are obviously looking to the future regarding the health of our planet and the fact that *fossil fuel* supplies are finite. With a budget of about $18 billion, the city will be run totally on renewable energy and is planned to have a zero *ecological footprint*.

At the moment, Masdar City looks like a university town because that is the only facility currently up and running. But it is still in development and will hold about 40,000 residents in a two-square-mile area, with the very latest in all forms of smart growth and green living.

ecofeminism

First coined in the early 1970s, this social or intellectual movement is very diverse, but central themes can be found throughout. One aspect is a strong concern for the environment and that women are inherently closer to the Earth and nature than men. Many in this movement regard a society historically controlled by men as one of the reasons for the continued degradation of our environment. Some see a link between the oppression of women and the exploitation of nature. Some within this movement believe women are more in tune with nature and more likely to want to protect it and therefore of central importance to protecting our environment. Vandana Shiva, Elizabeth Dodson Gray, Ynestra King, and *Carolyn Merchant* are prominent thinkers within this movement. [*ethics, environmental; environmental justice; eco-psychology; Women's Earth Alliance; women farmers*]

eco-labels

A label informing consumers that a company, or their products or services, has met specific environmental requirements. They are usually awarded by *NGOs* or government agencies. Companies use these labels to advertise their concern about their environmental impact (*ecological footprint*) and their *social corporate responsibility*. The first such eco-label was created in Germany in 1977, called the Blue Angel program. The United States and other countries began picking up on the idea in the 1980s, and by the 1990s, it was widespread. *Food eco-labels* and *organic food labels* have become especially popular.

Even though many eco-labels are valid ways to recognize environmentally conscious companies, others are nothing more than *greenwash*. Great care must be taken to ensure you are not simply listening to a new form of paid advertising. Luckily, many groups now provide lists of eco-labels and report on their veracity. One of these respectable *environmental rating systems* of eco-labels is The Ecolabel Index, a global directory that tracks more than 400 eco-labels from 197 countries. Others include the NRDC Label Lookup, ForestEthics, Consumer Reports Green Choices, Global Ecolabeling Network, International Trade Centre, and ISEAL Alliance. [*seafood eco-ratings*]

ecological footprint

A tool commonly used to measure how much of our planet's natural resources are required by some entity—such as a person or a country—to live. It was first developed by William Rees and Mathis Wackernagel. By resources, they include both the resources used (land, water, air) and the waste products produced that must be properly disposed of.

This measurement originally was based on how many hectares were needed per person to live. In 2003, it was calculated to be 2.2 hectares (about 5 1/2 square miles) per person, as a world average. However, possibly more important was the difference between developed and less developed countries. In Africa, it was 1.1 hectare per person compared to the United States, where it was 9.6 hectares per person, because of lifestyle differences.

In the United States—contrary to what many people think—those living in cities typically have smaller ecological footprints than those in the country. This is because a large percentage of city residents do not have *automobiles*, live in smaller homes, and don't have *lawns* to maintain—three important factors. Many reports suggest that residents of New York City have the smallest ecological footprint in the United States.

Possibly the most interesting way to use this measurement is to look at the ecological footprint of all people on Earth. In 1960 the global ecological footprint for everyone on our planet was about 50 percent of our planet, meaning we all used about half of the resources the Earth had to offer. During the 1970s our footprint was just about 100 percent, meaning we maxed out what the planet could provide for us on a sustainable basis. In 2013, humans have a footprint of about 150 percent, meaning to be sustainable we would need another half-Earth, which is not available at this time.

Today the ecological footprint concept has become a great way for anyone to get an idea about how much impact they, or their family, or a company, or a country have on the planet. A variation has recently become popular—the carbon footprint. This is a subset of an ecological footprint that only measures how much carbon someone produces as it relates to *climate change*. This too is useful although not as comprehensive. Numerous online calculators are designed to help you determine your ecological or carbon footprint. [*human population throughout history; population growth, limits of human; safe operating space; safe and just operating space*]

ecological restoration

An attempt to return an *ecosystem* to its original condition, including the native *biodiversity*. The restoration might be needed because of human activities such as *logging*, *dams*, and *overgrazing* or because of natural phenomenon such as hurricanes, floods, or fires. The science of restoration ecology is the study of ecological restoration via human intervention. [*virgin habitat; wilderness; no net loss*]

ecological studies

Ecology can be studied from many angles. It can be studied according to the *habitat* of interest (terrestrial, *marine*, and *freshwater*), by the types of organisms of interest (plants, animals, microbes), by the *level of biological organization* (from a single organism to an entire *ecosystem*), and by the methodology used during research. All of these approaches are usually used in conjunction with one another during ecological research.

ecological study methods
Ecology can be studied in many ways. These various methods can be grouped into one of three approaches or methodologies: *descriptive ecology*, *experimental ecology*, or *theoretical ecology*. Most early ecological research was descriptive (qualitative) in nature, but gradually changed to become primarily experimental and theoretical (quantitative). Most recently there has been a return back to descriptive ecology and a need for field scientists to study the *biodiversity* of the planet and all of the challenges facing life on Earth.

ecological succession
See *succession*.

ecological transparency
This refers to providing consumers with information about the environmental impact of products so they can make intelligent purchasing decisions. The premise of ecological transparency is if consumers know about the *life cycle* of a product and the harm it causes the environment, they will move toward more environmentally friendly products and can make a significant difference on a daily basis.

Ecological transparency is one of the driving factors behind the *sustainability imperative* and the growing number of *green consumers*. As consumers become more intelligent about products and their environment, companies must adapt and accommodate those consumers. [*green marketing; green products; green scam; environmental literacy, planned obsolescence*]

ecology
Ecology is the study of the relationships that exist between all the components of an *environment*. This includes interactions of organisms with other organisms and with the nonliving components of the environment, such as the geography and climate of a region. You can think of the environment as a set of dominoes and ecology as a study of the domino effect. [*eco-; environmental science*]

economics, ecological
This is considered to be an entirely new field of economics—not a subfield of it. It assumes the perspective that the conventional economic model is unsustainable, and it looks for other models, such as *steady-state economics*.

Ecological economists believe rising populations along with rising global consumption cannot continue without environmental degradation and destruction; an economy cannot be based on constant growth when our planet's resources remain fixed. The idea became popular with the publication in 1973 of E. F. Schumacher's book Small is Beautiful. [*Beyond GDP; CERES; climate capitalism; companies, top ten green global; corporate social responsibility; corporate sustainability rankings; economics and sustainable development; economics, environmental; economy vs. environment; Genuine Progress Indicator; green net national production; Sustainability Imperative; integrated bottom line; natural capital*]

economics, environmental

A subfield of economics that takes into consideration the value of nature, something conventional economics traditionally does not.

Environmental economics measures and analyzes the costs of using a natural resource and the environmental degradation that comes with its use. This model places value on natural services, often called *eco-services,* and applies them to their economic models.

This should not be confused with *ecological economics* that sounds similar but is very different. Where environmental economists believe the conventional economic system can be sustained indefinitely, as long as the value of nature is recognized and becomes part of the balance sheet, ecological economists believe an entirely new model is required. [*Beyond GDP; CERES; climate capitalism; companies, top ten green global; corporate social responsibility; corporate sustainability rankings; economics and sustainable development; economy vs. environment; Genuine Progress Indicator; green net national production; integrated bottom line; natural capital*]

economy vs. environment

Many people are under the false assumption that whatever is good for the environment is necessarily bad for the economy. Concerns about the quality of our environment have created its own booming global business. Professions in the environmental sciences—especially environmental cleanup, engineering, and renewable energy—are fast-growing sectors in the U.S. economy.

New technologies in environmental and related sciences are helping fill the decline in the aerospace and defense industries and continue to expand, producing many jobs.

The main problem is not the environment vs. the economy; instead, it is the age-old problem of change. People who earned a living tending horses or building carriages probably believed the *automobile* was bad for the economy. Environmental protection appears to be an enormous opportunity, not an impediment.

The following quote from *Thomas Berry* addresses the economy versus the environment folly. "The emerging *climate change* crisis arises from the simple question of whether economic profit or the integral functioning of the planet will be the normative value of guiding the human community into the future. Will the human economy be accepted as a subsystem of the Earth economy, or will the Earth economy be considered to be a subsystem of the human economy? The greatest folly of our times may be the setting of these two in opposition to each other, when obviously any conflict between the two is disaster for both." [*steady-state economics; Beyond GDP; CERES; climate capitalism; companies, top ten green global; corporate social responsibility; corporate sustainability rankings; Sustainability Imperative; economics and sustainable development; Genuine Progress Indicator; green net national production; integrated bottom line; natural capital*]

economics and sustainable development

Economic development is often accompanied by a decline in the quality of our environment. Until recently, most people had little concern for the negative environmental impact as long as the "progress" was good for the economy. Today, many people are trying to incorporate negative environmental impact into the economic picture. Removing a forest is good for the economy because it provides jobs, wood, and other materials, but if the land erodes and becomes unproductive for future generations, was it truly progress?

A new view of our economy is based on *sustainable development,* which takes the negative impact on the environment into consideration when assessing economic growth. Sustainable development has been defined as addressing present needs without compromising future needs. Instead of using economic indicators such as the *GNP (gross national product)*, new methods are slowly beginning to be used, such as the *Index of Sustainable Economic Welfare* (ISEW) that balances the economic worth of a product or service against environmental loss. [*Beyond GDP; CERES; climate capitalism; companies, top ten green global; corporate social responsibility; corporate sustainability rankings; eco-services; Genuine Progress Indicator; green net national production; integrated bottom line; natural capital; steady-state economics*]

ecopreneur

An entrepreneur who specializes in environmental products or services. [*green marketing; green products; green consumers; environmental literacy; greenwash; green scam*]

eco-psychology

A field that suggests a close relationship between the Earth's well-being and personal well-being; that the needs of the one are relevant to the other. [*Biophilia Hypothesis; Gaia Hypothesis; ecofeminism; ethics, environmental; ethnobotany; nature-deficit disorder; primitivism; self-concept and the human microbiome*]

eco-services

Our economic system has not provided, until recently, a way to place economic value on nature. When assessing the value of building a new housing development on wetlands, the economic value of the project will include things such as the jobs it will create, the business revenues it will generate, and the tax revenue the homes will add to the tax base. However, if the *wetlands* are destroyed, what natural services—now called ecological or eco-services—will be lost? That part of the equation was not taken into consideration until about ten years ago with the advent of eco-services—a way to assess nature's value.

One of many methods for doing this is to determine what will be lost and how much it will cost to replace. For example, if the wetlands protect the city from flooding after a storm, how much will it cost to build levies to prevent the flooding and how much will it cost to clean up after each flood? These costs are now considered the ecological or eco-services provided by nature and should be taken into account when economic growth occurs at the expense of nature.

If a lake becomes polluted due to *industrial waste*, what is the value of the eco-services that will be lost? Possibly, a million dollars in *commercial fishing* each year? The loss of recreation for the town's citizens and the revenue it brings in each year? If the lake is a spawning ground for fish that then go out to sea, how much is that loss worth?

Of course, this is a difficult task. What is the value of a city's clean water supply that might be jeopardized by *agriculture* or fresh air that might be threatened by industry? However, today many people are working on developing these numbers and how to apply them.

In spite of growing acceptance about eco-services, there is much criticism about assessing value on nature that comes free to all. *Aldo Leopold* described the problem many years ago. "One basic weakness in a conservation system based wholly on economic motives is that most members of the land community have no economic value. Wildflowers and songbirds are examples. Of the 22,000 higher plants and animals native to Wisconsin, it is doubtful whether more than five percent can be sold, fed, eaten, or otherwise put to economic use." [*tragedy of the commons; land grabs; risk assessment, environmental; risks, true vs. perceived; economy vs. environment; economics and sustainable development*]

ecosystem

A description of all the components of a specified area, including the living organisms and nonliving factors such as *air*, *soil*, and *water*, and the interactions that exist between all these components. These interactions result in a relatively stable assortment of organisms and involve the continuous cycling of *nutrients* between the components.

The area defined as an ecosystem is arbitrary. It can be a complex biological system, such as a *biome,* or a simpler *habitat,* such as a lake or *forest*. However, smaller entities, such as a rotting log, can be considered and studied as an ecosystem. [*environment; environmental science; ecology; anthrome*]

ecotage

An act of environmental sabotage, such as *tree-spiking*. [*terrorism, eco-; monkey-wrenching; eradicating ecocide movement; environmental movement*]

eco-terrorism

See *terrorism, eco-*.

ecotone

The boundary between two different *biomes*—acting as a transition zone. An ecotone often has greater *biodiversity* than the interior of either biome, due to the edge effect. This effect occurs for many reasons, but basically it is because individuals within this transition zone have the best of both worlds; they can take advantage of both as they move back and forth between the two. For example, birds often take advantage of the transition zone and are found at the edge of a forest or near a coastline. [*habitat; anthromes; succession; anthropogenic stress; ecological studies; bioregions; biosphere; zoogeography; zones of life*]

ecotourism

The business of nature travel. Areas that have unique natural wonders often center their economy on people willing to pay to see or participate in these attractions or activities. The entire economy of some countries revolves around ecotourism, such as in the Caribbean and Belize in Central America. Much of this tourism is based on attracting people from industrialized nations to visit *less-developed regions* where they can learn about local cultures, see beautiful natural resources, and enjoy unique activities. This can include everything from hiking, photography, and *wildlife* safaris to education about local art, music, dance, and so on.

Ecotourism has become a multibillion-dollar global business. But increased revenues are often accompanied by increased environmental degradation. Regions that are big on ecotourism must rigorously manage their natural resources to prevent the tourists they seek from destroying the attractions they come to see.

This model appears to work because regions that once had economies based on land destruction such as *clear-cutting* forests can now earn an income from tourism and do not need to destroy their land. Many villages have learned that one hunted lion shot in the wild might be worth a few thousand dollars to locals, but the tourism that lion can generate over many years will be worth many hundreds of thousands of dollars.

Some country's economies are already almost entirely based on ecotourism; others are trying to become ecotourism havens, such as *Laos* and Costa Rica. [*ecotourism, ten best; ecotravel; countries, ten greenest; geotourism; voluntourism, environmental; hotels, green; lodging, green; resorts, eco-; WWOOF; eco-kibbutz*]

ecotourism, ten best countries for

National Geographic ranks the top ten *ecotourism* countries based on efforts to protect ecological sites and species and promote sustainable tourism. They are 1) *Brazil*, 2) Dubai, 3) Canada, 4) Belize, 5) *Kenya*, 6) Gabon, 7) *Laos*, 8) Ireland, 9) Turks and Caicos, and 10) Greece. [*ecotravel; countries, ten greenest; resorts, eco-; hotels, green; lodging, green; voluntourism, environmental; volunteer research, scientific; WWOOF; eco-kibbutz*]

ecotravel

Traveling with a main focus of environmental awareness. Activities may be purely educational, such as studying *ecosystems* or indigenous peoples; hobby oriented, such as photo expeditions into exotic *habitats*; or thrill seeking, such as shooting the rapids or mountain climbing. Over a hundred private organizations in the United States specialize in ecotravel adventures. Travel agencies often have special ecotravel guides. [*ecotourism; resorts, eco-; ecotourism, ten best countries for; hotels, green; lodging, green; voluntourism, environmental; volunteer research, scientific; WWOOF; eco-kibbutz*]

eco-warrior

This was originally used for someone who attempts to protect the environment in a radical fashion, such as *ecotage* (eco-sabotage). It was popularized in the book Confessions of an Eco-Warrior, by Dave Forman in 1991. Today, however, almost anyone involved in environmental activism might be called an eco-warrior, including those who attend rallies or sign petitions. [*Earth First!; environmental movement; green consumer*]

ectoparasite

A *parasite* that lives outside its *host*, such as a tick. [*endoparasite; kleptoparasitism; hyperparasite; symbiotic relationships*]

edaphology

The study of *soils*, with a concentration on the use of the land for cultivation. [*agriculture; topsoil*]

eels and elver fishing

Eels are fascinating animals with complex life cycles—migrating between *fresh waters* and *marine* waters. Elvers are the immature eels. They are eaten as a delicacy in Asia—especially in *China* and South Korea—and are in great demand for sushi. They are raised in *aquaculture* farms in Asia but in recent years have been in short supply, forcing a search elsewhere. But with most of Europe having banned their capture to protect the wild populations, the search has moved to the United States.

The eels' life cycle means they spend only a limited time in U.S. coastal waters. The demand has driven wild populations into decline, forcing all but two states to make it illegal to fish for elvers. In Maine and South Carolina, harvesting adult eels (called sniggling)—as well as the elvers—is big business. During a shortage frenzy that occurred a few years ago, prices reached $2600 per pound! As expected, with demand that high, the eel population plummeted. Efforts to sustain the population, as well as the yield, meant these states had to begin requiring licenses for legal catches and increasing enforcement and fines for *poachers*.

As often happens, the *tragedy of the commons* could result in the local *extinction* of this species unless they are protected. The *U.S. Fish and Wildlife Service* has recently considered listing the American eel as a *threatened* or *endangered* species, which would end the hunting season in the United States.

effluent

The discharge and flow of liquid waste from a specific known source of pollution into the environment. Industrial effluent is one of the main causes of *water pollution* and was the prime target of the *Clean Water Act*. This type of pollution is called *point source pollution*. [*beach water quality and closings, U.S.; combined sewer systems;*

eutrophication, cultural; Great Lakes Water Quality Agreement; industrial water pollution; industrial water use]

egg farms

Factory farms for egg production are considered by many to be terrible places where the animals are treated inhumanely. There are about 270 million egg-laying hens in the United States alone. Pressure from groups such as the Humane Society, with support from the United Egg Producers, that represents most of these farms have (surprisingly) worked together to double the size of the living space allocated per hen plus a number of other measures to be phased in to make their lives a little less miserable. Not all in the industry are happy, fearing that this type of legislation can lead to similar initiatives for other livestock, such as pigs. [*CAFO; dairy cow output; farm animal cruelty legislation; rendering*]

Ehrlich, Paul R.

[1932–) Paul Ehrlich is a *population* biologist and *ecologist* who wrote about and popularized the concerns and effects of human *overpopulation* on the environment. Ehrlich's famous 1968 book <u>The Population Bomb</u>, coauthored with his wife Anne Ehrlich, warned the public about a forthcoming population crisis where *natural resources* would become exhausted and famine and instability reign. This thought is often called Neo-Malthusian after *Malthus,* who wrote in the 1700s about the *human population* overwhelming our natural resources.

Ehrlich continued this premise with other books, including <u>The Population Explosion</u>, published in 1990. He founded the *Zero Population Growth NGO* in 1968 to educate the public about this issue.

Some critics consider him to be a doomsayer and state that many of his predictions have not happened and population growth is leveling off naturally. However, many scientists and environmentalists continue to point out how our environmental problems are caused by our population's growth. [*baby boom; demographic transition, theory of; China's One Child Program; doubling time in human populations; human population throughout history; population growth, limits of human; population explosion, human; Population Reference Bureau; philosophers and writers, environmental*]

Eiseley, Loren

(1907–1977) Eiseley was an anthropologist, educator, and prominent nature writer. He has been called a modern-day *Thoreau.* His first and most famous publication in 1946 was <u>The Immense Journey</u>. This book demonstrated his skill to combine science with humanism. The book is a collection of essays, mostly about his early years living in Nebraska. He went on to write many books about the relationship between people and nature and became recognized globally for his work. [*philosophers and writers, environmental; anthropologists, environmental; Burroughs, John*]

electric power plants

Most electricity is generated by burning *coal*, the most polluting of all the *fossil fuels*. With recent newly found *reserves* of *natural gas* in the United States and the cost reductions that have accompanied it, a major shift is taking place. The amount of electricity created by coal-powered plants dropped by six percent from 2010 to 2011 and the amount created by gas-powered plants increased by three percent. A trend sure to continue, not only because of the price but also new legislation mandating reduced levels of *carbon dioxide* and pollutant *emissions*.

Some newer electric power plants, called *waste-to-energy power plants*, are fueled with *biomass* from *municipal solid waste*. [*alternative energy sources; energy consumption, historical; energy use; energy, A very brief history of U.S.; energy independence, U.S.*]

electric vehicle (EV)

EVs are not new. In the early 1900s about 50,000 electric vehicles were on the road, but the introduction of the internal combustion engine and cheap *gasoline* ended their popularity. EVs are once again becoming mainstream vehicles. Most *automobiles* today using battery technology are *hybrid vehicles*, but more and more are coming to market that are strictly electric.

Government pressure on car companies to reduce *emissions* and to improve mileage has spurred development, and high gas prices have made them appealing to the public. EVs produce *zero emissions* so they do not contribute to *air pollution* or *climate change*. However, the electricity they require for charging must be calculated into their *ecological footprint*, as well as their *cradle-to-grave* impact, which is different from regular combustion engines because of their batteries.

Improvements in battery technology are the driving factor to their recent success, because it is the battery that determines how far they can go on a single charge. The technology has been changing quickly over the past few years—today almost all EVs use lithium-ion batteries. However, this technology is thought to be maxed out and not likely to be improved any further. The next contender is probably the lithium-air battery. This would have much better energy density (how much energy can be packed into the weight of the battery) than previous batteries. Another possibility is the multivalent-ion battery, which uses magnesium as well as aluminum. [*automobile propulsion systems, alternative; batteries, automobile; CAFÉ; highways, electric vehicle; electric vehicles, charging; range anxiety; tire, high mileage]*

electric vehicle highways

See *highways, electric vehicles.*

electric vehicles, charging

Most *electric vehicles* on the road today are charged with the same electric outlets found in your home, using alternating current (AC). The problem is that the power stored in an electric car battery uses direct current (DC), so when an electric vehicle is charged, the AC power is converted to DC. This slows the process down.

Recent technological advances now make it possible to charge electric vehicles by using direct current (DC), charging an electric car in far less time. Both General Motors and BMW have announced plans to adapt cars for a DC charge. Some of the new *electric vehicle highways* are being fitted with this newer charging technology.

There is also a new, two-way car-charging system. The University of Delaware, in concert with the local utility company, has rolled out an experimental program called the Vehicle to Grid fleet of electric vehicles. This program provides electricity back to the *power grid* instead of only drawing power from it. When the grid is down or strained, these cars can be plugged into a special two-way charging station that actually takes the energy produced by the car and returns it to the power grid. Theoretically, you could power your home with your car, if your home loses power. [*automobile propulsion systems, alternative; batteries, automobile; CAFÉ; highways, electric vehicle; range anxiety; intelligent vehicle highway systems*]

elephant ivory, dating

A new technique has been developed that dates when an ivory tusk was removed from the *elephant.* This is helpful because ivory taken prior to 1989, when the *CITES* ban was enforced, is legal. Any ivory collected after that was illegally collected by *elephant poaching.* This method works by determining the exact amount of a radioactive substance, a specific form of carbon 14, present in the ivory. Carbon 14 became prevalent in our environment from 1952 to 1962 when atomic bomb testing was performed. After the testing was discontinued, the amount of this radioactive material declined in a known curve from that time to now. Testing how much of this substance is in the ivory determines within one year if it was taken legally or as an act of *poaching.* The method is now used to convict poachers. [*timber, DNA sequencing; endangered and threatened species lists*]

elephant poaching

Elephants continue to be slaughtered for their ivory at a sickening pace. In 2011, over 25,000 were killed and 30,000 in 2012. Today there are about 500,000 elephants left in Africa—only one-tenth of what there were 75 years ago. The illegal *ivory trade* (*poaching*) has exploded for primarily one reason: the insatiable desire by the Chinese, where an exploding population of middle class can now afford the high price of ivory. Over 80 percent of Chinese citizens recently surveyed stated they planned to buy ivory goods in the near future. This has driven the price of ivory at times to over $1000 per pound, making it a lucrative business for poachers who are willing to risk their lives for such large financial rewards.

The ivory trade has been banned by *CITES* since 1989, but that doesn't stop the illegal trade. In some countries, poaching goes on unabated. In the Central African Republic, which is currently a rogue state, elephant poaching is rampant. Elephant meat is sold in open markets and the *World Wildlife Federation* offices have been attacked and looted multiple times and the workers evacuated. The tense security situation prevents wildlife officers from even going out to check for carcasses, let alone protect those elephants that remain. [*rhinoceroses and their horns; polar bear trade; wildlife trade and trafficking; Wildlife Conservation Society; whaling; United Nations Convention on Biological Diversity; International Union for Conservation of Nature; debt-for-nature swaps; Convention on the International Trade of Endangered Species; biodiversity loss, prevention of; endangered and threatened species lists; extinction, sixth mass; species recently gone extinct; United Nations Convention on Biological Diversity; habitat loss; animal law; Lighthawk, the Environmental Air Force; conservation drones*]

elephants

Elephants are the largest land *mammals* on Earth. At one time, hundreds of species of elephants existed; today there are two—the African and the Asian. Studies show elephants to be among the most intelligent mammals, along with *dolphins*, porpoises, and primates. They are social animals and communicate extensively between individuals.

The African elephant is classified as an *endangered species* according to the *U.S. Endangered Species Act* and the *International Union for the Conservation of Nature and Natural Resources Red List*. Also the *Convention on International Trade in Endangered Species of Fauna and Flora* (*CITES*) has enlisted this animal on Appendix I, meaning they need protection.

All elephant populations are under extreme stress because of *habitat loss* but more so from *elephant poaching* for their ivory tusks. Entire herds are being slaughtered just to saw off their tusks for the illegal wildlife trade. In the 1970s and 1980s this slaughter was rampant. In 1985, CITES began to control the ivory trade, and in 1989, the United States and the European Union banned all ivory imports. However, the illegal ivory trade and the slaughter of elephants continue today, if not at a pace worse than ever before. Entire herds, including numerous infants, have been found slaughtered with their tusks removed.

E Magazine
See *E-The Environmental Magazine.*

emergent plants
Aquatic plants that are rooted at the bottom but protrude through the water's surface, often to flower. Water lilies, cattails, and bulrushes are all emergent plants. [*aquatic ecosystems*]

Emerson, Ralph Waldo
(1803–1882) Emerson was a writer, poet, environmental philosopher, and religious leader. He is considered the father of transcendentalism. He believed that the natural world is a revelation of the Divine and that experiencing nature is important to understanding the perfection of creation. Even though he was an advocate of protecting nature, he was not an environmentalist in the sense that we now consider one. His writings were more romantic and spiritual than of the real Earth. He was not one to enjoy the wild as did many of his contemporaries such as *John Burroughs* and *John Muir*. But neither of them attained Emerson's fame, either. [*philosophers and writers, environmental; Thoreau, Henry David; Berry, Thomas*]

emigration
Emigration is the movement of people out of their home country into other countries. *Immigration* describes movement into a country. Both emigration and immigration, along with fertility and *mortality*, dictate changes in the size of a population.

Emigration and immigration are typically only used for people. When other animals move, it is properly called *migration* such as birds migrating south for the winter. [*push-pull hypothesis; population stability; demography; population dynamics; population ecology*]

emissions
A general term used to describe any substance released into the *atmosphere* by humans. Pollutants such as *particulate matter*, gases such as *carbon dioxide* and carbon monoxide, compounds that contain nitrogen or sulfur, *heavy metals*, and *radioactive*

materials released into the environment from *incineration*, power plants, internal combustion engines, and other industrial processes are all considered emissions.

The term is usually used in a more specific fashion, such as *carbon emissions* and their relationship to *climate change* or smokestack emissions from an incinerator and their relationship to *air pollution*.

emissions trading

See *carbon markets*.

Encyclopedia of Life

This is the name of a robust online database used by scientists worldwide to share and study data about most species. Among other things, scientists use this to learn where specific species are located worldwide. However, anyone can use it by going to the site and typing in a species name. Text and multimedia links appear and give you a quick tour about what you seek. In the summer of 2012, the encyclopedia surpassed the one-million-species mark on its way to cataloging all life. Another database is currently being created called the *Map of Life*. [*biodiversity; Census of Marine Life*] {eol.org}

Endangered Species Act, U.S.

The U.S. Endangered Species Act became law in 1973 (it broadened the mandate of the original Endangered Species Conservation Act of 1969). This act does two important things. First, it empowered the *Department of the Interior* to create and manage *endangered and threatened species lists*. This is administered by the *U.S. Fish and Wildlife Service* (for all species other than marine life) and the *National Oceanic and Atmospheric Administration* (for marine species).

This act also gave the federal government jurisdiction over the management of any organism listed as an endangered species. It states that no government agency can perform any activity that would lead to the extinction of an organism on the list and that all government agencies must cooperate to prevent extinction. The original act had strong language and the clout to enforce the words.

Since its inception, however, the act has been underfunded and gradually watered down. In 1978 an Endangered Species Review Committee—also called the *god squad*—was established that could override this act if the economic advantages outweighed the negative ecological effects.

In addition, amendments to the Act made it harder to add species to these lists. As of 2013 there are about 250 "candidate species," meaning they are wait-listed. Almost 100 of them have been waiting for more than ten years and 73 have been waiting more than 25 years. While they wait, they are afforded no protection—so many will probably become extinct before they ever receive protection.

This act only pertains to U.S. species. The *International Union for Conservation of Nature (IUCN)* and the *Convention on International Trade in Endangered Species (CITES)* provide lists of species in jeopardy all around the world.

endangered and threatened species lists

Many organizations maintain lists of species in jeopardy of becoming extinct. The most important lists are the *International Union for Conservation of Nature* (IUCN) *Red List,* which covers species globally, and the *U.S. Endangered Species Act* list, which only covers U.S. species. In addition, the *Convention on International Trade in Endangered Species (CITES)* is an international agreement that lists species in jeopardy because of the *wildlife trade.*

These lists use similar but somewhat different ratings systems. For example, the IUCN uses the categories "endangered" and "vulnerable," whereas the Endangered Species Act uses the categories "endangered" and "threatened." [*biodiversity, loss of; biodiversity loss, prevention of*]

endemic

Something only found in a certain region. For example, many plants and animals are endemic to areas being deforested, or a disease may be endemic to a certain region. [*biodiversity hotspots; biodiversity, increased disease with a loss in; invasive species; native species*]

End of Nature

See *McGibbon, Bill.*

end-of-pipe technology

See *waste minimization.*

endoparasite

A *parasite* that lives inside its *host,* such as a tapeworm. [*ectoparasite; kleptoparasitism; hyperparasite; symbiotic relationship*]

energiewende

A *German* word meaning energy transition and used to describe Germany's aggressive plans to convert 100 percent of its energy to *renewable energy* sources. The plan's goals include being 40 percent on renewables by 2020 and 80 percent by 2050, with the final goal 100 percent. Although the word is German, many global players like to use it, because it conveys the world's need to move to renewable energy sources and to do so soon. [*energy, a very brief history of U.S.; energy sources, historical; unburnable carbon; biofuels, second generation; bridge fuel; soft energy path*]

energy, A very brief history of U.S.

Energy resources and *energy consumption* and are what drive civilizations. In the 1970s, the United States was a major importer of *oil* and *natural gas*. Oil prices continued to surge for many reasons but with an underlying belief that we had passed *peak oil* and supplies were going to decrease. Simple supply and demand made the price go up. This was also the case for natural gas. Large facilities were built to prepare to import large quantities of *liquefied natural gas*. Until about 2008, that is where the United States stood—dependent and paying for it.

Then, beginning around 2009, things changed when vast expanses of *shale formations* containing natural gas in the United States become available for extraction by new *unconventional technologies* called *fracking* and horizontal drilling. The change could not be more dramatic. Natural gas from the United States is now so plentiful that prices have plummeted. The facilities built earlier to import natural gas are now being retrofitted to export it to other countries. Unless fracking is stopped in its tracks—something that many environmentalists want to happen—the United States is much closer to reaching *energy independence* that it could have ever imagined only a few years ago.

In addition to natural gas, *tight oil* is now flowing within the United States, adding to our oil supplies. The result is a far more energy-independent country. However, many believe it a far more environmentally endangered one as well. These newer, unconventional extraction methods, such as fracking, are more environmentally harmful than conventional methods and there is an ongoing debate about whether it should continue. This brief history is still being written. [*unburnable carbon; biofuels, second generation; bridge fuel; soft energy path*]

energy consumption, historical

As civilizations became highly developed, the need for readily available fuels grew dramatically. Even though the world population doubled between 1900 and 1990, people consumed 12 times more energy during the same period. The twelvefold increase was not due to the increase in the number of people as much as a thirtyfold increase in the number of products produced during the same period.

About 80 percent of the world's energy supply today comes from *nonrenewable energy* sources—*fossil fuels*. With continued *population growth* and increased productivity, the demand for energy increases. *Climate change,* caused primarily from burning *fossil fuels*, needs to be addressed. Most environmentalists believe we must transition to *alternative energy* sources to prevent continued *global warming* and various forms of pollution and that, sooner or later, fossil fuels will become scarce.

Historically, the switch to an alternative energy source (once available), such as coal in the 19th century and oil and gas in the 20th century, takes about 50 years to be accepted, implemented, and put into large-scale use. Many people believe we do not have that much time for the transition from fossil fuels to renewable energy sources such as *solar power* and *wind power* before irreparable harm is done. [*unburnable carbon; biofuels, second generation; bridge fuel; soft energy path*]

energy content

See *vehicles, hydrogen fuel-cell.*

energy efficient states, top ten U.S.

The ACEEE (American Council for an Energy-Efficient Economy) publishes annual rankings of the most energy-efficient states. It uses criteria such as utility and public benefits programs and policies, *transportation* policies, building energy codes, heat and power policies, government-led initiatives, and many others. Their most recent ten most efficient states are 1) Massachusetts, 2) California, 3) New York, 4) Oregon, 5) Vermont, 6) Connecticut, 7) Rhode Island, 8) Washington, 9) Maryland, and 10) Minnesota. [*energy return on investment; energy independence, U.S.; energy use; power grid*]

energy (fuels)

Readily available sources of energy (fuels) are intimately related to the development of nations. These fuels have had a dramatic impact on the environment in the past and certainly will in the future. Sources of energy can be divided into two categories: 1) *fossil fuels*, which are *nonrenewable energy* sources and supply more than 80 percent of the world's energy demands and 2) *alternative energy* sources (alternatives to fossil fuels), which include many types of *renewable energy* sources and *nuclear power.*

energy independence, United States

For decades, the United States has talked, hoped, and worked toward energy independence—especially during times of unrest in the Middle East. We still have a long way to go before that happens, but sometime in 2014 the United States will export more *oil* than it imports—something unheard of only a few years ago. With newfound *reserves* of *natural gas* in *shale formations*, extracted by *fracking*, estimates have the

United States becoming energy independent somewhere between 2018 and 2030, depending on whose estimates you want to believe.

For this to happen, we would also have to find and extract more oil and gas in the Gulf of Mexico and with *Arctic drilling*—both actions that most environmentalists fiercely oppose because of the potential for catastrophic environmental harm. In addition, independence could only happen if *renewable energy sources* become a more important part of our total energy mix and if more energy is conserved—both actions that environmentalists applaud.

Many people had hoped our energy independence came about because of a shift from *fossil fuels* to renewable energy sources, but that is not turning out to be the case. Even though renewable energy sources are part of the equation, it turns out that new unconventional methods of fossil fuel extraction and new finds of fossil fuels will probably be the cause for energy independence. [*bridge fuel; energy, A very brief history of U.S*]

energy plantations
Farms that grow crops specifically for use as a source for *biomass energy*. [*biofuels*]

energy poverty
Refers to a lack of access to clean, affordable energy services for cooking, heating, lighting, communications, and productive uses. Estimates have 1.3 billion people globally without any electricity and another 1 billion without reliable access. Almost three billion are believed not to have access to modern fuels for cooking and heating and to rely on primitive *wood-burning cookstoves*. Almost 80 percent of people without access to electricity live in either sub-Sahara Africa or South Asia. The country with the largest number of people experiencing energy poverty is *India*, with 289 million without electricity. One of the simplest methods to reduce energy poverty is to provide improved cookstoves. [*energy sources, historical; methane digester; water shortages, global; food desert; brick kilns*]

energy pyramid
Energy pyramids illustrate how organisms obtain energy and how the total amount of energy changes as it flows through an *ecosystem*. Each level of the pyramid, called a *trophic level*, represents a different method of obtaining energy. The bottom level represents producers, called *autotrophs* (plants) that capture the sun's energy during *photosynthesis*, creating their own food. Each subsequent trophic level represents different types of consumers, called *heterotrophs*. The second level contains primary consumers, which eat the producers. The next level contains secondary consumers,

which eat the primary consumers, and this process proceeds as you go up the pyramid, trophic level by trophic level.

Organisms in each trophic level use most of the energy available to them. They use this energy for breathing, growing, moving, reproducing, making sound or possibly light, and all other of life's activities and functions. When a group of organisms at one trophic level eats those in another group, it only receives the energy that remains. About 90 percent of the energy that goes into each level is used up by that level or lost as heat. This leaves only ten percent of the energy for the next trophic level.

With less and less energy available at each level, the organisms at the higher levels have less food available to them. For example, a field of grass might contain hundreds of field mice, ten snakes, and one hawk.

Energy pyramids illustrate why it is more efficient to feed plant foods (grains) to people than it is to feed them meat. Assume a field yields 100 pounds of grain. Feeding people from this trophic level provides the entire 100 pounds of food. However, if the same 100 pounds of grain was fed to cattle, 90 percent would be used up, resulting in only ten pounds of meat left for people to eat. [*meat, eating; sustainable yield; food chain; food web; net primary production; niche, ecological*]

energy return on investment (EROI)

A newly created system used to determine the impact that different energy extraction methods have on our planet. It measures the amount of energy a fuel provides but subtracts the amount of energy it took to extract it—the higher the EROI, the more energy we get when taking into consideration the damage caused getting it. For example, *conventional oil* has an EROI rating of 16 to produce liquid fuels such as *gasoline*, while *unconventional tar sands oil* has an EROI rating of only 5. For another example, to generate electricity, the EROI rating for hydroelectric power is 40, for wind it is 20, and for coal, it is 18. [*energy, A very brief history of U.S.; energy consumption, historical; energy use*]

energy-saving apps and programs

Many *environmental apps* and websites are dedicated to helping you save energy in your home. In most cases, they link you, the home owner or renter, to their utility company and offer lots of energy-saving hints about their specific electric usage. The goal is to help you get your bills down by saving on energy costs. Examples include when to run appliances and when to replace them, adding energy saving devices, and using an energy company's *smart energy* program to reduce costs.

Some of these services, such as Earth Aid, use *social media* to share information and exchange ideas so the information not only comes from your utility company and the app itself, but also from your friends online. This same program uses incentives such as getting gifts for reducing your energy bills a certain percentage. [*energy, ways to save residential; negawatts; soft energy path; energy use; power grid; microgrids, power*]

energy, ways to save residential

There are many ways to lower power consumption in the home, thus saving money and reducing the size of your *ecological footprint*. Here are eight of the best: 1) add *insulation*, 2) *caulk* leaks, 3) use low flow showerheads, 4) use a *programmable thermostat*, 5) rethink your *lawn*, 6) upgrade an old, inefficient *refrigerator*, 7) use *phantom-load* power strips, and 8) start switching over to *LED* light bulbs. [*bathroom water-saving techniques; energy saving apps*]

energy sources, historical

Wood was the first fuel source for humans. It was used to cook food, heat living areas, and mold metals into utensils, tools, and weapons. As the demand for wood grew, the supply diminished in many parts of the world and alternatives had to be found. Western Europe began running out of wood in the 13th century, but the North American *forests* kept wood as the primary *energy fuel* in the United States through the mid-1800s.

As the supply of wood diminished, *coal*—a *fossil fuel*—became popular. During the early 18th century, countries with large supplies of coal participated in the *Industrial Revolution*. In 1850, 90 percent of the U.S. energy supply came from wood, but by 1900, 70 percent came from coal.

At the same time coal was taking the spotlight away from wood, *oil* wells were being drilled. From the mid-1800s to about 1900, oil was used primarily for lantern kerosene, while the *gasoline* component was considered a waste byproduct. In 1870, oil supplied only one percent of the U.S. total fuel needs, but by the early 1900s, the *automobile* and its thirst for gasoline made our dependence on oil surge. Today, about 40 percent of the U.S. total energy requirement is supplied by oil, with coal supplying about 23 percent.

Just as gasoline was originally considered a waste byproduct of crude oil, *natural gas* used to be burned off at oil wells as waste. Natural gas was gradually ushered in as a good source of energy, especially for heating. During the 1920s, about five percent of the U.S. energy supply came from gas, but it gradually increased for many years and recently has expanded dramatically with new *shale bed* reserves and *fracking* technologies. Natural gas recently surpassed the use of coal and now supplies about 24 percent of U.S. needs.

Hydropower also came into its own during the early 1900s. Over the past few decades, the percent of U.S. energy supplied by hydroelectric has been in steep decline. Currently only about four percent of U.S. energy comes from hydropower.

Nuclear power became popular during the latter part of the 1900s but has also been in decline, providing only about eight percent of the U.S. total energy need. Other forms of *renewable energy* are still in their formative development but are gaining ground quickly, including *solar power*, *wind power*, *geothermal energy*, and *biomass energy*.

Energy Star rating

An *eco-certification program* that designates products deemed energy efficient, run by the *Environmental Protection Agency*. The primary goal is to reduce *greenhouse gases* that cause *climate change*. Products that meet their criteria are awarded the Energy Star *eco-label*. The criteria are rigorous and the rating can be revoked if the products fall below the standards. Products that can attain an Energy Star rating include clothes washers and refrigerators; building products such as windows and doors; computers and other electronic devices; heating and cooling systems such as air conditioners, and dehumidifiers; lighting, fans, and light bulbs; and plumbing products, including water and solar heaters. There is also an Energy Star program for *green homes*. {energystar.gov}

energy use

Energy use can be divided into four major categories: 1) residential and commercial, where *air conditioning* and *thermal insulation* play a major role, 2) industrial, 3) *transportation*, and 4) electrical utilities, such as in *electric power plants*. In general, *less developed countries* use a greater percentage of the energy for residential purposes, while more developed countries use a greater percentage for industry and transportation.

The energy use of a country and its people can be profiled by looking at the amount of energy used per person for each of the categories. For example, the amount of energy used for transportation by each person in the United States is 100 giga-*joules*, compared to 40 in Denmark, 25 in Japan, 12 in Mexico, and 2 in Zimbabwe. [*energy saving apps; energy, ways to save on residential; Living Buildings; smart energy; energy consumption, historical*]

entomology

The study of *insects*. [*beetles; slave ants; ash tree demise; honeybees, pollination, and agriculture; honeybee sperm bank; Monarch butterfly decline; zombie bees*]

environment

An environment includes all the living and nonliving components and all the factors such as climate that exist where an organism lives. The plants and animals, mountains and oceans, temperature and precipitation all make up an organism's environment. Environment assumes the perspective of the organism being studied or discussed. For example, the rabbit's environment or dumping waste damages our town's environment.

This term is often confused with *ecology*, which is also the study of these components and factors, but more importantly, the relationships that exist between them. Ecology is the study of how the living parts interact with each other and with the nonliving parts, as well as how factors such as weather impact all the parts. You can think of the environment as an assortment of dominoes around you, and ecology as a study of the domino effect, or the impact of one domino on others.

Environmental and Energy Study Institute (EESI)

EESI began as the Environmental Study Conference. It was established by a bipartisan group of members of Congress in 1975, who wanted to call in outside experts to provide objective analysis of environmental, energy, and natural resource issues. As the group grew, they changed their name in 1982 to the Environmental and Energy Study Institute, and in 1984 it became an independent, not-for-profit organization. Although it always includes members of Congress, it is now controlled by a private board of directors. Its mission remains the same, though: "To provide information to policy makers as well as the public about *climate change* and *sustainable energy*." [*NGOs; Office of Technology Assessment, former; League of Conservation Voters; environmental literacy; environmental education; law and the environment*] {eesi.org}

Environmental Defense Fund (EDF)

The EDF was established in 1967 and now has about 750,000 members. It has broad interests but focuses on *ecosystems*, *climate change*, *marine* life, and human health issues. Its strategy is to present the science and to work with business and municipalities to find market-based solutions. It has always been at the forefront of legislative action when needed. Its unique approach of comingling litigation with market forces has been a huge success. The <u>New York Times</u> referred to the EDF as "one of the most powerful environmental organizations in the world." [*Group of 10; NGOs*] {edf.org}

environmental education

Almost all the major environmental organizations, including *NGOs* and government agencies involved in the environment, have separate education committees, offices, or programs. The *Environmental Protection Agency* and *National Oceanic and Atmospheric Administration* are two federal agencies with excellent programs, websites, and information. The popular nonprofit NGOs *World Wildlife Federation* and *Nature Conservancy* are just two of many with excellent educational services. Some organizations, such as The *North American Association of Environmental Education,* specialize in just environmental education, providing support and services for environmental educators. The *Environmental Media Association* offers information on educational films and other media. *David Orr* is a college professor, turned environmental writer, who has become one of our leading environmental educators. [*law and the environment; environmental literacy; philosophers and writers, environmental*]

environmental impact statement (EIS)

A report that attempts to predict the environmental consequences of a proposed project such as a building, highway, dam, or any project that might harm the local environment. These statements are required as part of The *National Environmental Policy Act* (NEPA) of 1969. The assumptions and predictions put forth in these statements are based on science, but whenever the future is projected, a good deal of conjecture is necessary.

Many people believe these required documents have protected the environment against countless destructive projects; others believe these statements are used to thwart progress. Many states have used the *NEPA* model to create their own reporting requirements in efforts to protect the environment from harm potentially caused by state and local projects. [*eco-services; cost/benefit analysis; economics and sustainable development; externalizing costs; industry self-regulation; triple bottom line*]

environmentalist

Someone who understands the *environment* and uses this understanding to help manage our planet. Managing the planet involves lifestyle changes, activism, voting, and many other actions.

Changing personal lifestyles may seem to have little impact on correcting major problems, but when multiplied by a million people, the impact can be awesome. Individuals who buy recycled or biodegradable products not only change their lifestyle but also are changing the way companies do business. *Green products* and *green companies* are springing up in response to this growing market demand. Sometimes called *green consumers*, by basing your everyday purchasing decisions on environmental issues, you can change the way business and economies work.

In a democratic society, not only can people take action personally, but they can also help create the laws that take action on a larger scale. Legislation such as the *Clean Air Act* and the *National Environmental Policy Act* were enacted in response to the pressures applied by constituents. You can exert pressure to pass legislation by communicating with your officials, either on an individual basis or by joining *environmental NGOs*. To do this effectively, environmentalists must be able to detect *greenspeak, greenscams,* and *greenwash.* Just because someone says it's good for the environment does not mean it is.

Possibly the best way to help manage Earth is to express your feelings at the ballot box by voting for representatives with views similar to yours on environmentally sensitive issues.

environmental justice

The combination of social justice and *environmentalism*. It can be viewed as both a form of law and a social movement that works toward advancing these laws. The U.S. *Environmental Protection Agency* defines environmental justice as "the fair treatment and meaningful involvement of all people regardless of race, color, national origin, or income with respect to the development, implementation, and enforcement of environmental laws, regulations, and policies."

"Fair treatment" means no particular group of people should have to deal with an unequal share of harmful environmental effects because of policies run by a businesses or a government. "Meaningful involvement" means those people who would negatively be affected by a policy must have an opportunity to participate in decisions about a proposed activity and this public participation should have an influence on the final outcome.

The aim is to study and uncover socioeconomic examples where there is no fair treatment and/or no meaningful involvement. When this occurs, it is usually called environmental racism.

Environmental justice (and environmental racism) first became an issue in 1987 when the United Church of Christ's Commission for Racial Justice released a report with serious social justice charges. Their report stated that people who are poor or of a minority group are disproportionately affected by hazardous chemical production and waste disposal. It found that three out of five African and Hispanic Americans lived near toxic dumps and that millions of farm workers, mostly Hispanic, were exposed to more *pesticides* than in any other occupation in the country. It also stated that environmental laws in wealthy neighborhoods produced fines that were five times higher than similar ones levied in poor neighborhoods. They alleged that *cancer alley* is an example of this social injustice. Many environmental organizations have incorporated environmental justice into their mission statements.

Environmental justice was first fully articulated by Dr. *Robert D. Bullard* in his book, <u>Dumping in Dixie</u>. [*environmental movement; law and the environment; LULU; schools and brownfields*]

environmental literacy

A basic level of understanding an individual should possess to make intelligent decisions about managing our planet—what many think of as *Spaceship Earth*. Why *recycle*, if you don't understand what it accomplishes? Why vote for or against a *landfill* or a *waste-to-energy electric power plant* about to be built in your town if you don't understand the advantages and disadvantages of each? Why be concerned if the fruits and vegetables you eat contain *pesticides*, the beef you eat has been shot up with *antibiotics,* or the planet is getting warmer—unless you know what these things mean?

Environmental literacy is essential in a democracy because we put the people in office who determine the fate of our spaceship. You trust the pilot while onboard a plane; but you don't vote for your pilot—pilots are trained and certified so you feel safe while flying. Elected officials are not trained or certified and in many cases not even educated about environmental issues. It will take an environmentally literate constituency to ensure we have knowledgeable leaders to pilot our spaceship in the right direction. The alternative is no less frightening than having an unqualified pilot on your next flight. [*argumentum ad ignorantiam; environmental education; environmental rating systems; eco-labels; eco-certification; EPEAT; environmentalist; precautionary principle*]

Environmental Media Association

A Hollywood-based *NGO*, created in 1989 and dedicated to urging major studios to use environmental themes in television shows, movies, and other media. They hold the Environmental Media Awards competition each year, where they recognize excellence in this field. [*Earth Communications Office; movies, environmental*] {ema-online. org}

environmental movement (1960 through 1990), the start of the U.S.

The modern environmental movement in the United States began in the 1960s but had its roots long before then. *Emerson* and *Thoreau* became literary greats writing about nature. Not long after, *Muir* and *Burroughs* continued nature writing, but for a wider audience, and they were joined by President *Theodore Roosevelt's* concerns about protecting the American *wilderness*. Still others, such as *Leopold*, continued to set the stage for what many considered the "official" kick-off of the modern environmental movement.

A few of the defining moments attributed to this kick-off include *Carson's* _Silent Spring_ in 1962, *Hardin's* _The Tragedy of the Commons_ in 1968, and a flurry of federal legislation people clamored for, such as the *Clean Air Act* and the *Clean Water Act*. In the same year, the *human population* became the focal point when *Ehrlich* published the _Population Explosion_.

But the actual starting line for this movement is usually reserved for the first *Earth Day* on April 22, 1970, and the creation of the *Environmental Protection Agency* in the same year. Environmental literature and legislation were joined by numerous environmental *NGOs* that rallied and organized people to join the movement. Older groups such as the *Sierra Club*, *National Wildlife Federation*, Defenders of Wildlife, *Nature Conservancy*, and the *Environmental Defense Fund* saw their numbers swell. Newer groups sprouted up, such as the *Friends of the Earth*, *League of Conservation Voters*, *Natural Resources Defense Council*, *Greenpeace,* and Save the Whales.

In 1972, _The Limits to Growth_ was published, triggering an ongoing debate about people and the planet that continues to this day. Also, the *United Nations Conference on the Human Environment* was held in Stockholm, beginning a long line of international attempts to balance the human population with the world's limited resources. The threat to *biodiversity* was addressed by the *Endangered Species Act* of 1973 in the United States and the *United Nations Environment Programme* at the global level.

In 1975 a more intense form of environmentalism took hold with *Edward Abbey's* _The Monkey Wrench Gang,_ and the *Sea Shepherd Conservation Society* created a year later. This new activism was fed with events such as the 1977 *Love Canal* debacle when a local resident, *Lois Gibbs*, made front-page news fighting against corporate toxic pollution.

As the 1970s closed and the 1980s opened, *Three Mile Island* brought *nuclear energy* to the front pages and *Earth First!* added another new twist to the environmental movement. As the 1980s progressed, the world watched a toxic chemical catastrophe kill thousands of people in *Bhopal, India*; the *Superfund* was amended to begin cleaning up all of the toxic mess we had already made; and *Chernobyl* caused another, far greater nuclear catastrophic event in 1986.

In 1987, the *Montreal Protocol* was signed—destined to become one of the few big environmental success stories—stopping the destruction of the *stratospheric ozone* layer, or the so-called *hole in the ozone. Our Common Future* was published, establishing the first serious link between *economics* and *sustainability*.

In the later 1980s, *Chico Mendes* was murdered in *Brazil* over the battle between corporations and local citizens defending their natural resources. And *environmental justice* became an issue in the United States as well. The decade closed with the *Exxon Valdez* running aground off the coast of Alaska, causing a massive *oil spill*.

It was a tumultuous 30 year period of successes and failures.

environmental nongovernmental organizations

See *NGOs*.

Environmental Protection Agency (EPA)

The EPA was established in December of 1970, soon after the first *Earth Day*. Its purpose was to consolidate federal environmental activities that were being carried out by numerous other agencies. The EPA was authorized to manage *air* and *water pollution* and *pesticides*. It was also charged to manage *municipal solid waste, hazardous waste*, and *nuclear waste disposal*. Its headquarters are located in Washington D.C., with 10 regional offices across the country. They have a budget of about eight billion dollars and 18,000 employees. [*Environmental Protection Agency, past and present*]

Environmental Protection Agency, past and present

Prior to 1970, the U.S. *Departments of Interior* and *Agriculture* handled our pollution problems at the federal level, and the U.S. Congress occasionally dabbled in environmental legislation. By the late 1960s, Americans had come to the conclusion that the health of our planet was intimately tied to human health and welfare and that federal dabbling simply wouldn't suffice. A few months after the first *Earth Day*, held on April 22, 1970, plans to create the *Environmental Protection Agency* (EPA) began. In January 1971 the EPA was established to "strive to formulate and implement actions which lead to a compatible balance between human activities and the ability of natural systems to support and nurture life." With a budget of three billion dollars, a work force of 7000 employees, and William Ruckelshaus as the first administrator, the EPA opened shop. This all occurred during the *Nixon* presidency.

The EPA's first decade was one of momentum and progress. Here are some of its accomplishments: In 1970, Congress amended the *Clean Air Act* of 1963, giving the EPA the responsibility of setting air quality standards for each pollutant. In 1972, Congress passed the Water Pollution Control Act that allowed the EPA to set water quality standards and regulate *water pollution*. The EPA aggressively reported offenders to the Department of Justice. In 1976, the *Resource Conservation and Recovery*

Act (RCRA) was passed by Congress, directing the EPA to get involved in preventing industrial *hazardous waste* problems.

In 1978, *Love Canal* introduced the American public to the magnitude of hazardous waste problems that could not be prevented because they already existed. The public began to realize there were thousands of other "Love Canals." In response, Congress passed *CERCLA (the Comprehensive Environmental Response, Liability, and Compensation Act)* in 1980 (usually referred to as the *Superfund*). This fund was to help pay for the management and cleanup of these hazardous waste sites. The EPA was no longer simply responsible for preventing pollution, but also for cleaning it up. A *National Priority List* of Superfund cleanup sites was created.

The EPA's progress was halted and often regressed, beginning in 1980 under a new administration. The attitude in the Reagan White House was that environmental well-being was contradictory to economic well-being. The Office of Management and Budget was instructed to review the economic impact of all EPA-proposed regulations, and their budget was cut by 50 percent and the staff cut by 25 percent. The number of environmental violations reported to the Justice Department dropped by almost 70 percent in just a few years.

In 1984, however, Congress re-enforced RCRA and in 1986 extended and increased funding for the *Superfund* and passed the *Clean Water Act*. During the late 1980s, the Bush Administration once again elevated the importance of economics over the environment. Many environmental regulations were reviewed and rewritten if they were considered unfair to business.

The role of the EPA again expanded during the late 1980s and 1990s. The *nuclear meltdown* at *Three Mile Island* got the EPA involved in monitoring *nuclear wastes*, and they began to control *hazardous wastes* and *dioxins*. The *hole in the ozone* layer and the *Montreal Protocol* started the phaseout of *ozone-depleting gases* by the EPA. In 1989, the *Exxon Valdez* spill resulted in the EPA fining the Exxon Corporation $1 billion—the largest criminal environmental damage settlement in history.

During the 1990s, the Pollution Prevention Act returned the agency's focus to prevention. In 1991 the agency created a voluntary industry partnership for energy-efficient lighting and later created the *Energy Star* program. In 1994, President Clinton ordered the EPA to ensure *environmental justice* become part of its mission, and in the late 1990s the EPA was dealing with a new threat—*global warming*.

By 2000 the EPA had become the federal government's largest regulatory agency. This size and power resulted in having critics on both sides of the fence. The EPA has many enemies in corporate America and especially among conservative politicians, who feel the EPA over-regulates and is anti-business. On the other hand, they also receive criticism from many environmental groups for not doing enough.

environmental racism

See *environmental justice.*

environmental rating systems

Eco-labels, *eco-certification programs*, and environmental rating systems all help consumers buy products and services in an environmentally conscious way. While the first two place products or services into categories, environmental rating systems assess one particular product or service.

These rating systems are probably the quickest, simplest way to understand the environmental impact caused by a specific product or service. With most of these ratings, you can look up a product by brand name to find out how environmentally friendly (or harmful) it is. Some of the well-respected environmental rating systems include the NRDC Label Lookup and the Consumer Reports Greener Choices.

Some environmental rating systems even rate eco-labels and eco-certifications, such as the Ecolabel Index (a rating system of eco-labels), ForestEthics (they rate forest certification programs), Global Ecolabeling Network (they rate various eco-labels), International Trade Centre (various eco-labels), and ISEAL Alliance (various eco-labels). [*green consumers; environmental literacy; green marketing; greenwash; greenscam; planned obsolescence; Skin Deep, EWG's; textile industry; tropical fish tanks; wine-bottle cork debate; wrapping paper alternatives*]

environmental resistance

See *S-curve.*

environmental science

To understand environmental science, one must first understand the term *environment.* When we talk about the environment, it is usually from the perspective of an organism. For example, a rabbit's environment is everything—living and nonliving—that affects the rabbit. Humans live virtually everywhere on the planet, so the human environment is—Earth.

Environmental science is the study of how human intervention, called *anthropogenic stress*, affects our environment. More specifically, it deals with the effect of *human populations* and technologies on our planet and how to resolve the problems they pose. It is an interdisciplinary study that transcends biology, geology, chemistry, politics, economics, and many other sciences. [*ecology; aerobiology; agroecology; agroforestry; applied ecology; autecology; synecology; computer modeling, ecological;*

cryptozoology; deep ecology; edaphology; entomology; industrial ecology; meteorology; oology; sociology; theoretical ecology; environmentalist]

environmental tobacco smoke (ETS)

Also called passive or secondhand smoke, this is the inhalation of cigarette smoke by nonsmokers. ETS contains many of the ingredients breathed in by the person doing the smoking. The American Heart Association estimates tens of thousands of premature deaths from heart disease caused by ETS each year. Nonsmokers exposed to breathing in ETS at home have a 25 percent higher risk of developing lung cancer or heart disease. When the exposure extends to the workplace and other public places, this risk goes up to as high as 60 percent.

But it harms children the most. Estimates state that ETS contributes to 150,000 respiratory infections in babies, triggers 8000 new cases of asthma in unaffected children each year, and aggravates symptoms in at least 400,000 asthmatic children. [*antibiotics and the human microbiome; autism studies and environmental factors; breast milk and toxic substances; geomedicine; medical ecology; nanoparticles and human health; puberty and the environment, early; spillover effect*]

Environmental Working Group

An environmental health research and advocacy *NGO*. They offer many online consumer guides that detail whether a product contains any toxic substances harmful to either you or the environment. Their well-known *Skin Deep* guide is for *cosmetics*, but others are for sunscreen products, *insect* repellants, cleaning products, produce and other foods, plus many others. [*eco-labels; environmental rating systems; EPEAT; dry cleaning; essential oils; lipstick and toxic substances*]

EPEAT (Electronic Product Environmental Assessment Tool)

An *environmental rating system* for anyone wanting to find out which electronic products are environmentally friendly and which are not. This website was developed by a grant from the *Environmental Protection Agency* but is managed by the Green Electronics Council who now manages the site and runs the entire EPEAT resource. As the site states, "Choose electronics that reflect your passion for protecting the environment. EPEAT-registered products are manufactured with less toxic content compared to devices that don't meet EPEAT criteria, and they're designed to be easily recycled. Also, because EPEAT-registered products meet the latest *Energy Star* specifications, they can help you lower your energy use and potentially your energy bill."

They rate electronic devices as Bronze, Silver, or Gold. Using such resources to check up on a product's *ecological footprint* prior to purchase is practiced by *green consumers*. [*LEED*] {epeat.net}

epiphyte

A plant that grows on another plant, using it for support. It causes no harm to the *host* (supporting plant), so it is considered a *commensal* relationship. Because epiphytes are not rooted to the ground to gain nourishment, they typically have small hair-like structures growing on their stems and leaves to absorb nutrients from the atmosphere or rainwater. Spanish moss is an example of an epiphyte in the southern United States. [*symbiotic relationship; communications between plants; mimicry; dientomophilous; niche, ecological*]

eradicating ecocide movement

Ecocide means destroying nature. This movement consists of a small group of people—mostly lawyers in the U.K. The term ecocide was coined during the *United Nations Conference on the Human Environment* in 1972 by Sweden's Prime Minister, who attacked the United States over its use of the defoliant *Agent Orange*, during the Vietnam War. It has even been suggested that ecocide be added to the four existing international crimes against peace, which include war crimes, genocide, crimes against humanity, and the crime of aggression. It is doubtful that crimes against the environment will ever become law, but that might not keep people from trying. [*law and the environment; environmental justice; environmental ethics; utilitarian environmental ethic; ecofeminism; "Land Ethic, The"; Douglas, William O.; Center for Health, Environment and Justice*]

erosion

The process of removing *soil*, sediment, or rock from its original location and depositing it in another location. *Water*, wind, and *glaciers* can all cause erosion. Erosion typically includes *weathering* when rock is broken into smaller pieces by natural processes and then transported. Water, wind, and glaciers move these materials and deposit them elsewhere during a process called *deposition*.

Erosion can dramatically alter an *environment*, such as by removing most or all of the *topsoil* from an area. Human activities greatly increase the likelihood of erosion. [*soil conservation; soil erosion; topsoil; Dust Bowl; plate tectonics; environmental science*]

essential oils

Essential oils have become popular for medicinal and therapeutic uses, as well as for aroma and flavor additives. They are naturally produced by the harvest and distilling of certain plant parts. Although they might be good for human consumption, the surge in popularity is becoming an environmental issue. Common plants used include rosewood, sandalwood, thyme, cedarwood, and gentian. The volume of plants needed to provide a commercial yield is large. For example, it takes 200 pounds of lavender to produce one pound of lavender oil and 10,000 pounds of rose blossoms for one pound of rose oil. When harvesting at these levels, sustainability is important. Many of these species are on *threatened species* lists, and some are harvested illegally with little concern for the environment or the local economies that depend on these plants.

When purchasing essential oils, like most products, you can avoid those causing harm by checking with reputable *eco-labels, eco-certification programs*, or *NGOs* that work to protect these plants. [*lipstick and toxic substances; nanoparticles in cosmetics and personal products; Skin Deep, EWG's; wine-bottle cork debate; natural food colorings*]

estivation

An adaptation that allows some organisms to enter a state of dormancy to survive extreme heat or dryness, as opposed to *hibernation* which does the same but for cold. This adaptation allows these organisms to remain in their habitat throughout these extremes as opposed to migrating elsewhere. [*torpor; cold resistant; quiescence; migration, natural selection*]

estuary

Estuaries are unique coastal *ecosystems* where *freshwater* from a river mixes with salt water from the sea. This results in a concentration of salts between freshwater and marine habitats called *brackish* waters. Because the degree of salinity, temperature, and other factors varies in estuaries with the tides, only certain kinds of organisms with wide *tolerance ranges* can inhabit this type of *ecosystem*.

Estuaries are among the most productive ecosystems on Earth, along with *tropical rainforests* and other types of *wetlands*. The constant flow of water from the river or other body of water into the estuary provides high concentrations of *nutrients* for organisms to thrive. The water is usually shallow, so sunlight can reach the bottom, allowing *emergent plants* and *algae* to grow abundantly. These producers (*autotrophs*) are the first link in *food chains* that include fish such as flounder and *crustaceans* such as shrimp. Many organisms use estuaries as a spawning ground and a nursery. The young have plenty to eat and protection from the open ocean. Once large enough, they move out to sea, making estuaries a vital link to much larger ecosystems. [*estuary and coastal wetlands destruction; marine ecosystems; neritic zone; eco-services*]

estuary and coastal wetlands destruction

Estuaries and coastal *wetlands* (bays, lagoons, salt *marshes*, and *swamps*) were thought by many to be worthless, mosquito-infested regions. Many of these areas were used as dumping grounds for waste, while others have been drained, filled in, and built on, and still others have had their source waters diverted for human use.

In fact, we now know estuaries, swamps, and marshes are the most productive of all *habitats* and similar in *net primary production* to *tropical rainforests*. These regions are spawning and nursing grounds to 70 percent of U.S. commercial fish and shellfish and they are breeding grounds and *habitats* for many waterfowl and other wildlife.

These areas filter out and dilute water pollutants from the rivers and streams that feed them before reaching the open sea. Some estimates value one acre of tidal estuary to be worth $75,000 in waste-treatment costs. These regions also act as an enormous buffer, protecting the inland from storm waves and absorbing vast quantities of water that would otherwise move onshore and cause flooding.

In the United States, more than 55 percent of all estuaries and coastal wetlands have been damaged or totally destroyed. Most have been filled for building developments because of the popularity of coastal living. Many others are contaminated with toxic substances because of either direct dumping or accumulation of toxic substances from the incoming waters.

These regions are still threatened by development. Although few people disagree with scientists about the importance of these habitats, many people still feel these areas should be developed. Some local, state, and federal agencies are trying to take the easy way out of this dilemma by redefining what a wetland is and offering a *no net loss policy*, where creating a so-called wetland elsewhere would replace the natural wetlands that are destroyed—a process most environmentalists believe to be more of a political solution than an environmental solution.

After many coastal storms, such as Superstorm Sandy in 2013, debate has renewed about reducing coastal development and even returning some areas back to their natural settings. However, it would be far easier and less expensive not to build in these areas in the first place than to try to relocate people out of an area and remediate an *ecosystem*. It will be hard enough to ask people to move out of their homes and close to impossible to perform the *geoengineering* required to re-create lost wetlands. [National Estuary Program]

ethanol

See *bioethanol*.

E - The Environmental Magazine

Called E magazine for short, this is an entertaining, easy to read *eco-magazine* and *green blog* that anyone with even a remote interest in our environment would enjoy. It contains many interesting, brief articles about environmental issues. These articles, although about our environment, are written with your interests in mind—whether it be about the foods you eat, your health, your home, your pets, or anything else about living in a green world. In addition to these brief articles, the magazine has full-featured stories on focused topics that are well-written, in-depth, and of value to hard-core environmentalists. {earthtalk.org}

ethics, environmental

Ethics is a branch of philosophy that attempts to determine right from wrong, without cultural influences. Environmental ethics is specifically concerned with defining right from wrong as it pertains to environmental issues. Environmental ethics, like any ethical problem, is defined in different ways, depending on the perspective of the individual espousing the beliefs.

Possibly the best definition of environmental ethics was written by *Aldo Leopold* in his essay "*The Land Ethic*" in 1949. "All ethics so far evolved rest on a single premise: that the individual is a member of a community of interdependent parts. The land ethic simply enlarges the boundaries of the community to include soils, waters, plants, and animals, or collectively, the land." [*tragedy of the commons; environmental justice; utilitarian environmental ethic; eco-psychology; environmental movement*]

ethnobotany

The study of how plants have been used by indigenous people throughout history and how they are used today. Some people call this field ancestral medicine or ancestral remedies. Some of these remedies were documented in ancient texts and others just handed down by oral history, with *China* being the most prolific contributor. These ancient remedies often make a comeback and become popular for a while before fading into history again. However, efforts to preserve these remedies are being made in many countries. For example, doctors in the islands of Palau and Pohnpei, in Micronesia, are writing island-specific manuals that cover basic health-care remedies for such things as stings, bites, colds, and diseases common to each island. In most cases, specific plants are used as the remedies. [*biopiracy; primitive technologies*]

euphotic zone

That portion of a body of water that receives sunlight. [*aquatic ecosystems*]

euroky

The ability of an organism to tolerate a wide range of environmental conditions. [*tolerance range; limiting factor; estuary; brackish water*]

eutrophication, cultural

Cultural eutrophication is an accelerated form of *natural eutrophication*. The natural process is accelerated when waste products produced by human activities enter bodies of water. Untreated *sewage, livestock wastes*, agricultural *fertilizers*, and many *industrial waste* products find their way into bodies of water and dramatically hasten the process of *nutrient enrichment*.

This sudden enrichment of nutrients produces a population explosion of organisms, especially *algae*. When these organisms die, *bacteria* have a population explosion of their own, feeding on the decomposing algae. This results in a reduction in the oxygen content of the water, causing the death of the lake or pond *ecosystem*.

Symptoms of eutrophic bodies of water (those with high levels of nutrients) include large amounts of shore vegetation, *algal blooms*, stagnant water, and the lack of cold water fishes. [*standing water habitats; biological oxygen demand*]

eutrophication, natural

Lakes are classified as either *oligotrophic* or eutrophic. Oligotrophic lakes are deep, clear, cold, and contain limited nutrients to support life. Eutrophic lakes have undergone the process of eutrophication and are usually shallow, warm, and cloudy, because they are rich with nutrients.

Bodies of standing water go through natural eutrophication—the process of *nutrient enrichment*—over long periods of time, usually hundreds or even thousands of years. Sediment is naturally washed into bodies of water from the surrounding *watershed*, and the *nutrients* within this sediment *leach* out into the water. High concentrations of nutrients results in expansive amounts of *algae* and other *producers*. Over time, enormous quantities of dead and decaying *organic* matter such as dead algae settle to the bottom, where it provides food for vast numbers of *bacteria*. As the bacteria decomposes the algae and other organic matter, they deplete much of the dissolved oxygen from the water, resulting in the collapse of the food web. Natural eutrophication is part of the natural process of *succession*, which is the gradual change from one type of *habitat* to another. [*standing water habitats; succession, primary aquatic; biological oxygen demand*]

Everglades

A unique wetland *ecosystem* in southeast Florida, roughly 200 miles long by 50 miles or less wide, by a few inches deep, found between the Kissimmee River and the Florida

Keys. Coined the "river of grass" by the late activist and protector *Marjory Stoneman Douglas*, the Everglades contain saw grass, tree islands, and *marshes*. Located within this region are the 1.5-million-acre Everglades National Park, the Big Cypress National Preserve, and the Loxahatchee National Refuge.

By 1970 about half of the original area had been lost to development and the quality of what remained had been degraded because of wastewater and *agricultural runoff* containing *fertilizers* and *pesticides*. What was once a *wilderness ecosystem* is now fragmented by 2000 miles of canals, many 20 foot-high-dikes on Lake Okeechobee to the north, and numerous channels cut through the Kissimmee River. A fifth of the original area became protected as the Everglades National Park, which is also a *United Nations Biosphere Reserve*, a *World Heritage Site*, and a Wetland of International Importance.

Even with this protection, however, the Everglades have been under environmental attack by surrounding private *agricultural* lands and development. Dairy farms produce animal waste runoff, and sugar cane plantations release fertilizers and pesticides in runoff as well. Because the water is controlled by canals and dams, it no longer maintains its natural cycles, causing indigenous plants and animal populations to go into decline.

Recent legislation has attempted to restore the ecosystem, including the Everglades National Park Protection and Expansion Act in 1989 that added more than 100,000 acres to the park, closed most of the park to airboats, and ordered water restoration to improve the health of the ecosystem.

Then in 2000, Congress approved the Comprehensive Everglades Restoration Plan (CERP) to restore, preserve, and protect the ecosystem "while providing for other water-related needs of the region." A heated debate followed as to whether this plan is designed to actually protect the ecosystem or the rights of private landowners surrounding it. While some environmental *NGOs* support the plan, others oppose it. This plan, if fully implemented, is supposed to be the largest environmental restoration project ever, but it continues to meet with opposition. In the meantime, the Everglades National Park is now not only a World Heritage Site but has been placed on the List of World Heritage Sites in Danger.

evergreen

Trees and shrubs that have leaves all year long, such as pines and spruces. [*deciduous; terrestrial ecology; deforestation; forest health, global; silviculture; timber, DNA sequencing of; tree farms*]

evolution

See *natural selection*.

e-waste

The waste produced when electronic equipment such as computers, tablets, and smart-phones are disposed of. These devices contain substances not commonly found in other products and thereby create a unique waste problem. The quantities of this waste are staggering. In the United States, more than three million tons are produced each year. Americans threw out 150 million cellphones in 2010. Globally, between 20 and 50 million metric tons of e-waste is disposed of each year. This represents about five percent of the total *municipal solid waste* produced worldwide. And it grows every year.

E-waste includes *heavy metals* such as *lead*, *cadmium,* and *mercury* as well as *PBDEs* and *rare earth metals.* These substances remain inert and harmless while you are using them in your electronic gadgetry. However, if not properly *recycled*, they become an environmental nightmare. Most of the e-waste that is not recycled ends up in *landfills,* where these substances can *leach* out, polluting the *soil* and *groundwater*. It can enter *food chains* and build up in organisms as a result of *bioaccumulation* and, sooner or later, might enter our food supply and possibly our bodies, becoming part of our *body burden*.

In the United States, about 25 percent of this waste (850,000 tons) is recycled; a growing business, because many of these waste substances are valuable and have a viable *postconsumer* market. It is usually less expensive to recycle some of these substances than mine for them. (Not to mention far better for the environment.) For example, one million cell phones contain about 35,000 pounds of copper, 750 pounds of silver, 75 pounds of gold, and 30 pounds of palladium.

E-waste recycling programs are expanding. In the early 2000s, only three states had specific e-waste laws; now they exist in 25 states. For example, Montgomery County, Maryland, has an e-cycling program that brings in 150 tons a month.

The recycling of e-waste is a dangerous process and must be performed in the proper manner so as not to harm the workers or the environment. Recycling performed in the United States and in Europe is well regulated; protecting those working in these facilities and, to some degree, the environment with what remains. The EU requires any remaining e-waste, after recycling, to be handled as a *hazardous waste*, but the United States does not.

Usually recycling programs are a feel-good topic. However, e-waste recycling has a dark side. Of the 25 percent that gets recycled in the United States, most of it—about 70 percent—is not done by legitimate recyclers within U.S. borders but is shipped to poor or developing nations. The EU does the same with about 65 percent of its recycled e-waste. Nations such as Nigeria, Ghana, *India*, and *China* accept this waste, but these countries have little or no oversight to safeguard workers—called *urban miners*—handling these *toxic* or *hazardous materials*.

In these countries, the e-waste is manually broken down and sold to the highest bidders. Much of the work is done by children or women, and the health risks are significant. Children smash computer batteries with mallets to gather flecks of *cadmium* and women work over steaming buckets of boiling water, cooking circuit boards in search of gold slivers, called *gold fingers*. After stripping these valuable materials and selling it, what remains includes toxic wastes that are simply dumped on surrounding areas or burned in open air containers, causing *air pollution*. Countries with few recycling regulations also have few landfills.

Shipping these wastes from developed nations to poor nations has become a serious global problem, and many countries have signed the *Basel Convention* to stop the practice. However, the United States has never ratified the agreement and therefore the practice is considered legal. The EU, however, has ratified the agreement, so the practice is illegal there. [e-*waste and you; e-waste product life cycle; e-waste recycling and disposal programs; beryllium in computers*]

e-waste and you

If you become familiar with the many environmental problems about *e-waste*, and you want to make an effort to reduce your e-waste impact, you can do many important things. The first is not popular—use less. If you get a new cellphone every three years instead of every two (22 months is the average), you just reduced your impact by half. Second—and everyone can and should do this—when you buy phones, laptops, and any electronic gadget, check the manufacturer's environmental ratings. Many such rating systems are out there. And don't just believe every rating you see. Be sure you are not a target of *greenwash* instead of facts. It sounds a bit strange, but you must rate the *environmental rating system*.

And third, don't throw your old devices in the garbage. Here, a few minutes of thought can make a big difference. If the device still works, you might be surprised that there are plenty of people that might want it and will give it another life. If it must go, numerous useful *e-waste recycling and disposal programs* will help you recycle it or at least find the appropriate place for disposal. [*e-waste product life cycle*]

e-waste product life cycle

The life cycle of a smartphone, computer, or any electronic device has three primary steps. Each of these steps can cause considerable environmental harm. First, the extraction of the materials needed to build the devices often involves various types of *mining* practices. Electronic devices require *rare earth metals,* making the extraction process more complex and often more environmentally harmful than for other materials.

Second, the manufacturing process involves the use of many harmful substances. Those at greatest risk are employees who manufacture these devices, exacerbated by the fact that many of these plants are located in countries that do not protect their workers. Statistics show that workers in these factories have higher than usual rates of certain diseases.

Finally, the "end of life" phase refers to how the product is disposed of once its useful life is over—which is something you, the consumer, decides. *E-waste* has its own unique problems, and most electronic waste products should be probably disposed of and not placed in the regular trash. Everyone should be familiar with some of the *e-waste recycling and disposal programs* available.

e-waste recycling and disposal programs

Special *recycling* programs help ensure your *e-waste* is properly disposed of and recycled. Best Buy, Office Depot, and Staples all accept e-waste products that they then send to responsible, third-party, verified service providers.

It is important to remember that any recycling program requires more than just turning in a product for recycling. A market must exist for that recycled material—a buyer who can then turn the product into another product. This is often the hard part of a recycling program.

Here is an example of what has gone wrong with an e-waste recycling program. *Lead* has long been recycled from the glass in old televisions and computer monitors (those with the big cathode ray tubes). Lead is valuable, so it has made sense to go through the expensive process of extracting it from these old screens. However, with the advent of flat panel televisions and computer monitors, everyone has been turning in these old devices. As a result, the value of those collected for recycling has plummeted, forcing many of the recyclers into bankruptcy. A few years ago, recyclers were paid $200 a ton for this glass; today they must pay $200 a ton to have it disposed of because of the glut in the recycled marketplace. [e-*waste and you; e-waste product life cycle; urban miners; gold fingers*] {electronicrecyclers.com}

exclusive economic zone

The area that begins at a country's coastline and goes out from the coast 200 nautical miles. A *United Nations Convention on the Law of the Sea* agreement states that this zone is for the exclusive economic use of that country. This means any *commercial fishing* done within this zone requires permission from that country, and any *natural resources* found within this region belong to that country. Most nations adhere to this agreement, but there are many ongoing national disputes about this limit. [*tragedy of the commons; National Marine Fisheries Service; no-take zones, marine; ocean farming; mariculture; United Nations Convention on the Law of the Sea*]

exobiology

The study of life from other worlds. [*outer space; near earth objects; panspermia; natural selection; extinction and extraterrestrial impact; space debris; space-debris cleanup; terraforming*]

exotic species

This is an organism introduced into a new area; one that is not native or indigenous to that area. Exotic species are sometimes introduced into an area for a specific purpose, such as *biological control* of *insect* pests. But in other circumstances, attempts are made to prevent exotic species from entering areas. For example, in many waterways, the *zebra mussel* and *Asian carp* soon overrun existing populations and damage stable ecosystems. [*alien species; invasive species; native species*]

experimental ecology

Experimental *ecology* became popular in the 1960s and continues to be important today. It focuses on studying the mechanics of an organism's environment by manipulating the environment (or organism) to see what happens when controlled factors are changed. Experimental ecology was preceded by *descriptive ecology* and was a predecessor to *theoretical ecology*. [*applied ecology; community ecology; deep ecology; ecological studies; industrial ecology; zoogeography*]

explosive remnants of war (ERW)

Every year many civilians are killed and injured by explosive remnants of war. Most are unexploded landmines, but they can be artillery shells, mortars, grenades, bombs, and rockets—all left behind from any one of the numerous armed conflicts across the globe.

In 2003, a large contingency of the international community adopted a treaty called The Protocol on Explosive Remnants of War, which took effect in 2006. As of 2013, it has been *ratified* by 84 member nations. Its purpose is to help reduce the human suffering caused by these devices and bring rapid assistance to those affected.

One of the better known and worst examples of ERW is the legacy of the Vietnam War—more than three decades after the war ended. The U.S. Department of Defense estimates there were more than one million tons of undetonated weapons in Vietnam. Estimates, by groups involved, report about 100,000 injuries and fatalities since 1975 because of ERW.

In addition to efforts by those nations that signed the Protocol, numerous humanitarian *NGOs* are working toward safe removal of these weapons. [*weapon dumps, abandoned; Agent Orange and the Vietnam War; Laos*]

extant
Organisms that are living during the present time; not *extinct*.

extended product responsibility (EPR)
Also called the polluter-pays principle. A policy that places the responsibility for *post-consumer waste* on the producer of the product. The original idea was developed by a Swedish professor who wrote about "waste-conscious product development." It was then endorsed and presented formally to the industrialized world by the *Organization for Economic Cooperation and Development.*

The first EPR law, called Green Dot, was passed in *Germany*. It required manufacturers to become responsible for any packaging materials they used for their products and the waste it generated. The results were impressive—dramatically reducing the volume of waste entering their *landfills*. It is now standard policy in the EU and Japan and beginning to be used in the United States.

Without EPR, a company is not responsible for the environmental costs of a product, such as filling up *landfills* or causing any form of pollution. (These are called *externalized costs.*) Instead, the consumer bears the cost of preventing a product from going into a landfill or of cleaning up pollution, probably in the form of taxes to pay for these services.

With EPR, these costs are the responsibility of the manufacturer and the cost becomes part of the product. For example, a *bottle bill* to *recycle* bottles or an electronic device return program are both financed by the manufacturer. The burden is removed from the consumer and placed on the manufacturer. (This is called *internalizing the costs.*) This is not only good for the consumer, but for the environment as well.

When EPR is carried out properly, it demonstrates how the competitive marketplace can work to the benefit of the environment—instead of always against it. When a cost becomes the responsibility of the manufacturer, the company figures out how to minimize this cost, typically also minimizing the extent of any environmental damage.

A good illustration of EPR at work is to compare packaging materials. In Europe, where EPR is the norm, companies use far less packaging material than they do in the United States, where EPR is not common—and this is for the exact same product. U.S. companies don't bother to reduce their packaging material, because they don't have to pay for its cleanup, but in the EU, where cleanup is their responsibility, these same companies figured out how to minimize the packaging material and the waste it causes.

The United States is moving in the right direction. In 32 states with product-specific EPR laws, products have various ways of being taken back at the end of their life, or the manufacturers pay for *recycle* programs. These include electronic device take-back programs, safe end-of-life disposal of products containing mercury, car battery return programs, and various beverage-container recycling programs paid for by manufacturers. [*cradle-to-grave; cradle to cradle; CERES, economics and sustainable development; Beyond GNP; life-cycle assessment; natural capital; slow money movement; triple bottom line*]

externalizing costs

When a product creates a cost, but that cost need not be paid by the company that makes the product. For example, a company makes a product that pollutes a town's *drinking water* supply or causes *air pollution*. If no laws are in place that require the company to pay for the cleanup of the water or air, these costs are considered *externalized*—the cost of cleaning up the mess made by the company or their products is external to the company. Usually, taxpayers are left to figure out how to clean up the environment and pay for it, as well.

In effect, you can think of these externalized costs as a subsidy paid by the city to the company. Today, many municipalities and states can no longer afford to continue this and are passing legislation that forces companies to *internalize* these types of costs. *Bottle bills*, take-back programs of electronic products and appliances, and *extended product responsibility* programs all internalize the cost, making the manufacturer responsible for the product's full *life cycle,* including *postconsumer waste.*

extinct

Organisms that are no longer in existence are said to be extinct (as opposed to extant, meaning they do exist). But the term is often used loosely and must be clarified. For example, many organisms called extinct are actually *extinct in the wild*; they remain an extant species because individuals exist in *captive breeding programs* in *zoos.* A truly extinct species will never be seen again other than stuffed or in a jar.

Species are forced into extinction when they cannot adapt to changes in their environment. The most likely organisms to become extinct are those that have a low population density, are found in a relatively small area, assume a specialized *niche*, and reproduce slowly. The *Northern Spotted Owl* is a good example of this type of organism. Rabbits, mice, and most insects have a high population density, are found widely, fill many different niches, and reproduce rapidly, so they are less likely to be forced to extinction, but this does not mean they won't become extinct.

These changes to the environment can occur naturally or be caused by human activities. As long as life has existed on Earth, there has been extinction. It is believed that 99 percent of all species that ever existed are now extinct. (After all, where are the dinosaurs today?) But the numbers of species forced into extinction today and for the past few decades has accelerated well above natural levels—so much so, that many say we are in the midst of a *sixth mass extinction*.

Some of the earliest warnings about the impact of human intervention on extinction were posed by Thomas Lovejoy in 1980, and his concerns have come to be. Today more than 10 percent of all birds, roughly 25 percent of all *mammals*, and about 40 percent of all *amphibians* face possible extinction.

Human activities responsible for extinction are numerous and include *habitat loss* and fragmentation; hunting (sport, commercial, and for local *bush meat*); introduction of *invasive species*; distortion of the *nitrogen cycle* because of agricultural *fertilizer* use; the *wildlife trade and trafficking,* including the *ivory trade* generated by *elephant poaching*; *pesticides*, toxic substances, and *pollution* in general; and *climate change* that causes unique forms of habitat loss, such as the reduction of *arctic ice,* which harms species such as *polar bears*, and *ocean acidification,* which kills *coral reefs*.

Some of the well-known classic examples of species forced into extinction by human activities include the *passenger pigeon* years ago and the *dusky seaside sparrow* more recently. It is very possible that in about 100 years, many of our large carnivores, including lions, tigers, and cheetahs, will only be found in zoos or protected wildlife areas, having become extinct in the wild.

Many organizations create and manage *endangered and threatened species lists*. [*biodiversity, loss of; polar bear trade; rhinoceroses and their horns; CITES; dodo bird; de-extinction; seed banks and vaults; species recently gone extinct; Lazarus Taxon; International Union for the Conservation of Nature; extinction and extraterrestrial impact; extinction, mistaken; Xerces Society*]

extinct in the wild

A *species* where individuals no longer exist in the wild and the only remaining members of that species are found in captivity, usually in some *zoos* that have *captive breeding programs*. [*extinct*]

extinction and extraterrestrial impact

Geological records reveal that throughout the Earth's history, five mass *extinctions* have occurred. A growing body of evidence shows that three of these mass extinctions could have resulted when an object from outer space collided with Earth, disrupting global climate. Much of this evidence is based on the find of unusually shaped "microtektites," tiny glass beads believed to have formed as a result of the collisions.

The most famous of the three extinctions is the annihilation of the dinosaurs and about two-thirds of all marine life about 65 million years ago. The other two occurred about 200 and 370 million years ago.

Various theories exist about how these impacts caused mass extinctions. Some center around dust clouds believed to have blocked out the sun enough to disrupt plant growth and destroy *food chains*. Also, the Earth might have cooled sufficiently to disrupt life as it existed at that time. Other theories propose that some of the extinctions were caused by *global warming*, when large amounts of *carbon dioxide* were released by limestone that melted during the collision.

Critics of these impact theories believe the mass extinctions were probably caused by natural factors such as volcanoes that spewed ash, blocking the sunlight and changing the climate. [*extinct; extinct in the wild; de-extinction*]

extinction, mistaken

For most species, once they are declared *extinct*, they are gone forever. However, extinction is a difficult state to declare definitively because it is based on an "absence of evidence" of their existence. This is especially true for species that live in remote, inaccessible areas. Many times, species have reappeared after being officially declared extinct. The coelacanths (pronounced seel-a-canths) are probably the most notorious such organism. It was believed to have become extinct along with the last dinosaurs about 65 million years ago, because only fossils were known to exist, but in 1938 live specimens were discovered, with many others after that time.

A more recent example is a freshwater snail called the oblong rocksnail, declared extinct in 2000 but recently found alive and well in its native *habitat* in Alabama.

extinction, newly identified species and

Occasionally, the discovery of new species and extinction happen almost simultaneously. Scientists have been discovering new species but soon realize their new finds are on the brink of extinction. A species going *extinct* is always unfortunate, but even more so when the species has not yet been studied. Recently, this has been the case with a huge spider found in Israel with a five-inch leg span. It is believed to only live in a small region that has been slashed in size through *habitat destruction* caused by *agriculture* and *mining* and may be on the brink of extinction.

A recently discovered small monkey in Colombia, called the Caqueta titi monkey and resembling a leprechaun, is already on the *threatened species* list. In addition to these and many other animals, numerous plants are believed to be in a similar state of new discovery at the same time they are in jeopardy of extinction.

Scientists estimate around 60,000 species of plants are yet to be discovered and are concerned that many of these—because they will most likely live in *fragmented* and fragile *habitats*—will be candidates for the threatened species list as soon as they are discovered.

extinction, sixth mass

Fossil records show five mass extinctions over the past 540 million years, with the last one wiping out the dinosaurs about 65 million years ago. Many scientists and environmentalists are saying we are in the midst of a sixth mass extinction, this one brought on by humans. Humans are believed to be responsible for increasing the normal rate of extinction by anywhere from 1000 to 10,000 times higher. They also calculate that somewhere between .01 percent and 0.1 percent of all species become *extinct* each year. This translates to between 200 and 2000 extinctions each year—no longer to ever be seen alive on our planet. Estimates are that another 20,000 species are in danger of extinction at any given time. The causes of *biodiversity loss* are numerous, but the end result is the same—extinction of epic proportions caused by humans.

extremely low-frequency magnetic radiation (ELF)

Power companies in the United States and Canada deliver electricity at a frequency of 60 hertz. Electromagnetic radiation created by currents at this frequency level is called extremely low-frequency (ELF) magnetic radiation. Studies about whether any public health concerns are caused by this form of radiation have been inconclusive and controversial. A few studies have indicated long-term exposure to high-power tension lines increases risks of childhood leukemia and cancers, but many other studies find no relationship. The general consensus today is that this form of radiation is not considered a health risk.

extreme weather and climate change

Weather and climate are not the same thing. According to the *National Oceanographic and Atmospheric Administration* website, "The difference between weather and climate is a measure of time. Weather is what conditions of the atmosphere are over a short period of time, and climate is how the *atmosphere* behaves over relatively long periods of time."

People are beginning to ask if the increase in extreme weather events is related to *climate change*. The answer according to most scientists is yes. The *Intergovernmental Panel on Climate Change*, in a 2012 report, stated that heat waves, coastal flooding, extreme precipitation weather events, and severe droughts show a strong correlation with change in our climate. However, increased numbers and severity of hurricanes and tornados is more questionable. But a few scientists disagree with this assessment, believing it is not possible to connect the two. [*drought, the latest; megafloods; warming hole; Global Reporting Initiative; global warming; Investor Network on Climate Risk; Keeling Curve*]

extremophiles

Organisms found beyond the so-called normal range of life within our *biosphere* are considered extremophiles. Not too many years ago, it was thought that all life on our planet existed within narrow limits up into the *atmosphere*, down into the *lithosphere*, (solid Earth) and *hydrosphere* (waters), and within a narrow range of properties such as temperature, *pH*, and salinity. But new discoveries have forced us to redefine where life exists on Earth. In most cases, these extremophiles are *bacteria* or other microbes.

Thermophiles live in regions too hot for most life. Some microbes thrive in temperatures as hot as boiling water (212 F) and may live in hot springs or in *hydrothermal vent ecosystems* at the bottom of the ocean where the vents pump out water more than 300 degrees F. These unusual ecosystems contain some of the most bizarre extremophiles.

Psychrophiles live in regions too cold for most life. Entire communities of microbes have been found living in frozen *Arctic* ice. Still others include acidophiles that thrive in highly acidic environments, alkaliphiles in highly basic (alkaline) environments, and halophiles in saline (salty) environments. Extremophiles demonstrate how *natural selection* fills any and every void with life on our planet.

Many of these newly discovered organisms are classified in a third major domain of life, called the *archaea*. [*weird life; cryptic species; aerobiology; chemoautotrophs; high-altitude ecosystems; hydrothermal ocean vent ecosystems*]

Exxon Valdez

On March 24, 1989, an *oil tanker*, the Exxon Valdez, went off course, hit a reef, and spilled 11 million gallons of *oil* into Alaska's Prince William Sound—creating the worst tanker *oil spill in U.S.* history. It remained the worst oil spill in U.S. waters until the 2010 *BP Deep Horizon* spill. (The *Amoco Cadiz* oil tanker spilled five times as much oil off the coast of France in 1978.) The Valdez incident polluted about 1000 miles of Alaska coastline and coated tens of thousands of animals with oil, many of which died. The degree of harm to surrounding ecosystems is still being studied. Follow-up studies show the damage to the region's habitats is long-lasting and continues to have a negative impact.

Exxon spent more than a year cleaning up the mess and established a damage claims program for the residents whose livelihoods were affected. The accident cost Exxon about 2.5 billion dollars for cleanup and damages. The Exxon Valdez was not a double-haul tanker, which could have prevented the accident. The following year, the United States passed the *Oil Pollution Act*, which requires all new ship hauls to be double-hulled. [*oil spills in history, ten worst; oil pollution in the Persian Gulf; ship-sinking program; Rigs-to-Reefs Program; pipeline spills*]

Facebook's carbon footprint

Facebook must be credited for being open about how they are doing environmentally. Beginning in 2011, they shared with the world their *carbon footprint* information by posting what types of energy sources they use to drive their business, including their *data centers* and other operations. In 2012 their energy came from the following sources: *renewable energy* sources such as *solar power* and *wind power*, 23 percent; *natural gas*, 17 percent; *coal*, 27 percent; *nuclear*, 13 percent; and other, 20 percent. The company continues to move toward being as green as possible by building a new green data center in Sweden that will take advantage of the natural cold climate to minimize energy consumption, keeping their cloud-computing hardware cool. [*blogs, green; crowd funding, green; crowd research, scientific; crowdsourcing, eco-; microblogs, green; social media, environmental; tweets, eco-*]

factory farms

Most livestock raised for slaughter comes from highly mechanized, high-tech, and high-volume factory farms. Factory farms have high-density populations in controlled environments. Animals are often kept in what many people consider to be inhumane conditions. Growth hormones and antibiotics are typically used extensively to ensure rapid, healthy, and substantial growth in the animals.

High-volume factory farms globally raise more than 1 billion pigs, 1.4 billion cattle, 2 billion sheep and goats, and 19 billion chickens. Eating meat is far more energy intensive when compared to eating plants. In addition, the animals on these farms—thanks to *belching* and flatulence—emit 16 percent of the global *methane* gas emissions that contribute to *climate change*. Also, the raw animal waste, when not properly managed, causes *water pollution*.

NGOs have raised awareness about these problems, but global per capita meat consumption has doubled since 1950 and continues upward—mostly because of the increased affluence in many developing nations.

Most of these large-scale factory farms are now handled in facilities called *CAFOs,* which exacerbates these bad conditions. CAFOs have become so prevalent that 72 percent of all poultry, 43 percent of all *egg* production, and 55 percent of all pork produced comes from CAFOs. [*antibiotics for livestock; bovine growth hormone; dairy cow output; farm animal cruelty legislation; industrial livestock industry; rendering; egg farms*]

factory ship

Sometimes called a mother ship. The term originally only meant a large ship used in *whaling*, but now it describes any ship equipped to not only catch but also process the marine catch. It might be the main ship in a fleet or it might be on its own at sea. These ships can catch, process, freeze, and store fish or *whales* while at sea before they are brought to shore and market. These ships contribute to *overfishing* and the decimation of many fish and sea mammal populations—so much so that on a global basis the number of factory ships has diminished, because there are not enough fish left in the sea for them to catch.

Most commercial fish populations have been reduced in size by 70 percent and some by 99 percent, so even these huge factory ships today catch about six percent of the quantity sailors did 120 years ago. Catching fish with modern technology cannot make up for fish that no longer exist.

Factory ships are still used for whaling by some nations and continue to be highly controversial. The *Sea Shepherd Conservation Society* plays an active role in reducing their effectiveness, often with great success. [*fishing, commercial; tragedy of the commons; fish populations and food; Magnuson Fishery Conservation Management Act; pirate fishing; bottom trawling; driftnets*]

fair trade certification

Fair trade is a form of trade designed to build a fair and long-term partnership between producers—usually in *developing countries*—with consumers in developed countries. It can be a valuable tool to help reduce poverty.

In a true fair trade agreement, the producers receive a minimum set price for their goods, and the laborers are afforded safe working conditions. This is usually accomplished by making agreements directly between the buyer and the producer, cutting out anyone in the middle. The agreement should also include some form of economic development for the producer communities and educational opportunities for their children. They should also receive some form of financial and technical support.

The consumers, at the other end of the agreement, receive quality products and know they are actively addressing poverty, helping protect the environment, and promoting an end to child labor.

As with any *eco-certification*, you must be sure the *eco-label* stating a product to be fair trade is legitimate. The *NGO*, Fair Trade USA, certifies that producers adhere to environmental and labor criteria. They put the farmer in direct contact with companies removing the middleman. Green Mountain Coffee uses Fair Trade USA to ensure it purchases fair trade coffee. Wine, flowers, cocoa, and nuts as well as coffee can all be fair trade certified, but coffee is by far the most prevalent.

Fair trade coffee buyers pay a community-development premium that goes directly into the producer's local farming communities to improve residents' lives. This is usually 20 cents per pound for conventionally grown coffee, all of which goes to the community, or 50 cents per pound for organic coffee, of which the extra 30 cents goes back to the farmer as an incentive to grow organic crops.

In 2011, about $17 million of fair trade premiums went to community-development programs to build new schools and health care facilities, among other projects. [*governments of the forest; coffee, environmental impact*] {fairtradeusa.org}

fallow field
A field that is left idle to restore its productivity. [*conservation agriculture; low-input sustainable agriculture; Conservation Reserve Program*]

farm animal cruelty legislation
There have been many undercover exposes about how farm animals can be treated inhumanly. Most of the time, these reports result in outrage and some form of criminal punishment or new regulation to prevent such treatment from continuing. Recently, the industry is fighting back, trying to pass legislation that makes it illegal to report about the mistreatment of farm animals. Laws have been passed in many states, called ag-gag bills by *animal rights* activists, making it illegal to covertly film such activities. Iowa, Utah, and Missouri have all passed such ag-gag bills, making it virtually impossible for animal rights activists to do any work in these states. [*CAFO; factory farms; cosmetics, cruelty-free; vivisection; dissecting frogs in school; greyhound dogs; circus elephant ban*]

farm bill
The first farm bill was part of the New Deal back in 1933. It created price-support assistance to help struggling family farms. Over the decades it has changed dramatically,

now supporting and enriching big industrial farms (*high-impact agriculture*) and agribusiness in general.

Besides this financial support, the bill has significant environmental importance because it plays an important role in land *conservation*, agricultural research, and food safety issues. The bill is rewritten every five years, so the funding for these environmental concerns is constantly changing up or down, depending on which politicians are in Congress at that time.

The bill can and often does provide assistance to farmers to install conservation measures, such as preventing *pesticides* from running off into rivers. The *Conservation Reserve Program* offers incentives to stop farming marginal lands that are prone to *soil erosion*. The bill can also encourage the use of *renewable energy* alternatives by providing cost-sharing for any farmers wanting to try *wind power*.

The most recent farm bill negotiations, in 2013, planned to cut ten percent from the Conservation Reserve Program. Environmentalists should take note whenever the farm bill comes up for renewal, because it's not just about farms, which is an environmental issue, but about the environment in general. [*sustainable agriculture; farmland lost*]

farmers' market

A market, usually outdoors, where farmers sell their products directly to the public—cutting out the "middle man." Most of these markets sell foods grown in or around the local community. This lessens the *food miles* travelled and, many believe, instills a sense of local civic promotion. In the past, most of these markets were a great way to get a great price. This is still the case in many places, but today they are often viewed as boutique venues that can demand a higher price. These markets often carry *organic foods*. Farmers' markets are one aspect of *community supported agriculture*. In 1994 only about 1700 such markets existed in the United States, but today there are thousands.

However, some so-called farmers' markets sell wholesale foods at retail prices or ship in food, grown elsewhere. Many people consider these to be *fake farmers' markets* instead of the real thing. [*locavore; seasonal foods; slow food movement; teaching farms; WWOOF*]

farmers' markets, fake

Farmers' markets are popular, so much so they have become the target of practices of deceiving unwitting consumers. Unless those running the market enforce what can and cannot be sold, the odds are good that consumers won't get what they expect.

Look behind some of the tables or booths and you might find the same packaged goods you find in your supermarket—coming from the other side of the country or overseas.

If there are no restrictions, expect the worst; it's no different than shopping at your local supermarket but at probably higher prices. If there are restrictions, ask what they are. They can be anything from all products sold must be 100 percent grown by local farmers to a certain percentage being grown by them, such as 50.1 percent from local farms. Also, check on their definition of what is local, because it might be local to the city or the state or something else. [*green scam; greenwash; greenspeak*]

farmland grabs
See *land grabs.*

farmland lost
Urbanization swallows up millions of acres of farmland in the United States annually. Some of this land is in *high market value farming counties*, meaning it is some of the most fertile farmland that remains. Some states tax land on its highest value use. The highest value of this land is often for development. Farmers who cannot afford to continue farming the land when it is taxed as if it has been developed are often forced to sell the land, so it does get developed. Many states have changed the tax laws to only tax on the existing use in an effort to save farmland. [*farm bill; land grabs; high-impact agriculture*]

farm-to-school movement
This movement can be thought of as a subset of the *farm-to-table movement*; in this case, the table is in the school cafeteria. The National School Lunch Program provides federally assisted meals to over 31 million school children in the United States every school day. The farm-to-school movement tries to make these meals healthy by working with school districts and food service companies to incorporate fruits and vegetables from local farms, *community gardens*, or *college quad farms*. [*community supported agriculture; food miles; slow food movement; teaching farms*]

farm-to-table movement
Food impacts both our health and the health of our planet as much as anything else. Farm-to-table refers to how food is grown, stored, shipped, prepared, and eaten, which is not only a health issue but also an environmental issue. Enough people are

concerned about this issue to have created what is now called the farm-to-table movement. You hear about it in the news and there are new restaurants and cookbooks dedicated to it. Farm-to-table means following and scrutinizing the path each step of the way and being as knowledgeable as possible about the process.

It typically includes *organic foods* from sustainable farms, which cause less harm to both us and the environment. Also, shipping distances are minimized to reduce transportation impacts, and storage times are shortened to improve nutrient value and decrease waste.

Those that take this belief to the max grow their own foods, harvest, and cook them. Some people call themselves *locavores* to emphasize these beliefs. [*farm-to-school movement; sustainable agriculture; community supported agriculture; food miles; slow food movement; teaching farms; organic farming and food*]

fauna

All the animals in a *habitat*. [*ecosystem; ecology; biodiversity; organism, most abundant; smallest vertebrate; insects; crustaceans*]

faunula

All the animals in a microhabitat, such as in *leaf litter* or under a rock. [*caliology; habitat; cryptozoology*]

Federal Food, Drug, and Cosmetic Act (FFDCA)

First passed in 1938 and regulated by the U.S. *Food and Drug Administration* (FDA), the Act was created in response to the public outcry caused by a large-scale poisoning episode. Its purpose is to keep harmful and mislabeled foods, *cosmetics*, medical devices, and drugs from being distributed. Since its creation, it has been amended numerous times—16 times just since 1980.

The FFDCA first determines if a product is considered a cosmetic, a food, or a drug. This occurs because they are regulated differently. Food products are regulated through the *Federal Insecticide, Fungicide, and Rodenticide Act*, which is primarily concerned with *pesticide residues* on food. The only other types of food controlled are *food dyes* and colorings, which are regulated by the FDA.

Drugs must go through a registration process. However, cosmetics have no approval process, only a labeling requirement. (For a company to claim that a cosmetic product has FDA approval is considered illegal, because the FDA doesn't approve the safety of cosmetics.) [*lipstick and toxic substances; nanoparticles in cosmetics; Skin Deep; body burden; toxic cocktail*]

Federal Insecticide, Fungicide, and Rodenticide Act (FIFRA)

Created in 1947 and amended six times (most recently in 1996), this law gave the *Environmental Protection Agency* authority to test and regulate *pesticides* in an effort to protect consumers, those who work with these chemicals (such as farmers), those who apply pesticides, and the environment. It was first managed by the *Department of Agriculture* but then transferred to the Environmental Protection Agency in 1970.

Under FIFRA, all pesticides must be registered with the EPA. Information about these substances must be provided on detailed labels that accompany the product, including use and safety precautions as well as disposal requirements. States have the right to mandate additional pesticide regulations and requirements.

Pesticide manufacturers must prove their products will not cause unreasonable harm to the environment or human health and that the product does what it claims to do (as opposed to the *Toxic Substances Control Act*, which does not). When the EPA weighs whether a product should be registered for sale, it takes into consideration the economic, social, and environmental costs, as well as its benefits. Almost 20,000 pesticides are registered with FIFRA at this time.

A related statute is the *Federal Food, Drug, and Cosmetic Act* (FFDCA), which is responsible for controlling *pesticide residues on food* in interstate commerce (including imports). FFDCA sets tolerance levels of how much of a pesticide can remain on a food product, such as an apple, when it is sold. Even though FFDCA sets the tolerances, it is FIFRA that regulates the use of the pesticide. For this reason, the EPA must coordinate the efforts of both. (The 1996 amendment did just that.) The tolerance levels established by FFDCA are based on a "reasonable certainty of no harm" by exposure to pesticide residues on the food.

The *Food and Drug Administration* (FDA) and the U.S. *Department of Agriculture* (USDA) are responsible for monitoring these pesticide residues and enforcing the established tolerances through inspection programs. (The USDA is responsible for inspecting meat and poultry, and the FDA inspects all other foods.) States can also monitor pesticide residues on foods sold within their borders. FIFRA does not prevent U.S. chemical companies from manufacturing and exporting pesticides not approved for sale in the United States. For this reason, many people believe there is a *circle of poison* problem.

FIFRA must follow regulations imposed by the *Endangered Species Act*. However, it does not have any considerations for the *Clean Water Act*. This means no studies need to be done under FIFRA to study the impact of pesticides entering *aquatic ecosystems*—something that routinely occurs.

Federal Trade Commission Green Guides
See *Green Guides, Federal Trade Commission.*

feed/conversion ratio
Modern livestock feed is high-tech big business. Livestock in *factory farms* and in *CAFO* facilities and seafood in *aquaculture* facilities are successful in part because of the efficiency of the feed/conversion ratio, which is the amount of feed needed to add body mass. For example, 8.7 pounds of feed are needed to produce one pound of beef, 5.9 pounds of feed to produce one pound of pork, 1.9 pounds of feed to produce one pound of chicken, and 1.2 pounds of feed to produce one pound of salmon. These numbers are much higher than they would be if these animals were free range or wild caught. [*dairy cow output; the Green Revolution; high-impact agriculture; genetically modified organisms*]

feedstock for energy production
Any substance used as an original source to produce energy is called a feedstock. For example, when corn or sugarcane is used to produce *bioethanol*, the corn or sugar cane is the feedstock. [*biofuels; corn to biofuel; sugarcane to biofuel; sunflower oil as a biodiesel fuel; plant oil fuels*]

Fertile Crescent
Even though the Middle East is one of the driest regions on Earth, it also contains some of the richest *soils*. The region along the Nile River and the Tigris-Euphrates river basin is where human *agriculture* is believed to have begun and is called the Fertile Crescent.

Today, this iconic region is making headlines for another reason. It is losing *groundwater* so fast as to be ranked second only to northern *India*. Its *water table* has dropped about one foot per year in recent years and portions of the Euphrates River only flows with 70 percent of previous capacity. This occurs because farmers remove more than *aquifers* can *recharge* and also because of ongoing droughts hitting the region.

Another problem is that the countries that share this basin—Turkey, Syria, Iraq, and Iran—all use the basin extensively and have no treaties or agreements on how to manage, use, and protect the water resources. Without some kind of international agreements, it appears the fertile crescent might be fertile not much longer. [*water shortages, global; water grabs; water diversion*]

fertilizer

Any substance added to the *soil* to supply *essential nutrients* required for plant growth is considered fertilizer. It can be an *organic fertilizer* or a *synthetic fertilizer*. Fertilizers supply the three primary plant nutrients: potassium, *phosphorus*, and *nitrogen*, called macronutrients. They may also contain substances required by plants in smaller amounts, called *micronutrients*, including manganese, boron, zinc, and many others.

fire blight on apples and pears

See *apples, antibiotics on organic*.

fireplaces, wood-burning

Conventional open wood-burning fire places are nice to look at, soothing to sit by, and—to many—the perfect thing for a cold winter's night. However, that is about all they are good for. They are terribly inefficient, with most of the heat going up the chimney—along with much of the ambient heat from the room. Plus, most of them contribute to *air pollution*. Only about ten percent of the wood's energy is converted into heat, producing 50 grams of *particulate matter* per hour.

A far more efficient method is to use an insert that slides into the fireplace. The inserts are designed to push the heat into the room instead of up the chimney and to burn the wood more cleanly. Inserts can convert as much as 80 percent of the wood's energy potential into heat and cut the smoke down to less than 8 grams per hour.

Another option these inserts provide is the flexibility of a fuel source. Some burn wood, wood pellets, gas, and many forms of organic waste, such as sawdust, woodchips, nutshells, corn, waste paper, and others. When burning what would normally be waste material, the *ecological footprint* of your fire can be negligible.

If you want to save further on heating bills and pollute less, consider a wood-burning stove that is EPA-certified. [*wood-burning stoves, primitive; thermal insulation, building; super insulation*]

Fish and Wildlife Service, U.S.

Created in 1940, this is the primary agency responsible for protecting species in jeopardy. The *Endangered Species Act* from 1973 empowered the *Department of the Interior* to create and manage *endangered and threatened species lists.* This is administered by the *U.S. Fish and Wildlife Service* (for all species other than marine life) and the *National Oceanic and Atmospheric Administration* (for marine species).

It also enforces the provisions laid out in the *Migratory Bird Treaty Act* and the *Marine Mammal Protection Act*. It manages the *National Wildlife Refuge System*, as well.

Environmentalists have been unhappy with how the agency does its jobs; particularly the agency's apparent preference for game species as opposed to worrying about nongame species and their *habitat protection*. They are also frustrated with the slow process of getting species designated as endangered or threatened, because there are long "waiting lists." Waiting to get on a list to prevent extinction often does not turn out well.

fisheries and seabirds

Seabird populations are harmed by *commercial fishing* in two ways. First, the depletion of the *fish populations* reduces the bird's food supply; basically, humans are competing for the same food source. Second, many seabirds dive for the baited hooks of *longlines* and become part of the unwanted *bycatch*. It is estimated that longline fishing kills more than 200,000 seabirds each year.

But changes are being made that should reduce these numbers. New devices weigh down the longlines below the surface so birds are less likely to go for them. Also, attaching streamers called tori lines scares some birds away. [*seafood eco-ratings; whaling and commercial ship collisions; marine ecosystems; barbless hooks; catch-and-release programs; fish populations and food*]

fish kills, nuclear power plant

Many industries cause *thermal pollution*, but none more so than *nuclear power* plants that cool the radioactive fuel rods with water pumped in from a nearby body of water, heated to almost 100 degrees, and then released back into the water—killing fish and much of the aquatic life. Just one such facility can suck in 2.5 billion gallons of water a day.

New Jersey and New York have passed laws requiring nuclear power plants to adhere to the *Clean Water Act* by installing modern cooling systems to replace the once-through cooling systems found on older power plants, which vacuum in fish and other sea life, killing them. These older designs kill staggering numbers of fish. Two facilities on the Hudson River are estimated to kill 26 billion fish each year. This includes the *endangered* Atlantic sturgeon, *sea turtles*, and probably trillions of newly hatched fish. [*fish populations and food; cooling ponds and towers; Hanford Nuclear Reservation; meltdown; nuclear reactor safety and problems*]

fish populations and food

Globally, fish populations are in decline. As of 2009, almost 60 percent of global fisheries were considered fully exploited, meaning they were at their maximum yield; almost 30 percent were considered overexploited; and only 13 percent were considered not fully exploited.

The primary cause of these declines is *overfishing*. Other causes include chemical *water pollution* caused by *nutrient enrichment, climate change,* and *ocean acidification. Aquaculture* is growing rapidly, because natural populations are no longer sustainable.

American fisheries had, like those in the rest of the world, been in serious decline during the 1980s because of *overfishing*. However, beginning about a decade ago, the *Magnuson-Stevens Fishery Conservation and Management Act* changed things by enforcing quotas to allow the depleted populations to rebound. The quotas were handled by scientists and taken out of the hands of the fishing industry—who preferred to self-regulate themselves out of business. This new management has resulted in good news for many U.S. fisheries.

A new report states that six commercial fisheries returned to health in 2011. There have been improvements for the past ten years, resulting in 86 percent of U.S. federally monitored fisheries—of which there are 250—are not considered *overfished* and 79 percent are actually considered healthy. America's fisheries are now considered the best managed in the world, along with Norway, Iceland, New Zealand, and *Australia*. The Act responsible for these improvements has been challenged in Congress on more than one occasion, even though the facts speak for themselves. [*fishing, commercial; Bluefin tuna; dolphin safe tuna; eels and elver fishing*]

fishing, commercial

Globally, commercially caught fish are in a desperate situation, with populations of the most popular species—cod, flounder, swordfish, and tuna—falling 90 percent in the past 50 years. *Overfishing* is the most obvious way to understand how we exploit our natural resources beyond their limits. Since the 1980s commercial fishing catch has remained the same, so the world's demand for fish has been supplemented by fish farms called *aquaculture*.

In addition to overfishing, *climate change* and *water pollution* take their toll on the world's fish populations. The world catch of 80 million tons per year is believed by many scientists to be well beyond sustainable levels. The emergence, growth, and necessity of aquaculture confirms this fact, because wild populations are no longer sufficient to feed the *human population*.

Roughly half of all fish caught worldwide are done so by *China*, Peru, the United States, Japan, Chile, *Indonesia*, and Russia. Other countries playing a major role include Spain, Poland, South Korea, Taiwan, Thailand, Mexico, Malaysia, and Vietnam. [*fish populations and food; bycatch; drift nets; longlines; pirate fishing; purse seine nets; turtle excluder device; gillnets; illegal, unreported, and unregulated fishing; tragedy of the commons*]

flagship species

These are species people love to protect from *extinction*, even if they are not the species in the most need. People love to love them because they remind us of us. They are large mammals with eyes that make eye contact (forward-facing eyes). Lions and tigers and polar bears, elephants and chimpanzees are among the flagship species. Because it is easier to get donors to open up their wallets for certain species, many organizations take the line of least resistance; raising money on flagship species that generate the most cash.

Unfortunately, this money-making method presents two problems. First, many of these animals are not nearly as endangered as some that seem less appealing but are more in need of help. Species listed as *vulnerable* are more likely to be the focus of a save-the-so-and-so campaign than species listed as *critically endangered*.

Second, these organizations often advertise their efforts to save the species—not the *ecosystem* the species resides in. This narrows the benefits that might be gained. Only 2.2 percent of the monies raised for flagship species went to protect the organism's ecosystem, and 61 percent went exclusively to protecting the species without any concern for its entire habitat. Some of the most endangered ecosystems, such as *estuaries*, have no flagship species to represent them. This is a terrible shortcoming considering how much *loss of biodiversity* will ensue without protection.

Many conservationists want to remedy this dilemma by urging *NGOs* and other conservation organizations to consider a new label along with flagship species—the Cinderella species: animals with traits somewhat similar to flagship species but not quite making the same cuteness grade. These Cinderella species are in greater need of conservation measures than most flagship species. So far, about 183 species have been designated as Cinderella species, including the pygmy raccoon, the Mindoro dwarf buffalo, the African wild ass, and many others.

In addition, many scientists want a broader, more encompassing effort to promote *biodiversity hot-spot* programs that raise conservation funds for entire ecosystems instead of just one species. [*biodiversity hotspots; charismatic megafauna; CITES; extinction, sixth mass; Lazarus taxon; Lonesome George; species recently gone extinct; Ugly Animal Preservation Society; Treaty on Biological Diversity; Wildlife Conservation Society*]

flame retardants, brominated (BFRs)

These substances first came out in the 1950s and, as is so often the case, were thought to be a great idea at the time, until their dark side was learned. These substances are used in electronic devices, furniture that contains older polyurethane foam, and clothes and other textiles to reduce the likelihood of fire. The primary problem is that most of them contain a substance called *PBDE,* which research shows is probably an endocrine *hormone disruptor.* These substances, once they work their way into the environment—usually as *municipal solid wastes*—tend to *bioaccumulate* within an organism's body and become more concentrated within organisms via *biological amplification.* Health and environmental concerns have resulted in many European countries banning them. These countries adhere to the *precautionary principle,* whereas the United States has no such regulations.

Products containing BFRs remain legal in the United States, but many manufacturers no longer use this substance in products such as polyurethane foam or clothes. Many electronics firms have promised to phase out the use of this substance in their devices. The *Environmental Working Group* has plenty of information about how to avoid these substances. [*BPA; chemical body cleanse; dioxin; nanopesticides; Toxic Substances Control Act*]

flaring

When drilling for *oil,* it is not uncommon to find *natural gas* pockets. If the well site is not designed to capture the gas, the only way to get to the oil is to release the gas in a process called flaring. Because natural gas is *methane* gas, the worst of all *greenhouse gases,* it is ignited instead of just being released. This converts the gas to *carbon dioxide* and water. However, methane typically is released prior to and after the flaring, making it still a damaging process—not to mention a waste of valuable fuel. (So valuable that many landowners are suing oil companies for lost revenues caused by flaring.)

In 2011, 210 billion cubic feet of natural gas was flared and wasted that would be enough to heat two million homes. A federal act has been proposed to stop the practice, but no action has yet taken place. [*green completion standards; the shale gale; stranded natural gas; fracking*]

flat screen televisions, green

Americans use more than 100 billion *kilowatt hours* of power watching TV, which exceeds seven percent of the total residential electric use. Your household *ecological footprint* can be minimized by buying the right TV. When buying a flat screen, here are some environmental factors to consider: The larger the screen, the more power used. Assuming similar size, LCDs use half as much power as plasma. LCDs with LED backlighting use even less power. Rear projection screens use even less power than the others. Finally, a new technology called OLED (organic light-emitting diode) will probably be the most efficient of all when it becomes readily available. [e-*waste; inkjet and laser printer cartridges; laser-printer ozone; laundry and the environment; couches and indoor pollutants; mattresses, recycling; refrigerator recycling; building related illness; glass recycling*]

flex-fuel engines

Engines that can run on *alternative fuels* such as *E85* (85 percent *bioethanol*) are called flex-fuel engines. They exist in the United States but in small numbers. They are the standard engine, however, found in some countries such as *Brazil,* where *biofuels* are commonly used. Flex-fuel engines get almost 30 percent fewer miles to the gallon when compared to regular gasoline engines. [*ammonia as a car fuel; automobile propulsion systems; diesel engine fumes; electric vehicles; hybrid cars; vehicles, hydrogen fuel-cell; alternative energy sources*]

flip-flops and preconsumer waste

When our summer flip-flops are mass produced, a great deal of waste material is left behind. Picture the shape of the flip-flop being stamped out of a large sheet of material. All of the remaining material is waste. This is a good example of *preconsumer waste*, as opposed to *postconsumer waste*, because you, the consumer, never used it. Many companies are trying to manufacture products in an environmentally friendly way, aiming for *zero waste*. This means everything in the manufacturing process must be used, with nothing left for disposal. At least one company uses the leftover materials that don't make it into the flip-flops to make colorful doormats. Using this preconsumer waste keeps it out of *landfills* and *incinerators*. [*waste stream; telephone pole disposal; tires, recycling / reprocessing; single-stream recycling; recycling, moving toward zero waste; downcycling*]

Flipper Seal of Approval

Earthtrust, an environmental *NGO*, has created the Flipper Seal of Approval that shows a cartoon likeness of the celebrity *dolphin*. Its purpose is to assist consumers in finding tuna that have not been caught in *purse seine nets* at the expense of dolphins. Only those canners using tuna captured following very strict guidelines set by the *Earth Island Institute* become certified to use the Flipper Seal of Approval. This *eco-label* is considered to be far more stringent in its definition of what *dolphin safe tuna* means, when compared to the federal (NOAA) definition. [*food eco-labels*] {earthtrust.org/fsa.html and noaa.gov/features/04_resources/tuna.html}

floating homes

Hurricane Katrina devastated much of New Orleans in 2005 and especially the low-lying 9th Ward. One of the many groups working to rebuild the city is Brad Pitt's Make It Right Foundation, which has been building unique floating houses. If another storm hits, these homes will lift up to 12 feet and will float. They are designed to safely break away from electrical and plumbing connections and have backup-power batteries to continue running appliances for up to three days. They even have an escape hatch in the roof in case things get out of hand. They are considered *green buildings* based on the materials used for construction.

These buildings made a lot of news not long after the hurricane. Today, they are still there and have been gaining in popularity and in numbers. Pitt raised $30 million dollars for the project. Enough of these odd but attractive homes have been built to become a destination for tour buses in the city. Some even offer Brad Pitt house tours. Whether they become a standard design for houses in flood-prone areas remains to be seen. [*climate change; sea level rise; Bangladesh and sea rise; extreme weather and climate change; hockey-stick graph; iceberg calving; mega-floods; melt ponds*]   .

floating wind turbines

The United States and Britain are working together on a proposed project for floating *wind turbines* that could be used in waters too deep for mounting the turbines on the sea floor and where the winds are typically stronger than around offshore *wind farms*. [*wind power; wind power brief history; wind power states, ten biggest; wind power transmission line; wind-energy areas; wind-speed vacuum*]

flocculation

The chemical or biological process in which clumps of solid waste (flocs) increase in size to expedite their removal. [*sewage treatment facilities; wastewater-to-energy power plants; water shortages, global*]

floodplain

The lowlands alongside a river, prone to flooding. Some may flood annually, while others only during catastrophic storms or any duration in between. *Urban sprawl* may use floodplains for building development, which often turns out to be a mistake.

[*wetlands; watershed; urbanization and urban growth; zoning, land use; eco-services; cost/benefit analysis; economics and sustainable development; growthmania*]

flora

Refers to the plant life in a given *habitat*. [*ecosystem; niche, ecological; imperfect flower; perennial plants; photosynthetically active radiation; phytoplankton; arborculture*]

Florida panther

A subspecies of the cougar (also called a mountain lion or a puma) that only lives in the forests of the *Everglades* National Park in southern Florida. It is one of the most *endangered species* of *mammal* and in steep decline. It is the top *predator,* with its only enemies being humans. At one time it was found throughout much of the southeastern United States and numbered in the thousands. A few wildlife refuges have been created to help this species survive, but today—primarily because of *habitat loss*— it numbers 100 to 150 in the wild and it is on the verge of extinction. [*apex consumer and ecosystems; biodiversity, loss of; poaching; endangered and threatened species lists; charismatic megafauna; habitat fragmentation*]

flowback

This is the mix of water, sand, and many chemicals that is injected into *fracking* wells and then "flows back" out of the wells, along with the *natural gas* being extracted. This flowback consists of contaminated water that must be properly handled, transported, and disposed of, usually at *hazardous waste* facilities.

flowers

The specialized reproductive organs of flowering plants (angiosperms), usually with female organs in the center, surrounded by male organs. [*pollination; annual plants; cacti; dientomophilous; National Wildflower Research Center*]

fluff

Parts of an *automobile*, other than steel, that can be recycled. [*automobile recycling; battery disposal in Mexico; recycling*]

fluorescent lighting

Fluorescent lights use less energy to produce more light, than conventional *incandescent bulbs*. A 40 watt fluorescent bulb produces 80 lumens of light per watt, while a 60 watt incandescent bulb produces less than 15 lumens of light per watt. Over a seven hour period, the fluorescent bulb in this example saves 140 watts of energy over the incandescent bulb. [*light bulb – recent history and legislation; light bulb technologies, domestic; compact fluorescent light bulbs; LEDs*]

fluvial

Organisms that inhabit streams and rivers. [*rivers in decline; running water habitats, human impact on; Clean Water Act; aquatic ecosystems; detergents, phosphates in; dredging; effluent; industrial water pollution*]

fly ash

Air emissions produced by the *incineration* of *municipal solid waste*, as opposed to *bottom ash* that remains at the bottom of the incineration facility.

flyway

The *migration* route of birds is called a flyway. In North America there are four major flyways: Atlantic, Mississippi, Central, and Pacific. Migratory birds such as ducks, geese, swans, and rails can fly thousands of miles between where they are hatched, often in Canada, and where they winter, often in Central America, such as Mexico. [*migratory bird protection; Migratory Bird Hunting and Conservation Stamp Act; Audubon Society; passenger pigeon; Monarch butterfly decline; biodiversity*]

follicular mites

Although it makes many people cringe, there is no denying our faces are flush with mites that live in hair follicles on the face. Mites are related to ticks and spiders of the class arachnids, so they have eight legs. Two species live on humans. They are only about a third of a millimeter long (a little more than 1/100 of an inch) and sausage shaped. Because they live on our body's surface, they are considered *ectoparasites* as opposed to *endoparasites* that live in an organism's body.

You are not born with mites; you contract them from others. So the older you get, the more likely you have them and the more you have. Later in life, it is almost a certainty that they inhabit your follicles. They are most common on the eyelids, face, and cheeks but can also be found on other parts of the face and even other parts of the body. They only live where there are natural oils, which is why the face is their favorite place and the summer is their best season.

They have pretty horrifying mouthparts that feed on the cells within the follicle. And they do actually crawl out of the follicle and move about on your skin, going from one follicle to another. They reproduce sexually and the female lays eggs that hatch in a couple of days. Their entire life cycle is only about two weeks. It is so short that they do not produce any waste; they don't even have an anus. Instead, the waste simply dissolves along with the decomposition of the entire organism upon death. [*parasite; hyperparasite; bedbugs and remedies; bird flu virus; microbiome and micro-biota, human; maggot medicine; nanoparticles and human health; stool transplants; skin bacteria; superbugs; zoonotic diseases*]

Food and Drug Administration (FDA)

A U.S. federal agency within the Department of Health and Human Services, whose mission is to protect the public health of consumers. They administer and enforce the *Federal Food, Drug, and Cosmetic Act* and related laws. They are responsible for preventing the spread of disease in our food and controlling the use of chemicals in both food and drugs. (Although the word *cosmetic* appears in the Act it enforces, no oversight is provided for cosmetics other than labeling.) They have the power to recall products deemed unsafe. [*lipstick and toxic substances; nanoparticles in cosmetics; Skin Deep; body burden; toxic cocktail; pesticide residues on food*]

food chain

The supply of food in any given *ecosystem* can be traced to create a hierarchy of "who eats what." This hierarchy, in its simplest form, creates a food chain, with green plants on land or *phytoplankton* in the oceans, called producers or *autotrophs* at the beginning of the chain performing *photosynthesis*. The next link contains animals that eat the plants, called primary consumers *(heterotrophs)*. They in turn are eaten by other animals, called secondary consumers. Tertiary consumers or possibly quaternary consumers may also exist. Food chains are the single lines or threads that form complex *food webs*. [*trophic levels; ecology; predator/prey relationship; symbiotic relationship; apex consumers and ecosystems*]

food desert

A phenomenon found in many low-income neighborhoods where there are few if any supermarkets or groceries. What does exist are high-priced convenience stores that offer little in the way of choice and healthy foods. In some regards, the local citizens are hostage to what they can eat. Many communities are making efforts to bring super-markets into these areas. New urban areas are being developed to include mixed zoning so groceries or supermarkets can be in the same area as residences. This is one aspect of what is called *smart growth* or smart cities. [*hunger, world; food miles; food waste; food's carbon footprint – best and worst; fregan; Good Samaritan Food Donation Act; imported food dangers; locavore; seasonal foods*]

food dye (artificial food color)

These are synthetic food colorings, made from *petroleum*-based substances. Many studies link these substances to health risks involving attention deficit hyperactivity disorder and a few to cancers, so much so that Britain has banned them completely and the European Union requires labels stating they may have adverse effects on activity and attention in children. In the United States, the FDA says there is not enough evidence to create regulations here.

Food dyes are found in numerous foods, from the obvious brightly colored candies to maraschino cherries, from packaged macaroni and cheese to salad dressings, and on and on. They are identified by numbers, such as Yellow 5 and 6 or Red 3 and 40.

An obvious alternative is *natural food colorings* that were used before the advent of these synthetic dyes. Strawberry sundaes sold in some fast-food chains in the United States use Red 40, but the same chain uses real strawberry juice color in Britain. So we get what our government allows us to get and it's up to the consumer to make smart choices.

Food dyes are not required on a package label but must be stated on the ingredients list. [*food eco-labels; organic food labels; pesticide residues on food; imported food dangers; irradiation; tomatoes—tasteless but red*]

food eco-labels

Eco-labels, *eco-certification* programs, and *environmental rating systems* all help consumers buy products and services in an environmentally conscious way. Using food eco-labels not only helps the environment but also usually means making a better choice for your health. However, all food labels are not of the same value, and many are nothing more than greenwashing. Here are some of the most commonly seen food labels and their meanings.

"USDA Organic," which is found on almost any food. These foods are produced without use of *pesticides*, *GMOs,* or *food irradiation*. Animals must be fed with 100 percent *organic* feed. The criteria for this label has recently been upgraded stating that animals must have spent at least four months per year roaming fields and feeding on grass.

"Certified Humane" is found on meat. The criteria states that the animals are fed, sheltered, and slaughtered in a humane manner.

"Hormone Free" for meat and dairy means what it says—but hormones are not used on pork or poultry, so it is only meaningful on beef.

"Non-GMO Project Verified" for cereals, breads, chocolates, pasta, and dairy means no more than 0.9 percent of the product contains GMO products. *GMO labels are not legally required as of yet, so this is one of the few ways to know if food contains GMOs. Also, the U.S. *Department of Agriculture* approved this label for meat and liquid egg products, so any of these products containing this label have not been fed any GMO feed.

"Marine Steward Council Sustainable" for canned fish and fish filets is a *seafood eco-certification* for sustainability.

"American Grassfed" for beef, bison, lamb, and goat meets standards set by the American Grassfed Association. The animals must be fed nothing but mother's milk and then fresh grass throughout its lifetime. It also means no *growth hormones* or *antibiotics.*

"Free-Range" for eggs and poultry only means the birds get some kind of access to a "range" of some sort—even if it's just a concrete slab. The only way to know if the birds are truly raised in an open field is to look into the *farm-to-table* movement, become a *locavore,* or at least visit the place to see how the birds live. "Cage-Free" for eggs and poultry only means the birds are not kept in cages, but it does not mean they are not crammed into a small shed. The only way to know for sure is to actually visit the farm.

"*Fair Trade*" for coffee, tea, sugar, bananas, and chocolates means it comes from farms with safe conditions and fair wages for its workers. It has nothing to do with organic, although they are often found together.

"All-Natural" is seen on just about everything except fruits and vegetables and is basically meaningless. There are no criteria and no one overseeing this statement. If anything, it probably is an indication that the product is not natural; if it were, it would have one of the other labels from this list.

Some other good food eco-labels are Aurora Certified Organic (various food products), Animal Welfare Approved (meat, diary, eggs), USDA Process Verified Grass Fed (meat), Soil Association Certified (produce), *Rainforest Alliance* Certified (produce, chocolate, coffee), Non-GMO Certified (produce, diary, processed foods), *Fair Trade Certified* (produce, coffee, chocolate, wine), and Fair Labor Practices and Community Benefits (produce, dairy, poultry, eggs, coffee, chocolate). [*organic food labels; seafood eco-ratings*]

food irradiation

Food irradiation refers to bombarding food products with gamma radiation to kill pests and pathogens such as salmonellae and extend shelf life. It is used in about 50 countries, including the United States. Some nations use it on virtually all types of foods, including grains, produce, and poultry, but others have approved only a few products, such as spices, for irradiation.

Irradiated food does not become radioactive, and almost all studies indicate the food is safe for consumption. A significant group of people contend that the long-term harmful effects have not been realized and urge banning of irradiated food. This might be why irradiation is not found in many U.S. supermarkets. Although food irradiation uses radioactive cobalt instead of *uranium* (as used in *nuclear reactors*), it results in *radioactive waste* that must be disposed of and therefore presents the same problems as other low-level nuclear waste. Irradiated foods are labeled with the *radura*.

food miles

The distance food must travel from the farm to your table. In the past, food *transportation* was reported to be one of the primary causes of *greenhouse gas emissions* and lessening this distance was expected to alleviate this issue. Recent studies, however, show this is a factor, but a minor one. If you look at all the environmental harm caused by agriculture, transportation is one of the least. However, it should not be dismissed, because the volume of food shipments still makes it important. For many, food miles is a more important indication of how local the food is.

Because fewer food miles indicates whether the food is locally grown, it is likely to taste better, be less contaminated (which often occurs in transit), better support your local community, and use less gas and cause fewer emissions to get to you. If you are interested in reducing your *ecological footprint*, especially your *carbon footprint*, look at your *meat consumption* habits. You would lessen your carbon footprint far better by not eating meat or by eating less meat in addition to reducing food miles. [*organic farming and food; food eco-labels; organic food labels; organic food supermarkets*]

food's carbon footprint – the best and worst

Which animal foods are responsible for the most *greenhouse gas* emissions? With this information you can modify your *carbon footprint*, if you should so choose. The *NGO, Environmental Working Group* published <u>The Meat Eater's Guide to Climate Change and Health</u>. They analyzed the *cradle-to-grave* impact of many types of foods to find out their carbon footprint. The results for the worst three (causing the most emissions) are 1) lamb, 2) beef, and 3) hard cheese (although a distant third). The best four are 1) yogurt and milk, 2) canned tuna, 3) eggs, and 4) chicken (although canned tuna has other environmental problems associated with it). [*organic farming and food; food eco-labels; seafood eco-ratings; organic food supermarkets; sustainable agriculture; wine, organic and biodynamic*]

food security

See *grain production and use.*

food waste

Estimates have the total global amount of food grown for human consumption to be about 1.3 billion tons—with about a third of that ending up as waste. Wealthy countries are the worst offenders, because people in poor nations cannot afford to waste the little they have. In the United States alone, people threw out more than 34 million tons of food in 2009. That is about 14 percent of the total *waste stream* that finds its way into *municipal solid waste landfills* or *incinerators.*

The only product we throw out more than food is paper. However, we *recycle* most of the paper—almost 60 percent—but only three percent of food waste is *recycled* (converted) into *compost.* This is a shame, because compost is valuable and can be used as an *organic fertilizer.* Only a few cities, such as *San Francisco,* Seattle, and Boulder, have residential curbside pickup of compostable waste materials.

Food waste is not just created after you, the consumer, are through with your meal. It begins with farmers. If the price for a crop drops too low, it is more economical for farmers to just plow it under instead of harvesting it. For example, a study found that 90 million pounds of broccoli are annually left un-harvested.

Then it's the supermarkets' turn to waste food. They routinely throw out tons of food that is not perfect in appearance when first received or begins to spoil, even if it is still edible. The United States *Department of Agriculture* requires supermarkets to throw out an entire carton if one egg is found broken, resulting in about five billion wasted eggs. Huge numbers of prepared foods, such as rotisserie chickens, are thrown out at the end of each day.

Many supermarkets are beginning to realize that consumers are willing to buy imperfect fruits and vegetables if discounted. Produce with a few blemishes might turn off some buyers but are usually just as tasty and healthy as those with no imperfections. Millions of tons of this imperfect produce could feed people instead of being thrown out or used to feed livestock. An estimated $15 billion worth of unsold produce is wasted annually in supermarkets. However, some supermarkets, in addition to selling discounted imperfect produce, routinely donate unsold food to soup kitchens and food pantries.

Restaurants have an effect, as well. Estimates are that about 15 percent of all food wastes entering landfills come from restaurants. One restaurant can routinely dispose of well over 10 tons of food waste a year and as much as 50 tons. Some of it comes from peeling and trimming before serving, much of it is simply never used, and much of it comes from leftover foods patrons don't take home in so-called *doggie bags.* Some restaurants routinely donate these foods to soup kitchens and food pantries, as do supermarkets, to feed the poor.

In the past, many retailers were concerned about the legal issues of donating food. In the past, if someone became sick from food they donated, it could have resulted in legal action. However, in 1996 President Clinton signed the Emerson Good Samaritan Food Donation Act that protects them when donating food products to nonprofit organizations.

And then—at the end of the food waste trail—are our homes. Any leftover foods—or any organic matter—can be *composted* to make some excellent *fertilizer* for a garden or *lawn.* Unless you live in one of the very few towns or cities that pick up compostable wastes, you will have to do the *composting* yourself. But it is not difficult to do. [*farm-to-table movement; food eco-labels; food miles; food's carbon footprint— the best and worst; fregan; grain production and use; hunger, world; seasonal foods; slow food movement; vegan and vegetarian diets and the environment*]

food web

Food chains are an easy way to learn about relationships that exist between organisms in an *ecosystem.* However, no food chain stands alone; all are connected to other food chains. These intertwined chains create a complex food web that shows all of the feeding relationships that exist within an *ecosystem.* For example, a single type of plant might be eaten by five types of *insects,* each leading to its own food chain, or a mouse might be eaten by a fox or a hawk, each leading to its own food chain.

The two basic kinds of food webs are grazing and decomposing (also called detritus) webs. The examples in the previous paragraph are grazing webs, in which the energy harnessed by the green plants is used as food to build larger and more complex organisms. In decomposing food webs, however, dead organisms are broken down and decomposed by *insects, bacteria,* and *fungi.* For example, a dead mouse might be eaten first by scavenging insects. Once the insects decompose the mouse sufficiently, it may be attacked by bacteria and other microbes that complete the decomposition, returning the nutrients to the soil or water.

In grazing food webs, organisms eat each other. But in decomposing food webs, all the organisms are eating the same thing—the decomposing mouse—but at different stages of decomposition.

forage fish

Also called bait fish, these small fish, low in the ocean *food chain,* usually eat *plankton*—some are filter feeders. They in turn are eaten by larger predatory fish and other organisms, such as seabirds and *marine mammals.* Because they are small, they find safety in numbers and often swim in large schools. They include sardines, anchovies, menhaden, and herring. Forage fish play an important role in *oceanic zone ecosystems*, transferring the energy from the *producers* of the sea, the *phytoplankton*, along to the larger fish that feed on them, such as tuna, salmon, and cod.

Forage fish populations are dwindling because they have become an important harvest in recent years—mostly as feed for *aquaculture*, or artificial *fish farms*. Almost 90 percent of all the forage fish caught are for nonhuman use. For example, most of the sardines caught off the U.S. West Coast are used for Bluefin tuna fish farms or as bait for commercial *long-line* fleets in search of tuna.

The increase is dramatic. In 1950 only 8 percent of commercial fish caught were forage fish, but today it is 37 percent. As aquaculture grows at an explosive pace, so too does demand for forage fish. The commercial value of these fish is estimated to be 5.6 billion dollars per year. However, studies determining the *eco-services* provided by these fish are estimated to be more than 11 billion dollars per year.

Like many other populations of wildlife on our planet, we are seeing the *tragedy of the commons* repeat itself. Without quotas and enforcement, these fish will disappear. Many people are calling for stronger regulations to make the forage fish harvest sustainable. [*fish populations and food; fishing, commercial; Magnuson Fishery Conservation Management Act; overfishing*]

forams

Much of our planet's limestone is made of calcium carbonate, some of which comes from *coral reefs* but most from a single-celled organism called foraminifera, or forams for short. There are 6000 species of these small *marine* organisms that produce beautiful shells of calcium to protect their bodies. Their ornate shells have earned them the name "nature's masons."

When these free-floating organisms die, they float to the bottom of the sea and build up in the sediment below. Over millions of years, this became limestone deposits, still harvested today for its building qualities just as it was in ancient Egypt when it was used to build the pyramids.

Forams play an important role in removing *carbon dioxide* from the *atmosphere*; a process called *carbon sequestration storage*. When the gas dissolves in the oceans, carbonate is formed—and this is what forams use to build their shells. Vast quantities of these forams float along with other plankton in the upper layers of the ocean and store (sequester) about 25 percent of all the carbonate that is created in the oceans each year. This removal of carbon dioxide from the atmosphere and storage of it in their shells is considered an important *ecoservice*.

Unfortunately, if the concentration of carbon dioxide continues to rise—as is happening with *climate change*—the seas will become more and more acidic. Instead of increasing carbonate production, as one would expect, the opposite occurs. More acidic water reduces the amount of carbonate in the water, making it harder for foraminifera to build their shells. This results in less carbon being stored in the oceans and more in the atmosphere. Recent studies show that their shells are now 30 percent thinner than before the *Industrial Revolution*. [*ocean acidification; carbon sink*]

forest

Among the many different types of forests, such as *temperate deciduous forest* and *tropical rainforest*, the common feature is the dominance of trees. Forests are among the most important *biomes* on Earth. They provide some of the most essential *ecosystem services* of all regions, such as protecting *watersheds* and acting as *carbon sinks* that store vast amounts of carbon, keeping it out of the *atmosphere*. Wood remains, in many parts of the world, the prime fuel source (called *biomass* energy).

Estimates have almost 90 percent of all land-based organisms living in tropical rainforests alone, and the majority of medicines derived from plants come from forest *ecosystems*.

Deforestation is one of the more pressing environmental problems facing our planet today. [*forest health, global; forests, sustainable; forest management, integrated; afforestation; reforestation; clear-cutting; deforestation and climate change; forest, pop-up; governments of the forest; logging, illegal; timber, DNA sequencing of; Tongass National Forest; dual-use policy; multiple use policy; Forest Stewardship Council*]

forest fragmentation

Forests are cut down in many ways. If not totally cleared, as in *clear-cutting*, they often are cleared piecemeal, leaving small parcels of forest like a patchwork of forests, called forest fragmentation. Although some of the forest remains, it destroys the *forest ecosystem* that once thrived. Fragmented forests do not provide animals with food, water, mates, and other factors required to exist. If the food is located in one fragment and water in another and a mate in still another, and there are no trails to connect them all, a fragmented forest spells doom for many of the residents. [*anthromes; habitat fragmentation, human-caused; deforestation*]

forest health, global

The global health of *forests* around the world is measured by both density and area. A recent report analyzed data from 1953 to the present and shows that forest density—how many trees grow within existing forests—improved everywhere except Asia. The researchers assume the improvement is probably because of better *forest management* in many countries—except in some parts of Asia, where *deforestation* has been extensive. *China* was a major exception to the problems in Asia; forest density actually increased there because of large-scale *afforestation* programs.

Forest area expanded in Europe, remained the same in the United States, and decreased in South America and Africa.

Forest ecosystems play a vital role in our planet's health, and because forests act like a huge *carbon sink*, healthy forests mean less carbon dioxide in the *atmosphere*. [*forest management, integrated; forest, sustainable; forest, pop-up*]

forest management, integrated

Timber in *forests* is a *natural resource* that humans have always exploited. The wildlife that live within forests are also a natural resource, and the entire forest *ecosystem* provides valuable *eco-services* we cannot afford to lose or even reduce. Balancing our economic needs with environmental protections is called integrated forest management.

Integrated forest management techniques can minimize the negative environmental impact of *logging* and result in *sustainable forests*. But improper techniques often damage or destroy entire forest ecosystems. (And the loss of *ancient forests* can never be replaced.)

Some methods, such as *clear-cutting,* are economically beneficial to the logging industry but often devastating to the forest. However, s*elective harvesting, reforestation,* and *afforestation* can result in sustainable forests, even while they're being harvested.

forest, sustainable

There is a difference between a sustainable yield of timber within a *forest* and sustaining a forest. The former has historically been the focus, but today emphasis is placed on the latter.

Forests provide enormous *eco-services* for our planet. You cannot envision a healthy planet without healthy, sustainable forests. Efforts to this end include the creation of important *eco-certification programs* such as the Sustainable Forest Management certification and its international arm, the Programme for the Endorsement of Forest Certification. These certifications allow companies to place *eco-labels* on their products to assure consumers that the products come from sustainable forests.

As always, these certification programs and eco-labels only work when people choose to consider the environment when making purchasing decisions.

forest, pop-up

Tundra on Russia's northwest Arctic coast is showing unusual signs of *climate change.* The tundra typically is inhabited with low shrubs such as willow and alder. They remain low because harsh conditions limit their growth. But with warming from climate change, the tundra has become greener and the low-lying shrubs are growing into trees, creating what appears to be—for lack of a better term—pop-up *forests.*

This is not what scientists expected to happen. They thought changes in temperature would allow trees from a neighboring biome, the *boreal forests* to the south, to migrate north. However, instead of nearby species migrating into the tundra as it warmed, the original inhabitants adapted to the change. Scientists are just beginning to learn about the rapid changes caused by climate change. [*natural selection; climate change and biodiversity*]

Forest Service, U.S.

In 1891, the first U.S. Forest Reserves were created and managed by the *Department of Interior.* In 1892, *Gifford Pinchot* thought these *forests* were not managed properly and persuaded Congress to turn them over to the Bureau of Forestry, within the *Department of Agriculture,* which he managed. In 1905, under the presidency of *Theodore Roosevelt*, Pinchot changed the name of the Bureau of Forestry to the Forest Service. In these early years, *Aldo Leopold* worked for the agency, contributing to their *conservation* efforts. However, much of its history has seen controversy over its *dual-use policy*—preserving these forests as well as using them for economic purposes. The stated goal is to make the *forests sustainable* for future generations—an endgame that includes both uses.

Over the years, many acts and amendments have been passed to change how the Forest Service operates. Some have been in the interest of preserving the land, while others gave forest managers complete control over timber sales, with little regard to protection. The battle goes on, but the forests appear to be losing and the logging industry appears to be ahead.

The Service in more recent years has favored timber cutting and road building over all other forest uses. The most recent act—Restoring Healthy Forests for Healthy Communities Act—is still in debate; it mandates that 50 percent of the forest can be used for logging and includes language that will make it difficult to challenge in the courts. Former President Clinton stated the following about the Forest Service: "These unspoiled places must be managed through science, not politics.Many people, including even those who don't call themselves environmentalists, find it appalling that the Service routinely loses money each year, selling valuable logging rights to private companies. However, Forest Service managers receive budgetary rewards for selling timber, even if they lose money. In past years, the Service spent far more money managing its timber sales program and building logging roads for the timber industry to use than it received from the sale of rights to cut down the trees. When the government loses money, the American taxpayer picks up the tab. Therefore they are contributing to the destruction of their own national forests.

The Forest Service celebrated its 100th anniversary in 2005 and currently manages 155 national forests, 20 national grasslands, and more than 192 million acres.

Forest Service Employees for Environmental Ethics (FSEEE)

A group devoted to changing the current U.S. *Forest Service's* values to reflect a greater ecological understanding. Members include current, retired, and former employees of the Forest Service. Their strategies are to provide an open forum for expression of the facts about public land management, provide a support system for Forest Service employees, and educate the public on effective ways to practice good land management. {fseee.org}

Forest Stewardship Council (FSC)

The FSC is an *NGO* founded in 1993 and primarily funded by the *World Wildlife Fund*. Its mission is to improve *forest management* to promote *sustainable forests*. Members include the forest industries, *eco-certification* organizations, indigenous peoples, local community groups, and environmental organizations. There are more than 550 member groups from 67 countries, including both developed and *developing countries*.

The FSC is itself an eco-certification organization that accredits other groups that certify forest products. Buying a product with the FSC-certified *eco-logo* assures you it comes from a well-managed sustainable forest. {fsc.org}

formaldehyde

Formaldehyde is a *volatile organic compound* (VOC) used in many building materials, such as particleboard, plywood paneling, and fiberboard. But it is also used extensively in *plastics*, textiles, *pesticides*, and even *cosmetics*. Its ability to bind dissimilar substances together to produce new, complex compounds led to its manufacturing use. Billions of pounds of this colorless substance are produced in the United States each year.

Formaldehyde "evaporates" out of these products by a process called *outgassing*. Most outgassing occurs within the first few months after production, releasing substantial quantities of the substance into the air. Formaldehyde has been found inside buildings in fairly high concentrations.

The most obvious health problem associated with formaldehyde is upper respiratory tract irritation producing symptoms similar to a cold, such as irritated eyes and sinuses, coughing, and running nose. Some people get headaches. It has been identified by the National Toxicology Program's Report on Carcinogens as a probable human carcinogen.

Because it is found in many products, many types of alternatives have been developed. Consider formaldehyde-free *cosmetics* and cleaning products. Look for carpeting that has few or no VOCs. Many manufacturers of products such as shampoos and deodorants are removing formaldehyde from their products on a voluntary basis because of public pressure and *green consumers*. (There are no regulations preventing their use.) Possibly one of the best ways to learn which products contain this and other toxic substances is to check out the *Environmental Working Group's* online consumer guides.

formations

Scientists have divided the Earth into large regions containing distinct groups of plant life called formations. For example, all the plants in a grassland are considered a formation. [*biomes; anthromes; zones of life; bioregions*]

fossil fuel

Fossil fuels include *coal*, *oil*, and *natural gas*. These fuels provide the majority of the world's energy supply. They are considered *nonrenewable energy* sources because deposits of these substances are no longer being replenished and will eventually become depleted. [*energy consumption, historical; energy use; energy, A very brief history of U.S.*]

fossil-fuel reserves

Reserves are the identified quantities of *fossil fuels* available for immediate extraction from the Earth—that portion of the total amount believed to exist (called resources) that are economically feasible to extract. For example, *natural gas* reserves in the United States have dramatically increased in recent years with the advent of new extraction technologies such as *fracking* and horizontal drilling. [*energy consumption, historical; energy use; energy, A very brief history of U.S.*]

fossil-fuel resources

Fossil-fuel resources are the projected amount of a *fossil fuel* that exists within the Earth. The total resource may not ever become available for use, because it might not be economically feasible to extract, or the technology to do so might not exist. [*fossil-fuel reserves*]

fossil fuel subsidies

A subsidy is money given by a government to a recipient. The recipient can be a person, a group, or an entire industry. The money is usually in the form of a tax credit. Traditionally, subsidies are to be provided to a recipient to relieve a financial burden and to be in the public's best interests. Examples of subsidies include welfare payments, first time buyer housing loans, student loans, farm subsidies, and *fossil fuel* subsidies.

Many people feel strongly that fossil fuel subsidies have little to do with the public's best interests and far more to do with politicians and corporate best interests—but at the people's expense. Subsidies provided to *oil*, *coal*, and *gas* (fossil fuels) companies are among the most hotly debated environmental issues of our day.

There are two kinds of fossil fuel subsidies. Production subsidies lower the cost of production by methods such as beneficial tax treatment for these companies. Consumption subsidies lower the price consumers pay for the energy by offering tax credits.

A large number of people believe subsidizing oil, gas, and coal companies – some of the richest companies in the world—is nothing less than ludicrous. The total amount of fossil fuel subsidies globally is about 520 billion dollars per year (2011).

These subsidies help keep fossil fuels entrenched in our societies and thwart efforts to advance *renewable energy* sources. Burning fossil fuel adds tremendous costs to society, from both environmental degradation and public health problems. A reduction in these subsidies would reduce fossil fuel use and result in estimated savings of $120 billion in pollution and health-related costs each year, globally.

From 2002 to 2008, U.S. fossil-fuel companies received about $70 billion in subsidies, while renewable-energy companies received a little more than $17 billion. Many people believe far more subsidies should go toward renewable energy resources to help transition our use in the right direction. Additionally, fossil-fuel subsidies are such a time-honored tradition that they are in our tax code, meaning they are permanent, whereas renewable subsidies are approved only when Congress deems they are needed and for a limited time.

It's not that the world is unaware of this issue. People have been talking about stopping these subsidies at many levels for a long time. At the *Rio+20* meetings, an organized anti-fossil fuel subsidies *tweetstorm* was unleashed toward the delegates. In 2008, at the highest levels of government, G20 nations pledged to phase out these subsidies. This looked great in the press and made many of their constituents happy;

however, as of 2013, not one of the 20 nations has made much progress doing so. In many cases, these nations do not even bother reporting these numbers to the G20, thereby avoiding bad public relations.

Fossil-fuel subsidies remain a major obstacle to making a smooth transition from fossil fuels to renewable energy because, with these subsidies, companies have little financial incentive to change. Eliminating them, or even significant reductions, would move renewables into the forefront far more quickly, with the endgame being reduced *greenhouse gas* emissions and social change on a major scale.

Fossey, Dian

(1932–1985) Fossey was an American biologist and wildlife activist who spent her life working with *endangered* mountain *gorillas* in Rwanda and Zaire. Her work became famous with the release of her book <u>Gorillas in the Mist</u> (1983) and a subsequent movie of the same name (1986). While doing field work, she used techniques similar to *Jane Goodall*'s work with chimpanzees.

After one of her favorite gorillas was killed by *poachers*, she became a spokesperson and leader of the anti-poaching movement. She was murdered in Rwanda in 1985 in what many believe was retaliation for her work fighting poaching. The Dian Fossey *Gorilla Fund International* is still in operation and an important *NGO* working on their behalf. [*biodiversity, loss of; animals experiencing grief; elephant poaching; primate research*]

frackademia

Describes universities that perform research on *fracking* but have financial ties to the *oil* industry. The implication is that research is supposed always to be unbiased and strictly science-based but it cannot be so if the research is funded by an industry that would benefit from a particular result—in this case, the result that fracking is not environmentally harmful.

fracking

The common term for hydraulic fracturing, an *unconventional natural gas extraction* technology often used with horizontal drilling. The process extracts *natural gas* and *oil* from geological formations that, in the past, were impossible or too expensive to retrieve because the gas or oil had low permeability (wouldn't flow easily enough for removal.)

Three such types of geological formations where low permeability is an issue and fracking required are *shale beds*, tight sands, and *coal-bed methane*. Two enormous reserves of shale beds have recently opened up for fracking in the United States: the *Marcellus Shale* bed and the *Bakken Shale* bed.

Fracking has changed the energy situation in the United States and probably the world—for better or for worse is not yet known. It is a high-impact process that most environmentalists oppose. At the federal level and in many states, a debate is currently raging over the future of fracking. The choices are to allow fracking to continue unabated, allow it with strict regulations, or to ban fracking outright because of environmental concerns.

Many in favor of the process believe the *energy independence* provided by fracking is worth the *environmental risk* caused by it. But almost all environmentalists believe the *environmental concerns of shale-bed fracking* far outweigh the advantages. Environmentalists are also concerned that these new, vast sources of natural gas will slow the progress made in advancing *renewable energy* sources.

Even with the ongoing debate, fracking is more tolerated in the United States than in Europe. In the United States, landowners own the rights to all minerals found beneath their property. (A good example of *YIMBY FAP* which stands for, "yes in my backyard, for a price.") However, in Europe, they do not, so they will not reap the financial rewards if they allow fracking on their property. *NIMBY* plays a big role when there is little financial incentive. Most European nations have restrictions on fracking, and France has banned it outright. [*Clean Water Act and fracking; flaring; flowback; frackademia; Gasland, the movie; mortgages and fracking; natural gas loophole; shale gale; stranded natural gas*]

fracking, alternatives to shale bed

Many people want to stop *fracking* completely because of *environmental concerns about fracking*, but others believe the process might become acceptable if companies would use less damaging techniques. Two major concerns are the *toxic chemicals* injected into the fracking wells and the amount of contaminated water produced.

Some companies now minimize the toxic chemicals used during the fracking process. They use no chemical additives or, at least, nontoxic additives. Other companies now store the contaminated water in "lined" *open pits* so they do not leak. In a few locations, the *flowback* water goes directly into tanks to be transported away from the site, bypassing all together the problems associated with temporary storage in open pits.

Another solution being considered is to filter the contaminated flowback water removed from a fracking well so it can be used again at another site. But possibly the best solution would be waterless fracking. Research is ongoing to see if some gels can replace water completely. These gels would not only save water but would not collect natural contaminants, as water does, meaning it would be less contaminated when retrieved.

Another technique being researched is a "fracking water sponge" that soaks up the contaminants in the water once retrieved from the fracking process. The sponge cannot absorb water but is designed to absorb substances found in water, such as fracking contaminants. So when the sponge is placed in the used fracking water, it soaks up the harmful substances, increasing in size eightfold; none of it is water—only contaminants.

fracking, initial environmental concerns of shale bed

Early concerns about *fracking* for *natural gas* focused on two problems: contaminated *groundwater* and earthquakes. Fracking takes place thousands of feet below the *water table*. However, the bore and pipes must obviously travel through the water table, and leaks can occur. In many cases, *groundwater* supplies have been found contaminated with gas from these wells. Even though the natural gas industry states fracking cannot contaminate groundwater, evidence is beginning to show the contrary.

Studies have shown groundwater wells in Pennsylvania and New York were contaminated with excess methane, and there was a direct correlation between the distances from the wells to the fracking sites. Refuting this, industry spokespeople often state that this methane was in the water to begin with and is not caused by fracking. However, further research shows that the methane found in contaminated groundwater wells has a signature known to be from the fracking process.

The most recent reports and studies suggest that possibly some of this methane contaminating groundwater wells comes from the fracking operations above ground, such as from the stored wastewater sitting in *open pits* and ponds, which is part of the *shale bed fracking process*.

Reports about earthquakes generally agree there is little likelihood of earthquakes caused by fracking, although they have been known to occur. Some research shows an increase in the number of small earthquakes near areas where *deep injection wells* were drilled to store the contaminated water.

Now that fracking has become established and much more is known about it, scientists have learned with greater certainty about the *ongoing environmental concerns of shale bed fracking* and how to deal with them.

fracking, ongoing environmental concerns of shale bed

The United States does not follow the *precautionary principle* as does most of Europe. We often learn about serious environmental and public health threats as they unfold—somewhat like a real-world science experiment. This is what is happening with *fracking* for *natural gas*—routinely used throughout much of the United States but banned in many other countries.

Many environmentalists have serious concerns about fracking for natural gas; recent research shows why. The *initial concerns about shale bed fracking* have not turned out to be the most pressing problems. We now know the biggest problems come from the water used and then removed.

In the United States, about 100 billion gallons of water is used for fracking each year and rapidly increasing. This is roughly the amount of water used by 75 cities of 50,000 people for a full year. One fracking well site uses anywhere from 2 to 10 million gallons of water. This water use poses many environmental issues, the most obvious being the possibility of local water shortages.

But many opponents believe a bigger problem is managing the disposal of the large volumes of contaminated wastewater produced by fracking. A typical fracking site that injected four million gallons of water would use roughly 100 tons of chemicals—most of which are toxic. They typically include acids, *biocides*, buffers, stabilizers, corrosion inhibitors, friction reducers, gelling agents, *VOCs,* and others.

After this mix of water and chemicals does its job during the *fracking process*, it is removed (called *flowback*) from the well and stored in temporary open pits or impoundments—called frac ponds—until it is trucked out to *hazardous waste* sites or simply sent to regular *landfills*. In some locations, the waste is placed in *deep well injection* sites.

In Texas alone, 290 million barrels of contaminated wastewater is disposed of in deep well injection sites every month. This is equal to about 18,500 swimming pools. Texas only requires the water be injected 250 feet down into impenetrable rock, separating the contaminated water from any *groundwater aquifers*. The environmental impact of these wells is not yet known.

From the time the contaminated water is removed from the well, stored in pits or impoundments, injected underground, or trucked to disposal sites, there is a danger of spills, leaks, and accidents when toxic substances can make their way into the environment and the local *ecosystem*.

Many of these chemicals are known to be toxic to humans and wildlife and some are known *carcinogens*. Even though the number of fracking wells is growing rapidly, the number of hazardous waste disposal sites is decreasing because many cannot properly treat this type of waste. Some municipalities find it difficult to handle the waste produced.

Some fracking wells are in areas that produce *radioactive* waste flowback and are required to be disposed of in special facilities. However, there have been cases where radioactive wastewater was released illegally into public waters.

Spills and leaks have occurred many times, both of chemicals used for fracking while in transport to a site and of wastewater stored on or leaving a site. Spills have been reported of chemicals entering ponds and lakes, resulting in fish kills. Open pits have also overflowed and contaminated waterways and the soil.

Air quality in some neighborhoods near wells has degraded because the contaminated wastewater stored in open pits contains chemicals that are volatile and readily evaporate. Many of these substances are called *hazardous air pollutants*. Some examples of *air pollution* caused by these pits were so bad residents had to sue to have the pit moved farther from their homes.

Another environmental concern is not knowing which chemicals are used in the fracking process, making assessment of the risks impossible. Only recently have some states passed legislation requiring companies to divulge the chemicals they used. The following states have passed laws requiring companies to disclose what these chemicals are: Arkansas, Colorado, Montana, Oklahoma, Pennsylvania, Texas, and Wyoming.

The most recent concerns coming to light about fracking include the growing demand for a proppant—typically sand—that has resulted in the creation of numerous *frac sand mines* in the United States, a form of *mining* with its own set of problems.

But possibly the biggest concern coming to light is about recent studies showing leaks from these wells. *Methane* is a far more potent *greenhouse gas* than *carbon dioxide*, so it would not take much of a release to negate any advantage of gas over *coal*.

A major report about fracking concludes that fracking makes economic sense, but only if done properly—meaning safety must be the top priority. Doing things properly means additional cost and the report concludes that by adding seven percent to drilling costs, it could be done safely. The problem is whether the companies are willing to pay and if the public is willing to accept the end result.

fracking process, shale bed

Fracking is a technology used to extract *natural gas* from *shale bed* formations and works as follows: A drill pad is created, similar to the concrete slab foundation under a house but about 3.5 acres in size. Pipes are installed, roads constructed, holding ponds dug, and outbuildings set up for tanks, compressors, generators, and such. A huge tank reservoir—often one million gallons—is trucked in to hold the needed water. A drill rig about 100 feet high is brought in to drill a borehole into the Earth about 8000 feet. The bore not only goes down, but can turn 90 degrees and run horizontally for up to a thousand additional feet.

Small blasts create perforations along the lateral portion of the borehole where the gas is known to be located. The borehole is injected under pressure with large quantities of water, a *proppant* such as sand, and a mixture of various chemicals. In a typical well, 2 to 10 million gallons of water, roughly 4 million pounds of proppant, and at least a ton of chemicals is injected into the well.

This pressurized liquid mixture is forced into the borehole, through the perforations, and into the shale, causing it to fracture. At this point, the injection process is stopped and the pressure released, so the fluids begin to flow back out of the borehole, called *flowback*. The flowback contains not only the liquid that was injected but also natural gas that was unlocked from the shale.

The sand (proppant) keeps the cracks open so as much of the gas as possible is released. The natural gas and contaminated water are removed from the well separately. The natural gas is then transported to be sold and the contaminated liquids must be disposed of. The contaminated water is typically stored in open pits or impoundments to be trucked to *hazardous waste treatment* facilities. However, in some states they simply go into *landfills* or are placed in *deep injection well pits*.

The fracking process does not begin until a few months after the start of construction. However, the actual fracking process only takes a few days. Once complete, the well could extract gas for 30 to 50 years.

Of all the *fossil fuels*, natural gas is far more environmentally friendly than *oil* and *coal*. Converting from coal use to natural gas has long been a goal of many environmentalists. However, now with new found *reserves* of natural gas becoming available, many environmentalists believe an expansion of natural gas use will be to the detriment of the advancement of *renewable energy* development.

There are many *environmental concerns about fracking* and there are *alternative fracking processes* that minimize some of the environmental problems.

fracking regulations, states and

Fracking has been used for more than 30 years, but not in combination with horizontal drilling and not for the current application. This new use is so controversial that a raging debate is ongoing about whether it should or shouldn't be allowed. *Shale beds* are located in many states. Some, such as Pennsylvania, have welcomed it. About one-quarter of the entire state has already been leased by companies for fracking—about seven million acres of public and private lands, including state forest properties. Many people studying fracking believe Pennsylvania's having welcomed the process with open arms so quickly is a textbook case of how not to proceed—with few regulations.

Other places have outright banned the process, pending further scientific review. Then there are states that simply cannot make up their minds, such as New York. Each week they appear to make a new decision about what to do or not do. At the moment, New York has a moratorium on fracking. Some states are okay with the practice but concerned about specific aspects such as the storage of fracking waste. For example, New Jersey has banned the treatment or storage of fracking waste.

Regulations for fracking on public lands are controlled by the *Bureau of Land Management* (BLM). They have recently upgraded these rules, much to the displeasure of environmentalists, who say the rules do not go far enough and favor protecting the industry as opposed to residents. The new rules still allow companies to keep secret the ingredients of their chemical cocktail injected with water into the site—a major concern. But it does state that the BLM can require this information if deemed necessary. Most environmentalists believe it is always necessary. Many states do as well and are passing state laws requiring this disclosure.

Franken-salmon

Some people opposed to *genetically modified organisms* (GMO) use this spin on Frankenstein to describe a GMO salmon recently approved for U.S. sale. Although we all eat lots of GMO because it is in our processed food, we did not have any true GMO species on the market until this salmon. A Massachusetts-based biotechnology company plans to market this new, fast-growing species. They combined genes from the Pacific Chinook salmon with more common species to produce a salmon that reaches maturity much faster.

In a heated ongoing *debate about GMOs,* this salmon has become a focal point. Although many people are opposed to GMOs, many highly respected scientists firmly believe GMOs are safe and beneficial. Many people are upset not as much because these foods are becoming available but because there are no *GMO label* requirements, meaning you won't know what you are eating.

Free Range label

See *food eco-labels.*

fregan

A person who believes in the exact opposite of *conspicuous consumption.* Such people practice conspicuous nonconsumption and try not buying anything new. They *walk* or ride a *bicycle* instead of using a *automobile*, barter instead of making a purchase, and even participate in events such as dumpster dives looking for useful trash. [*locavore; community gardens; community-supported agriculture; green consumer; group coupons, green; planned obsolescence; Small is Beautiful; sneakers, green; tubeless toilet paper; wedding dresses, reused; wine bottle corks, recycled; microgreens and urban gardening*]

Freon

See *CFCs*

freshwater ecosystems

Aquatic environments can be divided into *marine ecosystems* (saltwater) and freshwater ecosystems. Freshwater contains relatively low concentrations of dissolved salts and therefore provides a very different *habitat* from marine bodies of water. Freshwater ecosystems can be divided into two types: *standing water habitats,* which include ponds, lakes, and reservoirs, and *running water habitats,* which include streams and rivers. Limnology is the study of all types of freshwater habitats.

friability

A soil's ability to crumble. Friability is necessary for good plant growth. Friability is usually determined by the *soil texture* and the amount of moisture in the soil. For example, "sandy" soils are friable but "clay" soils are not. [*topsoil; humus; loam; erosion; soil erosion*]

Friends of the Earth

Founded in 1969 by *David Brower*, this *NGO's* early goals were to stop the first supersonic transport plane, called the SST and help restore the Florida *Everglades*. Today their efforts are wide-ranging, including resource conservation and public health issues. They work with grassroots environmental groups, lobby members of Congress, litigate when needed, conduct workshops, and provide advice and information to the public. Friends of the Earth International is one of the largest networks of grassroots organizations in the world, with 70 national organizations, 5000 local groups, and 2 million members. [*environmental movement*] {foe.org}

freshwater springs

See *artesian springs.*

frugal science

Scientists in *developed countries* are trying to make it easier for people in developing countries to create low-tech, inexpensive versions of high-tech devices—something called frugal science. For example, an MIT lab called Little Devices creates medical nebulizers from bike pumps and converts a pressure cooker into an autoclave to sterilize medical equipment. The logic is that these important devices can be constructed without a lot of expertise and out of common parts and products. [*utilitarian environmental ethic; biopiracy; ethics, environmental*]

fruit waxing

Conventionally grown (nonorganic) fruits and vegetables often have a coat of a wax-like substance applied to the skin to prevent moisture loss, protect them during shipping, and extend shelf life. Some of these waxes are natural substances such as carnauba wax (from a palm tree), beeswax, and natural shellac (from a beetle). However, the vast majority of these waxes are petroleum-based (made from *oil*). This oil-based wax is allowed by the *Food and Drug Administration* (FDA) and is technically considered a form of edible "modified atmosphere packaging" (MAP). MAP controls the atmosphere the food resides in during transport to your table. In the industry, it is not called a wax but a film. These films are used by the "fresh cut" industry to create the MAP to "protect" your food.

In addition to chemicals that compose the oil-based film, other substances are often included, such as *fungicides* to prevent mold, *dyes* to enhance color, and other substances to improve overall consistency of appearance. In addition, many opponents say that any harmful *bacteria* on the produce when it was waxed will remain locked on the food when eaten.

Whether these films are harmful to human health or to the environment has been an ongoing debate for a long time. Of course the industry says it is perfectly safe, but other groups disagree. There has not been a great deal of independent research on the topic. The only way to avoid eating this brew of chemicals is to peel off the skin, which is unfortunate, because many of the most valuable vitamins and minerals lie right beneath the skin.

You can avoid these films totally by buying organically grown fruits and vegetables. Most *organic food labels* allow organic waxes as mentioned earlier, which most people (even the critics) consider harmless. Some organic growers state they don't wax their fruits and vegetables at all. You can always ask your grocer about the wax status of your produce, even when purchasing organic.

The nonorganic fruits and vegetables most commonly waxed include cucumbers, bell peppers, eggplant, potatoes, apples, lemons, limes, and oranges.

Waxed produce is required by federal law to be marked as such. Although these signs rarely appear, stores can be fined $1000 for failure to do so. Consumers can request their supermarkets to post these signs. If this doesn't work, report the violation to the Food and Drug Administration (FDA) or to your state Attorney General's office. [*pesticide residue on food; circle of poison; Federal Insecticide, Fungicide, and Rodenticide Act; Toxic Substances Control Act; POPs, dirty dozen; Dirty Dozen—fruits and vegetables; body burden*]

fuel cell vehicles, hydrogen

Many engineers, scientists, and environmentalists believe and hope that fuel-cell technology might be the fuel of choice for vehicles of the future, but they have been saying this for a long time. It remains in early stages of development and appears to have a long way to go, although enthusiasm runs deep.

Hydrogen fuel cells generate electricity by combining hydrogen and oxygen; there is no combustion, no pollution, and the only waste products are water with heat. These vehicles would not run down and need a recharge. Experimental fuel-cell vehicles are on the road today.

The problem is getting the hydrogen. This sounds odd, because hydrogen is abundant on our planet, but it is bound to other elements such as water (H_2O). Because a lot of energy is needed to break, or crack, the hydrogen from the rest of whatever it is attached to so it can be used in a fuel cell, this technology is not economically feasible—at least not yet. [*automobile propulsion systems, alternatives; ammonia as a car fuel; electric vehicles; greenest cars, top ten; hybrid cars; natural gas vehicles*]

fuel wood

Also called firewood, this remains the most popular source of energy for cooking and heating; about three billion people in the world rely on wood. After the *Industrial Revolution*, most of the *developed countries* began using *fossil fuels*, but many of the less developed countries never moved away from wood. The wood is usually burned in primitive *wood-burning cookstoves* that pollute the homes, causing a great many health problems. The use of fuel wood is also one of the major causes of *deforestation* in these nations. [*biomass energy; alternative fuels*]

Fukushima nuclear power plant disaster

On March 11, 2011, Japan was hit by a 9.0 earthquake that physically moved portions of the country 15 feet to the east and raised the sea floor near the coast (where it occurred) as much as 230 feet. The quake caused a devastating tsunami, killing 20,000 people along the country's eastern coast. The Fukushima Daichi *nuclear power* plant that housed six *nuclear reactors* was destroyed and three of them suffered reactor *meltdowns*, releasing *radiation*. About 80,000 people living within 12 miles of the plant were evacuated and remain so today. As opposed to the *Chernobyl* reactor meltdown, where the site was and remains abandoned, the Japanese are attempting to decontaminate the site hoping to make it once again habitable.

In 2012, leaks were found in the large underground pool designed to hold contaminated radioactive water. More than 30,000 gallons of radioactive water leaked out of a lining but it was not known if it had reached the soil.

Following the accident, Japan closed all of its 54 nuclear reactors. The leadership has gone back and forth about its nuclear future. They have ranged from plans to have it provide 50 percent of the country's energy needs (it now provides about 33 percent) to scrapping all nuclear plants. Recent plans include a gradual phasing in of the existing plants and even plans for some new ones. However, there have been large protests because much of the public is opposed to *nuclear power* ever since the disaster. In late 2013, none of Japan's nuclear power plants were operating. [*Fukushima radioactive water leaks; cooling ponds and towers; nuclear reactor safety and problems; nuclear waste disposal; nuclear waste dilemma, the global; Three Mile Island nuclear accident; fish kills, nuclear power plants*]

Fukushima radioactive water leaks

Two years after the *Fukushima nuclear power plant disaster*, radioactive water continues to leak into the Pacific Ocean. In August 2013, more than 300 tons of tainted water reportedly leaked from a storage tank into the ocean. Reports indicated that the radioactive water had not yet been completely contained and continues to pose a serious threat.

Fuller, Buckminster

(1895–1983) Fuller was a technology optimist—and a genius. Although many *environmentalists* might disagree with his belief that technology could solve most of our problems, Fuller certainly saw the need for sustainability before most and knew how to apply technology to attain this goal.

He was an architect, mathematician, inventor, and designer, among other things. Back in 1927, when few even considered such things, he built a house that used a *graywater recycling* system, recycled *solid waste* for *fertilizer*, used *solar energy* for heat and low-flow shower heads to conserve water—things that are just today becoming mainstream. He invented the geodesic dome and many other, somewhat outlandish designs.

In his book, Operating Manual for *Spaceship Earth* (1969), he compared our planet to a spaceship to point out how fragile our existence is. The following quotes from this book speak of the man and of our planet.

"Spaceship Earth was so extraordinarily well invented and designed that to our knowledge humans have been on board it for two million years, not even knowing that they were on board a ship. And our spaceship is so superbly designed as to be able to keep life regenerating on board."

"Our little Spaceship Earth is right now traveling at sixty thousand miles an hour around the sun and is also spinning axially which. . . adds approximately one thousand miles per hour to our motion. Each minute, we both spin one hundred miles and zip in orbit one thousand miles. That is a whole lot of spin and zip." [*Ward, Barbara; philosophers and writers, environmental; Dubos, René*]

fungi

Fungi are primitive plants incapable of *photosynthesis*. They reproduce *asexually* by spores. Because they cannot produce their own food, they are either *parasitic* on other organisms (the fungus that causes athlete's foot) or *saprophytic*, meaning they feed on decaying plants (such as mushrooms on dead wood).

Recently, fungi have been found to be a part of the *human microbiome* found in humans. Hundreds of new fungal species have been found in the guts of mice and men, but their role here is not yet known. [*decomposer; detritus; ecosystem; niche, ecological; chytrid fungus; white-nose syndrome*]

fungicide

A *pesticide* that kills *fungi*. Fungi help decompose dead organisms or tissues naturally. Most fungicides are applied as fumigants to protect stored food products from spoiling.

Gaia Hypothesis

Gaia was the ancient Greek Mother Earth deity. The Gaia Hypothesis, proposed in 1972 by Lovelock and Margulis, suggests that all life on Earth acts together in a way that causes the planet to be self-regulating. The hypothesis states that the planet possesses a life force that monitors and maintains conditions at a level best suited for the continued survival of life on Earth. Some people believe this means any single species threatening the survival of life in general on the planet would naturally be forced into extinction.

Many environmentalists view this hypothesis as a metaphor instead of a theory, one that emphasizes the interdependencies of our global environment and the need for a holistic approach to environmental issues. [*Biophilia Hypothesis; eco-psychology; primitivism*]

gall

An abnormal growth, usually a swelling on the leaves or twigs of a plant, caused by the invasion of microbes, *fungi*, or *insects*. The invaders use the structure for protection and as a food source. The plant is usually unharmed. [*symbiotic relationship; mutualism; niche, ecological; host; parasite; synecology; caliogy*]

garbage

The term garbage has a few meanings. The *Environmental Protection agency* defines it as *food waste*. Many people consider garbage to be waste from a kitchen. Others, however, use the term for any substance that is no longer needed and must be disposed of. It can be anything from leftover food to an old refrigerator. It typically refers to solid waste.

Because most garbage in a city finds its way to a collection site, it is commonly referred to as *municipal solid waste* (MSW for short) and is usually *incinerated* or sent to a *landfill*.

Another way to define garbage is by its location. Garbage is waste "where it is supposed to be." For example, in a garbage bin or in a landfill. *Litter*, on the other hand, is waste "where it is not supposed to be." For example, the litter along the roadside, littering the beach, or blowing in the wind.

Garbage is similar in many ways to a *pest* or a *weed* because all three can be considered useful if found in another place at another time. One person's garbage might be another person's food, just as one person's weeds might be another person's bouquet of flowers. [*toxic waste; hazardous waste; e-waste; junk mail*]

garbage patch, marine

These are huge regions of ocean filled with vast quantities of small pieces of floating *plastic* waste. One of these zones, called the Great Pacific Garbage Patch (a.k.a., Eastern Garbage Patch) is spread across a vast swath of ocean located in the North Pacific *Gyre*, where the vortex of wind and water accumulates the waste over time. Some reports state this patch is twice the size of Texas, but it is probably smaller. And the analogy is not a good one, because the waste is spread out as if in was in a soup. There is also a Western Garbage patch off of the coast of Japan.

The plastic waste is near the water's surface, where most marine life feeds. Much of the plastic is decomposed into tiny pellets resembling food particles and these areas contain more plastic that resembles food than actual food—*zooplankton*. Small fish eat the plastic, and it starts its way up the *food chains*. Environmentalists are concerned about the impact on *oceanic zone ecosystems*, and health experts are concerned about the impact of so much plastic entering into our food supply.

Some far-fetched ideas have been proposed for cleaning up these patches. One thought is to build large, manned sailing catamarans that would skim the surface, collecting the plastic. However, the only real solution will be getting control of our *plastic waste* dilemma. [*ocean pollution; ship pollution; plastic pollution; deadtime; Plastic Pollution Research and Control Act; marine ecosystems; Ocean Dumping Ban Act*]

garden hose, drinking from

It is not uncommon to see someone sipping water from a garden hose on a hot summer day, but it is not recommended and can be harmful to one's health. Many garden hoses contain *toxic chemicals* and the metal fittings can contain *lead*. During research, water from a garden hose that sat outside in the sun for a few days was found to contain 4 times the safe standard of *phthalates*, 18 times the standard for *lead*, and 20 times the standard for *BPA*.

Alternatives exist, such as hoses made from polyurethane or natural rubber with lead-free fittings. However, these researchers warned that microbial contamination is probably the biggest worry about water sitting in hoses, and this occurs no matter what kind of hose is used. So it's simply better to not drink from a hose and leave it at that. [*lead poisoning*]

gardens, rain

Gardens designed to collect rainwater *runoff* from surfaces such as roads, parking lots, and rooftops. They can be placed anywhere including in dense urban settings. Their primary purpose is to help keep storm water runoff from overpowering a town's *sewer system*, which can result in polluted water flowing into bodies of water. Some municipalities encourage citizens to build these gardens and offer support and workshops on how to do so. [*pointless pollution; Alliance for a Paving Moratorium; smart growth; straw-bale gardening; urban gardens; microgreens and urban gardening; combined sewer systems; aquifer; groundwater*]

gasification

One method of using *biomass energy* involves converting *biomass* (plants and animal wastes) into liquid gas fuels called *biofuels*. If the fuel produced is in the form of a gas, the process is called gasification. One example of gasification combines heat, steam, and small amounts of oxygen to produce a mixture of carbon monoxide and hydrogen called *syngas*—short for synthetic natural gas because it has many of the same properties as natural gas. [*alternative fuels; alternative energy sources*]

Gasland (the movie)

This highly controversial 2010 film about *fracking* was nominated for an Academy Award for the year's best documentary. Gasland II, a sequel, came out in 2013. The movies add fuel to the fire of debate about *ongoing environmental concerns about fracking*. The controversy about the movies centers around whether the movie was factual enough to be considered a documentary, as it was labeled.

The New York Times calls Josh Fox's documentary "an example of melancholy muckraking: somber in tone and vivid in its images of fouled landscapes and flaming faucets. The weight of Mr. Fox's evidence is convincing, though his less-than-rigorous methods leave him open to attack by corporate and government naysayers."

Although most of the movie is based on facts, some of it—including a scene with flames coming out of a faucet—was not. Follow-up investigations showed the person's water well had tapped into a natural gas pocket, resulting in a flame-throwing sink. Some believe the movie would have been more helpful in restricting fracking or ensuring better regulations if it presented a more balanced approach and depended less on theatrics. But others feel it was the sensationalism that made it so successful.

[*news literacy; environmental literacy; movies, environmental; Earth Communications Office; Environmental Media Association*]

gasohol

A blend of *gasoline* mixed with 5 (E5) to 25 (E25) percent *bioethanol*, a biofuel, is often called gasohol. The most common gasohol has 10 percent bioethanol (E10). It is used in many regions of the United States and throughout the world. *Biofuels* come with environmental advantages as well as disadvantages. [*alternative fuels; flex-fuel engines; syngas; alternative energy sources*]

gasoline, saving on auto

The best way to save money on gas and reduce your *ecological footprint* is to walk, take *mass transit*, or *carpool*. But when you do drive, you can save gas by doing any or all of the following: a) Slow down—driving at 55 mph versus 65 mph saves up to 15 percent of gas; b) unload useless waste—getting rid of 100 pounds of excess weight (either on the car or on your occupants) improves mileage up to 2 percent; c) keep tires properly inflated—this saves gas and improves handling; d) tune up the engine—faulty firing dramatically reduces mileage; e) start slowly—if you have a car with a readout that shows mpg used, watch it during acceleration to see how dramatic the difference is on mileage when accelerating slowly versus quickly; f) keep the air filter clean because it too impacts mileage; g) get a car with the highest mileage possible—for obvious reasons. [*tire, high-mileage; automobile propulsion systems, alternative; corporate average fuel economy; electric vehicle, charging; greenest cars, top ten; idling your car; remote-control car starter*]

genetically modified organisms (GMOs)—basics about

GMOs are organisms (plant or animal) whose *DNA* has been modified by *biotechnology*. This includes organisms that have been genetically engineered (manipulated) by changing the genes within a specific species or by transferring genes from one species to another. The most common method is to take genes from *bacteria* that have some unique quality and insert them into the DNA of certain crop plants. For example, some bacteria produce chemicals that some insects won't eat. The bacteria's gene—responsible for this trait—is inserted into a plant; this new genetically modified organism is a plant that is now resistant to certain insect pests.

Some odd combinations of species have also been used to create GMOs. For example, a certain fish gene that keeps the fish from freezing has been inserted into a tomato plant to produce a GMO tomato plant that is frost resistant. The possibilities are endless.

These organisms are created by scientists working in the field of *synthetic biology*. Some of their research includes creating GMOs that clean up *pollution*, create *biofuels*, and create medicines to fight diseases such as malaria. Others include developing livestock resistant to mad cow disease, sheep with high yields of wool, and cows with high amounts of omega-3 acids in their meat. Some GMOs planned for the very near future include fast-growing salmon and apples that don't turn brown.

The expansion of this technology has been so fast that most of us missed seeing it happen. First grown experimentally in 1988, today about 90 percent of all corn, cotton, and soybeans in the United States are genetically modified crops. Many of these crops are used to feed livestock, but about 75 percent of the processed foods we eat contain GMOs.

Even though GMOs have become common in the United States (with most of us not knowing it), people in other parts of the world don't like the idea at all. Austria, Hungary, France, Luxembourg, *Germany*, and Greece have all banned the use of GMOs completely. Seventy-five percent of all Europeans, when asked, say they don't want GMOs in their food. In most countries—but not the United States—food containing GMOs must be labeled as such. Some African nations such as *Kenya* will not accept U.S. food-crop imports unless they are assured they are GMO free. Now that the American public has been made aware of GMOs in our foods, a large anti-GMO movement and an even bigger GMO-labeling movement have become established. [*GMOs—the debate about; GMOs you routinely eat; GMOs—resistance in; GMOs—animals; GMOs—labels; GMOs in your food, the five most common; selective breeding*]

genetically modified organisms (GMOs)—debate about

If ever a topic evoked the phrase "there are two sides to every story," GMOs are it. The first side is: A few decades of research have come up with no conclusive evidence that GMOs pose a health risk. But the other side of the story is that many scientists do not believe GMOs have been thoroughly enough tested; the *Food and Drug Administration* doesn't regulate GMOs and foods are not required to have GMO labels, making them difficult to identify and then test.

Proponents say GMOs are nothing more than copying what occurs in nature through natural processes and through *selective breeding* that people have performed for hundreds of years. However, most of these genetic modifications would never occur naturally. For example, genes from bacteria do not naturally jump into the genes of tomato plants.

In 2013, a new GMO corn specifically for human consumption and containing *pesticides* genetically built into its DNA was released to the public. It won't be mixed or processed with anything else, so it will be the first true human GMO food. (Not that you will know about it, because it need not be labeled as such.) The first completely new GMO animal food for humans was going to be a rapidly growing salmon, but it has never been approved and remains on hold.

Supporters praise GMO crops for their ability to increase yields and help feed the 7.2 billion people on Earth. Some think of GMOs as the next *Green Revolution*. Just as the first one helped feed the masses, beginning in the 1950s, some think the advent of GMOs will help feed the next few billion people who populate Earth over the next 40 years.

Many critics, especially in Europe, worry about making "alien" species that didn't evolve through normal means and might carry risks we don't know about (yet). Some feel we are messing with Mother Nature. Others believe these new organisms might breed with the "normal" species and create who knows what. However, at this time, we have very little evidence that anything harmful has actually happened.

Some *NGOs* have worked successfully to form a growing anti-GMO movement. They use tactics ranging from traditional legal actions to *guerrilla labeling* to civil disobedience such as destroying GMO test plots or actual crops. These tactics are primarily used in Europe. An even larger movement is brewing in the United States and demanding that, at least, GMO products should be labeled as such. [*GMOs—basics about; GMOs you routinely eat; GMOs—resistance in; GMOs—animals; GMOs—labels; GMOs in your food, the five most common; DNA sequencing technology*]

genetically modified organisms (GMOs) you routinely eat

Americans eat almost 200 pounds of genetically modified organisms each year! These are mostly in processed food that contain beet sugar, corn syrup, soybeans, and corn-based foodstuffs that have come from GMO crops.

The two most common GMOs we routinely consume are crops that have had their *DNA* altered. The first are plant crops that have been implanted with a gene that makes them tolerant to a specific *herbicide* called *glyphosate*. This means the entire cropland can be sprayed with this herbicide, and it kills everything (all the *weeds*) but will not harm the GMO crop.

The other is a GMO that produces a natural *insecticide* called *Bacillus thuringiensis (Bt)*. Bt is bacteria that naturally produces a toxin that repels insects. The gene from the bacteria responsible for repelling insects was inserted into many crop species. The resulting GMO crops won't be attacked by *insects*.

Resistance in genetically modified organisms is beginning to appear, which can limit the value of these crops. [*GMOs—the basics; GMOs—the debate about; GMOs—resistance in; GMOs—animals; GMOs—labels; selective breeding; GMOs in your food, the five most common*]

genetically modified organisms (GMOs)—resistance in

One well known problem with *GMOs you routinely eat* is resistance. GMO crops "immune" to a certain *herbicide* (*glyphosate*) worked great—at first. The crop grew, and the weeds did not. However, after a few years, weeds have become resistant to this herbicide, and are now called *superweeds*. Today about 75 million acres of U.S. cropland has superweeds. What is the solution?

The original advantage of this GMO crop was supposed to be that fewer herbicides would be needed. But the result has been the exact opposite. The superweeds created by the GMO crops now require more and more herbicides, not less—the same vicious cycle we experienced with *pesticides*. [*GMOs—the basics about; GMOs—the debate about; GMOs—animals; GMOs—labels; selective breeding*]

genetically modified organisms (GMOs)—animals

Most *genetically modified organisms* are crop plants. However, there are GMO animals—for example, a GMO mosquito. Scientists have modified their genes so once they reproduce, all of the offspring die. These GMO individuals will compete with normal mosquitoes when released, with the hope that the overall mosquito population will decline. The hope is to control dengue fever that makes 100 million people sick every year. This is being tested primarily in *Brazil* but also in the Cayman Islands and Malaysia. The company responsible for these tests has requested U.S. approval but has so far been denied.

Another example is a GMO Atlantic salmon that grows to full size in 18 months instead of three years. It was created by inserting genes from a Chinook salmon with the Ocean pout (an eel-like fish). The result is a fish that continues to grow during its entire life instead of—as is normal—only during certain periods of its life. The company that produces this GMO has requested *Food and Drug Administration* approval for its use in *aquaculture* farms.

An active campaign is ongoing against GMO animal releases, with a great deal of debate as to whether releasing GMO animals is wise. In the case of the salmon, some fear it will escape and breed with the natural species, forcing their population into further decline. [*GMOs—the basics about; GMOs—the debate about; GMOs you routinely eat; GMOs—resistance in; GMOs—labels; selective breeding; biological control*]

genetically modified organisms (GMOs)—labels

Although many people are against the creation of GMOs, they are here now; we use them, eat them, and put them on our faces and bodies—like it or not. So trying to stop GMOs in the United States is a bit "after the fact." Many more people want to know what products are or contain GMOs, especially in the foods we eat. Companies that manufacture these foods are in full battle mode, trying to prevent any legislation that would require such labeling.

More than 60 nations enforce labeling of foods containing GMOs, but the United States hasn't passed any laws to this effect. Even state mandates have been thwarted by big companies spending a fortune to defeat such legislation. Some large natural-foods supermarkets, such as Whole Foods, now label GMO foods on a voluntary basis. You can also use the *Non-GMO Project Verified eco-label* to help you identify GMOs, or you could grow your own food.

California came close to mandating a law in 2012 with Proposition 37, called A Mandatory Labeling of Genetically Engineered Food Initiative. However, the biotech industry spent more than $45 million on a campaign to fight it and won by a margin of a few percentage points. Besides the biotech companies, many manufacturers also contributed to the fight against labeling.

President Obama promised in the 2007 election to support such a law stating, "Americans have the right to know what they're buying." Little to nothing has been heard about this since that time. Going back to 1992, when GMOs were in their infancy, they were given GRAS (generally regarded as safe) status, meaning manufacturers can simply inform the *Food and Drug Administration* that a new GMO product is being produced—without needing to prove it is safe. Most of Europe follows the *precautionary principle,* requiring independent scientific studies before any GMOs are sold or they are banned completely.

This battle, however, has just begun. Dozens of GMO-labeling campaigns are being waged in 20 or more states. Some businesses are siding with consumers wanting labeling. Stonyfield Farm, maker of yogurt products, has initiated a Just Label It campaign, and many others are joining the fray.

Instead of spending their money on campaigns to prevent labeling, companies should be educating the public as to why they should not fear GMOs and the benefits that might come from them.

Connecticut became the first state to mandate labeling of GMO foods but with many caveats, including a requirement that at least four other states do so as well and that one of them must border Connecticut—something that has not yet happened. [*GMOs—the basics about; GMOs—the debate about; GMOs you routinely eat; GMOs—resistance in; GMOs—animals; selective breeding*]

genome

The group of genes contained in a single set of chromosomes is called its genome. It is basically a map of a species' entire genetic structure. Having this map allows scientists to understand the traits of the species—how and why they are what they are.

The genome can help scientists determine why genetic diseases might occur and possibly how to prevent or cure them. Once scientists know which genes do what, they can also change these genes with genetic engineering, producing *genetically modified organisms*. [*DNA sequencing technology; biotechnology; synthetic biology; Human Microbiome Project*]

Genuine Progress Indicator (GPI)

The GPI is an alternative method of evaluating the overall health of an economy and the inhabitant's well-being. The economic health of a country is typically measured by the *gross national product* (GNP). The GNP measures the total output of goods produced and services provided, but it does not take into account the depletion of *natural resources* or the negative impact they have on the *environment* or its people.

For example, *irrigating* fields raises productivity by producing crops but reduces the *water supply* for the next generation. Burning *coal* raises productivity but pollutes the air. The GPI, however, incorporates this environmental degradation into its calculations. It takes into account the depletion of *nonrenewable* resources, the loss of *farmland* and *wetlands*, and the cost of *water* and *air pollution*. It also includes long-term factors such as the impact of a product that releases *greenhouse gases*.

Since the 1990s, the GNP has increased substantially, but the GPI has remained flat. This would indicate that the economic growth experienced by the United States has not been accompanied by an overall improvement in the welfare of its citizens.

This indicator is currently only calculated for the United States and seven other countries. The GPI may prove to be a transitional indicator, helping the world find better ways to measure the health of economies and the well-being of populations now and for the future.

The GPI was first created by *Herman Daly* and John B. Cobb in 1989 and called, at that time, the Index of Sustainable Development. A few years later, they made significant revisions to the calculations and it became the GPI. Other, similar indicators can be used—for example, the United Nations' Human Development Index and a few lesser known models such as the *Happy Planet Index*. The *Beyond GNP Movement* is working to establish other viable alternatives. [*economy vs. environment; slow money movement; triple bottom line; steady-state economics; Sustainability Imperative*] {genuineprogress.net}

geocaching

A high-tech new form of treasure hunting. An environmental group often sponsors the hunt where someone hides caches outdoors along a trail, within a park, *nature center*, or *zoo*. The cache contains a nature-related activity that must be performed, such as locating a certain plant and photographing it or writing down five *insects* found at the site. Each trail typically has an environmental theme, such as *biodiversity*, and might have three or four or many more caches.

This is typically a family affair, with children searching for the caches using clues provided via a smartphone or a GPS. Once one cache is found and completed, they move on to the next until they have completed the trail. If you have not heard of this activity, you will be surprised to learn over five million people participate in finding almost two million caches placed outdoors worldwide. The *National Wildlife Federation* has over 80 *Ranger Rick's* geocache trails across the United States. [*apps, eco-; crowd research, scientific; theme parks, greening of; zoos, future; volunteer research, scientific; crowd research, scientific; voluntourism, environmental; WWOOF*]

geoengineering

Technologies that make changes on a global scale. This might be something humans have done accidentally such as *climate change* or the dramatic increase in *biodiversity loss*. Humans are quite good at geoengineering, when done by accident, causing harm.

The term however, could and probably was meant to mean something humans try to do on a global scale with a goal of improving our planet. This type of geoengineering is mostly in the research and development phases. One of the more well-known types of geoengineering is cloud-seeding, in the hopes of making rain when nature doesn't deliver. But lesser known examples—hopefully more successful—might be various methods of *carbon sequestration storage* or dumping vast amounts of limestone into the oceans to stop *ocean acidification*.

One geoengineering project that is up and running is occurring in Mongolia. They are drilling large holes in the frozen surface of the Tuul River, pumping up water and releasing it on nearby land, where it freezes into large chunks of ice, called naleds in Russian. These naleds thaw much more slowly than the river ice and can be used to create cool areas in the summer and provide water as they melt. [*terraforming; synthetic biology; iron hypothesis*]

Geological Survey, the

An agency located in the U.S. *Department of Interior*, responsible for collecting, analyzing, interpreting, publishing, and disseminating information about *natural resources* found on our *public lands*. The research they perform and results they publish are important in managing the nation's resources for energy, mineral, land, and water resources. [*Bureau of Land Management; mineral exploitation*]

geological time scale

Most people find it difficult to understand the concept of time when it goes well beyond our life span and the life span of numerous generations—spans of time so vast that the concept of human generations becomes meaningless. Perhaps the main reason most people cannot understand the process of evolution and *natural selection* is that it is so difficult to comprehend the time spans necessary for the process to occur. This vast scale of time is considered a geological time scale, meaning it is more in tune with the span of rocks, mountains, and seas than it is to human life. Geological time measures time since the Earth was formed.

Many attempts have been made to illustrate how long geological time is and how short a period of time humans have been part of it. One way is to use clocks that condense all of geological time into a one-hour period, showing that the first human-like people would not even appear until the last one second and modern man would not appear until the last one-tenth of the last second.

A different twist was created by Berkeley University representing time as a book, with one page per every 10,000 years. That book would contain 460,000 pages and be almost 80 feet thick. Page 1 would be when the Earth first formed, 4.6 billion years ago. Life would not appear until page 70,000 (3.9 billion years ago). Dinosaurs don't become extinct until page 453,500 (65 million years ago). And the very first humans (hominoids) won't appear until page 459,600. (Remember, the book has 460,000 pages.) [*opinions and science; science; pseudoscience; natural selection; extinction and extraterrestrial impact; panspermia; Pangaea; supercontinents*]

geology

The study of the Earth, especially regarding the history of rocks. [*geological time scale; Pangaea; supercontinents*]

geomedicine

An informal term for a new generation of high-tech health devices, including products, Web sites, and apps that link an individual's health to local environmental conditions. Some are on the market now but many more are on the drawing board. For example, there are asthma inhaler devices with built-in *biosensors* that indicate a person is having an asthma attack and transmits the person's location to a central database to help determine what might have triggered the attack. In the future, biosensors might analyze pollution levels to inform people of health risks, while others might have sensors that read vital signs, recognize danger, and transmit a call to a 911 dispatch. [*apps, eco-; best available technology; biomaterials; biomimetics; bioplastics;*

biotechnology; crowd research, scientific; satellite mapping; Wide-Field Infrared Survey Explorer Satellite]

geothermal energy

Geothermal energy, along with *solar power, wind power, hydroelectric,* and *biomass energy* are all called *renewable energy sources,* because the supply is considered inexhaustible. The inner core of the Earth consists of a molten mass. In some areas in the United States and throughout the rest of the world, the intense heat within the Earth rises near the Earth's surface and heats underground water, forming hot water or steam.

If these reservoirs are close enough to the surface—about 10,000 feet—wells can be drilled to tap the steam and hot water and then use it to drive turbines to produce electricity.

About 24 countries use geothermal power, generating electricity at about the same cost as using *coal.* The largest geothermal plant is in northern California and supplies energy for the city of San Francisco. The United States is the largest producer of this form of energy (although a tiny percentage of our energy supply), followed by the Philippines and *Indonesia.* Reykjavik, the capital of Iceland, uses geothermal power to heat all of its commercial buildings, and about half of all energy in Iceland comes from geothermal power. However, from a global perspective, less than one percent of the world's energy supply comes from geothermal.

Geothermal power is clean and the supply is considered *baseload power,* meaning it is available around the clock—unlike solar or wind power that comes and goes and therefore requires *time shifting* to smooth out the power supply.

Some minor environmental problems are associated with geothermal power. In some locations, the facility can cause *air pollution,* because it might emit hydrogen sulfide and ammonia and might even emit some *radioactive* substances. It also can cause *water pollution* from dissolved solids that might come to the surface. Pollution control devices are used to prevent this environmental damage. A geothermal plant produces less than three percent of the *carbon dioxide* emissions when compared to a *fossil fuel* plant.

A new form of geothermal power under research shows great promise in expanding the number of sites. Most conventional geothermal sites are in regions where the thermal sites are close to the surface, but this new method—called enhanced geothermal systems—can use sites where the heat is thousands of feet deeper.

Another problem associated with these sites is finding them. Of all geothermal sites investigated, 20 to 50 percent turn out to be unusable, making the search costly.

However, a recent find of extensive geothermal data will make the search far more specific and hopefully productive. This data, collected from the 1970s through the 1990s and sitting idle for decades, is being reviewed by a coalition of universities and agencies and might boost geothermal activity in the United States substantially.

Also, some believe that many *fracking* wells could become *geothermal fracking* energy sites.

geothermal fracking

A technology called *fracking*, used to extract *unconventional natural gas*, is also being used to generate *geothermal energy*. Many of the sites used for fracking appear to be good sites to produce *geothermal energy*. These sites release their natural gas and also release lots of heat that can be tapped as a geothermal energy source. Some people believe this could become a substantial form of energy production within the United States in the not too distant future.

geotourism

Most tourism has a negative impact on the local environment, but geotourism means the environment is sustained or even enhanced by visitors. The concept is that tourism can be a positive for both the economy and the environment. Where *ecotourism* focuses on minimizing the environmental damages caused by visitors (do no harm), geotourism is more about sustainability based on the specific cultural or aesthetic aspects or heritage of the area. The goal is to turn these positives into a sustainable business model—and the model cannot continue unless these features are sustained. [*ecotravel; hotels, green; lodging, green; resorts, eco-; theme parks, greening of; ecotourism, ten best countries for*]

Germany

Germany has a long tradition of environmental awareness and concern, dating back to the mid-1800s. A 1972 amendment to the German constitution (Bund) gave the national government more control over environmental policy, including oversight for *waste* management, *air pollution* controls, *noise* abatement, *radiation* protection, and criminal law as it pertains to environmental protection. A federal Environmental Agency was created in 1974. In the late 1970s through to 1990, an environmental movement grew, and in 1979, a *Green Party* formed and some members were voted into office. In 1986, the Environmental Agency became part of a Federal Ministry for the Environment, Nature Conservation and Nuclear Safety. Environmental policy in Germany is based on the *precautionary principle* and the *polluter pays principle*.

ghost fishing

Discarded or lost *commercial fishing* nets of many types continue to catch fish at the bottom of the oceans, a process called ghost fishing because no one is there to harvest the catch and the marine life die when captured in the useless nets. The quantity of these nets is surprising. About ten percent of all ocean trash comes from abandoned and lost fishing gear. It sometimes washes up on shore but most of the time remains at the bottom of the sea—ghost fishing. [*fisheries and seabirds; pirate fishing; fish populations and food; overfishing*]

giant squid in native habitat

Giant squids are elusive *cephalopods* that had never before been photographed deep in their native habitat near the bottom of the *ocean*. In 2013, high-definition equipment 2000 feet down in the Pacific, off the coast of Japan, took amazing pictures of a 40-foot-long *bioluminescent* creature. The dinner-plate-sized eyes, the largest in the animal kingdom, are clearly visible in the photos. [*deep sea trenches; benthic organisms; blobfish; camouflage, squid; Ocean Conservancy; ocean pollution; biodiversity; oceanic zone ecosystems*] {deepseanews.com/2013/01/first-stills-of-the-giant-squid}

Gibbs, Lois

(1951–) In 1978, Lois Gibbs discovered a *toxic waste* dump site at an abandoned chemical plant three blocks from her home in Niagara Falls, New York, in a neighborhood known as the *Love Canal*. She believed this dump site might be the cause of the severe asthma, seizure disorders, and other illnesses her children and others in the area were experiencing. Although she passed out petitions, gave speeches, and lobbied politicians to clean up the site, her plea was ignored. Finally she took a few *EPA* agents hostage, getting the attention of then President Jimmy Carter and leading to the government's evacuation of the neighborhood.

Gibbs is a great example of how one person with a conviction can make a huge contribution to society. Currently, she is the Executive Director of the *Center for Health, Environment and Justice*, a grassroots environmental crisis center that offers resources and training to community groups to help them protect their neighborhoods from exposure to *hazardous wastes*.

She authored several books about the Love Canal story and toxic wastes. Her story was dramatized in the TV movie <u>Lois Gibbs: the Love Canal Story</u>, in 1982. [*cancer alley; cancer villages; POPs, dirty dozen; hazardous waste; persistent organic pollutants; dirty dozen POPs; Safe Chemicals Act; schools and brownfields; silent killer; Superfund; toxic cocktail; Rachel Carson*]

giga-

A prefix meaning a billion.

gigatons of carbon dioxide (GTCO2)

A unit of measurement commonly used to quantify the amount of *carbon dioxide* emitted into the air when discussing *climate change*. [*ton of carbon dioxide look like?, What does a*]

gillnets, commercial fishing

These are fishing devices used in coastal waters to catch large numbers of tuna and salmon. Unsuspecting fish swim into these walls of invisible netting, entangling their gills. In addition to the intended catch, the nets entangle large quantities of fish and other marine life not sought after. This unintentional catch—called *bycatch*—is thrown back into the sea with most of the animals dead or dying. These nets are similar to *driftnets* but not as large because they are used in coastal waters.

A recent report states that at least 400,000 seabirds are killed each year by gillnets. The report states that "The status of seabird populations is deteriorating faster compared to other bird groups, and bycatch in fisheries is identified as one of the principal causes of declines." A few states have banned the use of gillnets, but there are no federal mandates.

An estimated 2500 ships put out more than 50,000 miles of driftnets and gillnets every night. These nets devastate populations of many fish species and kill *dolphins*, *whales*, seals, and *sea turtles* among other sea life. [*fish populations and food; fishing, commercial; fisheries and seabirds; purse seine nets; overfishing*]

glacial lake outbursts

Glaciers and ice sheets are receding, most likely because of *climate change*. As they do, they often leave behind lakes of meltwater that become trapped by large mounds of rock and debris that had formed as the glacier advanced. These mounds, called moraines, are highly unstable and often break down from the water pressure of the melting ice. On numerous occasions, these moraines have burst, releasing vast quantities of water that can destroy whatever lies below. These dramatic releases are called glacial lake outbursts. The water can travel at speeds estimated to be 60 mph and wash away everything in their path. These outbursts occur in many locations, including Iceland, Alaska, Pakistan, Nepal, Tibet, and elsewhere. The volume of water, rock, and sediment can be huge. In 2000 an outburst in Tibet wiped out 10,000 homes and caused ten million dollars damage. [*Arctic ice, changes in; archeologists, ice-patch; Petermann Glacier ice detachment; melt ponds*]

glaciers, mountain

Glaciers are large, slow-moving masses of snow and ice that originate on mountain ranges. They are smaller than continental ice sheets but can be hundreds of miles square and reach depths of a thousand feet. Glaciers form and grow when more snow falls in the winter than melts in the summer. As the mass of snow builds up, the lower layers are compressed and turn to ice. As the mass increases in size, gravity pulls the mass downward at a few feet a year. However, if more snow and ice melt during the warm season than fall during the cold season, glaciers recede. Over *geological time*, the glaciers and ice sheets have advanced and receded many times. However, human-induced *climate change* appears to be reducing many of the world's glaciers at an alarming rate—far more rapidly than during natural cycles. [*Arctic ice, changes in; Antarctica; Third Pole, The; archeologists, ice-patch*]

glass recycling

Made from natural materials including silica, sand, soda ash, and limestone, glass can (theoretically) be recycled indefinitely. Some countries recycle 85 percent of their glass. Most glass containers today contain over 35 percent *postconsumer* glass product. Some estimates state glass can take one million years to decompose, but whatever the definitive time, *recycling* prevents it from ending up as *landfill* waste and uses less energy than producing new glass.

Even though glass can be recycled many times, it goes through a process called down-cycling, in which the quality diminishes with each recycle. Once its quality is below a certain minimum, the glass can no longer be used for containers and often becomes construction and road material. [*preconsumer waste; recycling metals; tires, recycling/reprocessing; telephone pole recycling; soda can recycling, aluminum; single-stream recycling; paper recycling; motor oil recycling; grass cycling; composting*]

Global Environment Outlook (GEO-5)

The *United Nations Environment Programme* produces these comprehensive reports that can be thought of as health-checks for the planet. The 5th edition of GEO-5 was published in 2012, just in time for the United Nations *Conference on Sustainable Development (Rio+20),* with the hope it would offer direction to the meeting. It urges governments to create more ambitious targets for addressing *climate change*, reducing *overfishing* and *desertification* among many others. It warns, "If current patterns of production and consumption of natural resources prevail and cannot be reversed and *decoupled*, then governments will preside over unprecedented levels of damage and degradation."

The report states there has not been a great deal of success addressing most environmental issues and expresses hope the next meeting (*Rio+20*) would do so. However, now that Rio+20 has come and gone, we know that few of the suggestions made in the report were taken seriously.

Global Reporting Initiative (GRI)

A joint venture of *CERES* and *United Nations Environment Programme*, created in 1997, this initiative is meant to improve the credibility of companies reporting on their *sustainability* efforts. It urges them to produce sustainability reports with the same rigor as when they write financial reports. The GRI is composed of 100 organizations, including corporations, *NGOs*, consultants, and universities that produce reporting guidelines for many industry sectors. Over 1500 corporations use these guidelines to write *triple bottom-line* reports that include economic, social, and environmental information. The GRI is used to advance *corporate social responsibilities*.

global warming

This phrase is often used inaccurately as being synonymous with *climate change*. Although climate change does produce a general global warming, it includes other conditions, such as more erratic weather patterns and even colder temperatures in some areas. So global warming is only one aspect of climate change—the general warming of the planet due to *greenhouse gases*.

Global Wind Day

June 15 is officially deemed Global Wind Day, coordinated by the *wind power* industry and observed since 2009. In Europe, it is typically celebrated with the opening of new *wind farms*, but it is not well known in the United States. [*Cape Wind; floating wind turbines; wind power brief history; wind power states, ten biggest; wind power transmission line; wind turbine; wind-energy areas; wind-speed vacuum*] {globalwindday. org}

glyphosate (GMO and super weeds)

This is the generic name for the most commonly used *herbicide* on the planet. The problem with all herbicides is they don't know the difference between the *weeds* you don't want and the crops you do. Farmers have to be careful when applying the herbicide or they will end up with a dead field to harvest.

In the 1990s, genetically modified corn, cotton, and soy crops were created in the lab that are resistant to the herbicide glyphosate. If farmers plant a crop that cannot be harmed by a specific herbicide, they could spray that herbicide as much and often as they wish over the entire field and know it will not harm the crop—just the weeds. (Much of our food supply is said to be genetically modified because it comes from varieties of corn or soy resistant to glyphosate.)

However, continued use of glyphosate has built up resistance in many types of weeds, called *super weeds* because they cannot be killed by it. At least 12 species of weeds are now considered super weeds—resistant to glyphosate. Because these super weeds resist the normal application of glyphosate, farmers use more of it or revert to some of the older, more dangerous herbicides such as 2, 4-D. [*genetically modified organisms—resistance in; natural selection; pesticides*]

GMO

See *genetically modified organisms.*

GMOs in your food, the five most common

The *Non-GMO Project* and the Cornucopia Institute provide information about the most common foods containing GMOs. Their data shows the top five are 1) soy, 2) canola, 3) corn, 4) milk, and 5) sugar beets. [*genetically modified organisms—the basics*]

god squad

In 1978, the *Endangered Species Act* was amended, allowing the government to make exceptions to the original intent of the law. (The Act prevents the government from doing anything that would increase the likelihood of forcing a species toward *extinction.*) The amendment provided exceptions to this Act by creating a special review committee that could override the law. The committee contains seven federal cabinet-level heads and was informally called the god squad because they wield the power to decide which species might be forced into extinction or not. It has only met on a few occasions over the years and has not been as contentious as first thought it might become. [*biodiversity loss, prevention of; CITES; Convention on Biological Diversity, The United Nations; endangered and threatened species lists*]

Golden Rice

A *genetically modified organism* that was destined to become the first to be sold as a new food, not a food product that is mixed or processed with other foods. This rice, developed in the 1990s, was created by incorporating three genes into one variety of rice: two from *bacteria* and one from the daffodil flower. The rice produced beta-carotene, which creates vitamin A, was of high yield, and had a built-in *insect* repellant. It also had a yellow tint, which gave it its name.

The goal of the inventor is to feed this rice to children in *less developed nations* where it could help prevent blindness of children (thanks to the vitamin A). The rice was such big news, <u>Time Magazine</u> featured it on the cover of an issue in 2000. However, at the moment, it is not being sold anywhere. The anti-GMO movement has stopped much of its progress to make it to market. Golden Rice is not owned by any of the giant companies that manufacture the other GMOs currently used in most of our foods. It is owned by a nonprofit group called the International Rice Research Institute. The status of Golden Rice's future remains an unknown. [*genetically modified organisms—the basics*]

gold fingers

The connecting pins found on computer chips that contain precious metals such as silver, gold, and palladium. This and other parts of *e-waste* contain roughly 40 times, by weight, the amount of metal than is found when mining the ore directly from the ground. This makes the waste of great value for *recycling*. [*Basel Convention*]

gold mining

Gold mining causes the same environmental harm as other types of mining operations, such as *strip mining*. In most cases, it creates large open pit mines where the rock is blasted out in search of the gold. The removed *overburden* must be placed somewhere, and this, along with its accompanying *acid mine drainage*, is often the biggest environmental issue.

However, mining for gold and some other precious metals is often done in poor countries with little to no environmental protection or worker protection. These mines can destroy the environment and people's health or even lives. Efforts similar to the certification program that identifies blood diamonds exist for gold. Because 80 percent of gold mining production is used for jewelry, efforts such as No Dirty Gold can help. Many large jewelry retailers have signed this *eco-certification* campaign, stating they will try to purchase gold from wholesalers dealing with mines that follow criteria that protect workers and the environment. [*mineral exploitation (mining); rare earth metals; Pebble Mine; gold fingers; Basel Convention*] {nodirtygold.earthworksaction.org}

golf courses and the environment

A golf course is very similar to an agricultural *monoculture*. It is a large expanse of one crop—not a crop such as corn but one of grass. It has the same environmental issues as do all monocultures. Golf courses use lots of water to keep the crop growing (well over 100,000 gallons per day on a typical course) and lots of synthetic *pesticides* and *herbicides* to be sure nothing eats the grass crop and no weeds interfere with the crop. They also use lots of *synthetic fertilizer* to be sure the crop remains healthy and beautiful.

The environmental issues are also similar to monocultures. The extensive water use often draws down local water supplies, and the chemicals used often pollute local bodies of water. The fertilizers cause *nutrient enrichment,* often resulting in *eutrophication.*

And just like the monocultures grown on farms, golf courses are here to stay, so efforts are underway to make them environmentally less damaging. These include growing varieties of grasses that require less water and using *organic fertilizers* and *integrated pest management* solutions to reduce the use of toxic *pesticides.* There is a new movement to create some of the first organic golf courses. The United States Golf Association (USGA) "recognizes the importance of taking proactive steps to minimize our environmental footprint and integrate environmental considerations into all aspects of our activities." An Environmental Institute for Golf is working toward this end as well.

The sincerity of these efforts will become evident in time. Many newly proposed golf courses throughout the world have met with strong local opposition based on environmental objections. The future growth of the popular sport might depend on how successful golf courses are at reducing their *ecological footprints* in their communities. [*water shortages, global; agriculture, high-impact*]

Goodall, Jane

(1934–) Dr. Jane Goodall has studied chimpanzees for more than fifty years. In 1960 she began working with Dr. Louis Leakey on a study of wild chimpanzees near Lake Tanganyika in *Kenya.* Their studies revolutionized anthropology and the definitions of what characteristics are considered strictly human.

She established a program called the ChimpanZoo to improve the lives of chimpanzees in *zoos* and worked to eliminate the use of chimps in scientific research. In 1977 she established the Jane Goodall Institute to promote initiatives to protect chimps in the wild.

Currently she spends her time at the Gombe Research Center and takes speaking tours to raise money for continued research on chimpanzees and protecting the environment in general. She has received numerous awards for her humanitarian and environmental work. [*primate research; vivisection; Fossey, Dian; animals experiencing grief; biodiversity, loss of*]

Good Samaritan Food Donation Act, Emerson

See *food waste.*

Gore, Al

(1948–) Gore's interest in the environment began with his concerns about *soil erosion* on his family's farm. His political career begin in 1978, and a significant aspect of his agenda has always been environmental issues, much of which was documented in his book <u>Earth in the Balance</u> (1992). In Congress, he helped lead the drive to establish the *Superfund* Act.

Gore was the 45th U.S. vice president, serving under President Bill Clinton from 1993 to 2001. After losing the 2000 presidential election, he delved back into environmental issues. He won the Nobel Peace Prize in 2007 for his work on *climate change* (shared with the *IPCC*). His book and documentary film—titled <u>An Inconvenient Truth</u>—and accompanying speaking engagements helped spread the word about climate change. [*McKibben, Bill; hockey-stick graph*]

gorillas

The largest and rarest of the great apes, gorillas only live in equatorial regions of *Africa*. Four subspecies live in distinct regions: the western lowland, mountain, eastern lowland, and cross river gorillas. All are endangered or critically endangered. Probably no more than 100,000 western gorillas, about 17,000 eastern lowland gorillas, fewer than 900 mountain gorillas, and fewer than 300 cross river gorillas are left in the wild.

Bushmeat hunting, *habitat destruction*, the *wildlife trade*, and diseases are responsible for the gorillas' reduced populations. Hunting gorillas is banned throughout Africa, and *CITES* prohibits all international trade of the animal or animal parts. Unfortunately, *poaching* and illegal wildlife trade continue. Rwanda has several gorilla refuges, including one established by *Dian Fossey*. [*Goodall, Jane; primate research; vivisection; biodiversity, loss of; animals experiencing grief; elephants; dolphins and porpoises, true*]

Gorilla Fund International, The Dian Fossey

This *NGO* works to save the 900 or so remaining mountain *gorillas* by educating the public and offering an adopt-a-gorilla program. [*Fossey, Dian*] {gorillafund.org}

governments of the forest (extrativistas)

Since the 1990s, a successful global movement—called governments of the forests—has saved *forests* and helped the people that live in them. The idea is to grant ownership rights to the communities that live in the forests. Studies show that where these autonomous governments are established, *deforestation* has been reduced—and a livelihood provided to the indigenous people.

The most successful and longest running such government in the forest is in Acre, *Brazil*—the former home to *Chico Mendes* and an area with a long history of working with *environmentalists*. The people living in these forests and participating in this form of government are called the extrativistas.

When people both live and make their livelihood in the forest, they protect it as well; sustaining the forest sustains them. The locals are provided with technical guidance from *NGOs* or the government, and the products they produce often get subsidies to ensure a market, at least at the onset. These products can be some sort of harvest or crafts from the forest or services such as *ecotours*.

Some governments in the forest have failed elsewhere because they lacked an essential component. The indigenous people must not only be given rights to use the land but also the right to prevent outsiders from moving in and destroying it. In Brazil, extrativistas can legally keep outsiders out. And that is why they have been successful. Many nations in Africa have tried these arrangements but failed because the locals have no legal right to keep outsiders out. Locals have little incentive, knowing that someone can come in and undo what they initiated. In many cases, *land grabs* occur, where the locals begin a successful sustainable business just to lose control to others.

grain production and use

There are about 50,000 species of edible plants, but two-thirds of everyone on Earth gets their food from only three of them: corn (maize), wheat, and rice. In 2011, over 570 million tons of these grains were consumed worldwide.

Since the 1960s, the land used to grow these grains has increased only 25 percent, but the amount of grain produced on this land has increased almost 270 percent. This is because of the *Green Revolution*, which has managed to continue feeding the ever growing *human population*. But scientists are concerned about having so many people depend on so few crops. Food security, as it is called, has become a major concern. If something, such as *climate change*, causes dramatic drops in any one of these crops, there will be a catastrophic impact. Many experts believe we must diversify our food crops to ensure food security in the future.[*heirloom plants; vegetable seeds, heirloom; tomatoes, tasteless but red; genetically modified organisms—the basics*]

Grand Canyon

The Grand Canyon, located in northwestern Arizona in the United States, is considered one of the natural wonders of the world. The exposed rock was eroded by the *Colorado River*. The age of its origins is hotly debated, with most people believing it was created about six million years ago, while others believe it to be much older—about 75 million years. That would mean dinosaurs might have enjoyed the view.

It was originally protected in 1908 by *Theodore Roosevelt* and became a *national park* in 1919. In 1975 it was expanded to encompass more than 1.2 million acres of land. Millions of people visit the park each year.

John Burroughs called the canyon the "divine abyss." Roosevelt said of it, "Leave it as it is. You cannot improve on it. You can only mar it. . . . What you can do is keep it for your children, your children's children, and for all who come after you, as one of the great sights which every American if he can travel at all should see." [*National Park System; national parks, global; Hetch Hetchy*]

grant-making foundations

In 2011, almost 300 billion dollars was doled out by philanthropic organizations in the United States and about eight percent of that went to environmental causes. The environmental *NGOs* that received the most of it were *The Nature Conservancy*, the *World Wildlife Fund*, and *Ducks Unlimited*. Just a few of the largest donors were the Gordon and Betty Moore Foundation, The David and Lucile Packard Foundation, The William and Flora Hewlett Foundation, the Bill & Melinda Gates Foundation, The Ford Foundation, Ted Turner, and Michael Bloomberg.

grass cycling

One of the most obvious forms of *recycling* is overlooked by millions of individuals who mow their lawns and remove the clippings, only to later apply *synthetic fertilizer* to replace the lost *nutrients*.

Dead and decaying organisms are the primary source of nutrients for future generations of plants. Lawn clippings have a fertilizer value of 5-1-3, (*nitrogen, phosphorus*, and potash.) It takes about two pounds of fertilizer per thousand square feet to replace the nutrients removed by the clippings. When left on the lawn, clippings decompose and become useable to the growing grass in about one week. This reduces the amount of synthetic fertilizer required by at least 25 percent. Grass cycling keeps clippings out of *landfills* and lowers the need for synthetic fertilizers which come with a host of environmental problems.

To recycle grass clippings, the grass should be cut often because the clippings must be one inch or shorter and the grass should be left about two inches tall. (Remember, you don't have to bag, so it takes less time.) The grass should not be damp. Tall grass, short clippings, and dry grass assure that the clippings fall within the remaining grass instead of smothering it.

Another alternative is to collect the clippings but place them in a *compost* heap to create *organic fertilizer* which can then be used at a later time instead of synthetic fertilizers. Both leaving the clippings on the lawn and composting them let the cut grass act as a *carbon sink*, significantly lowering *carbon emissions* that contribute to *climate change*. [*biogeochemical cycles, human intervention in; recycling metals; glass recycling; gold fingers; paper recycling; soda can recycling; tires, recycling*]

grassland, temperate

One of several kinds of *biomes*. Temperate grasslands are also known as steppes or prairies. They receive only 25 to 75 centimeters (10 to 30 inches) of precipitation per year. They are usually windy environments with hot summers and mild-to-cold winters. Grasses are the dominant plants, making up about 75 percent of the total plant life, because they require less water than trees.

Grazing animals (*herbivores*) from small mice and ground squirrels to *bison*, wild horses, sheep, cattle, and wildebeest are often abundant. There are also many species of *insects*, reptiles, and birds. Grassland soils are very rich and fertile and are therefore being converted at an alarming rate into farmland. Some of the drier grasslands have been converted into grazing land for domesticated animals.

Grasslands typically do not match the *biodiversity* found in *tropical rainforests*. However, recent studies in Eastern Europe, *Argentina*, and the Czech Republic have shown that grasslands match or even surpass the *biodiversity* found in some rainforests—a very surprising discovery. [*farmland grabs; farmland lost; Fertile Crescent; anthromes*]

gray air smog

Synonymous with *industrial smog*.

gray water

Water that comes from washing machines, sinks, showers, and tubs. It is not water from toilets or anything that comes in contact with feces. It is perfectly safe to be used for watering plants. There are gray-water projects that show how to build drainage systems from sink or laundry drains directly to the outdoors to water plants. [*detergents, phosphates in; water use for human consumption; toilet-to-tap drinking water; toilet paper, eco-; laundry and the environment; tub bath vs. shower; biosolids and wastewater; domestic water use; water treatment*]

grazing

Foraging for food by *herbivores*. The food eaten is either plants such as grasses in terrestrial habitats or *algae* and other *phytoplankton* in *aquatic ecosystems*. [*food chains; ecosystems*]

Great Barrier Reef

The only life form visible from space, it is typically considered one of the seven wonders of the natural world and has been designated a *World Heritage Site*. It is not a single *coral reef* but a collection of thousands of reefs located off the east coast of *Australia*. The beautiful colors are produced by the *mutualistic* algae that help feed the coral. Dead coral has no *algae* and is white, resulting in the phrase *coral bleaching*.

The Great Barrier Reef is home to over 1500 species of fish, 4000 types of mollusk, 200 species of birds, and numerous other species, such as the loggerhead turtle.

This reef, like all coral reefs, is in decline. Studies show that even though some portions of this reef have expanded, the overall size of the live reef is in dramatic decline. In the past 27 years, studies show it has lost about 50 percent of its live coral.

A recent study shows the decline has the following causes: tropical storms are responsible for about 50 percent; coral bleaching, probably resulting from *climate change*, is about 10 percent; and roughly 40 percent is caused by a *predatory* starfish called the crown-of-thorns starfish, which is native to the reef but has been enjoying a large population explosion in recent years. Although they are not sure, scientists believe this increase in starfish populations resulted from agricultural *runoff* that causes *algal blooms* the starfish *larvae* feed on. There are ongoing efforts to control the starfish and the agricultural runoff to save the reef from further decline. [*biodiversity, loss of*]

Great Britain

Europe's first *Green Party* was founded here in 1973. Environmental regulation at the federal level is handled by the Department for Environment, Food and Rural Affairs (DEFRA). In 1995, the Environment Agency for England and Wales and the Scottish Environment Protection Agency were created. Great Britain ratified the *Kyoto Protocol* in 2002.

The Open Spaces Society—the oldest *conservation* organization—was established in 1865. *Environmentalism* was still alive and well in 1912 when the Society for the Promotion of Nature Preserves helped establish many protected areas. The society is still in existence but has evolved into the Royal Society of Wildlife Trusts (2004). The first of the *World Wildlife Federation* national organizations started here in 1961.

GMOs and *pesticides* have met with very strict regulations in Great Britain and the entire European Union.

Great Green Wall of Africa, The

Forty percent of Africa is affected by *desertification*. *Drought* and *climate change* are expected to increase the expansion of the Sahara Desert to engulf more and more of the continent. Eleven African nations have a plan to thwart the desert's encroachment. They hope to build a green wall of trees some 4300 miles long and 9 miles wide from Senegal to Djibouti. African leaders hope this green wall will stop the sands from advancing farther. [*shelter belts; windbreaks; assisted migration; Bangladesh and sea rise; climate change and biodiversity; Convention on Climate Change; drought, the latest; deforestation and climate change*] {thegef.org/gef/great-green-wall}

Great Lakes Environmental Assessment and Mapping Project (GLEAM)

The Great Lakes hold 21 percent of our planet's fresh *surface waters*. Their importance to trade, commerce, and recreation is vast, but they, like most *ecological* systems, are under threats. They have some of the highest levels of pollutants in the United States, including: *mercury*, *PCBs*, and agricultural *runoff*.

GLEAM provides a recently released map that should help protect the lakes. It is a comprehensive map of all the environmental stressors impacting the lakes. These stressors are organized into seven categories: changes to *habitat*, *climate change*, coastal development, *fisheries* management, *invasive species*, *nonpoint source pollution*, and *toxic pollutions*. Lawmakers and resource managers use this interactive map to help guide the lakes' future. [*Great Lakes Water Quality Agreement; water pollution*] {greatlakesmapping.org}

Great Lakes Water Quality Agreement

In 1972 the United States and Canada signed an agreement to reduce pollution in the Great Lakes, which is the largest body of fresh *surface water* in the world. This agreement was amended in 1978 and 1987 when a new protocol was implemented turning over the stewardship of the lakes to the *International Joint Commission*. This commission reports progress of the lakes' status to both governments on a biennial basis. The commission has made progress cleaning up the lakes, especially reducing *cultural eutrophication* and *toxic substances* entering the lakes, but much remains to be done. [*Great Lakes Environmental Assessment and Mapping Project; water pollution*]

green

The word green, just like the prefix *eco-*, is used as a reference to the environment or something good for the environment. There are *green blogs* and *green jobs, greenways* and *green parties*, the Green Seal and the *Green Guides,* to mention very few. It is used to describe things that want to be green but are not, such as *green scam, greenspeak*, and *greenwash*. There is even *green fatigue* for those sick of hearing the word green.

Once a word becomes synonymous with its meaning—such as green and good for the environment—some people will think consumers believe adding the word to a product's name provides magical powers that make it good for the environment—which, of course, it does not.

However, overuse doesn't make the word less useful. It just emphasizes the importance of the word and what it represents. Environmentally literate consumers and *green consumers* know when the word green is used in a meaningful way and when it is a green marketing gimmick. [*environmental literacy; eco-labels*]

GREEN: A Blog about Energy and the Environment

A New York Times blog that many believe "was" the best environmental blog available. The Times discontinued Green in March of 2013, much to the dismay of numerous fans. [*blogs, green; tweets, eco-; microblogs, green; apps, eco-; crowdsourcing, eco-; DOT Earth blog; Yale Environment 360 blog; magazines; eco-; news literacy*]

Green Belt Movement

One of the most successful grassroots environmental movements, *Wangari Maathi* founded this as a women's organization in 1977 to stop *deforestation* in her native land of *Kenya*. The program relies on women planting tree seedlings in green belts in both urban and rural areas. Donations are used to purchase the seeds and pay a small amount to those doing the planting. In Kenya, over 30 million trees have been planted by this group. The group uses this simple act as a springboard for a wider philosophy of good governance, environmental stewardship, and human rights for all. Maathi was awarded the Nobel Peace Prize in 2004 for her work, which has been emulated in many other countries. [*environmental justice; ethics, environmental; ethnobotany; governments of the forest; Women's Earth Alliance; women farmers; Great Green Wall of Africa; ecofeminism; United Nations, global environmental issues and the*]

Green Climate Fund

During the *Intergovernmental Panel on Climate Change (IPCC) conference* in Copenhagen in 2009, promises were made that *developed countries* would help developing nations pay for the costs of reducing *carbon emissions* to lessen *climate change.* At the following IPCC conference at Cancun in 2010, the vehicle for this transfer of funds was created and called the Green Climate Fund. At the moment, the fund is working on procuring the monies from the developed countries before it can determine which developing countries are to receive them. [*United Nations, global environmental issues and the*]

green completion standards

It is not uncommon for *oil* and gas companies to burn off, or flare, *natural gas* that is often found while drilling for oil. Even though the gas is of value, if there are no facilities to capture it, it is simply flared off. However, because natural gas is *methane*, a potent *greenhouse gas*, this release contributes to *climate change*.

To prevent companies from *flaring* natural gas, new federal regulations require it to be collected and sold. These new mandates are called green completion standards. [*fracking; natural gas as an energy source; natural gas extraction, unconventional*]

green consumer

See *consumer, green*.

Green Cross certification

See *eco-certification*.

Green Dot program

See *extended product responsibility*.

greenest cars, top ten

GreenerCars.org publishes a list of the greenest *automobiles* based on criteria such as fuel economy, emissions, and others. Their best ten list for 2013 is 1) Toyota Prius C, 2) Honda Fit, 3) Toyota Prius, 4) Toyota Prius Plug-in Hybrid, 5) Honda Civic Hybrid, 6) Honda Insight, 7) Volkswagen Jetta Hybrid, 8) Mercedes-Benz Smart for Two, 9) Scion IQ, and 10) Ford Focus Electric.

ConsumerReports.org publishes a list of the greenest cars based solely on fuel consumption. Their best ten list is 1) Ford Focus Electric, 2) Nissan Leaf, 3) Tesla Model S+, 4) Chevrolet Volt, 5) Toyota Prius, 6) Toyota Prius V Three, 7) Ford Fusion SE Hybrid, 8) Toyota Camry Hybrid XLE, 9) Volkswagen Golf TDI, and 10) Ford C-Max Hybrid SE. [*automobile propulsion systems, alternative; batteries, automobile; battery disposal in Mexico, automobile; diesel engine fumes; electric vehicles; gasoline, saving on auto; hybrid cars; idling your car; natural gas vehicles; range anxiety; ammonia as a car fuel; vehicles, hydrogen fuel-cell*]

green growth

Most of today's developed nations, including Europe and the United States, became developed the old-fashioned way, where economics and the environment often collide with one another and economic growth usually means environmental destruction. The mantra has been "grow first, clean up later."

Many of today's developing nations are trying to learn from past mistakes and take a new path called green growth, where development is green from the start—not after the fact. This new mantra includes *eco-services*, with nature's value part of the equation for growth. [*economy vs. environment; Beyond GDP; CERES; economics and sustainable development; economics, environmental; economics, ecological; Genuine Progress Indicator; integrated bottom line; steady-state economics*]

greenhouse effect

A natural phenomenon that occurs when gases in the atmosphere, called *greenhouse gases*—including *carbon dioxide*, water vapor, *methane*, and others—regulate the amount of energy trapped within the Earth's *atmosphere*. These gases allow solar energy to pass through to reach Earth, but trap the heat (radiant) energy, preventing it from being reradiated and lost into space. The amount of these gases in the atmosphere determines the amount of heat trapped by the greenhouse effect. Without greenhouse gases and the greenhouse effect, life would never have formed as we know it on our planet.

As long as these greenhouse gases remain at a constant concentration along with other climatic factors, the temperature on the planet remains relatively steady. Increased concentrations of greenhouse gases resulting from human activities increases the greenhouse effect and causes *global warming* and *climate change*.

green fatigue

Most people (57 percent in a recent survey) do not believe the so-called green *eco-labels*, *eco-certifications*, and various other claims companies make about themselves. This general feeling of mistrust is now called green fatigue. Many efforts are underway to reduce *greenwash* and make these claims more meaningful. [*environmental rating systems; green consumer; green growth; green marketing; greenspeak*]

greenhouse gases

Gases that keep our planet warm by trapping heat—much like the glass in a greenhouse keeps the interior warm. The higher the concentration of these gases in the *atmosphere*, the more heat is trapped.

Greenhouse gases occur in nature, but human-induced emissions are upsetting the natural mix, increasing concentrations and causing *climate change*. These gases include *carbon dioxide* from burning *fossil fuels*; *methane* gas from *landfills*, feedlots such as *confined area feed operations*, and probably leaking *natural gas* wells; *nitrous oxides* from *synthetic fertilizers* and burning fossil fuels; and *CFCs* from old *air conditioners* and refrigerators.

The amount of human-induced greenhouse gases has steadily risen over the past few decades and with it, increased global temperatures. This *global warming* and *climate change* are projected to be destructive to our way of life—so destructive, in fact, that worldwide efforts have been ongoing, attempting to rein in the production of these greenhouse gases. The *Kyoto Protocol*, which took effect in 2005 and expires in 2015, is the most important international effort to control these gases. Many countries have introduced their own regulations to control greenhouse gases, including the use of *carbon markets* that try to reduce *carbon emissions* with market forces.

Carbon dioxide accounts for more than 70 percent of all greenhouse gases. One of the longest lasting of these gases, it continually builds up in the *atmosphere*. There is a direct correlation between the amount of carbon dioxide in the atmosphere and increased temperatures, called *climate sensitivity*.

Green Guides, Federal Trade Commission (FTC)

As more and more people became environmentally concerned *green consumers*, many companies responded with *greenwash*—the unfortunate practice of saying they too are green when in fact they are not. (Creating a *green marketing* campaign is far less expensive than actually becoming green.) However, many companies truly want to clean up their act and need to know the proper way to convey these actions to their consumers—and avoid greenwash. The FTC Green Guides do just that.

When last revised in 1998, these guides were considered weak, accomplishing little. However, this document had a major overhaul in 2012 and now has become an important document for business and industry. It provides guidelines about what it means to be truly green and offers insight into how businesses can prove that they are as green as they claim to be. The FTC can levy fines on companies that don't follow these guidelines but to date have only issued warnings. These federal guidelines can provide help to smaller businesses so they won't need to hire outside consulting firms; something routinely done by large corporations.

Many environmentalists hail these new guidelines as a major improvement, while others continue to say they remain far too weak to do any good. [*eco-labels; eco-certifications; environmental rating systems*]

green jobs

The Bureau of Labor Statistics defines these as jobs that produce goods or services that help the *environment* or protect *natural resources* or jobs that make the processes of a company more environmentally friendly or decrease the use of natural resources. There are many places to look for green jobs. [*climate capitalism; economics and sustainable development; economy vs. environment; ecopreneur; green net national product; investing, green; microfinance, green; microlending; natural capital; nurture capital*] {eco.org}

Greenland

See *Arctic ice.*

green manure

Green manure is one of three types of *organic fertilizers.* Green manure is any plant that is plowed into the soil to improve fertility, increasing the amount of organic matter available to the next crop grown in the field. Typical types of green manure include any weeds that may have grown on an uncultivated field and turned over into the soil, rangeland grasses turned over into soil to be cultivated, or *nitrogen-fixing* plants such as alfalfa, planted to enrich the soil. [*conservation agriculture; cover crops; fallow field; guano; intercrop; low-input sustainable agriculture; night soil – biosolids; organic farming and food; sustainable agriculture*]

green marketing

Many manufacturers are taking advantage of the surge in consumer interest about environmental issues. Marketing studies show that consumers prefer to buy products that are environmentally safe and are even willing to pay more for them—usually called *green products* and services. Advertising campaigns that target these issues are called *green marketing* or green advertising. More recently they have also been called sustainable marketing.

Much of this marketing is based on fact; many companies have become serious about their environmental impact, so their green marketing campaigns are justified. On the other hand, much of it has little if any basis of fact and is called *greenwash,* among other things.

Green marketing has become big business. But you cannot have true green marketing without true changes in how a company operates. The marketing must reflect changes made to a company's business model. (Otherwise, it would just be greenwash.) These changes are designed to lessen a company's *ecological footprint,* including the *reuse* and *recycling* of *pre-consumer waste, cradle-to-grave* and *cradle-to-cradle* tracking of their products, adhering to *corporate social responsibility,* and *extended product responsibility* tenets. Some have taken it as far as making *zero waste* or *zero carbon emissions* their goals. Valid green marketing reflects these kinds of serious changes and commitments. [*CERES; climate capitalism; companies, top ten green; corporate acquisitions of small green companies; corporate sustainability rankings;*

ecopreneur; investing, green; microfinance, green; microlending; money market funds, environmental; sustainable marketing; green consumers]

green net national production (GNNP)

This is one of many suggested alternatives to the *GDP* (gross domestic product), which is a measurement of a nation's economic output related to the nation's income. It is an indication of the health of a nation's economy. The GNNP considers factors the GDP does not measure, such as the depletion of natural resources and environmental degradation.

For example, filling in a wetland to create a housing development would be viewed by the GDP as only a positive, because all the products and services that went into this project were good for the economy. However, the GNNP would do the same, but would also include in the equation the negative aspects of the *wetlands destruction* and all the *eco-services* it provides that would be lost. [*Beyond GDP; Genuine Progress Indicator; gross national happiness; nurture capital; slow money movement; steady-state economics; triple bottom line*]

green parties, political

These are political parties that support a platform centering on environmental issues. In some countries, they have been somewhat successful in placing representatives into office at various levels. The first such parties started in the early 1970s in *Germany*, where today they remain active.

At various times, many countries, primarily in Europe, have had green party candidates, some with success but most without. The United States has not had any successful green parties over the years. Americans tend to put their environmental efforts and hopes into very successful *NGOs* that carry these causes to the electorate. [*greens, the; League of Conservation Voters; environmental movement*]

Greenpeace

Greenpeace is a worldwide NGO with more than three million members. It aggressively and sometimes dramatically promotes public awareness about environmental problems. Greenpeace is primarily involved in defending *marine ecosystems* and controlling the spread of *toxic substances* and *nuclear weapons*. They have been relentless in their pursuit to stop the use of *driftnets* and the indiscriminate death of sea mammals and fish.

Their ship the *Rainbow Warrior* was often used to protest activities such as *whaling*, *driftnet* operations, and nuclear testing. The Rainbow Warrior was destroyed during one of these protests.

Greenpeace does not accept corporate, government, or political funds and supports itself solely through donations. Its claim to fame is its use of nonviolent civil disobedience and confrontation. Through the years their goals have expanded to include protecting *biodiversity* and *old-growth forests* and encouraging sustainable trade. Greenpeace continues to be one of the premier worldwide environmental organizations. [*environmental movement; greenwash; whaling*] {greenpeace.org}

green products

An informal term meaning a product is in some way "good for the environment." It has little value without some form of verification, such as being backed by a credible *eco-certification program* or *eco-label*. Green products might be made from *recycled* materials, or they themselves are recyclable or *biodegradable*. They might have less *packaging materials* or be more *energy-efficient*. They might have specific ingredients omitted such as *phosphates in detergents* or BPA in *plastic bottles*.

Green products have become big business. Although many companies manufacture green products because they truly believe in protecting the environment and people's health, others do so because it is good for business. Surveys show consumers want products and services that are environmentally friendly. About half of them say they will pay more for these products if needed. Many people routinely express this concern every time they take out their wallets and are called *green consumers*.

Even though the best way to ensure what you buy is truly green is to check for valid eco-labels, an older, less used method still might help. In the early 1990s, when green products were new to the market, they were classified as either being "deep green" or "greened up." Deep-green products are primarily sold through companies that specialize in only selling green products. These products were created from "scratch" to be environmentally friendly. Most of these companies go out of their way to substantiate their green marketing claims with appropriate eco-labels.

Greened-up products were originally sold as regular products and then converted into so-called green products. It is harder to make an existing product environmentally friendly. These products are more likely to be sold in large chains. But the lines between the two are becoming fuzzy. When it comes to making a purchase, the end justifies the means, and the only important thing when selecting green products is to be sure the claims can be substantiated. [*corporate acquisitions of small green companies; food eco-labels; organic food labels; green marketing; green scam; green wash; greenspeak; green fatigue; nurture capital; companies, top ten green global; corporate sustainability rankings; investing, green; shareholders, corporate; sustainable marketing*]

Green Report
See *green marketing*

green revolution, the
The green revolution definition surprises many people because it does not mean what they think. Back in the 1960s, scientists, using *conventional breeding methods,* developed a few varieties of wheat and rice that were better adapted to grow in tropical regions and produce larger crop yields than the standard varieties. These new varieties dramatically increased the total amount of food grown on a worldwide basis. The most important breakthrough was cross-breeding a variety of dwarf wheat with a high-yield variety, producing a new type of wheat that could produce a high yield and withstand the weight of lots of wheat grain clusters without toppling over.

These new cross-bred varieties of our most important crops resulted in what is now called the green revolution, and it is the primary reason people were wrong who predicted mass starvation as the *human population* increased.

These new varieties, however, required a new type of farming, often called *high-impact agriculture*, which began a spiral of associated problems. With the *monocultures* required by these farms came the need for synthetic *pesticides*. As production increased, soils became depleted and there was also a need for *synthetic fertilizers* to replenish the soil.

The green revolution did prevent food shortages, but with it came numerous environmental affronts we are dealing with today. Many people are looking for ways to return to the past, before the green revolution, when growing food did not require a barrage of synthetic chemicals. *Organic farming*, crop diversity and *organic fertilizers* such as *compost* are all pre-green strategies. On the other hand, some believe *genetically modified organisms* (GMOs) might provide a second coming of a green revolution. Only time will tell. [*International Maize and Wheat Improvement Centre*]

green scam
As defined by the *League of Conservation Voters*, a "crime committed by politicians who portray themselves as environmentalists but consistently vote to pollute our air and water and destroy our planet."

But the term can go beyond politicians. Any statement that misleads the public about environmental issues, products, or services can be considered green scam. It is often used by groups with names that mislead the public about their agendas. The best way to get to the heart of a group's agenda is to see who provides funding. A group that declares the environmental virtues of *fracking* is probably funded by the *natural gas* industry. A group called Save the Planet might actually mean save the planet from environmentalists. [*corporate acquisitions of small green companies; food eco-labels; green marketing; green; green wash; greenspeak; green fatigue; nurture capital; companies, top ten green global; corporate sustainability rankings; investing, green; shareholders, corporate; sustainable marketing*]

Green Seal

See *eco-certification*.

greenspeak

Political rhetoric about being pro-environment when the politician's voting record is clearly anti-environment. [*League of Conservation Voters; green scam; Office of Technology Assessment, former; science; pseudoscience; opinions and science*]

green tax

An informal term used to describe taxes imposed on products or activities that pollute, deplete, or degrade the environment. These taxes act as an incentive to reduce the abundance of these products or activities, urging alternatives to be found, and also help pay for the cleanup of harm already done. Green taxes can be assessed on the consumer or on the manufacturer. Although the taxes placed on the manufacturer are often passed along to consumers at first, the idea is that companies will begin to find new ways to produce their products that do not carry the burden of this extra tax. This lets market forces work to improve products to the benefit of the environment.

One especially important green tax is based on the amount of carbon produced by burning *fossil fuels*, called a *carbon tax*. Green taxes have been used successfully for many years in Europe. [*carbon markets; carbon offsets; Green Climate fund; Investor Network on Climate Risk; Regional Greenhouse Gas Initiative; REDD+*]

greens, the

In general terms, "the greens" are any political party that makes the environment its principal concern. However, over 50 parties throughout the world are actually called The Greens. The party originated in West *Germany* and stressed environmental issues. The American Greens were organized during the mid-1980s. [*NGOs; environmental movement; countries, ten greenest; countries, ten most energy-efficient; environmentalist; environmental literacy; Native American environmentalism*]

greenwash

A combination of the words green and whitewash, it refers to a corporation, or any entity, portraying themselves as being environmentally responsible when in fact they are not.

Beginning around 1990, companies realized consumers wanted to patronize companies that respected the environment. Many companies made serious efforts, changing processes and products to be less harmful to the environment and working to adhere to *eco-certification programs* and obtaining various *eco-labels* proving these efforts. Their goal was to lessen their *ecological footprint*. Many other companies made less than serious efforts and simply changed their marketing and advertising tactics to appear as if they were environmentally conscious—something now known as greenwash.

The term became popular when *Greenpeace* published a book in 1992 titled <u>The Greenpeace Book of Greenwash</u>. Many environmental *NGOs* report on greenwash. CorpWatch writes a bimonthly letter called the Greenwash Awards and closely follows corporate greenwash. Many consulting firms specialize in helping companies ensure they avoid greenwash.

Perhaps the most important improvement in preventing greenwash is the U.S. Federal Trade Commission (FTC) publication called the FTC *"Green Guides,"* recently revised. [*greenscam; green marketing; green fatigue; League of Conservation Voters*]

greenways

These are linear parks usually containing a trail for *bicycles* or hiking. These narrow parks are often built along parkways or rivers or on converted railroad tracks. Besides providing recreation, some are used as commuting routes. [*Rails-to-Trails Conservancy; bicycling cities, ten best; golf courses and the environment; lawns; National Park Service; national parks (global); outdoor recreation; outdoor recreation and motorized vehicles*]

greyhound dogs

Many people believe greyhound dog racing is a cruel and inhuman sport and should be banned. These dogs can be overbred, treated poorly, and discarded when their racing days end. Organizations actively try to ban this enterprise, stop state subsidies and other incentives, and provide a process for caring for and finding homes for these dogs. [*animal rights movement and animal welfare; cosmetics, cruelty-free; vivisection; farm animal cruelty legislation; CAFO; cage-free label; teaching farms; factory farms; fisheries and seabirds*] {ngap.org and grey2kusa.org}

GRI

See *CERES*.

gross domestic product (GDP)

The most commonly used indicator of economic success—and therefore so-called progress—in the world. It measures the total market value of all goods and services produced in a country in a given year. The United States releases this number each quarter, and its increase or decrease is considered the bellwether of our economy and how well we as a nation are doing.

GDP has been the key indicator of whether a country is an economic success or failure and is also supposed to be an indicator to the health and happiness of its citizens. However, many environmentalists, some economists, and others believe it is no longer valid and is even counterproductive. The problem is that the GDP considers everything sold to be a good thing. Building a shopping mall over a *wetland* contributes positively to the GDP, even if the end result is massive flooding whenever it rains because the wetlands are gone. Cleaning up the massive flooding is good for the GDP because it employs lots of people to clean up the mess every time it floods. The *BP Deepwater Horizon oil spill* was good for the GDP because of all the money spent cleaning it up. Because no consideration or accounting is given for the types of products or services included in this value, many people believe alternatives are needed.

New methods have been developed to properly incorporate environmental and societal aspects of what is sold. *Eco-services* and new ste*ady-state economic* systems better measure our so-called progress. A *Beyond GDP* movement is forming that builds human and societal health and the environment into the equation. [*climate capitalism; decoupling; economics and sustainable development; economy vs. environment; economics, environmental; economics, ecological; growthmania; integrated bottom line; natural capital; nurture capital; slow money movement; gross national happiness; Sustainability Imperative*]

gross national happiness (GNH)

In 1972, Jigme Singye Wangchuck became king of the Himalayan nation of Bhutan. He made popular the concept of replacing the universally accepted indicator of economic success—the *gross domestic product (GDP)*—with a new measure called the Gross National Happiness index. Wangchuck believed that economic growth does not necessarily lead to contentment, so he focused on what he called the four pillars of GNH: economic self-reliance, a pristine environment, the preservation and promotion of culture, and good governance in the form of a democracy.

In more general terms, the GNH pushes public policy more toward the well-being of a people than the well-being of an economy. While it sounds quaint to many, some people have tried to advance these ideas elsewhere. There are efforts to devise a new economic index that would measure well-being gauged by things such as satisfaction with personal relationships, employment, and meaning and purpose in life.

The GNH is primarily concerned with the well-being of people. Another, possibly more tangible index is concerned with the well-being of our environment. The

Beyond GDP concept has met with some success in the United States and other developed countries.

gross national product (GNP)

A newer economic metric very similar to *gross domestic product* (GDP). The primary difference pertains to what the value represents. The GNP is supposed to indicate how well (or poorly) the individuals living in a country are doing, while the GDP is supposed to indicate how well (or poorly) the country as a whole is doing. From many perspectives, it doesn't matter if one uses the GDP or the GNP—both are seriously flawed because neither considers *eco-services* or human, societal, or environmental concerns. Alternatives—collectively called *Beyond GDP*—are being developed.

ground-level ozone

See *ozone, ground-level.*

ground-level ozone, ten U.S. cities with the worst

According to the American Lung Association's annual State of the Air Pollution report (2013), the following are the ten U.S. cities with the worst levels of *ground-level ozone*, a major component of *photochemical smog:* 10) El Centro, CA; 9) Washington D.C.–Baltimore–Northern VA region; 8) Dallas–Fort Worth, TX; 7) Houston–Baytown–Huntsville, TX; 6) Sacramento–Yuba City, CA; 5) Hanford–Corcoran, CA; 4) Fresno–Madera, CA; 3) Bakersfield–Delano, CA; 2) Visalia–Porterville, CA; and 1) Los Angeles–Long Beach–Riverside, CA. [*air pollution; air pollution in Beijing, China; personal, portable air-pollution sensing device; tweets, eco-*]

groundwater

Water found on the surface of our planet is called *surface water.* When surface water is absorbed into the ground and stored in underground *aquifers,* it is called groundwater. Only 2.5 percent of all water on Earth is fresh water (the rest being salt water). Of this, only 0.5 percent of all fresh water is groundwater, but it supplies much of the world with fresh *drinking water.* In the United States, for example, about 50 percent of all drinking water comes from groundwater. Groundwater becomes depleted in a process called *groundwater mining* and degraded from *groundwater pollution.* [*artesian wells; artesian springs; domestic water pollution; Kenya water find; Ogallala aquifer; potable water; Safe Drinking Water Act; toilet-to-tap drinking water; water conservation; water table; water use for human consumption*]

groundwater mining

This pertains to *groundwater* that is pumped out of the Earth for human use faster than natural processes can replace it, lowering the level of the *water table*. The water is removed for residential or commercial use but primarily for *agricultural irrigation*. A recent study found that almost 800 of the world's large underground *aquifers* are experiencing water mining. In many cases the water is being removed over three times as quickly as nature recharges it (naturally refills it).

In some areas of the San Joaquin Valley in California, the water table has dropped over 300 feet and the land itself is dropping a few fractions of an inch each year. In parts of west-central Kansas, large regions of once irrigated land have now run completely dry. A 100-mile stretch of the High Plains aquifer is totally depleted and no longer capable of irrigation.

It is estimated that water mining is occurring in 35 of the 48 contiguous states in the United States. When aquifers run dry, they will take hundreds of years to recharge. [*water grabs; domestic water pollution; Kenya water find; Ogallala aquifer; potable water; Safe Drinking Water Act; toilet-to-tap drinking water; water conservation; water table; water use for human consumption*]

groundwater pollution and depletion

Groundwater in the United States is a vital resource that is being polluted and depleted. It is polluted in four ways. 1) *Agricultural* products, such as *synthetic fertilizers*, animal feedlot (*CAFO*) wastes, and especially *pesticides* have already contaminated groundwater in many states. 2) *Landfills* produce *leachate* that can carry waste products to our water supply. New landfills capture the leachate, but many older *landfills* do not, possibly allowing waste to pass into the groundwater. 3) *Hazardous waste* disposal sites leak contaminants into the groundwater. Millions of underground storage tanks hold hazardous substances and many of them are known to be leaking their contents (called *LUST* for leaking underground storage tanks), with some of it reaching groundwater. Waste is also stored in more than 100,000 pits and lagoons in the United States. Almost all are unlined and many have yet to be inspected. Some are believed to be leaking and contaminating groundwater. 4) Well designed, properly functioning septic tanks should not cause harm, but many don't function properly, thereby contaminating groundwater.

But this is not just a U.S. problem. In *China*, about 70 percent of the population get their *drinking water* from groundwater, and much of it is contaminated. The country's weak environmental regulations have led to 90 percent of their groundwater being polluted. More than 35 percent is in such bad condition that even with treatment facilities, it still cannot be used for drinking. Some people point to China to illustrate what can happen when a nation does not protect its groundwater. It is estimated that 190 million Chinese become ill and 60,000 die because of drinking-water pollution. The Chinese government has finally decided to take action and begin investing in a groundwater monitoring system and attempting to prevent pollution and clean up the waters.

Groundwater is also depleted. This process is called *groundwater mining*. [*water grabs; domestic water pollution; Kenya water find; Ogallala aquifer; potable water; Safe Drinking Water Act; toilet-to-tap drinking water; water conservation; water table; water use for human consumption*]

group coupon sites, green

Groupon was the first to create this type of online discount website. Many of the products and services on Groupon pertain to *green consumers*, but that is a small part of all their offers. There now are discount coupon sites that focus entirely on environmentally sensitive products and services. [*apps, eco-; double-value coupon program; online petition sites and environmentalism; blogs, green; microblogs, green; tweets, eco-*]

Group of 10

The Group of 10 is an informal name used, often critically, to refer to ten of the most influential environmental *NGOs*. The ten members vary depending on who is listing them. Here are 12 organizations most commonly included on some of these Group of 10 lists: Defenders of Wildlife, *Environmental Defense Fund*, National *Audubon Society*, *National Wildlife Federation, Natural Resources Defense Council, Friends of the Earth, Izaak Walton League, Sierra Club, Wilderness Society, World Wildlife Fund, Greenpeace,* and *Nature Conservancy*. [*environmental movement*]

Group of 77, The

An organization consisting of and representing the poorest nations on the planet. They were very vocal in their demands at the *United Nations Conference on Sustainable Development*, emphasizing how Europe and the United States have a historic debt to pay for eating up so much of the globe's natural resources since the *Industrial Revolution*. [*United Nations, global environmental issues and the; Clean Development Mechanism*]

growthmania

A phrase coined by economist E. J. Mishan to describe the belief that bigger is always better. [*conspicuous consumption; planned obsolescence; green growth; Small is Beautiful; decoupling; economy vs. environment; triple bottom line; nurture capital*]

guano

The accumulation of large deposits of bird, *bat*, or other animal excrement. Guano is rich in *organic* nutrients and can impact the *ecology* of the *habitat*. [*organic fertilizer; animal manure; green manure; night soil-biosolid; toilets and biosolids*]

guerrilla gardening

When people illegally plant seeds or seedlings in urban spaces such as vacant lots, highway medians, abandoned lots, and even commercial land, some people call it guerrilla gardening. It is not popular in the United States but is an issue in Europe and especially Great Britain. Although their actions are illegal, its activists believe in nonviolence and work toward a goal of urban renewal. Some members try to work with local authorities to advance their goal of renewal. Their efforts are to protect *biodiversity* and advance environmental stewardship of public spaces.

In addition, the term can also refer to people in *developing countries* who use other people's land to plant food for sustenance. [*community gardens; community-supported agriculture; campus quad farms; teaching farms; gardens, rain; interim use; straw-bale gardening; Transition Towns; urban gardens; food desert; food miles; food's carbon footprint – the best and worst; fregan; locavore; seasonal foods; slow food movement; WWOOF; microgreens and urban gardening*]

guerrilla labeling

The practice of applying unauthorized labels on food packaging to identify products that contain a *genetically modified organism (GMO)*. People have taken to this illegal activity because corporations have resisted regulations that would require labels for foods containing GMOs—the vast majority of foods we eat. GMO labeling is required in most of Europe. [*GMO – labeling; GMOs in your food, the five most common; Golden Rice; Franken-salmon*]

guild

All the species within an *ecosystem* that utilize the same resource in the same way are described as a guild. For example, many different types of *insects* feeding on the leaves of trees of a certain species, form a guild and therefore can be studied as a single unit. [*habitat; levels of biological organization; competitive exclusion principle of*]

gully reclamation

Soil erosion occurs quickly on sloping land without sufficient plant cover. *Runoff* water forms gullies that wash away the soil. Gully reclamation—planting these slopes with fast-growing crops such as oats and wheat—returns this eroded land to productivity.

If the water movement is too great, mini-dams are created and allowed to fill with sediment and then planted. Fast-growing shrubs, vines, and trees can also be used to stabilize the soil. [*conservation agriculture; integrated multi-trophic aquaculture; intercrop; low-input sustainable agriculture; organic farming and food; regenerative agriculture; sheet erosion*]

gym movement, green

A new type of gym is sprouting up where working out means you not only get in shape but contribute to the *power grid*. All of the treadmills, bikes, and other apparatus are hooked up so the power you generate doing your workout is sent to the local utility company and added to the local power grid. These gyms also emphasize using reusable water bottles, and some have green equipment such as floors and mats made from *recycled* materials. [*bicmaqiunas; frugal science; microgrids, power; buildings, green; coffee, environmental impact of; ecological footprint; NBA arenas, green*]

gyre

The spiral movement of ocean currents. For example, the South Atlantic Gyre is the major anticlockwise circulation of surface waters in the South Atlantic Ocean. [*National Oceanic and Atmospheric Administration; ocean currents; oceanic zone ecosystems; Ocean Conservancy; ocean thermal energy conversion; ocean-current power; upwelling*]

habitat

The place where an organism lives. The habitat must fulfill the needs of the species if it is to survive, including sufficient nourishment, water, sunlight, proper temperature, and so on.

Habitats can be categorized by the types of ecosystems they reside in. For example, *aquatic ecosystems* are divided into *freshwater ecosystems* and *marine ecosystems*. Freshwater contains *standing water habitats,* which include lakes, ponds, and bogs, and *running water habitats,* which include rivers, streams, and springs.

When organisms live in restricted areas such as the confines of a leaf, under a rock, or in a pile of dung, the term microhabitat is often used. [*biomes; extremophiles; giant squid in its native habitat; migration; oysters and climate change*]

habitat fragmentation, human-caused

A form of *habitat loss*. Habitat loss usually refers to a large continuous region being destroyed by human intervention, but habitat fragmentation is caused by human activities that fragment or divide up what was previously one area. For example, building many roads through a wilderness area or many nonadjacent housing developments within a forested area results in a fragmented habitat. The result is a mosaic of natural and artificial areas. Although fragmentation doesn't seem as bad as wholesale habitat loss, the result is usually the same. Most species require continuous areas of habitat to survive. Many species cannot cross roads (or never make it across roads) or will not find enough food or mates or other requirements for survival in heavily fragmented habitats. [*biodiversity, loss of; biodiversity loss, prevention of*]

habitat loss

The greatest contributor to *biodiversity loss* is lost *habitat*. When human activities destroy where organisms live—on such a large scale that we are driving numerous species toward *extinction*—this is called habitat loss.

These activities include the conversion of natural habitats to *agriculture*—usually *high-impact agriculture*—where land that once contained a great amount of biodiversity has been converted into a *monoculture* of one crop with little else. Whatever else tries to exist in these new human monocultures is terminated with *pesticides* and *herbicides*. About 40 percent of the planet's entire land surface, once covered in natural ecosystems with a great deal of biodiversity, is now converted to agriculture.

Conversion to *rangeland* or pasturelands changes natural ecosystems to favor only those species of livestock *grazing* the land. Because many of these rangeland areas are semiarid or arid, overuse results in making them more desert-like—a process called *desertification*.

Deforestation is another cause of habitat loss. Wood is a form of *biomass energy* and has long been harvested. When cut down, the *food webs* that once existed in a forested ecosystem are obviously gone.

Urbanization also destroys habitats. As cities grow and *urban sprawl* expands, natural habitats are destroyed. Much of the land becomes paved over, making it impossible for any life to exist and even impeding the natural *biogeochemical cycles*.

Aquatic ecosystems also experience habitat loss. *Dams* are built on rivers, and the river waters are redirected, controlled, and drawn down, destroying habitat. *Point source* and *pointless pollution* degrades aquatic habitats.

All of these actions make habitat loss responsible for the majority of *threatened* and *endangered species*.

habitat management

See *wildlife management*.

hair dyes and PPD

Almost 99 percent of all hair dyes contains a compound called para-phenylenediamine, or PPD for short. Darker dyes have more of the substance than lighter shades. More than 100 million Americans use hair dyes, so any medical symptoms caused by this substance must be viewed as a public health concern. PPD has been accused of causing everything from mild to fatal allergic responses and dermatitis.

The simplest way to avoid problems is to follow the manufacturer's suggestion to patch-test the dye in a small area for 48 hours before actually using the product. However, because PPD is also considered a possible *carcinogen*, for which a patch test is useless, the best solution might be to look for alternatives.

Many companies now offer green alternatives. Look for dyes that specifically state they do not use PPDs or other irritants such as ammonia or resorcinol. [*BPA; BPAF; chemical body cleanse; hormone disruptors, endocrine; pesticide residues on* food; beryllium in computers; POPs, dirty dozen; fruit waxing; outgassing; PERC; Safe Chemicals Act; silent killer; THM; VOCs; toxic cocktail; body burden; couches and indoor pollutants; HEAL]

halons

A substance that was commonly used in fire extinguishers; it is an *ozone-depleting gas*. Halons are *hydrocarbons* that have had some of their hydrogen atoms replaced by bromine atoms and others by other halogen atoms such as chlorine, fluorine, or iodine.

Although produced and released into the *atmosphere* in much smaller quantities than *CFCs*, halons are far more damaging to the *ozone layer*—three to ten times more potent. The U.S. *Clean Air Act* (in compliance with the *Montreal Protocol*) banned the production and import of halons beginning January 1, 1994. However, recycled halon and existing inventories produced before January 1, 1994, were exempt. The only remaining use in the United States is called "critical use," where fire protection outweighs the environmental hazards, such as on airplanes.

handshake buildings

Urbanization takes many forms. In Shenzhen, China, not far from Hong Kong, hundreds of 10 story-high apartment buildings are illegally built so close together you can reach out and shake the person's hand across the way. Called handshake buildings, they are stacked next to one another, with about three feet of separation. China plans to remove these handshake buildings, as well as many other illegally built structures. [*urbanization and urban growth; eco-cities; homes, green; buildings, green; megacities; slum; small house movement; smart growth; tract development; Transition Towns; urban sprawl; zoning, land use*]

Hanford Nuclear Reservation

The United States military has produced vast quantities of *nuclear waste* as a by-product of manufacturing bombs. About two-thirds of it (65 million gallons) is stored at the Hanford Nuclear Reservation near Richland, Washington. Most of it is stored in large tanks and drums, but much was simply dumped into pools and lagoons as late as 1970. Dangerous levels of *radiation* have been found along the Columbia River all the way to the Pacific Ocean, 200 miles away. Recent studies have revealed that radioactive iodine used to reprocess spent *nuclear reactor fuel rods*, was released from the plant during the late 1940s.

Nuclear waste that hasn't made its way into the *soil* and *groundwater* is waiting for a permanent *nuclear waste disposal site*—something that does not yet exist anywhere in the world. The Department of Energy performed a cleanup effort, conservatively estimated to cost taxpayers about 30 billion dollars. Hanford is one of 15 locations referred to by the Department of Energy as part of the Nuclear Weapons Complex. All of these sites contain nuclear waste and require cleanup.

In 2013, storage tanks at Hanford were found to be leaking radioactive waste. The liquid portion in most of these tanks was removed a few years ago as part of the cleanup process, leaving a toxic sludge that is now leaking. First estimates had it at a few hundred gallons per year, but it is now believed to be higher. There are 177 tanks now holding nuclear waste and most of them are single-lined with a life expectancy of 20 years—but we are well beyond that now. At this time, no one seems to know how to clean this up, and the leaks continue. In time, the contaminants could reach *groundwater aquifers*. [*hazardous waste, military; hazardous waste, new technologies; hazardous waste disposal, historical; LUSTs; remediation of hazardous waste; Love Canal; brownfields; schools and brownfields*]

Hardin, Garret

(1915–2003) Hardin wrote what is probably the most read and cited environmental paper of all time. The article was published in <u>Science</u> in 1968 and called *The Tragedy of the Commons*. He was an academic who wrote numerous articles and books. His expertise was in human ecology, especially the intersection *of human population* and environmental destruction. His later writings, such as "Living on a Lifeboat" (<u>Bioscience</u>, 1974) inflamed many, stating wealthy nations should not assist poor nations and everyone should fend for themselves for the stability of our species and planet.[*philosophers and writers, environmental; environmental justice; ethics, environmental; utilitarian environmental ethic; human population throughout history; overfishing*]

hard pesticides

Pesticides that do not decompose into harmless chemicals readily. Hard pesticides can remain in their original state for 2 to 15 years. [*chronic toxicity; Agent Orange and the Vietnam War; DDT; Federal Insecticide, Fungicide, and Rodenticide Act; insecticides; pesticide dangers; Toxic Substance Control Act; Rachel Carson*]

Hayes, Denis

(1944–) Hayes is best known as the national coordinator for the first *Earth Day* held in 1970. He later made the event go global by creating and expanding a group called the Earth Day Network to include most of the nations of the world. He later turned his efforts to *solar power* technology and currently is the chief executive officer of a Seattle-based environmental *NGO* called the Bullitt Foundation. [*environmental movement*]

hazardous air pollutants (HAPs)

Also known as air toxics. In 1990, the U.S. Congress amended the federal *Clean Air Act* to address many air pollutants now known to pose a risk to human health or cause environmental harm. The initial list of 188 pollutants covered in this amendment are now known as hazardous air pollutants (HAPs) These are substances that were not included as the original *criteria pollutants* listed in the Clean Air Act.

HAPs are known or suspected to cause cancer, birth defects, and other serious ill effects or adverse environmental effects. They include benzene, found in *gasoline*; perchlorethlyene (*PERC*), emitted from some *dry cleaning* facilities; and others, such as *dioxin*, *asbestos*, and *heavy metals* such as *cadmium*, *mercury*, and *lead*.

Most of these chemicals originate from human-made sources, including *non-point pollution* sources such as *automobiles* or *point pollution* sources such as factories and *electric power plants*. The *Environmental Protection Agency* works to control HAPs by targeting the source of these pollutants, which is a different method than the one taken to control criteria pollutants. For example, new regulations will require gasoline sold in the future to reduce the release of *volatile organic compounds*. [*air pollution; air pollution in Beijing, China; brick kilns; fireplaces, wood-burning; ozone, ground-level; couches and indoor pollutants; HEAL*]

hazardous waste

Includes all substances that pose an immediate or long-term danger to the health or well-being of humans or to the environment, during transport or storage of these substances. These dangers are defined by the *EPA* in four categories based on their characteristics: 1) "ignitability" refers to waste that can easily catch fire, 2) "corrosiveness" refers to waste that requires special containment because it corrodes normal materials, 3) "reactivity" means the waste could easily explode, and 4) "toxicity" means the waste can cause physiological harm to humans or other organisms. This final category is often separated from the others and classified as *toxic waste*. *Radioactive waste* is a form of toxic waste, but typically discussed separately because of its unique characteristics.

Hazardous wastes come from many sources. The hazard might be a product itself, or a by-product of a manufacturing process. Some common hazardous wastes are organic chlorine compounds from *plastics* and *pesticides*; *heavy metals* and various solvents from medicines, paints, metals, leather, and *textiles*; and salts, acids, and other corrosive substances from *oil* and *gasoline*. The majority of these wastes are produced in *developed countries*, where most manufacturing occurs.

The management of hazardous wastes has become a pressing environmental dilemma. More than 250,000 *hazardous waste disposal sites* are located in the United States, all posing a threat to our *aquifers*, our health, and the environment in general.

Because no one wants this waste, wealthy nations often pay poor nations to accept shipments. This jeopardizes the health of those who work with this material, because many of these poor countries have few if any protections for workers. The *Basel Convention* tries to prevent this form of environmental injustice. [*solid waste; hazardous waste disposal, historical; hazardous waste, military*]

hazardous waste disposal, historical

For decades *hazardous waste* was disposed of by simply dumping it virtually anywhere. Although much has been done to improve the situation with legislation such as the *Resource Conservation and Recovery Act*, we continue to dispose of this waste, but only in a more regulated way. And many *less developed nations* continue to just dump it anywhere.

Huge amounts of hazardous waste have been dumped into or onto the *soil* in thousands of unregulated, industrial, *landfills* and in hundreds of thousands of pits, ponds, and lagoons. In addition, countless storage facilities have thousands of corroding steel drums—many leaking—containing hazardous wastes. Finally, *deep-well injection sites* deposit the waste 20 feet to several thousand feet in the Earth.

All of these soil-based sites are likely candidates to contaminate *groundwater aquifers* that supply much of our *drinking water*. Some of our aquifers are already contaminated with hazardous wastes, and the number is rising.

Those hazardous wastes that are not deposited on or in the soil, historically have been dumped directly into the *surface waters.* Wastes were routinely discharged into rivers, streams, and even *sewage treatment facilities*, where it still makes its way to the open water. This waste damages, if not destroys, many *aquatic ecosystems*.

Hazardous wastes—and especially the subset of *toxic wastes*—are gradually infiltrating our environment, and the price to clean them up is staggering. The U.S. government's attempt to clean up past hazardous waste mistakes falls to the *Superfund*.

Almost all of the methods just described continue to be used today, but with stricter regulations. In addition, n*ew technologies for hazardous waste disposa*l have improved the situation somewhat, but the risk remains. [*solid waste; hazardous waste disposal, new technologies for; hazardous waste, military*]

hazardous waste disposal, new technologies for

Hazardous waste disposal usually means dumping the material into a hole in the ground or into a body of water, resulting in contaminated *soil, water pollution, groundwater* pollution, and health hazards. New regulations have resulted in alternatives including waste destruction (or *incineration*), waste immobilization (or stabilization), waste separation, and neutralization.

Waste destruction means incinerating the waste at very high temperatures and is used primarily on organic hazardous wastes. This process can destroy 99.9 percent of the waste by breaking it down into harmless substances. Waste immobilization places hazardous substances into a form that can be easily disposed of and less likely to leak into the environment. For example, the waste can be broken down to a very small size and encapsulated in glass—a process called *vitrification* (or glassification). It can then be buried with little likelihood of contamination.

Waste separation is used to separate hazardous materials from nonhazardous materials, thereby reducing the volume of the hazardous waste. And finally, neutralization is changing the chemical composition of waste under controlled conditions, such as neutralizing acids to be less harmful.

Many environmentalists feel the best solution for hazardous wastes is to produce less of them in our society to begin with. [*solid waste; hazardous waste disposal, historical; hazardous waste, military*]

hazardous waste, military

The U.S. military generates more *hazardous wastes* each year than most of the biggest chemical companies combined. And they store more hazardous wastes than anyone, including locations such as the *Hanford Nuclear Reservation*. About 100 military properties are on the *Superfund's National Priorities* list, as well as thousands of military installations that do not comply with federal environmental laws. Many grassroots organizations take an active role in trying to get the government to clean up these sites and prevent more from being developed. [*hazardous waste disposal, historical; hazardous waste disposal, new technologies for*]

HCFC

See *CFC*.

HDPE (high-density polyethylene plastic)

HDPE is an acronym for high-density polyethylene—one of the most common types of *plastic*. It is used to make rigid containers such as milk and motor oil containers.

HDPE is one of the more commonly recycled types of plastics, with about 30 percent recycled in the United States. [*plastic recycling; plastic pollution; nurdles; plastic shopping bags; Plastic Pollution Research and Control Act, Marine; ocean pollution; garbage patch, marine*]

HEAL

The Human Ecology Action League (HEAL) was incorporated in 1977 by physicians and citizens concerned about the harmful effects of chemicals in the environment and their threat to human health. HEAL publishes a quarterly newsletter "The Human Ecologist," which contains information on environmental health hazards, legislative matters, and sources for safe food, clothing, and household supplies. [*medical ecology; nanoparticles and human health; puberty and the environment, early; superbugs*]

heating, ventilation, and A/C systems (HVAC)

Beginning in the 1970s, many buildings were constructed with nonopening windows to save on energy costs. The indoor *environment* of these buildings is controlled by HVAC (heating, ventilation, and air conditioning) systems. Fresh air is introduced, but much of the air simply recirculates throughout the building. When these systems are not working properly, pollutants and contaminants from office equipment, building materials, furnishings, carpeting, and smoke are trapped within the building. *Indoor air pollution* has become a major environmental and public health concern. [*air conditioning; energy saving apps; energy, ways to save on residential; smart energy; energy consumption, historical; homes, green; buildings, green; superinsulation; indoor ecology; insulation; laundry and the environment; R-Value; thermal insulation; weather stripping*]

heavy metals

Heavy metals are natural elements such as *lead, mercury, cadmium*, and nickel, mined from the Earth and used in numerous manufacturing processes and countless products. Chromium and nickel are used to electroplate other metals to withstand corrosion and heat. Mercury is used in paint because it acts as a *fungicide*. Cadmium is used in plastics to stabilize colors. Lead is used in gasoline to boost octane, in *marine* and industrial paints as preservatives, and in car batteries. Thirty-five heavy metals pose a risk to human health and the environment.

Lead, the most prevalent heavy metal, has been phased out of most gasoline and was banned from most paints, but still exists in the paint found on millions of homes. Mercury is still allowed in most water-based paints and is known to cause symptoms and illnesses similar to lead poisoning.

Industry dumps its heavy metal wastes directly into bodies of water or into *sewage systems*, which then make their way into our natural bodies of water. Shellfish in many regions are contaminated with heavy metals. Lead leaking out of disposed car batteries into the groundwater is the primary reason many *landfills* have been closed.

When materials containing heavy metals are incinerated, most of the metals go up the smokestack, resulting in *air pollution*. Others, placed into landfills, *leach* out into the soil and have been detected in *groundwater* used for drinking.

Heavy metals can enter the body through the air when we breathe; as *particulate matter* in food or water; or by absorption directly through the skin. They accumulate in the body, especially in the brain, liver, and kidneys. If ingested, they can cause neurological or liver damage or stomach cancer. Contact with these metals causes rashes or ulcerations, while chronic inhalation can result in respiratory problems that range from coughing to lung cancer. [*lead poisoning; mercury poisoning; sunsetting; Superfund; body burden; recycling metals*]

heavy-water nuclear reactor (HWR)

Most *nuclear power* plants today use the *light-water nuclear reactor* design that uses common water (light water). Some reactors, however, are designed to use heavy water, which contains a different form (isotope) of hydrogen called deuterium. Heavy water works better than light water in controlling *nuclear fission*, so the operating costs of an HWR are less than that of an LWR. HWRs have been gaining popularity. *India* recently announced plans to build more than a dozen HWRs during the next few decades. Both light- and heavy-water reactors have the same *nuclear reactor* safety issues, with the biggest problem being *nuclear waste disposal*. [*nuclear reactors; nuclear reactor safety and problems*]

heirloom plants

When talking about *biodiversity loss*, people are usually referring to animals becoming *extinct*. But it also pertains to plants. Today's *high-impact agriculture* and the *monocultures* they grow use only a few varieties of crops, and this too results in lost biodiversity. For example, almost 90 percent of all U.S. crop varieties used during the past 100 years are no longer commercially available.

Apples are a great example. Orchards in North America used to grow about 15,000 varieties of apples. Today there are no more than 3000 and most all of them are grown on a few small orchards. They are on the verge of disappearing, leaving only a few standards grown by the largest growers.

What has happened to the great diversity of plants that existed in the past? The answer is they are gradually being lost. But a relatively new movement is trying to save the old varieties of plants, now called heirloom plants. These heirloom plants not only provide consumers with more choices but they are also ecologically important. Genetically diverse crops are more resistant to pests, blights, and *climate change*. Groups of people interested in heirloom plants are working to find and cultivate rare varieties on the verge of being lost forever and reintroduce them to consumers. [*vegetable seeds, heirloom; seed banks and vaults; seeds and climate change, Project Baseline; organic farming and food; Non-GMO Project Verified label; rice, contaminated; seasonal foods; tomatoes—tasteless but red; WWOOF; wine, organic and biodynamic*]

heliostats

Huge mirrors used in *solar thermal power plants* (a.k.a. concentrated solar farms). [*solar power; solar thermal power plants*]

herbicide

A *pesticide* that kills vegetation. Herbicides make up about 65 percent of the total pesticides used in the United States. They are used to kill everything from a *weed* on your lawn (when they are commonly called weed killers) to all plant life along a proposed highway or railroad track or under high power tension lines. Their most common use, however, is now on farm crops to rid unwanted vegetation from stealing water or sunlight from the crop to be harvested. Many crops are now *genetically modified organisms (GMOs)* to be resistant to certain herbicides.

Herbicides are usually categorized according to how they kill and by which method they enter the plant. Most herbicides kill by inhibiting some process, such as *photosynthesis*, cell division, respiration, *chlorophyll* production, or leaf growth. Like all pesticides, herbicides do not only kill the *target species*. They typically cannot tell the difference between a weed and a beneficial plant. For this reason, some call all pesticides *biocides*.

However, there are herbicides that only kill the target species. They do so by using the specific weed's growth hormones, called *auxins*. Herbicides that use these growth hormones promote excessive growth, forcing the plant to grow abnormally large, killing the plant because it cannot support its own weight. These target-specific herbicides are far safer for the environment, but only make up a very small portion of all herbicides used.

Herbicides enter a plant and kill by one of the three following methods. 1) Contact herbicides kill on contact and include the well-known paraquat. 2) Systemic herbicides kill by being absorbed through roots and include *alar* and *Agent Orange*. 3) Soil sterilants, which include trifluralin, are placed in the soil, where they kill microbes essential to the plant's growth. [*integrated pest management; biological control; biological control methods; body burden; nanopesticides; pesticide residues on food; Toxic Substances Control Act*]

herbivore

Animals that only eat plants are called herbivores (primary consumers or *heterotrophs*). Grazing animals such as cattle, sheep, rabbits, and grasshoppers are all herbivores. Human herbivores could be called *vegetarians*. [*vegan and vegetarian diets and the environment; meat, eating; Meatout Mondays; pesticide residue on food; fruit waxing; food chains; ecosystem*]

hermaphrodite

An animal that contains both male and female reproductive organs or a plant that contains both types of organs in a single flower. For example, the *earthworm* is a hermaphrodite. [*sexual reproduction; asexual reproduction; decapitating flies; honeybee sperm bank; imperfect flower; pollen; parthenogenesis*]

herpetology

The study of reptiles and *amphibians*. [*amphibian decline; Amphibian Ark project; alligators; keystone species; vertebrates*]

Hetch Hetchy

In the early 1900s, Hetch Hetchy was the site of one of the biggest environmental controversies of its day—one that environmentalists lost. Hetch Hetchy was a beautiful valley in *Yosemite National Park*. In 1906, after the San Francisco Earthquake, the city concluded they needed a better water supply. Plans were made to create a large dam in the Hetch Hetchy valley to provide water to the city. *Gifford Pinchot*, at that time Chief of the U.*S. Forest Service*, argued for the dam and *John Muir* argued against it. The O'Shaughnessy Dam was completed in 1923, backing up the Tuolumne River to form a reservoir, Hetch Hetchy was submerged, and San Francisco got its water.

Today a group is working toward the removal of the dam and the return of the valley to its natural state. [*hydropower, traditional; hydropower, nontraditional; dam removal; dam, hydroelectric; dams, ten largest*] {hetchhetchy.org}

heterotrophs

Organisms can be categorized according to how they obtain energy to survive. Animals that must consume plants or other animals to obtain energy are called heterotrophs (a.k.a. consumers). Animals that eat plants are primary consumers and include cattle, rabbits, and grasshoppers. Animals that eat the primary consumers are secondary consumers and include *predatory* birds, *fish*, and *insects*. Tertiary consumers eat secondary consumers and can include lions, *sharks*, and hawks. Which types of organisms fill the role of producers and consumers depends on the type of *ecosystem* in which they are found. These relationships are illustrated in *food chains* and *food webs*. [*niche, ecological; trophic levels; energy pyramid; autotrophs; chemoautotrophs*]

HFC
See *CFC* and *ozone depletion meets climate change*.

hibernation
Some warm-blooded animals (a.k.a. endotherms) have the ability to become dormant during extreme cold periods. This adaptation allows them to remain in a region that would otherwise kill them during certain seasons. During hibernation, the individual's metabolic rate dramatically declines so it conserves energy and can survive long periods without food. The amount of oxygen going to the brain can drop down to two percent of what is normal in some animals. A squirrel's heartbeat rate goes from about 300 beats a minute down to three or four per minute. Some fish in cold regions hibernate.

Some hibernating animals must wake up for short periods of time to warm up their bodies and then return to "sleep." Black bears can give birth to their young while in hibernation. The young can suckle the mother even in this state. [*cold resistant; estivation; quiescence; torpor*]

high-altitude ecosystems
The *atmosphere* is not typically thought of as being full of life, but it appears it is. *Bacteria, algae,* and *fungi* are found at levels as high as 20 miles up. An estimated two million tons of bacteria and 55 million tons of fungal spores make their way up into the atmosphere each year. Exactly what they are doing up there is not known, because this is a newly discovered *ecosystem* that received little attention in the past. What is learned about this high-altitude ecosystem can be an indication of what life might be found on Mars, because the environment is similar. Research is being done to see if these microbes in the sky play a part in our weather; particularly rain- and ice-making. It is believed microbes may drive nucleation; the process of ice crystal formation.[*aerobiology; extremophiles; hydrothermal ocean vent ecosystems; atmospheric rivers*]

high market value farming counties, U.S.

This typically refers to counties that are in the top 20 percent of agricultural production for each state. More than half of these counties are inside or adjacent to metropolitan areas and are threatened by *urban sprawl*. Millions of acres of high-quality agricultural lands have been converted to other uses in this expansion each year. [*farmland lost; urbanization and urban growth; farm bill; land grabs; high-impact agriculture*]

high-temperature, gas-cooled nuclear reactor (HTGCR)

Most *nuclear power* is generated by *light-water nuclear reactors (LWRs)*, but a few are either heavy-water nuclear reactors (HWRs) or HTGCRs. The main difference between these three is the substance used as a coolant. The HTGCR design uses *carbon dioxide* gas as opposed to regular (light) water in the LWR and a heavier form (isotope) of water in the HWR.

highways, electric vehicle

If *electric vehicles* are ever to go mainstream, either they need to go farther on a single charge or there must be sufficient infrastructure, such as charging stations all along major highways, just as we have gas stations now. Many people say they won't buy electric *automobiles* because of *range anxiety*—not knowing if they can make it to their destination before running out of power. But electric vehicle highways are becoming a reality in some places. The West Coast Electric Highway is a 200-mile stretch of I-5 that boasts fast charging stations every 25 to 60 miles. The plan is to extend these stations the entire 1300-mile route from Canada through California. [*automobile propulsion systems, alternative; autonomous automobiles; electric vehicle, charging; greenest cars, top ten; intelligent vehicle highway systems; natural gas vehicles; vehicles, hydrogen fuel-cell; ammonia as a car fuel*]

Hill, Julia "Butterfly"

Redwood forest destruction was rampant in the 1980s, spurring a 1990 effort by *EarthFirst!* to initiate civil disobedience in the redwood forests to stop the *logging*. The movement was called Redwood Summer. During this time, Julia Hill climbed one of the 1000-year-old, 180-foot-tall *coast redwoods*, built a small tree house, and did not come down for two years. She drew worldwide attention to the ongoing environmental destruction. She was in danger because logging went on all around her; a fellow activist died in a logging incident that was considered an accident. She finally came down after getting an agreement from the logging company to save the tree, now called *Luna*, and a three-acre buffer zone around it. She continues to be an activist and educator and is the youngest inductee in the Ecology Hall of Fame. [*environmental movement*] {ecotopia.org/ecology-hall-of-fame}

hip-hop, green

An emerging music genre in which the lyrics are pro-environmental. Music, as always, can inspire, educate, and provide a medium for activists. Whether this is a flash in the pan or a long-term trend; only consumers will decide over time. [*jazz, Earth*]

HMS Beagle

Charles Darwin traveled aboard the ship HMS Beagle as the official naturalist. He was 22 when it left port in December of 1831 and they did not return for five years. The ship traveled up and down the eastern coast of South America, through the Straits of Magellan, and into the Pacific, heading back by way of *Australia* and Cape Town. The most important part of the trip was a five-week visit to the Galapagos Islands, where Darwin studied a variety of species, including finches, giant tortoises, and black lizards. These observations later led him to his world-changing theory of *natural selection* and evolution, which was published decades later in his <u>Origin of the Species</u>. He wrote specifically about his travel onboard this vessel in <u>The Voyage of the Beagle</u>, published in 1839. [*science; pseudoscience; opinions and science*]

hockey-stick graph

A now famous and controversial chart used in the 2001 "*Intergovernmental Panel on Climate Change*" *report* and then by *Al Gore* in his documentary movie <u>*An Inconvenient Truth*</u>. The chart illustrates the *global warming* trend in North America over the past 1000 years. The rapid increase in the most recent years makes the trend line appear similar to a hockey stick lying down with its blade pointing upward.

The chart illustrated the first research done to reconstruct the global warming trend by using proxy data to represent temperatures. (Proxy data uses things such as tree rings and glacier records to determine temperatures hundreds of years before people were recording this information. Proxy data has always been a scientifically accepted technique for many types of research.)

Because of questions about how the proxy data were used in the graph and later an event some call *Climategate*, the chart became the target of a group of *climate change deniers*. They claimed that not only was the chart false but also it proved the entire premise of *climate change* is false—actually not only false but a conspiracy or worse.

Unfortunately, some members of Congress joined with the climate change deniers, giving them a louder voice. But this resulted in a flurry of additional analysis and research that once again confirmed the science behind the chart and climate change in general.

In 2006, the National Academy of Sciences was asked to report on the veracity of *global warming* research and the resulting climate change. They stated in their report—and it has been repeated in countless additional independent studies—that climate change is real and the hockey-stick chart reflects this reality. Ridiculing the original hockey-stick graph, even deleting it from history—never to be seen again—does not change the facts about climate change. [*science; pseudoscience; opinions and science; Agenda 21 conspiracy*]

Holdridge Life Zone System

See *zones of life*.

holiday waste

Here is an often cited statistic: Americans produce 25 percent more *municipal solid waste* during the holiday season between Thanksgiving and Christmas than at any other time of year. Some news articles question this number because it was calculated long ago and there have been no new studies to confirm or deny its veracity. However, the consensus of those who study these things is that today the percentage is higher, because the production of solid waste has grown worse over the years, not better. What is known is that Americans produce 243 million tons of waste per year, which comes to almost 4.5 pounds per person per day. [*municipal solid waste disposal; wrapping-paper alternatives; postconsumer waste; product packaging; recycling*]

homes, green

Homes can be built to reduce the occupants' *ecological footprint*. The most important aspect of a green home and reducing the footprint is to minimize the amount of energy a home uses. *Eco-certification* programs help consumers understand how to do so. Builders can state they build homes that meet these certifications, but in most cases third-party verification is required to ensure the builder has done what is expected to meet the certification requirements.

Certification programs include the *Energy Star* for Homes certification, which includes a thermal envelope consisting of the proper insulation, seals, and certain types of windows, heating and cooling, lighting, and appliances to meet specific levels of efficiency. Some states offer tax incentives when homes are built to this level.

A more robust program is the *LEED* for Homes certification, which has silver, gold, and platinum levels of efficiency. To meet this certification, builders must meet stringent criteria, such as lot design, energy resource and water efficiency, and indoor air quality, to name just a few.

The most efficient homes are those considered to be *passive homes,* where most of the heat comes from *passive solar* energy from the sun. These homes are almost airtight, so all air entering and leaving the home is controlled. [*Living Buildings; zero-energy buildings; buildings, green; indoor ecology*]

honeybee sperm bank

The *colony collapse disorder* is devastating honeybee populations and has many people worried and trying some unusual and interesting ways to ensure their future success. One such plan has scientists storing sperm from United States and European honeybees to foster research on cross-breeding of honeybees. They hope to develop hardier bees that are not susceptible to the disorder. [*honeybees, pollination, and agriculture; robotic bees; doomsday seed vault*]

honeybees, pollination, and agriculture

Most people do not realize the importance the honeybee and other pollinators play in our daily lives. One method scientists are now using to assess the value of nature and natural processes is called *eco-services,* and pollination is one such service. Eco-services research estimates that pollination just by honeybees alone is worth about 14 billion dollars per year in the United States.

About 70 percent of all flowering species require pollination to reproduce, and more than 150 crop species in the United States require pollinators for their survival. Estimates show that about one-third of everything we eat is directly because of the pollination services provided by honeybees and wild pollinators such as *insects, bats,* and birds.

This is one reason the dramatic decline in honeybee populations—caused by a disease known as the *colony collapse disorder*—is so important. [*honeybee sperm bank; robotic bees; zombie bees*]

horizontal drilling

When most people talk about *fracking*, they are actually talking about two technologies: the actual fracking process and another called horizontal drilling that allows the bore hole to make a 90-degree turn even though it is thousands of feet underground. The debate about fracking is really about the process of fracking and horizontal drilling.

hormone disruptors, endocrine

Endocrine hormone disruptors are chemicals found in many commonly used products that mimic natural hormones and are believed to interfere with the body's endocrine system. Some of these substances are found naturally, but most are produced synthetically. They include drugs, *dioxin*, *PCBs*, *DDT* and other *pesticides*, and *BPA* (bisphenol A).

Hormone disruptors can be found in *plastic* bottles, metal food cans, *detergents*, *flame retardants*, food, toys, *cosmetics*, and pesticides. These substances are known to cause health problems in some types of *wildlife*, and research is ongoing to look for links to human health. They are thought to cause harmful developmental, reproductive, neurological, and immune effects in humans and wildlife. This includes lowering sperm count and possibly increasing the incidence of endometriosis and some cancers.

Many of these substances are persistent, meaning when disposed of, they do not break down into harmless chemicals but instead can build up in the environment. Some of these substances are found in our bodies and have become part of our *body burden*. [*autism studies and environmental factors; breast milk and toxic substances; medical ecology; nanoparticles and human health; puberty and the environment, early; Center for Health, Environment and Justice*]

Hormone Free label

See *food eco-labels*.

hormone weed killers

These are synthetic organic *herbicides* that control the growth of *weeds* by producing an effect similar to the plant's natural *growth hormones*, called *auxins*. This means it is one of the few *pesticides* that truly kills only the *target organism*—in this case, the weed. [*natural pesticides; biological control; biocides*]

host

An organism that provides food or shelter for another organism, such as a *parasite* living on a *host*. [*hyperparasites; microbiome and microbiota, human*]

hotel load

On rest stops, long-haul truck drivers used to typically leave the *diesel engines* running throughout the stop period, which could be the entire night, to power the cab's heat or cooling, appliances, and other electrical devices. The amount of power required to run these necessities is called the hotel load. About half a million of these trucks on

U.S. roads, idling for thousands of hours a year, uses a lot of energy and produces a lot of pollutants. In newer vehicles, the hotel load is handled by auxiliary power units that provide power for the hotel load without the engine running. [*energy use; energy-saving apps and programs; energy, ways to save on residential*]

hotels, green

Many hotels and motels have become green lately. Many have cards in their bathrooms asking guests to reuse towels if possible and not have the linens changed every day for those staying more than one night. This is a good example of how what's good for the environment is good for business. Just by using this little card, a typical 150-room hotel can save about 72,000 gallons of water, almost 500 gallons of detergent, and 17 percent of hot water costs.

These hotels use the services of groups such as the *NGO* Green Hotels Association and the World Travel and Tourism Council, which provide these marketing tools as well as many other services to save energy and water costs. [*three Rs; green consumer; credit cards, green; ecotourism; resorts, eco-; kibbutz, eco-*]

Hudson River School art movement

This art movement—which flourished throughout the 19th century—is considered the first truly American school of art. It is also credited with helping to shape the new country's love of nature and *wilderness*. Some consider the movement to have started earlier, but most attribute *Thomas Cole* with formalizing the movement in the first half of that century.

Members of this movement are said to have listened to *Ralph Waldo Emerson*'s call to "ignore the courtly Muses of Europe" and find a unique vision for American art. They typically painted large canvasses with vast landscapes of the Hudson River and its valley, illuminated with what many considered heavenly light. (It later evolved into Luminist art.) Although still considered the Hudson River School, many of this group traveled, later capturing the open spaces of the new west.

Their artwork appeared to be closely tied to the Transcendentalist ideals of Emerson and *Thoreau,* who wrote about nature and God. These Transcendentalist writers were later joined by nature writers loved by the masses, such as *John Muir* and *John Burroughs,* just as the Hudson River School painters were later joined by the mass-produced nature art prints of Currier and Ives. They all led to an America ripe for environmentalism and later an *environmental movement.* [*Beirstadt, Albert; Moran, Thomas; Audubon, John James; Caitlin; Porter, Eliot; Roosevelt, Theodore; preservation*]

human microbiome

See *microbiome and microbiota, human*.

Human Microbiome Project

The *human microbiome* refers to the collection of *bacteria* and other microbes that live on and in humans. Research on the human microbiome is ongoing in many countries, being led by the Human Microbiome Project in the United States—a five-year, federally funded research program that started in 2007. It is discovering new ways to treat illness and disease.

The study first identified a control group of "healthy" people to determine what the normal human microbiome consists of—a baseline. Using the latest in *DNA sequencing technology*, they determined which types of bacteria existed in these people. They now are comparing these microbiomes with those of "sick" people. The results are just beginning to be reported and appear fascinating. [*antibiotics and* the *human microbiome; self-concept and the human microbiome; bacteria; stool transplant; poop pill; virome*] {hmpdacc.org}

human population throughout history

Roughly 7.2 billion people exist on Earth today. How many people have ever lived on the planet since our ancestors evolved about 50,000 years ago? The *Population Reference Bureau*—the leading organization about all things concerning the people population—estimates it to be about 108 billion. This means the current population is about six percent of all the people who ever lived. [*population; population explosion, human; population growth, limits of human; biodiversity*]

humidifier fever

A respiratory illness with symptoms much like the flu, caused by microorganisms that inhabit and flourish in humidifiers and *air conditioning* units if they are not properly maintained and cleaned. [*indoor air pollution; heating, ventilation, and A/C systems; indoor ecology*]

hummingbirds and flight

Some consider hummingbirds in flight one of nature's wonders. Their flight patterns are unlike those of any other birds. They can fly forward, backward, and sideways effortlessly. They have been studied, flipping over and flying backwards and upside down for a brief time. They have been shown to remain in a hover for almost 50 minutes, if needed. They accomplish this, in part, by the speed and motion of their wings.

Their wings do not go up and down but round in almost a rowing motion. They complete one full wing-beat cycle (up and down) in 1/500 of a second, which is why their wings look like a blur in flight. [*biomimetics; bumblebee senses; camouflage, squid; communications between plants; sixth sense in animals; smell sensitivity in albatross; stereoscopic smell in moles; whistle names, bottlenose dolphin*]

humus

A dark brown, soil-like substance formed by the partial decomposition of plants and animals, found mixed with the surface layer of *topsoil*. Humus supplies new plants with most of the *nutrients* needed for growth. It changes the texture of the *soil* and increases its ability to absorb water. Rich topsoil usually contains a large amount of humus. [*soil types; soil particles; soil profile; soil organisms*]

hunger, world

Hunger is the want or scarcity of food in a country. World hunger pertains to an aggregate view of all nations. This is considered by many to be the number one global environmental issue.

The United Nations Food and Agriculture Organization estimates that nearly 870 million people— one in eight—suffered from hunger (chronic undernourishment) in 2010 to 2012. Out of this total, 852 million lived in developing countries and 16 million lived in *developed countries*. Hunger is obviously a far greater problem in poor nations.

Many regions of the world made significant progress in reducing hunger during these same years. In Asia and the Pacific, the number of hungry decreased by 30 percent, thanks to socioeconomic advances. Also, improvements were seen in Latin America and the Caribbean, but to a lesser degree. Africa, however, added about 20 million more hungry people. Developed countries also saw an increase of those hungry, from 13 million in 2004–2006 to 16 million in 2010–2012.

The environmental question often asked is whether the world can grow enough food to feed everyone, and the most common answer is yes. Beginning with the *green revolution* that began in the 1960s and other advances in agriculture, there should be enough food to feed everyone. The primary reasons many don't have enough food to eat is poverty. For those without enough land to grow food or money to purchase food (both linked to poverty), the result is hunger.

In developing countries, this is usually exacerbated by unstable political and economic forces within many nations, meaning the hungry are not going to receive aid from their governments. In addition, and more recently, *climate change* is believed to be adding to these problems, making it more difficult for those living in marginal regions to grow food.

World efforts targeting poverty have made significant progress. The *Millennium Development Goals* that come to a close in 2015 are considered responsible for most of the recent reductions in hunger in many parts of the world.

Many people believe the world will require a second "green revolution" to continue supplying enough food to the world's seven plus billion people. The first green revolution used selective-breeding techniques to increase crop yields. Many believe this second revolution might use *genetically modified organisms* as the new technique to feed billions of more people. [*safe and just operating space; water shortages, global; water grabs; land grabs; food waste; grain production and use; meat, eating*]

hybrid cars

Hybrid cars have become common, and fully *electric vehicles* (plug-in) are beginning to make inroads. The hybrid automobile became main stream with the 1997 release of the Toyota Prius. Although still a small segment of the total *automobile* market, hybrids are the fasting growing segment, with almost all car manufacturers producing at least a few models. Most environmentalists believe hybrids are a good thing but not a solution to the overriding problem of the environmental cost of so many cars burning *fossil fuels*. Hybrid cars are one of many *alternative automobile propulsion systems*. [*ammonia as a car fuel; CAFÉ; intelligent vehicle highway system; natural gas vehicles; tire, high mileage; vehicles, hydrogen fuel-cell*]

hybrid plants

See *selective breeding*.

hydrocarbon emissions

Hydrocarbons are one of the five primary pollutants that cause *air pollution*. Hydrocarbons are released by the incomplete combustion of *fossil fuels* or by simple evaporation of fuels such as *gasoline*. Hydrocarbons by themselves are not a problem, but when they react with other primary pollutants, they form dangerous *secondary air pollutants*.

Positive crankcase ventilation (PCV) valves and gas-cap air pollution control valves (APC), required on cars sold in the United States, help reduce hydrocarbon emissions. *Catalytic converters* also help reduce hydrocarbons (along with other primary pollutants) spewing from the tailpipe.

hydrologic cycle

See *water cycle*.

hydroponic aquaculture

Growing crops and fish together in closed, circulating systems. The fish are raised for food and the *hydroponic* vegetables or other plants are grown in the water on the waste from the fish. Some companies use hydroponic aquaculture to grow fish and vegetables to be sold to *organic* food supermarkets. [*aquaculture; integrated multitrophic aquaculture; hydroponic aquaculture; tuna aquaculture*]

hydroponics

Growing plants without soil. These plants are grown in an artificial liquid environment, using an artificial nutrient solution and providing artificial support for the plant. This technology allows crops to grow in areas where they would not normally be found. As long as all mineral nutrients required for growth are provided within the liquid solution, almost any type of plant can be grown hydroponically. [*hydroponic aquaculture*]

hydropower, nontraditional

Traditional hydropower facilities use *dams* to create reservoirs. Newer, nontraditional hydropower technologies involve capturing energy from moving water in rivers, such as in *run-of-the-river hydropower* facilities. Rivers are the main focus of new hydropower development. Smaller facilities do not necessarily require dams. They use a series of pipes with turbines inside that are turned by the current and have a less negative impact on the local ecosystem.

Many rivers considered feasible run-of-the-river sites have been put off limits for future development. The 1968 National Wild and Scenic Rivers Act prohibits development on virgin rivers and streams and protects about 40 percent of those sites identified as feasible for a hydro plant. This will limit the potential for this type of hydropower. Small-scale projects are believed to cause little harm to the environment, at least when compared with the larger projects that are environmentally destructive.

Many of these new plants are so-called small hydropower plants that generate less than 10 *MW* of power and even micro-hydro plants that create less than 1 MW. These small plants are often built in remote areas and create electricity for the immediate areas. About 1500 such plants have been built in the U.S., and more than 80,000 of these small hydro plants exist in China.

Many other nontraditional methods have also been used to harness water power, including *wave power*, *ocean thermal energy conversion*, *solar ponds*, *ocean current power*, and *pumped storage hydropower*.

hydropower, traditional

Also called hydroelectric power, this is the generation of electricity produced by water movement. It has been used for more than 100 years, ever since the invention of the electric generator. The traditional form of hydropower is to construct *dams* that form large reservoirs. The stored water is then released in a controlled manner to spin turbines and generate electricity. They are also built on natural waterfalls, with no need for a dam.

Hydropower dams are found in more than 100 countries, five of which produce more than half of all the hydropower produced globally. They are (in descending order): *China*, *Brazil*, United States, Canada, and Russia. About 24 percent of the world's electricity and 12 percent of U.S. electricity is generated by hydropower. (Hydropower produces almost half of all the *renewable energy* generated in the United States.)

The first commercial hydroelectric power plant was built in 1882 on the Fox River in Appleton, Wisconsin, to provide about 12 *kilowatts* of power that lighted a paper mill and a home. The first large hydroelectric plant was built on Niagara Falls in 1878.

Large hydroelectric *dams* that were built in the United States during the 1930s—such as the Hoover Dam (1455 *MW*) and the Grand Coulee Dam (6800 MW)—are dwarfed by new *megaprojects* such as the *Three Gorges Dam* in China (22,500 MW). Most environmentalists agree that although dams produce no *greenhouse gases* or *air pollution*, they are environmental disasters.

Hydropower has become cost competitive with *fossil fuels*. They are also easy to control and can be turned on and off, depending on need. However, major changes in weather, such as drought, can reduce production. A 1988 drought caused a 25 percent drop in hydroelectric power for the year in the United States.

The Federal Energy Regulatory Commission has identified thousands of sites within the United States that are feasible for hydropower development, capable of producing more than 150,000 MW of power. Many have been built and others are under development.

Hydropower, along with *geothermal power*, provides *baseload power*, meaning it is consistent, without the cycles of wind and solar power. Baseload power is important to utility companies providing electricity.

Water can be used for energy in many other ways, as well, called *nontraditional hydropower.*

hydrosphere

All forms of water on our planet constitute the hydrosphere. The majority of the hydrosphere consists of *surface waters,* which cover 74 percent of the entire Earth's surface. This includes *oceans*, fresh water and saline lakes, rivers, ice caps, and *glaciers*. *Groundwater* and moisture in the soil are also considered part of the hydrosphere.

Human intervention is affecting the quality of water in many ways, including *acid rain, ocean acidification,* and groundwater contamination. The extent of this contamination can be seen by the increase in the amount of *lead* found in layers of ice at the polar icecaps. [*biosphere*]

hydrothermal ocean vent ecosystems

In 1977 scientists discovered a new form of life—one that does not use the sun as its ultimate source of energy. These organisms get their energy from chemicals dissolved in the water as it emerges from deep sea (hydrothermal) vents. The chemical-rich water often looks like black smoke when it emerges from these vents, hence the nickname "black smokers." The water emerging from these vents is hundreds of degrees Fahrenheit. The *autotrophs* living in these regions do not perform *photosynthesis*; instead, they perform *chemosynthesis* to build organic compounds for life. The most common life forms living in these unusual *ecosystems* are tube worms.

New research on these vents continues to redefine life on Earth, especially as it pertains to *extremophiles*. Recent research in the Southern Ocean discovered many new species, including yeti crabs, living near waters produced by these vents more than 700 degrees F. In addition to these strange-looking crabs were two-dozen new species, including unique starfish, barnacles, and sea anemones.

As scientists study these vents, they are finding different varieties of these ecosystems. A recent discovery, also in the Southern Ocean, found one with no tube worms as found in the Atlantic and Pacific vents. Some scientists believe there are six distinct vent ecosystems, while others believe there might be as many as eleven.

Scientists studying these unique environments are fearful that *mining* companies might begin exploiting the high concentrations of minerals, such as copper, that shoot out of these vents. One such company is thought to have plans to start a mining operation on a vent off the coast of New Guinea. These strange and highly unique *ecosystems* could be damaged or destroyed before even being studied. They might be gone before we thoroughly understand what they are. [*extinction, newly identified species and; weird life*]

hyperaccumulators

Plants or animals that thrive on contaminated soils. Some plants can absorb contaminants from the soil, thereby decontaminating it. The process is called *bioremediation*. Research is ongoing with plants that hyperaccumulate *heavy metals*. After the plants have accumulated the metals, they can be disposed of as a *hazardous waste* or possibly processed to remove the heavy metals and *recycle* them for further use.

Plants such as common ragweed and hemp dogbane remove *lead* from soil, while others have been found to remove zinc, *cadmium*, and nickel. Once the plants have been harvested with high concentrations of the metals, they act like a *bio-ore,* which is then processed—or "smelted"—like mineral ores for their metal content. Hyperaccumulators within *Superfund* sites and around *nuclear reactor disasters* have decontaminated the soil there to some degree. [*phytoremediation; bio-ore; remediation of hazardous wastes; Chernobyl*]

hyperparasites

Parasites that parasitize other parasites. For example, some birds are parasitized by blowflies. In this situation, blowfly larvae parasitize a bird, feeding on its tissue. The blowfly larvae within the bird are also parasitized by a hyperparasitic wasp.

This type of relationship can continue when a *bacteria* parasitizes the wasp, meaning the bacteria are also hyperparasites, although the term is not typically used for bacteria.

But we are not finished. The bacteria can have their own parasites, as well. *Viruses* can parasitize bacteria and are called *bacteriophages*. This is quite common. Finally, *viruses* have even been found to have their own parasites. Recent research shows tiny bits of genetic matter called *transposons* living within some viruses. [*parasitoidism; decapitating flies; follicular mites; zombie bees*]

hypoxia, environmental

Low levels of oxygen within *aquatic environments* often caused by *nutrient enrichment*. When oxygen levels fall too low, it becomes a *limiting factor*, forcing organisms to move out of the area and even causing death. Extensive hypoxia can result in large *dead zones* in the ocean.

ice ages

Over the past 700,000 years, there have been eight great ice ages, in which huge sheets of ice moved down from the polar ice cap, blanketing large portions of North America, Europe, and parts of Asia. These dramatic transformations of our planet's environment resulted from gradual climate changes lasting thousands of years. Each ice age lasted up to 100,000 years, after which gradual warming allowed the ice caps to recede.

The periods between the ice ages, called *interglacial periods*, only lasted about 10,000 to 12,000 years each. The last ice age drew to a close about 10,000 years ago. The difference in temperature between an interglacial period and an ice age is only a few degrees. The difference in temperature between the last ice age and the current interglacial period is only about nine degrees F. [*climate change*]

iceberg calving

See *Petermann glacier ice detachment.*

ichthyology

The study of fishes. [*fish populations and food; fishing, commercial*]

idling your car

Automobiles that are left running while not being driven wastes gas and contributes to *air pollution.* People often think that letting their car idle uses less fuel than starting a car. However, leaving a car idle for 10 minutes typically uses more gas than simply turning it off and then back on when needed. Also, idling for 10 minutes emits about one pound of *carbon dioxide* into the air. Of course, it also is a waste of gas. [*gasoline, saving on auto; remote-control car starter; tire, high-mileage; right-to-dry laws; bathroom water-saving techniques; Christmas tree lighting; energy, ways to save residential; energy-saving apps and programs*]

igneous rock

Rock formed from cooled, hardened molten magma. [*plate tectonics*]

Illegal, unreported, and unregulated fishing (IUUF)

IUUF is a global problem threatening sustainable fisheries. These illegal catches typi-cally come from areas that have ineffective oversight. IUU fishing could include fish-ing in illegal waters, such as a *marine protected area* (MPA), but it most often violates quotas or *bycatch* limits that have been established under international agreements. In 2010, almost 20 percent of all seafood was caught in this illegal manner, making protecting seafood populations and *marine habitats* difficult, if not impossible.

This can be viewed as a typical case of *the tragedy of the commons*. [*fishing, commercial*]

immigration

Immigration refers to the movement of people into a country, and *emigration* refers to the movement out of a country. Immigration and emigration, along with *natality* and *mortality*, dictate changes in the size of a nation's population. [*human population throughout history; doubling time in human populations; population explosion, human; sex ratio; China's One-Child Program; migration*]

imperfect flower

A unisex flower that contains only the male (pistil) or female (stamen) reproductive organs. For example, the scrub oak produces imperfect flowers. [*sexual reproduction; pollen; hermaphrodite*]

imported food dangers

Roughly 10 to 15 percent of all of American's food is imported. But some sectors are much higher—about 66 percent of fruits and vegetables and 80 percent of seafood is imported. These numbers have risen dramatically and are projected to continue to rise. Only about two percent of all imported food is ever checked by *Food and Drug Administration* inspectors. Independent research shows a significant percentage of this imported food is contaminated in one form or another. Recent reports state imported raw produce and seafood are considered to be "of high risk" of contamina-tion. The most common finds are illegal *pesticide residues*, *pathogens*, and filth.

Roughly half of all imported seafood comes from *aquaculture,* which has addi-tional risk of contamination because these confined artificial habitats are prone to bacterial outbreaks, and any antibiotics and antifungal agents used can remain as residues on the food. [*food miles; food's carbon footprint – the best and worst; seasonal food; circle of poison; fruit waxing; organic farming and food; food eco-labels; organic food labels; organic food supermarkets; sugar, sugar alternatives and the environment*]

incandescent bulbs

These have been the traditional light bulb of the 20th century, using a tungsten filament to produce light. It is very inefficient compared to other *light bulb technologies*; only about 5 percent of the energy is converted into light, with the other 95 percent lost as heat. Newer incandescent bulbs have been developed using materials other than tungsten, which improves their efficiency. However, they still cannot compare to the newer *compact fluorescent bulbs (CFL)* and especially the most recent *LEDs.* In the United States, recent *light bulb legislation* is phasing out incandescent bulbs.

incineration

One of four methods of *municipal solid waste* disposal, in which incinerators burn *garbage* to reduce its volume. Some cities use the heat to generate electricity in *waste-to-energy power plants.*

Burning solid waste reduces its volume between 60 and 90 percent, meaning there is far less waste to be placed into a *landfill.* However, incineration is associated with significant environmental problems. [*incineration problems; cremation, human*]

incineration problems

Incineration is one method to dispose of *municipal solid waste.* The biggest advantage of incineration is that it reduces the volume of solid waste, meaning less needs to go into *landfills.* But problems are associated with its use. Where does the incinerated waste go? Most goes up the smokestack, causing *air pollution.* Air emissions from incinerators contain *fly ash* that often contains *hazardous* and *toxic* substances such as *dioxin,* furans, *heavy metals*, and acid gases. Pollution control devices such as scrubbers and electrostatic precipitators filter and trap substances, dramatically reducing these substances but not eliminating all of them.

The solids that remain after incineration are called *bottom ash.* It must be placed in *landfills* and may contain concentrated *toxic* substances that didn't go up the smokestack. Bottom ash from some incinerators must be handled as a *hazardous waste* and cannot be simply dumped into landfills.

Opposition to new incineration facilities has stopped the building of many planned incinerators because of the *NIMBY* syndrome. [*municipal solid waste disposal; waste-to-energy power plants; wastewater-to-energy power plants; cremation, human*]

India

India and *China* have similar population and environmental challenges. They are the only countries with a billion-plus population, and both are moving quickly from *developing* to *developed country* status. India has almost 20 percent of the world's population but resides on less than 2.5 percent of the world's land mass. India's economy has expanded rapidly, taking millions of citizens out of poverty and making them middle-class consumers. Their burgeoning population and newfound affluence are dramatically increasing the nation's *ecological footprint* and taking a toll on our planet's *environment*.

Some of their most pressing environmental issues include *deforestation* in the Himalaya *watershed* region, caused by the need for *fuel wood* and clearing for *agriculture*. This deforestation has resulted in chronic flooding. More than half of all existing cropland has become damaged or unusable by *overgrazing* and nonsustainable, *conventional agricultural* practices. Estimates have between 5 and 10 billion tons of *topsoil* lost each year due to this poor farming use.

From a public health perspective, 70 percent of all *surface waters* are polluted due to wanton dumping of *industrial waste,* such as *toxic* chemicals as well as *fertilizers* and *pesticides* from agriculture. Many people living in cities and most of those in rural areas do not have clean *drinking water.* Exacerbating the problem is the lack of *sewage treatment facilities*, found in only a few cities. More than 100 cities routinely dump raw sewage into rivers, including the Ganges; also polluted by *human cremation* remains. Downstream, residents use this contaminated water for drinking, cooking, and bathing.

Air pollution is worst in the cities, primarily from old, poorly running *diesel fuel* vehicles and human cremation fires. But primitive *woodburning cookstoves*, *brick kilns,* and the general use of biomass as a fuel source (such as animal dung) cause serious air pollution in rural regions as well.

Most of their *electric power plants* (found in cities) run on *coal*, but recent years have seen a move toward *natural gas.* More than 40 percent of the country's total energy needs come from coal. Natural gas provides less than 10 percent of the total, but is growing. Rural regions still primarily rely on *biomass energy* such as wood and waste, comprising more than 20 percent of the country's total. And *oil* remains important, especially for *transportation*, exceeding 20 percent of the total.

Wildlife trade and trafficking is a serious problem in India, with its abundance of *endangered species* and its paucity of wardens to defend *wildlife refuges* from *poaching*. Although many regions have strict laws, allowing wardens to shoot poachers on sight, little can be done to stop the wanton killing.

The country has made efforts to clean up and protect the environment. Governmental committees have existed since the 1970s to look for solutions, and in 1980 a Department of Environment was established. In 2006, a new National Environment Policy was adopted to further guide the country's environmental direction.

How India and China manage their environmental problems might not seem all that important to those of us on the other side of the world; but national borders are meaningless when it comes to billions of people living on our one and only planet. [*aquaculture-producing countries, top ten; Fertile Crescent; irrigation, agricultural; organic food and farming; brick kilns; carbon emissions; cremation, human; e-waste; municipal solid waste; toxic cocktail; weapon dumps, abandoned; heavy-water reactor; biochemical conversion, biofuels; breeder reactors; energy poverty; methane digesters; nuclear power; solar ponds; wave power; Bhopal; companies, top ten green global; Women's Earth Alliance; megacities; population explosion, human*]

indicator species

Some species can be used to indicate the overall health of an *ecosystem* and are called indicator species. For example, krill has been found to be sensitive to *ozone*. Krill is also the most important food source for marine life in *Antarctica*. When the hole in the *stratospheric ozone* was at its worst, the krill population dropped. As the ozone situation improves, as it is doing, the krill populations are improving. Scientists now monitor the health of the krill as an indicator species for stratospheric ozone. [*Landsat; satellite mapping; biosensors; crowd research, scientific; volunteer research, scientific*]

indigenous

Refers to organisms, including humans, that live naturally in an area, as opposed to being introduced from elsewhere. For example, Native Americans are indigenous to North America, and the buffalo is indigenous in the American West. [*alien species; invasive species; exotic species*]

Indonesia

Indonesia is an archipelago of about 17,000 islands. It has some of the world's largest *coral reefs* and *tropical rainforests*. Roughly 17 percent of all species that exist on our planet are found here. *Deforestation* is a major problem that results from *illegal logging*. The number of *automobiles* has increased, leading to *air pollution*. The country's attempts to curb their environmental problems have waned and ebbed based on the political leanings at the time. Vast forest fires ravaged some of the islands between

1997 and 1998, with estimates of 38,000 square miles burned. It is believed that most of the fires were intentionally lighted to clear land for *palm oil* plantations.

indoor air pollution

Most people worry about outdoor *air pollution*, and rightly so. Far too few people, however, think about the quality of air indoors, where they spend 90 percent of their time.

Two factors play an important role in creating indoor air pollution: increased use of *toxic* chemicals and synthetic materials in building construction and furnishings as well as the move toward energy efficiency by building tightly sealed buildings.

Substances such as *formaldehyde* are released from building products into the air by the process of *outgassing*. *Asbestos* particles, along with numerous other harmful substances, are still found in many building materials and furnishings. New carpeting alone emits (in the first few months) formaldehyde, ethyl benzene, toluene, xylene, and other *volatile organic compounds* (VOCs), all of which are toxic. Other substances commonly found in the home and office include *PVCs*, *phthalates*, chlorine, and others.

The fact that buildings are often sealed shut concentrates these and other substances. *Pesticides*, *heavy metals* from road dust, naturally produced *radon*, and pathogenic microbes can accumulate indoors and thrive in *air conditioning* systems. Still, one of the biggest indoor air problems remains smoke from cigarettes, including *passive smoke* breathed in by nonsmokers.

These indoor pollutants can cause short-term *acute* symptoms such as dizziness, headaches, and fatigue. Long-term exposure to some of these toxic substances can cause *chronic* serious illnesses.

Solutions can be difficult, because you usually do not have a choice about the construction materials used to build your place of residence. However, you can make wise purchasing decisions to purchase products that are environmentally friendly. Carpeting and furniture, for example, can be found that are free of VOCs or other pollutants, but be sure their *eco-labels* or claimed *eco-certification programs* are valid and not just a *green scam*. You can check the indoor air for *radon* with self-test kits that work well.

Another good way to reduce indoor environmental pollutants is to use a HEPA vacuum cleaner commonly found in retail stores. HEPA stands for high-efficiency particulate air. HEPA filters remove particles as small as one-third of a micron at close to 100 percent efficiency. (A human hair is around 90 microns thick.) If that's not enough, you can purchase an ULPA vacuum (ultra-low penetration air) that filters particles down to 0.12 microns; these are used by professionals during *hazardous waste* cleanups.

Don't forget that one of the main reasons indoor air pollution can become serious is that homes and offices are built tight these days, with a lot of insulation to keep the heat or air conditioning working well and the expense down. Opening windows allows the polluted indoor air out and fresh air in. [*couches and indoor pollutants; humidifier fever; inkjet and laser printer cartridges; laser printer ozone; mattresses, recycling; refrigerator doors; indoor ecology*]

indoor ecology

We spend most of our time indoors; a place with its own environment, its own pollutants, its own environmental problems. Only recently has the *ecology* of the indoors become mainstream science. Entire *ecosystems* surround us while indoors, very different from those outside. As the science advances, the goal has been changing. At first it was meant to seek out and destroy all forms of life indoors, trying to make it a sterile, "healthy" environment. But now science has come to realize we must understand the unique *biodiversity* of the indoors environment because we spend so much time there.

Every breath we take indoors contains millions of *bacteria* and other microbes. This new science is learning that—at least to a certain degree—it might be better to live and let live instead of spending all of our efforts trying to kill the indoor ecosystem.

Places such as shower curtains and showerheads are full of life, mostly bacterial. That scum you feel is more appropriately called microbial biofilm, because it is alive. The steam that emanates from the shower head is not just moisture but a mist full of life. While most of these microbes are benign, some are not. Therefore, we have much to learn about indoor ecology. Just as new studies are revealing the importance of our *human microbiome*, the importance of the indoor ecosystem is also in its earliest stages.

An interesting new study is combining both indoor ecology and the human microbiome. The Hospital Microbiome Project at the University of Chicago is studying the human microbiome of patients in hospitals, plus the indoor ecology of the hospital within which the patient resides. The goal is to lower infection rates and deaths that accompany hospital visits. [*indoor air pollution*]

industrial agriculture

See *agriculture, high-impact.*

industrial ecology

A new field that attempts to identify and quantify the impact of manufactured products on the environment. It integrates chemistry, physics, engineering, and *ecology* and is closely related to environmental health as well. You can get an idea of when a field of science has become mainstream by looking at the release of its own journal. The Journal of Industrial Ecology did not publish until 1997, indicating this to be a relatively new science. [*couches and indoor pollutants; humidifier fever; inkjet and laser printer cartridges; laser printer ozone; indoor ecology*]

industrial livestock industry

See *CAFO* and *factory farms*.

industrial smog

Smog with high levels of *sulfur dioxide* and *sulfuric acid* produced by burning *coal* containing sulfur impurities. Pollution-control devices, such as scrubbers, can eliminate enough of these contaminants to make industrial smog rare in developed countries. Also called gray smog. [*air pollution*]

industrial waste

See *industrial water pollution*.

industrial water pollution

Industrial water use causes *water pollution* because its final destination, just like *domestic water*, is streams, rivers, and finally the ocean. Some industrial waste is combined with *domestic wastewater*, but most is handled separately. New industrial facilities are required to treat their wastewater before releasing it, but older plants often simply dump the wastewater directly into a river or stream.

Industrial wastewater may contain *oil*-based products, *heavy metals*, acids, salts, or *organic* substances. In many cases these substances are highly *toxic* and should be handled as *toxic wastes*. *Thermal water pollution* is another form of industrial water pollution, caused when water is used for cooling. Discharging this hot water into natural bodies of water changes the water temperature and impacts the *aquatic ecosystem*.

industrial water use

One of four ways people utilize water. The majority of industrial water is used for cooling processes. Most *electric power plants* use water as a coolant. Paper mills and many manufacturing processes require vast amounts of water. Some manufacturers *recycle* the water once used, thus dramatically reducing the volume needed. Water used during manufacturing processes often becomes contaminated, causing *industrial water pollution*. [*water use for human consumption; fracking, ongoing environmental concerns; water shortages, global*]

industry self-regulation

This is an alternative to governmental regulations that protect the public or the environment from offensive products, services, processes, or any corporate actions that cause harm. It is also called command-and-control regulation. Self-regulation can be voluntary or mandatory. Some believe this is a meaningful way to create *corporate social responsibility* and protect the environment, but others—probably the majority of environmentalists—feel it is nothing more than *greenwash* and advertising hype. [*CERES; companies, top ten green global; corporate sustainability rankings; economics and sustainable development; economy vs. environment; integrated bottom line; internalizing costs; steady-state economics; Sustainability Imperative*]

infectious disease

Any disease transmitted without physical contact. [*pathogens; skin bacteria; bubonic plague; Lyme disease; maggot medicine; nanoparticles and human health; spillover effect; superbugs; zoonotic diseases*]

infrastructure, green

The *Environmental Protection Agency* defines this as "weaving natural processes into the built environment." Green infrastructure includes the use of trees and other plants to reduce *urban air pollution*. Recent research confirms that green infrastructure—when taken seriously—can significantly improve air quality. It includes well placed grasses, climbing ivy, and other plants that reduce the amount of *particulate matter* and *nitrogen oxides* in the air. Green walls—well-placed ivy-covered walls—appeared to be the most efficient way to naturally clean the air.

Other forms of green infrastructure include porous pavement, plants on rooftops, painted *white rooftops*, curbside vegetation, and *rain gardens*. It can also include the use of *biomimetics, community gardens, greenways, living buildings*, and smart cities that grow thanks to *smart growth, soft energy*, and *terraforming*. Green infrastructure is the opposite of gray infrastructure, which is the norm.

inholdings

Parcels of private land surrounded by public lands. [*inland wetlands*]

inkjet and laser printer cartridges and the environment

These cartridges are made of *plastic*, meaning they are made of refined forms of petroleum (*oil*). Estimates show that one laser cartridge requires one gallon of *fossil fuel* oil to produce and about 2.5 ounces to make one new inkjet cartridge.

At the other end of this product's *life-cycle assessment* are the three hundred and fifty million ink and laser cartridges estimated to be thrown out each year, globally. Some people call these cartridges, technology's Styrofoam when it comes to waste.

In addition, the ink contains *toxic* components. Not that it is harmful to those using it, but when disposed of improperly—as is often the case—they can contribute to pollution if they end up in the wrong *landfills*. The ink residue contains *volatile organic chemicals* (VOCs).

Less environmental harm is done if you purchase recycled cartridges and turn in old ones for *recycling* or proper disposal. (Most places that sell them will take them back for these purposes.) [*laser-printer ozone; indoor air pollution*]

inland wetlands

Wetlands are divided into two types: *coastal wetlands* and inland wetlands. About 95 percent of all remaining U.S. wetlands are inland. Wetlands play a vital role on our planet. They provide a *habitat* for numerous fish, waterfowl, and wildlife. They help control flooding by storing vast quantities of rainwaters that are then used to recharge (replenish) *groundwater,* which supplies much of our *domestic water*. Wetlands filter out sediment and dilute pollutants from incoming streams that would otherwise end up in our water supply. Crops such as rice, blueberries, and cranberries are cultivated in these areas.

More than 50 percent of the wetlands in the continental United States have been destroyed. Of the remaining inland wetlands, only about 20 percent are under federal protection. The other 80 percent are on private land. Many local laws are insufficient to prevent individuals from developing this land. [*estuary and coastal wetland destruction*]

inorganic

Substances that are not alive and do not come from recently decomposed organisms. They do not contain carbon and are not *organic*.

insecticides

Most people would think that with a name like insecticides, it would be a substance that kills *insects*, and that is the case—with one important correction. They also kill many other forms of life. Most *pesticides* (which includes insecticides), are nonspecific, meaning they don't know one organism from another. This is why *Rachel Carson* suggested we call them *biocides*—they kill life.

However, there is no question that one of the primary reasons we can feed the number of people we feed today is because insecticides are so heavily used. The flipside of that is we are also contaminating our soil and water with these chemicals.

The vast majority of insecticides are synthetically produced, and these can be placed into one of four categories: *chlorinated hydrocarbons, organophosphates, carbamates*, and *pyrethroids*. But alternatives to these synthetic insecticides do exist. *Biological pesticides* use natural and *biological control* methods. Both use natural relationships such as *predators, parasites,* and microbes such as *bacteria*, to control insect pests. In addition, *genetically modified organisms* have been developed for use as insecticides. The advantages to most of these alternatives are that they target the specific pest and they are far safer for the environment. [*pesticide residues on food; biological amplification; DDT; circle of poison; nanopesticides; pesticide dangers*]

insectivore

An organism that feeds on *insects*. Insectivores include numerous birds and fish; some *mammals*, such as shrews and anteaters; and many *predatory* insects. Also, a few plants such as the pitcher plant and the well-known, Venus fly trap can ingest insects and are therefore considered insectivores.[*carnivore; omnivore; coprophagous; apex consumers and ecosystems; chemozoophobous; chledophyte; decomposer; food chain; predator; saprophyte*]

insects

Of the approximately 1.8 million identified species of plants and animals on our planet, close to one million of them are insects. And most of these are *beetles*. The number of insect species that probably actually exist is estimated to be between 2 and 15 million. New insects continue to be routinely discovered. No other type of multicellular organism even comes remotely close to the *biodiversity* of insects. Of course, this is just an educated guess, but it is suggested that at any given time there are ten quintillion insects on Earth. (That is 1 with seven zeroes.) They probably provide more *eco-services* than any other group of organisms.

The typical insect body is divided into three regions (head, thorax, and abdomen), and the thorax typically contains three pairs of legs and often two pairs of wings. [*ant slaves; beetlecam; bumblebee senses; bumblebee decline; colony collapse disorder; decapitating flies; honeybees, pollination and agriculture; Monarch butterfly decline; neonicotinoid pesticide and honeybees; queenright; traumatic insemination; zombie bees*]

insect sterilization

Insect sterilization is an innovative way of controlling insect pests. It is an alternative to synthetic *insecticides* and has been used as part of an *integrated pest management* program. Insect sterilization involves mass-rearing a pest *insect* and sterilizing the males with either chemicals or radiation. The sterilized males are released at the appropriate time in the infested area to mate with unsuspecting females. Because they are competing with wild, nonsterilized males, they must far outnumber the virile males by ten to one. The release is usually done multiple times and only works on species that mate just once.

The screwworm fly—an often fatal parasite on cattle, goats, and deer—has been virtually eliminated in many parts of the United States, Mexico, and Central America by using this method. [*biological control; biological control methods; integrated pest management*]

in-stream water use

One of four ways people use *water*. In-stream refers to using a flow of water for human activities and includes *hydropower*, navigation, and *outdoor recreation*—boating, swimming, and fishing. Water recreation requires clean, unpolluted waters. To sustain this, the public using the water must be held accountable to keep the water clean.

Navigable waterways often require *dredging* or widening, which often cause environmental damage to the *aquatic ecosystem*. Today most projects that affect waterways require *environmental impact statements* that analyze the damage that might be caused. [*water use for human consumption; running water habitats, human impact of; rivers in decline; blue-jean pollution; Clean Water Act*]

insulation

See *thermal insulation*.

integrated bottom line

Combining financial, environmental, and social costs and benefits into a single reporting measure of a business. Standard goals such as profitability, competitive advantage, efficiency, and growth are only considered successful if they are compatible with environmental and social goals pertaining to *biodiversity*, sustainability, equity, community support, and the general well-being of its stakeholders.

The integrated bottom line approach is similar to the *triple bottom line* method. However, the triple bottom line uses three separate balance and income sheets, whereas the integrated bottom line uses one. [*companies, top ten green global; CERES; corporate social responsibility; corporate sustainability rankings; economics and sustainable development; economy vs. environment; internalizing costs; extended product responsibility; life-cycle assessment; steady-state economics*]

integrated multi-trophic aquaculture (IMTA)

As *aquaculture* farming becomes more prevalent, these farms are beginning to cause the same environmental problems found with other forms of *high-impact agriculture. Nutrient enrichment* occurs when uneaten food and wastes collect in the water and cause *cultural eutrophication.* Efforts have begun to make aquaculture more sustainable by reintroducing old practices such as raising many different species in the same aquaculture farm—an approach called integrated multi-trophic aquaculture (IMTA).

Just as small farms used to raise livestock that produced manure that was then used as fertilizer to grow crops, IMTA uses a similar approach. For example, a single aquaculture facility is raising salmon in pens but also places blue mussels downstream to consume the organic waste produced by the salmon. Even further downstream, kelp is grown, which extracts *inorganic nutrients*, and finally sea urchins and sea cucumbers on the *sediment* below feed on large waste particles that settle out.

The salmon, mussels, and kelp are all sold as part of the harvest. Thus, instead of just raising salmon and polluting the surrounding habitat, they are raising many species in concert with one another and keeping the habitat clean and the entire process sustainable. Various combinations of species are being used in IMTA in such diverse countries as Chile, Turkey, South Africa, Israel, Norway, Ireland, Scotland, and *China.* [*vertical farms; permaculture; conservation agriculture; green manure; low-input sustainable agriculture; regenerative agriculture; organic farming and food; organic fertilizer; sustainable agriculture; slow food movement*]

integrated pest management (IPM)

Integrated pest management is an alternative to exclusively using synthetic *insecticides*, with the accompanying environmental harm. IPM uses a variety of techniques in a carefully planned program designed to reduce our dependence on dangerous, contaminating *pesticides.* These techniques include *insect sterilization, sex attractants, resistant crops, cultural* or *organic farming* techniques, *natural insecticides, biological control*, and some forms of *genetically modified organisms,* as well as the selective use of synthetic pesticides.

IPM is used to reduce the use of traditional insecticides, but its use has been very limited. This is due in part to the absence of coordinated programs sponsored by local, state, and federal governments and the funding required. Also, many farmers are hesitant, because it is not the norm and takes more effort and time and possibly expense. The continued use of dangerous pesticides as our primary form of pest control, in spite of the availability of alternate existing technologies, is a management and education problem that needs to be addressed further.

intelligent vehicle highway systems (IVHS)

Millions of research dollars are spent each year in the United States on ways to create "intelligent" highways. Optimistically, these technologies hope to reduce the average commute in crowded areas by 50 percent. Some of the simpler aspects of this technology are already in operation in some areas. For example, some roads monitor traffic flow with magnetic induction loop detectors embedded in roads at half mile intervals. The data is collected by roadside devices that pass the information along to a computer. The computer controls the traffic lights and message boards in an effort to relieve congestion.

More advanced plans for IVHS include sending traveler advice directly into navigation receivers in *automobiles* and someday—and many think soon—controlling an "intelligent" car's speed, braking, and steering. [*autonomous automobiles*]

intercrop

Growing two or more crops in the same plot of land. [*monoculture; high-impact agriculture; conservation agriculture*]

interglacial period

A warm period that exists between two *ice ages*.

Intergovernmental Panel on Climate Change (IPCC)

The IPCC is a group of about 800 climate science experts from all over the globe. The panel was convened in 1988 by the *United Nations* and has issued five major IPCC Reports since then. Their mission is to "assess the scientific, technical and socioeconomic information relevant for the understanding of the risk of human-induced *climate change*." The panel writes in-depth reports considered to be the most important information about global climate change.

The panel briefly damaged their credibility by releasing some erroneous analysis in past reports. Although it was quickly retracted, they have been the focus of attacks from *climate change denier* proponents. However, the IPCC remains the most important worldwide group helping the world figure out what must be done to control climate change. The latest report—the *IPCC Fifth Assessment*—was released in October 2013.

Intergovernmental Panel on Climate Change (IPCC) Conferences

The *IPCC* has annual meetings where thousands of representatives from around the world come to discuss *climate change* and what to do about it.

The past climate conferences have had mixed results. In Copenhagen 2009, the *Copenhagen Accord* was accepted. In Cancun 2010, the *Green Climate Fund* was developed. In Durban 2011, they agreed to extend the *Kyoto Protocol* that expires in 2012 and create a replacement for that protocol by 2015. In Doha 2012, they agreed—once again—to have a new protocol ready before the France 2015 meeting.

Intergovernmental Panel on Climate Change (IPCC) Fifth Assessment Report

One of the most important pieces of information people want to know from the *Intergovernmental Panel on Climate Change Reports* is how sure we are that *global warming*—since the mid-20th century—results from human causes (*anthropogenic*). This latest report states we are 95 percent certain people are the cause of this warming.

The most recent 30-year period has probably been the warmest in the past 1400 years. Many critics, however, point to a short-term leveling off in increasing temperatures. The panel explains they look at long-term trends—not short-term "wiggles"—in the increase. Questioning why we should worry when things don't look so bad at this moment is understandable. However, just as most of us have a hard time understanding *geological time* as needed to clearly understand *evolution*, it is hard to imagine *climate change* time periods that must be studied over many hundreds and thousands of years.

These reports always give high and low estimates based on assumptions about how much is being done to mitigate future *carbon emissions*. This latest report has a worst-case scenario of global temperatures rising 4 degrees C (7.2 F) by 2100 to a best-case scenario of rising 2 degrees C (3.6 F).

Other information in the report is about *sea levels*, ice sheets, and *oceans*. The report states that the seas are rising more rapidly than first expected, with a range between 1 meter (3.3 feet) and 3 meters (9.8 feet) by 2300. Ice sheets in *Greenland* and *Antarctica* are getting smaller, *glaciers* are shrinking worldwide, and the *Arctic* sea ice and all of the Northern Hemisphere snow cover is decreasing.

The Atlantic Ocean's current circulation (the Gulf Stream) is weakening and estimated to be between 12 and 54 percent reduced by the end of this century. Weakening *ocean currents* are considered one of the *tipping points* some scientists fear will cause nonreversible damage. Also, the *ocean's acidification* continues to increase.

Intergovernmental Panel on Climate Change (IPCC) reports

The *Intergovernmental Panel on Climate Change* releases in-depth reports on *climate change* every five or six years. It issued reports in 1990, 1995, 2001, and 2007, and they released the newest report in 2013. These reports are considered the most important scientific assessments of climate change and are respected across the globe. (With the exception of *climate change deniers* that love to use these reports for target practice.)

The fact that they release these only once every few years has many calling for a change in their reporting structure. Climate change is moving at a pace that many feel should be reported on in a more timely fashion. Therefore, the panel is considering shifting from these tome-like thousand-page reports to smaller, focused, news-like briefs that are generated in a more timely fashion.

Intergovernmental Science-Policy Platform on Biodiversity and Ecosystem Services (IPBES)

This new group works under the auspices of the *United Nations* and will write regular assessments about the state of global *biodiversity*, primarily for policy-makers. They are headquartered in Bonn, Germany. The group is similar in design and purpose to the *Intergovernmental Panel on Climate Change* (IPCC), but where the IPCC focuses on *climate change*, the IPBES focuses on biodiversity.

interim use

A new phrase that describes ways to make unused inner-city space once again productive. Portions of many inner cities have become abandoned in the wake of economic crisis. This space is often made available for any type of interim use that might help bring a city back to life. Some suggested interim uses include *urban farms*, short-term businesses, entertainment venues, and even a canvas for artists. The hope is that interim use might gradually evolve into permanent use and a rebirth of some inner cities. [*urbanization and urban growth; brownfields; schools and brownfields; slum; smart cities; urban gardens*]

internalizing costs

See *extended product responsibility* and *externalizing costs*.

International Convention for the Prevention of Pollution from Ships

See *MARPOL*.

International Joint Commission (IJC)

The IJC is composed of three members from the United States (appointed by the President) and three from Canada (appointed by their government). This organization carries out environmental studies and makes recommendations to policy makers in both the United States and Canada about bodies of water shared by both countries—for example, the *Great Lakes*.

International Maize and Wheat Improvement Centre (CIMMYT)

Scientists at this organization, centered in Mexico, are given credit for starting the *Green Revolution* of the 1960s, when new varieties of wheat and corn were developed that enabled a rapid advance in yields. The development of these new varieties was done with *selective breeding*. (*Genetically modified organisms* had not yet been developed.)

The organization remains at the forefront of developing new varieties of wheat and corn (maize). Many crops grown today could be wiped out as *global warming* increases due to *climate change*. The CIMMYT hopes to develop new varieties to once again keep the world from starving. Interestingly, although most of their research and development is done with hybrid and selective breeding methods, they do not rule out the use of GMO varieties as well. [*hunger, world; grain production and use; dairy cow output; subsistence farming; United Nations, global environmental issues and the; safe and just operating space*] {cimmyt.org}

International Organization for Standardization (ISO) 14000

Eco-certification is one way to demonstrate that a company, process, or product is following specified criteria to be considered environmentally conscious. Adhering to international standards is another way. The ISO is an *NGO* based in Geneva that creates standards on numerous technical and managerial processes. It has no regulatory capabilities and is strictly voluntary. The World Trade Organization recognizes ISO standards, making it respected worldwide. ISO 14000 is a series of environmental standards and can also be considered when determining if you want to use a company's products or services. {iso.org}

International Union for Conservation of Nature (IUCN)

Founded in 1949 (as the World Conservation Union), this is the oldest, largest, and one of the most respected conservation networks in the world. It works with hundreds of governments, agencies, *NGOs*, and the *United Nations* to advance its mission of promoting *biodiversity* and sustainability of our planet's resources. It does so with research, education, and political discussions. The IUCN concentrates on helping *developing nations* establish strategies for *sustainable development*. It produces many publications, including the world-respected IUCN *Red List of threatened species*. {iucn.org}

International Whaling Commission

See *whaling*.

invasion

The mass movement of a *population* into a new area, often displacing an existing population. [*immigration; emigration; migration; population ecology population stability; push-pull hypothesis*]

invasive species

Species that are introduced to a new area because of human intervention are considered *alien species*. If these alien species overwhelm the native species, they are typically considered an *invasive* species.

Alien species might be introduced accidentally, such as an individual hitching a ride in a ship's ballast, or intentionally, such as has happened during ill-conceived research or the release of exotic pets. For the alien species to become an invasive species, it goes beyond just reproducing and establishing itself; it also wreaks havoc in the *ecosystem* because it has no natural predators or because other species simply cannot compete with it. This lets the invaders grow and reproduce unchecked. Invasive plants are often very easy to spot, because they engulf entire areas, creating what appears to be a *monoculture*—something always attributed to humans.

Invasive species typify how human intervention upsets the natural order. Adaption (through *natural selection*) and evolution are long, gradual processes. Transporting an organism from the other side of the world and releasing it where it would never be normally found is an immediate action that ecosystems cannot adapt to.

Some examples include the introduction of rabbits to the Australian island of Norfolk in the early 1900s, where they have no natural *predators*. Rabbits without predators are not a good thing—their numbers grew wildly, forcing many species of plants into extinction. Today, that original species has evolved into more than a dozen species, and rabbit control remains a problem.

In Florida, pet owners have been releasing Burmese pythons into the Florida *wetlands* after they get too large to be pets any longer. Examples of invasive plants include phragmites, which have overtaken habitats around ponds, and the Japanese barberry, which engulfs many wooded trails.

Invasive species contribute to *biodiversity loss*. [*Asian carp; ship ballast and invasive species*]

invertebrates

Animals that don't have backbones. Support is provided by other means, such as *insects* having an exoskeleton and *jellyfish* floating in water. [*ant slaves; beehive sound research; beetlecam; bumblebee decline; bumblebee senses; colony collapse disorders; cephalopods; camouflage, squid; earthworms; follicular mites; forams; giant squid in its native habitat; crustaceans; decapitating flies; hyperparasites; Monarch butterfly decline; nematodes; oysters and climate change; zombie bees*]

investing, green

See *money market funds, environmental.*

Investor Network on Climate Risk (INCR)

See *CERES.*

iron hypothesis

Carbon sequestration storage (CSS) is a hot topic these days. Most of these CSS methods of trapping and storing vast amounts of carbon to reduce the amount of *greenhouse gases* in the *atmosphere* are controversial. The iron hypothesis is a suggested method of carbon sequestration that would increase the amount of carbon stored in the oceans. Our oceans are the largest *carbon sink* on the planet, storing about 25 percent of all carbon emitted into the air.

The iron hypothesis is a plan to add vast amounts of iron salts into barren regions of the oceans to act like a *fertilizer*, boosting *marine* plant life, especially *diatoms* that are a type of *plankton*. When plankton die, they float to the bottom and remain there for long periods of time, increasing carbon sequestration in the seas.

Some research has shown that the risks outweigh the possible advantages, because this process might cause certain types of *algal blooms* where toxic diatoms become prolific, causing more harm than good. [*metal organic frameworks; karst carbonate; peatlands; permafrost, Siberian; Intergovernmental Panel on Climate Change reports*]

irradiation

Exposure to *radiation*. [*food irradiation; nuclear power*]

irrigation, agricultural

Human water use is divided into four categories: *domestic water* use, *industrial water* use, *in-stream water* use, and agricultural irrigation. Irrigation uses 70 percent of all water used by humans, worldwide.

Agricultural irrigation involves transporting or diverting water from its source to regions in need of water. More than 311 million hectares of land are irrigated globally. Countries with the largest acreage irrigated are *India*, *China*, and the United States. In the United States, for example, 35 million acres of crops grown in arid regions of the west are irrigated with water piped from hundreds of miles away.

The vast amounts of water used for irrigation do produce results, however. Irrigated lands provide about 40 percent of the world's food, which is grown on about 20 percent of land.

In *developing nations*, much of the water for irrigation comes from *groundwater aquifers,* because it is inexpensive to draw water from a well and use it for irrigation as opposed to diverting streams or other surface waters. This, however, results in *groundwater mining*—a lowering of the water table—which could result in *drinking water* shortages in these regions.

The most common form of irrigation is *flood irrigation*, where the land is simply flooded with water. More than two-thirds of the water used during this type of irrigation evaporates or runs off, never reaching the plants in need, but new methods of irrigation have been developed. Five major types of irrigation are in use: flood, furrow, sprinkler, *drip irrigation,* and sub-irrigation. [*precision agriculture; sustainable agriculture*]

ivory trade

One of the saddest affronts to the animals we share our planet with is the senseless slaughter of magnificent two-ton *elephants* and *rhinos* for nothing more than their tusks and horns. Entire families are often killed outright. Ivory refers to the tusks of elephants but is also informally—and incorrectly—used for teeth, horns, and tusks of a variety of animals, including rhinos, walruses, and *whales*.

In the 1930s, Africa had more than five million elephants. By the mid-1970s, little more than one million remained, and by the late 1980s, half of those were gone as well. At the height of this reign of terror, during the mid-1970s, almost 90,000 elephants were slaughtered each year just to cut off their ivory tusks to be sold on an international market, with most of the clients from Western countries.

During the 1980s, conservation organizations got the upper hand to help stop the slaughter. Attempts to prevent this slaughter started with The *Convention on International Trade in Endangered Species (CITES),* which in 1986 got a successful international ban on the ivory trade. *Kenya* had a shoot-on-sight policy against poachers. At the time, the battle between protection and poachers was described as war-like.

Unfortunately, this story does not have a happy ending. Because the elephant populations in Africa have rebounded, CITES has allowed the ivory trade to resume in southern parts of Africa. This opened the door to widespread killing of elephants once again. The worst part of this change is it is now legal to sell ivory to *China*, where a new-found middle class can afford and relish ivory trinkets. Japan is also a hotbed of demand for ivory.

The results have been the creation of international criminal cartels who are well armed and determined to get their booty—even if it means slaughtering thousands of elephants. In Kenya, where the average income is less than a dollar a day, a hunter can earn $2,500 killing one bull elephant.

Examples of this onslaught are numerous: The war has returned—just last year, two Kenyan protection rangers were killed by poachers, and in January 2013, 11 elephants were killed and poached in the worst single incident in Kenya's history. In July 2013, customs officials in Hong Kong seized 4400 pounds of ivory in the biggest single haul since 2010. It was worth 2.2 million dollars, being shipped from China to Togo, and consisted mostly of ivory from baby elephants.

Today, the latest numbers show there are fewer than 400,000 elephants in Africa. [*elephant poaching; poaching; elephant ivory, dating; polar bear trade; rhinoceroses and their horns; wildlife trade and trafficking*]

Izaak Walton League of America

This *NGO* was established in 1922 by a small group of concerned anglers. It is named after the 17th century author of probably the most famous book about fishing ever written, <u>The Compleat Angler</u>. The League is dedicated to protecting the nation's *soil*, air, woods, *waters*, and *wildlife*, with an emphasis on protecting areas for recreation and improving recreation-landowner relations. They publish newsletters and a magazine that contain articles on *conservation* and recreation. Their member pledge says it all: "To strive for the purity of water, the clarity of air, and the wise stewardship of the land and its resources; to know the beauty and understanding of nature and the value of wildlife, woodlands, and open space; to the preservation of this heritage and to man's sharing in it." {iwla.org}

jactitation

Some plants disperse their seeds by a jerking motion that tosses them from the plant. This process is called jactitation. [*pollination; pollination, wild; dientomophilous; imperfect flower; traumatic insemination; sexual reproduction; asexual reproduction*]

James Bay Power Project

A huge *hydroelectric dam* project in central Quebec, Canada, divided into two phases, the James Bay I and James Bay II. The project diverted and dammed nine rivers and flooded an area the size of New York State. The tiered spillway that was built to generate power is three times the height of Niagara Falls. It met with fierce opposition by environmental groups and indigenous peoples. Phase II was never fully completed.

The *habitats* of thousands of species were destroyed by the project, and the breeding grounds of millions of *migratory* birds were damaged. The toll the project might have on the larger *ecosystem* of the entire region is still being studied.

Proponents say the power created reduced the number of *coal*- and *oil*-fueled *electric power plants* and *nuclear power* plants needed, which would have caused far more harm to the environment than the hydro project. [*hydropower, traditional; hydropower, nontraditional; dam, removal; Aswan High Dam; dams, ten largest; pumped-storage hydropower; run-of-river hydropower; Three Gorges Dam; tidal power, wave power*]

jazz, Earth

Earth jazz, also called Earth music, "celebrates the cultures and the creatures of the whole Earth," according to Earth jazz musician Paul Winter, who helped create the genre and who has won seven Grammies. Animal sounds become part of the music—such as in "Wolf Eyes," which is a duet with a timber wolf.

Earth music has evolved into a category commonly known as world music. This usually refers to musical genres such as traditional or folk music and might be played by indigenous musicians. Some purists consider Earth music to only include traditional styles such as Japanese koto, Indian raga, and Tibetan chants. [*hip-hop, green; Seeger, Pete*]

J-curve

Theoretically, the *population* of any organism could continue to grow until it took over the entire planet. For this to happen, it would need unlimited resources, such as food and shelter, and ideal conditions, such as temperature and available water. If you plotted such a curve on a graph, it would look something like the letter "J"—starting off with a small population in the beginning (bottom) and ending with a large population at the end (top).

However, resources are not limited and conditions are not always perfect, so this ideal curve doesn't usually occur in nature. Once the population reaches the limits of what the environment can support (called the *carrying capacity*), growth levels off, creating a curve that looks more like the letter "S"—experienced by virtually all organisms except humans.

The human population has yet to level off and is still on the upward swing of the J-curve. Technology has so far forced our planet's human carrying capacity to support our numbers, but in many regions of the world, it appears to be pushed to the limit. For example, the infant mortality rate is far greater in the *less developed countries* (LDCs) than in the more developed countries (MDCs). Only time can tell when the Earth's carrying capacity for the human population will be reached and exactly what will happen. [*limiting factors; organism, most abundant; resilience, population, S-curve; r-strategists; population ecology; population growth, limits to human*]

jellyfish

Ocean acidification, *overfishing*, and *cultural eutrophication* are always mentioned as causes of population decline and the demise of many *marine ecosystems*. And it is true—except when it comes to jellyfish. Surprisingly, these same environmental problems increase the populations of the jellyfish because many of the fish in decline are jellyfish predators or competitors of jellyfish for the same food supply. The result has been population explosions of these *coelenterates* and a growing imbalance in many marine ecosystems.

The future of these stressed ecosystems is unknown as *climate change* continues. Commercial problems have already begun. Fisheries from the Gulf of Mexico to the Sea of Japan often find their nets fouled by masses of jellyfish, driving up prices. Jellyfish stings have risen dramatically in vacation areas such as southern Florida. Power plants have temporarily closed when jellyfish swarms clogged the vents, and *aquaculture* farm pens have been attacked.

Joint Implementation

This international agreement expands the *Clean Development Mechanism* (CDM). Whereas the CDM only allowed an investment in *carbon emission* reduction projects created in *developing nations*, the Joint Implementation also allows for these projects created in *developed nations*.

joule

The quantity of energy that can be generated from a fuel such as *oil* or *natural gas* is measured in units called joules. One joule equals the energy necessary to raise one kilogram of weight ten centimeters in height. Large quantities of energy are measured in megajoules, MJ (one million joules) and gigajoules, GJ (one billion joules).

jungle

Habitats characterized with very high precipitation and usually found in tropical regions. [*tropical rainforest; ecosystems; biomes; anthromes*]

junk mail

Unsolicited mass-distributed marketing materials sent through the mail. It is considered by many people an incredible waste of *natural resources*, because millions of trees are required annually to produce the volume of junk mail generated. More than five million tons of junk mail end up just in U.S. *landfills* each year. The typical family of four receives almost 1000 pieces of junk mail per year. Estimates state that about 100 million trees are cut down annually just in the United States to create this junk mail, and less than half ever gets opened. Paper is the largest single component found in today's *municipal solid waste*. To equate it to another environmental issue, junk mail produces the same amount of *greenhouse gases* as nine million *automobiles* on the road.

The most amazing aspect of this is that most people want it to stop, whether for personal or environmental reasons. Unfortunately, companies that want to keep sending it seem to have the upper hand.

karst, carbonate

Many *karst* formations contain vast amounts of calcite, a crystalline form of calcium carbonate; a common mineral substance. When water flows through the karst terrain, a chemical reaction occurs with the calcite, producing bicarbonate ions—a stored form of carbon. Studies are underway in *China* to see if these karst formations might be a natural *carbon sink*. Because about 15 percent of the Earth's land surface has karst formations, these formations have the potential to be a way to naturally perform *carbon sequestration storage*—one method being considered to reduce *greenhouse gases.*

karst terrain

When water makes its way into the soil, it sometimes passes through regions containing limestone or gypsum. The water mixes with acids in the soil, which dissolves, cracks, and otherwise breaks down the rock. The deteriorated substrate is called karst terrain and contains many underground caves and surface *sinkholes.*

Keeling Curve

Carbon dioxide is a *greenhouse gas,* and adding more of it into our *atmosphere* is the primary cause of rising temperatures and *climate change.* The more carbon dioxide, the warmer it gets.

Carbon dioxide in our *atmosphere* is measured in parts per million (ppm). Many people watch the ppm of carbon dioxide in our atmosphere the way others watch the stock market. While stocks are tracked by your brokerage house, ppm of carbon dioxide in our atmosphere is tracked by the Keeling Curve.

In 1958, Charles D. Keeling, a chemist, began tracking carbon dioxide in the atmosphere. He did so from high up on a mountain in Hawaii, which was remote enough to not be influenced by local industrial emissions. He later began tracking in many other locations, as well.

He found a seasonal cycle where the concentration of the gas dropped in the summer when plants took in the carbon dioxide and increased in the winter when much of the plant life ceased growing and decayed, releasing carbon dioxide. However, over a span of 50 years, he noted a distinct trend upward, which is now called the Keeling Curve in his honor. (He died in 2005.)

When he began, in 1958, the carbon dioxide reading was 312 ppm. In 2013, it broke 400 ppm, which is considered by many an important threshold indicating our failure to limit these emissions.

Many scientists believe we must keep the global warming below a 2-degree C (3.6 F) limit to prevent catastrophic results. To do so would require no more than 450 ppm of carbon dioxide. From the lack of global action—such as the results (or lack of results) of the *Kyoto Protocol*—all indications are that our atmosphere will reach 450 ppm by 2037.

Kenya

Kenya has been a leader in environmentalism, with initiatives to manage its extensive *biodiversity* through establishment of many *national parks* and promotion of *ecotourism*. It was an early adopter of *debt-for-nature swaps* back in the 1980s and their *Green Belt Movement* is noteworthy.

Problems, however, include serious *water pollution* in their large lakes and in their *drinking water* supplies. They have joined with surrounding countries to establish the Lake Victoria Environmental Management Project (1995) to promote *sustainable development* of the basin. *Elephant poaching* has also taken a turn for the worst recently in Kenya.

Kenya water find

About 40 percent of all Kenyans do not have steady access to clean, fresh *drinking water*. A new discovery of five large *groundwater aquifers* might help alleviate this problem. The source holds 250 trillion liters of groundwater—enough to serve about 40 million people. Further tests are needed to ensure the water is *potable*. [*water shortages, global*]

keystone species

If a single species is of critical importance to the stability of an *ecosystem*, it is called a keystone species. Removal of this species could result in the collapse of the existing ecosystem. The *alligator* in the southeastern United States is an example. Alligators dig holes that accumulate water that later becomes a *habitat* for many forms of aquatic life. They also build nesting mounds that turn into nests for herons and egrets.

Alligators eat large amounts of *predatory* fish, which stabilizes the populations of the *prey*, such as bass.

When the alligators were almost hunted into *extinction* in the 1950s and 1960s, the entire ecosystem in the region was altered. The aquatic organisms lost their habitat, birds lost their nesting grounds, and populations of fish dramatically shifted because the predatory fish were no longer kept in check by the alligators.

Other organisms considered keystone species are as diverse as *sharks* in some *marine habitats* and mistletoe (a *parasitic* plant on trees) in some woodlands. [*apex consumers and ecosystems*]

Keystone XL pipeline

A pipeline currently exists that carries diluted bitumen *oil* from Alberta, Canada, to refineries in Illinois and Oklahoma. The proposed Keystone XL pipeline would be a sister line and would double U.S. imports from the Canadian *tar sands*. It would carry 900,000 extra barrels of oil per day of this heavier, more viscous form of bitumen oil from these tar sands, through six states, to refineries near the gulf coast of Texas.

This proposed project is one of the most divisive environmental issues in recent years. It has been up and down many times and is still in limbo as to whether or not it will be built. In January of 2010, President Obama rejected the application for the pipeline. In November of 2011, thousands of protestors marched around the White House demanding it be stopped. The *Environmental Protection Agency* first said it was safe and then backtracked. Proponents say it will provide jobs and energy security. And the battle continues.

The oil itself is considered to be one of the dirtiest that exists. It is thicker, more corrosive, and more acidic than refined oil. Because of its thickness, it must be mixed with volatile natural gas liquids and pumped at high pressure to keep it flowing. The concerns are that this corrosive, thick, and acidic flowing oil under high pressure is a leak waiting to happen. The route the pipeline takes crosses much of the *Ogallala aquifer*, the largest *groundwater aquifer* in the United States, and the danger of leaks worries many environmental groups. This aquifer provides *drinking water* to 2 million people and supports 20 billion dollars of *agriculture*. A serious leak would cause devastating damage to both health and the economy. The first pipeline has had many leaks. In North Dakota in 2010, shortly after it first opened, it spewed 21,000 gallons of oil.

Those that favor the pipeline speak of the jobs and value to the economy. However, most predictions show the buildup of jobs would be temporary and not a long-term boost to the economy.

Many consider the most important aspect of this argument not to be the risk of a leak but the acceptance of a long-term extension of *fossil fuel* use—especially this particular polluting type of oil—instead of *renewable energy* sources that would fight *climate change*. Environmentalists ask if an investment of $7 billion to build a fossil-fuel infrastructure make sense when almost all science shows we should be moving away from fossil fuels and toward *alternative energy* sources that do not contribute to climate change. [*oil recovery or extraction; oil spills in U.S.; pipeline spills*]

kibbutz, eco-

A kibbutz is a close-knit communal farm or settlement in Israel, where all work for their common good. They are typically owned and operated by those living there. Some kibbutzim concentrate on creating a small *ecological footprint* with *renewable energy* sources, high-yield *organic farms*, and a recycled water supply. Some are *ecotravel* destinations where visitors lodge, eat, and enjoy a nature-friendly vacation. [*resorts, eco-; lodging, green; hotels, green; ecotourism, ten best countries for*]

kilowatt (kW)

A kilowatt is used to measure the generating capabilities of an *electric power plant*. A small *wind turbine* might generate ten kilowatts of power, while a *coal*-burning power plant may have a 1000 kilowatt capacity.

kilowatt hour (kWh)

One *kilowatt* of electricity supplied for one hour. An average household consumes between 500 and 1000 kWh per month. [*energy use; negawatts; energy, A very brief history of U.S.*]

kingdoms and domains

Up until the 1960s, the classification of organisms was pretty simple—life was divided into two kingdoms: plants and animals. Then in 1969, most scientists agreed that system was too simple and reclassified life into five kingdoms: Monera, Protista, Fungi, Plantae, and Animalia. This method used the classification system originally created by Linnaeus, which was based on characteristics such as anatomy, morphology, embryology, and cell structure.

Then, beginning in the late 1970s, scientists agreed that the traditional five-kingdom system didn't do enough to address evolutionary relationships that exist between organisms and could not address the discovery of new organisms—many of them considered *extremophiles*. In the 1990s, scientists agreed on the addition of a new level above kingdoms, called domains.

There are three domains: Eukarya (kingdoms include Protista, *Fungi*, Plants, and Animals), *Bacteria*, and *Archaea* (an ancient group of single-celled organisms, many of which are *autotrophs* capable of *chemosynthesis* and found in extreme conditions—therefore called extremophiles).

kleptoparasitism

A form of *parasitism* where an individual of one species steals food from another species to feed its own young or itself. The most common example is seagulls that harass other bird species and take away their food. If given the opportunity, they harass people and take away our food, as well. [*parasite; parasitoidism; host; ant slaves*]

K-strategists

Organisms that are usually large, have relatively long lives, and produce few offspring are called K-strategists. These organisms invest a great deal of energy in assuring the survival of the few offspring they have. Most large *mammals,* such as horses, deer, and humans, are K-strategists. The *population* size of these organisms is defined by *density-dependent limiting factors*. This means the *population* grows until the density of the population limits any further increase.

For example, the population of *predatory* birds such as the hawk continues to increase until there are too many hawks and too few snakes, mice and other *prey* available for the young hawks to survive. Food becomes the *limiting factor* and the hawk population stops increasing. K-strategists have populations that stabilize at the *carrying capacity*. [r-*strategists; S-curve; population dynamics; population ecology; populations stability*]

Kyoto Protocol

This is the most important legally binding (but easily broken) international agreement addressing *climate change*. It came about as an offshoot of the *United Nations Conference on Environment and Development* (the Rio Summit), in 1992. This summit created the *United Nations Framework Convention on Climate Change,* which was given the responsibility to develop a document that would lower *greenhouse (carbon) emissions* on a global scale.

The document they created was signed in Kyoto, Japan in 1997 and became known as the Kyoto Protocol. It went into force in 2005, but the United States did not ratify it. Today, 160 countries have ratified the agreement representing more than 60 percent of the total emissions. (Since then the United States signed but still never ratified the protocol.)

The protocol is complex but basically requires countries to reduce their *greenhouse gas* emissions by certain amounts on certain deadlines. The amounts and deadlines vary, depending on which category (called Annexes) the countries were placed into. These categories provide some countries more time than others, based on numerous variables. After all these years, the reductions in emissions produced by this agreement have been slight.

The agreement expires in 2015 and nothing has yet been written to replace it. The recent meetings of the *Intergovernmental Panel on Climate Change* (in Cancun 2010, Durban in 2011, and Doha in 2012) have resulted in nothing more than pronouncements that the Kyoto Protocol needs to be replaced. Delegates have stated there will be a new document by the time the 2015 meeting takes place. But even if that does occur, it would not take effect until 2020, leaving the world without any international agreement on carbon emissions for many years.

Many people in the United States do not believe carbon emissions should be legislated but that they should be controlled by using a market-based approach, such as *carbon markets*. On the other hand, most environmentalists and *NGOs* do not believe markets will ever produce the needed reductions in carbon emissions. [*ratified vs. signed*]

Ll

Lady Bird Johnson Wildflower Center

Formally called the National Wildflower Research Center, the Lady Bird Johnson Wildflower Center, in Austin, Texas, is the only institution in the nation dedicated to conserving and promoting the use of native plants in North America. This research center studies flowers not only for their beauty but also for the economic and environmental benefits they provide. Their clearinghouse and reference library contain materials on native plants, propagation and landscaping, and wildflower identification. The organization was first created by Lady Bird Johnson in 1982. {wildflower. org}

land acquisition, wilderness

Two major forces are involved in the acquisition of *wilderness* areas to protect and preserve *natural resources*: the federal government and environmental *NGOs*. The government procures lands with the Land and Water Conservation Fund, the Migratory Conservation Fund, and the North American Wetlands Conservation Fund.

The Land and Water Conservation Fund is funded by royalties collected from *oil* and *natural gas* drilling on federal lands. Most of the lands acquired are near or within the borders of existing *national parks* or *forests*. Over the past years, more than 35,000 projects have benefitted.

The Migratory Bird Conservation Fund was established to restore and enhance migratory bird *habitats,* with some of its funds coming from the sale of duck stamps, through the *Migratory Bird, Hunting, and Conservation Stamp Act*, and others from entrance fees at refuges and firearms import duties. About four million acres of *migratory bird* habitat have been acquired through this fund.

The Wetlands Conservation Fund, passed in 1989 by Congress, authorized the purchase, management, and restoration of *wetlands* in the United States, Canada, and Mexico. Determining exactly what is a wetland and saving these areas has been more of a political football game than a scientific endeavor in the past few years.

A few environmental NGOs, such as the *Nature Conservancy* and the Trust for Public Land have had great success in saving wilderness areas. [*land grabs; land trusts*]

"Land Ethic, The"

This is an environmental essay of great importance, both from a historical and a modern-day perspective. It was written by *Aldo Leopold* and appeared in his book titled A Sand County Almanac, published in 1949. It was the first call for people to consider our environment within our definition of ethics.

There is no better explanation than the following direct quote: "All ethics so far evolved rest upon a single premise: that the individual is a member of a community of interdependent parts. His instincts prompt him to compete for his place in the community, but his ethics prompt him also to co-operate (perhaps so there may be a place to compete for). The land ethic simply enlarges the boundaries of the community to include soils, waters, plants, and animals, or collectively: the land." [*ethics, environmental; philosophers and writers, environmental; governments of the forest; land trusts; tragedy of the commons*]

landfills

About 70 percent of all U.S. *municipal solid wastes* are dumped into landfills. Another 25 percent is *recycled* or *composted,* and the remaining 5 percent is *incinerated.* Other countries use landfills to varying degrees: *Germany* has no landfills at all, Bulgaria puts 100 percent of their waste into landfills, and other nations fall somewhere between these two.

The size of many landfills is truly staggering. Puente Hills, which takes most of the waste from Los Angeles, California, has more than 130 million tons already stashed away. It covers about 1400 acres and looks like a slow-growing (very unnatural looking) large hill.

Today's modern landfills, also called sanitary landfills, are quite sophisticated. They contain liners to collect fluid (called *leachate*), which is pumped away for treatment, and gas collection systems to remove *methane* gas as it is produced. This gas is usually then used as a fuel to generate electricity in *landfill gas-to-energy power* plants. Monitoring devices that surround the site test the *groundwater* beneath, checking for contamination and also for methane gas escaping above.

The waste goes into huge cells or holes in the ground. As the waste fills each cell, it is covered with a layer of clay or other material. All of these cells form what looks like a vast honeycomb. Each one is built on the previous one, so the site gradually grows upward.

Contrary to popular belief, the material in a landfill is not expected to *decompose*, and that is why "fill" is part of the name. Landfills are kept dry to prevent the leachate from contaminating the groundwater and only a little oxygen enters deep into the pile of waste. Without water or oxygen, not much decomposition can occur. Once a landfill is "full" with "fill" or is being closed for some other reason—such as *NIMBY*—it is graded, capped, and usually turned into some form of recreational area such as a park, *golf* course, or athletic field. However, even once capped, the *landfill problems* associated with an operational landfill usually continue for decades. [*landfill mining*]

landfill gas-to-energy power plants

When *organic* matter within *landfills* decomposes, *anaerobic bacteria* generate *methane* gas—lots of it. Estimates state that for every million tons of *municipal solid waste*, about 430,000 cubic feet of landfill methane gas is produced daily. If this gas simply is released into the *atmosphere*, it becomes a potent *greenhouse gas*. Methane is about 20 times more potent than *carbon dioxide* as a greenhouse gas. Because of this potential harm, large landfills were required, starting back in 1996, to capture and flare (burn) it so it never makes it into the atmosphere. They must do so while the landfill is active and for 30 years after it closes.

But another option exists: to use it as a fuel to generate power. After all, the *natural gas* we use for energy is also methane. So instead of drilling for it, why not use the natural gas produced from our landfills? Even though the number of landfills in the United States is diminishing, the number of them using landfill gas to produce energy is growing. Currently, more than 500 landfills are using gas-for-energy. It can be burned to generate electricity (the most common use) or used to create hot water or steam, or both. It is believed another 500 sites could begin using landfill gas for energy. It is a win-win, because power is produced and the atmosphere doesn't get an extra dose of potent greenhouse gas. [*landfill problems; landfill mining; waste-to-energy power plants*]

landfill mining

Landfills are filled with waste, but that waste is loaded with what were once valuable metals and other minerals, and being in a landfill doesn't mean they are no longer of value. Theoretically, a landfill could be mined for these resources and that potential is called landfill mining. At the moment, no such facilities exist.

Organic waste decomposes, but metals and minerals do not. If the time comes that some metals and minerals cost a fortune—even more so than now—it might make sense for people to go back into landfills and remove what we previously threw out.

Besides metals and minerals, the same might be true of *plastics,* which are manufactured from *fossil fuels.* When we run out of fossil fuel—which will happen someday—how will we make plastic products? Most environmentalists would agree we will probably be better off without them, but some entrepreneurs might disagree. Because plastic decomposes at such a slow rate, it will still be in those landfills long after all the fossil fuels are depleted. [*landfill problems; recycling metals; gold fingers; e-waste*]

landfill problems

At first glance it appears that *landfills* in the United States are going away. Their numbers are dropping precipitously; from 18,500 landfills in 1980, to 6500 in 1988, and to less than 2000 today. But a huge asterisk marks this statistic—the overall capacity of those that remain is increasing. The truth is we have fewer landfills, but those we do have are much, much bigger. So the problem isn't really getting better as some would have you believe; it is getting worse. Landfills today are massive, city-like edifices with their own *ecosystems* and even regional wind patterns.

Today's modern landfills are far better than they used to be. But regardless of how many safety features are installed and how high-tech they may be, landfills are environmental disasters waiting to happen (if they haven't already).

The biggest problem with landfills is that they produce *leachate* and *methane* gas. Leachate is a toxic brew of water and all the contaminates from the waste. It filters (leaches) through the fill and into the soil and often into the *groundwater aquifers* below, which supply *drinking water.* The *methane gas* is produced *anaerobically* deep within the fill. Methane is one of the most potent of the *greenhouse gases.*

Modern landfills have liners to collect the leachate and a pipe system to collect the gas, (which can then be used to generate electricity in *landfill gas-to-energy power* plants). However, most existing landfills were built prior to the requirements for liners and pipes. If the liner and pipes were installed after the fact, the leachate that formed beneath the add-on liner continues to seep into the ground. Almost all older landfills and some of the newer landfills have been found to leak contaminants into the groundwater supply.

In addition to *groundwater pollution* and the contribution of greenhouse gases, the various forms of *aesthetic pollution* produced are obvious. Even with the covering applied each day, there is a *malodor pollution* problem (they stink), and they cannot prevent *visual pollution,* with *litter* flying all around.

Another problem and probably one of the reasons there are fewer (albeit bigger) landfills is the *NIMBY syndrome*—not in my backyard. Plenty of physical land exists for more landfills, but they must be relatively close to the municipality producing the garbage to be economically feasible, and people don't want *garbage* and contaminated water in or even near their communities.

Even though these are formidable problems, there appears to be little push in the United States to move to other forms of *municipal solid waste disposal*. Other nations have been moving away from landfills and using other options. Many countries place less than five percent of their total waste into landfills, including Japan, *Germany*, the Netherlands, Austria, Sweden, Belgium, Denmark, and Switzerland. These nations recycle and incinerate most of their wastes.

land grabs

Wealthy nations without enough land to grow crops or raise livestock to feed their people or enough water to quench their thirst have been buying or leasing vast quantities of land in developing nations. Between 2000 and 2010, roughly 270,000 square miles of agricultural land was purchased or leased. Most of these transactions are made by governments, but some are made by private enterprises from wealthy nations.

Most of the land is in Africa and most of it is acquired by investors from *China* and the Middle East. The losers in these transactions are the local residents, who often go hungry as outsiders take and use their land, as well as small-scale farmers who lose their livelihood. In addition, the *high-impact agricultural* farms that replace the small farmers causes far more environmental harm to this land.

More recently, many of these land grabs are actually water grabs. Countries in dire need of freshwater buy up land not to grow crops but to secure supplies of water that come with the land rights. For example, some Saudi Arabian companies have bought tens of thousands of square miles of land in countries such as Ethiopia, primarily for its *groundwater* supply.

In many situations, local residents have become activists against these incursions, but their efforts are usually thwarted by government officials, often time resulting in incarceration and even worse for the activists. [*water shortages, global; hunger, world*]

land mines, abandoned

See *explosive remnants of war (ERW)*.

Landsat

A series of satellites that orbit the Earth, taking pictures of the surface by using thermal technology. The pictures are used to monitor the health of our planet and can indicate such things as *urban sprawl,* farm and land use, *water pollution, rainforest destruction,* and wildfire hazards. [*indicator species; drones, conservation; Great Lakes Environmental Assessment and Mapping Project; satellite mapping; near earth objects*] {landsat.gsfc.nasa.gov}

land trusts

Private or government entities or *NGOs* that acquire land for the sake of protecting and conserving it are called land trusts. These lands are typically those deemed scenic or important *wildlife habitats. The Nature Conservancy* is the leading organization performing this important environmental role. [*land acquisition, wilderness; land grabs*]

Laos

For those old enough to remember (or for those who took a U.S. History course), Laos conjures up thoughts about *Agent Orange and Vietnam War.* But today, it conjures up a lush tropical *habitat.* Perhaps because it is one of the least developed of the Asian nations, 70 percent of the country remains *tropical rainforest* and it has become a booming *ecotourism* location. Numerous nature preserves and parks exist to protect many *endangered species.* Ecotourism, as is typically the case, has been a great economic success story. Many who would work in *rice paddies* have other opportunities to earn an income. Travelers typically rave about the hospitality found in Laos from the local people. [*ecotravel; ecotourism, ten best countries for*]

laser-printer ozone

Older laser printers emitted ozone (*ground-level ozone*)—especially if their filters were not routinely cleaned or replaced. However, most new laser printers use a different technology that does not produce ozone at all. To be safe, if yours has a filter, be sure to replace it as needed. Ground-level ozone can cause respiratory ailments, nausea, and headaches. [*indoor air pollution; couches and indoor pollutants; ink-jet and laserjet printer cartridges; e-waste*]

laundry and the environment

Four hundred loads of laundry that use 3,000 or more gallons of *water*—that's what the typical U.S. household goes through each and every year. And you're not finished until it's dried. The dryer is the second biggest electricity user (next to your *refrigerator*) of any household appliance. Doing the wash costs both you and the environment a significant price.

To save money and reduce your *ecological footprint,* follow some of these simple steps: Most energy used in the wash comes from making the water hot, so use cold water when possible. Even though most machines have small load settings, you waste water when the machine is not full, so load it up. Washing-machine detergents are much better than they used to be—when *phosphates* were common ingredients—but today there are still green choices better than others. (*Environmental rating systems* can help you find them.) Clean the lint filter religiously in the drier because a clogged one uses far more energy to get the same job done. If you have more than one load to dry, do them in quick succession, because energy is wasted heating up at the start but the dryer will already be hot after that first load.

And finally, consider going back to basics and try a *clothes line dry,* outside. It was once common but became old fashioned; it is once again cool because it is green. The clothes really do smell fresher when they dry outside, and it's a natural fresh— not one concocted by an artificial freshener. Some cities even have *right-to-dry laws.*

When buying a new washer or dryer, be sure to check their environmental ratings. Look for *Energy Star-rated* machines that will save you plenty of money on energy and water bills. And, if possible, buy a front-load washer, which uses about half the water as a top-loader. [*domestic water conservation; domestic water use; tub bath versus shower*]

law and the environment

Unfortunately, the environment and litigation often go hand and hand. For our judicial system to work, prosecutors, defense attorneys, and judges need to be environmentally literate about topics such as *wetland* protection, *air pollution*, *hazardous waste disposal*, *risk assessment*, and all of the areas likely to find their way into the court system.

A number of organizations, some nonprofit and some for profit, provide *environmental literacy* for these professionals. The Environmental Law Institute in Washington D.C. is a leader in this field. The Institute was created in 1969 as a nonprofit that provides well-respected environmental law education programs and materials. [*SLAP; SLAB back; environmental justice; ethics, environmental; risk communication; Douglas, William O.*]

lawnmower pollution

Lawnmowers were not regulated until the mid-1990s, but even now they cause far more *air pollution* than you might imagine. A gasoline-driven lawnmower produces far more *carbon monoxide, nitrous oxides*, and *volatile organic compounds* than does a car driven for the same period of time. Although much smaller and with fewer horsepower, the lawnmower is far less sophisticated in design and less likely to be tuned than cars. Estimates have about five percent of all U.S. air pollution being caused by lawnmowers. The *Environmental Protection Agency* also believes that about 17 million gallons of gasoline are spilled each year when people fill up their lawnmower tanks.

Electric lawnmowers are alternatives, but like *electric vehicles*, they transfer environmental problems to *electric power plants*. However, they are considered less harmful to the environment than gasoline mowers.

lawns

It's hard to dislike a beautiful, well-manicured, green lawn—unless you stop and think of how unnatural a *habitat* it really is and the environmental toll it takes. In some aspects, it is very similar to *monocultures* grown on farms. Lawns account for about one-third of the total *domestic water use* in the United States, which comes out to roughly seven billion gallons of *water* per day. Maintaining lawns uses 800 million gallons of gasoline each year and tons of synthetic *pesticides* and *fertilizers*. There are ways to minimize this damage. You can decrease your use of synthetic pesticides and fertilizers. Use organic replacements or some of the *biological control* methods for pests. You can investigate types of grasses that do best in your location and use the least water. Local agricultural extension services at universities usually have lots of helpful information.

A more extreme alternative is to discard the lawn idea completely and use *native* plants. Why try to plant and maintain grasses that would not normally grow when you can plant vegetation that have *evolved* to live there? Using native plants saves water and, once established, they are easier to maintain. Some garden centers specialize or at least carry native plants. [*golf courses and the environment; water shortages, global; irrigation, agricultural; xeriscape*]

Lazarus taxon

Taxa (taxon singular) are categories, such as those used to classify life: *kingdom*, phylum, class, order, and so on. The Lazarus taxon doesn't really exist—it describes an informal group of species that have been declared *extinct* but later reappeared. The name comes from biblical times and refers to "raising from the dead."

The most famous species, long thought extinct, is the coelacanth—an ancient fish that was known to exist at the time of the dinosaurs. When the last of the dinosaurs became extinct 65 million years ago, these fish were thought to have died out with them. However, a few were found alive and well in 1938. And in many other cases, species thought extinct were found to still exist. [*extinction, mistaken; species recently gone extinct, ten; de-extinction*]

LD50

Commonly used unit to measure the toxicity of a substance. LD stands for lethal dose, and 50 refers to 50 percent. LD50 measures the dose required to kill 50 percent of the individuals in a sample *population* exposed to the substance. It is measured by body weight. For example, the substance that causes botulism (a food poisoning) has an LD50 of 0.0014 milligrams per kilogram of body weight. Say the animals average a weight of 100 kilograms and have 0.14 milligrams of the substance in their bodies (one milligram for each kilogram of body weight). This would result in the death of 50 percent of the animals. [*toxic waste; radiation; tailings, uranium mine; weapon dumps, abandoned; plutonium pit disposal; toxic cocktail; toxic pollution; lead poisoning; mercury poisoning; POPs, dirty dozen*]

LDPE

LDPE is an acronym for low-density polyethylene, a type of *plastic* commonly used to manufacture a variety of plastic items, such as *grocery shopping bags*, bread bags, frozen food bags, squeezable bottles, shrink-wrap, garment bags, and many other products. These products are stamped with the number "4," surrounded by a triangle formed of arrows. [*plastic pollution; plastic recycling; nurdles; garbage patch, marine*]

leachate

As rain or snow falls on a *landfill*, water gradually passes through (percolates) the fill, becoming a soup of decomposing waste and microbes called leachate. It often contains a variety of *hazardous* and *toxic waste* substances, including *heavy metals* and *organic compounds*. Regulations call for new landfills to be lined to catch the leachate so it can be treated before being released into the environment. However, older landfills have no liners, and even those with liners have, on many occasions, have been found to leak leachate, which will eventually contaminate *groundwater*. [*municipal solid waste disposal*]

leaching

As water flows over or through *soil* or rock, it removes chemicals and carries them elsewhere. This process is called leaching. For example, many of the *nutrients* found in *topsoil* will leach down to the lower levels of soil. [*soil horizon; leachate*]

leaded gasoline

The story about how *lead* got into our car's gasoline is emblematic of many of our environmental problems. In 1921, a form of lead was found to improve the performance of the internal combustion engine that powers our *automobiles*. Because lead was already a known poison, many argued against adding it to gasoline. However, by the 1930s, the economic advantages of this additive had won the argument over the health concerns and almost all gas in the United States was leaded. Soon other nations followed suit. By the 1950s, research began to show increased levels of lead in people, and *lead poisoning,* with its associated problems, began to rear its ugly head.

It took decades before lead would be phased out of our gasoline, and it was finally discontinued in 1975. This occurred when all new cars sold in the United States were required to have *catalytic converters* (to reduce *air pollution*), and these devices could not tolerate any lead in the gas.

This did not prevent the United States from exporting leaded gasoline to other countries that still used it. It was not until only a few years ago that lead-free gas became the norm worldwide. However, a few countries continue to use leaded gas, including Afghanistan, Algeria, Iraq, Myanmar, North Korea, Sierra Leone, and Yemen. [*precautionary principle; flame retardants; brominated; mercury poisoning; silent killer*]

lead poisoning

Exposure to the element *lead* and its accompanying health effects. Lead poisoning has occurred since ancient times, when Romans were chronically poisoned by drinking out of lead goblets. Problems have persisted in the United States—albeit not from wine goblets—but because lead was a common additive in *automobile* gasoline up until 1975.

Lead poisoning occurs by the inhalation of fumes containing lead or the ingestion of particles containing lead, such as from lead paint. Leaded gasoline, along with many other products, has resulted in lead contamination permeating throughout the environment.

Lead has been phased out of many of these products—most notably *gasoline,* paints, and toys for more than a decade—but because many of these products are still around, such as old toys and houses painted before 1978 (when it was banned in the United States), lead is still causing harm.

Public health concerns remain so great that even with the phaseout of lead in many products, the Centers for Disease Control, in 2012, lowered the recommended allowable blood level concentration for children aged one and two.

Lead can damage the nervous system (including the brain), as well as the kidneys and reproductive organs. It is most damaging to a developing fetus, so expecting women are at most risk. Lead poisoning can cause *acute* illness, such as vomiting, or *chronic* illness as lead accumulates in the body over long periods, such as permanent damage to the nervous system. Because it affects the nervous system, it can cause cognitive and behavioral problems, including criminal behavior. Since the U.S. ban on leaded gasoline, it is estimated that IQ scores have increased several points and 58 million crimes did not occur.

Research has shown that children in *urban* areas have a spike in lead levels in their blood because of playing outdoors where lead remains in the soil where it was used years ago in gas, plumbing, and *pesticides*.

Even with a phaseout of lead, it is still used in many products. It is found in some computer circuit boards, where it causes no harm to users but becomes a *municipal solid waste* problem. It can also be found in products most people wouldn't expect. For example, you can avoid ingesting lead that might be on wine bottles by washing off the lip of the wine bottle where the foil cap comes in contact with the bottle. The wine can pick up lead (originally in the foil) as it flows over the lip. Another unusual situation can occur if you reuse *plastic* food packaging, such as bread wrappers. When reusing, don't turn the plastic bag inside out, because lead can be released from the painted labels and be absorbed into the bread. [*toxic waste; sunsetting; toxic cocktail; e-waste*]

leaf litter

Plants and plant parts that have recently fallen to the ground and are only partially decomposed are called leaf litter. Leaf litter often forms the surface layer in many *habitats*. [*soil horizons; duff; biogeochemical cycles; caliology*]

League of Conservation Voters (LCV)

The LCV is an *NGO* that works to help elect pro-environment candidates to office by endorsing and providing financial support to those with good environmental track records. It was founded by *David R. Brower*. They "educate the public, lobby Congress and the administration, build coalitions, promote grassroots power, and train the next generation of environmental leaders."

They publish the annual National Environmental Scorecard, which provides objective, factual information about important environmental legislation up for consideration, along with the corresponding voting records of members of Congress so everyone knows who they are voting for (or against).

With its goal of changing the balance of power in the U.S. Congress to reflect the pro-environment sentiment of the electorate, the LCV is one of the most important of all environmental organizations. [*environmental literacy; environmental education; Student Environmental Action Coalition; law and the environment*] {lcv.org}

Leaping Bunny certified

See *cosmetics, cruelty free.*

least developed countries

Defined by the *United Nations*, this refers to the world's poorest countries—those with a per capita income of less than $905 per year and in a poor state of development and economic vulnerability. [*less developed countries; developing countries*]

LEDs (light-emitting diodes)

See *light-emitting diode bulbs.*

LEED Certification

LEED stands for Leadership in Energy and Environmental Design. This is an *eco-certification program* run by the *U.S. Green Building Council*, an internationally recognized and respected nonprofit that assesses the *ecological footprint* and overall *sustainability* of buildings, including business, residences, schools, manufacturers, and just about all others. LEED certification began in 1994 and had its roots in the *National Resources Defense Council*.

To be LEED certified means a building meets specific criteria for things such as site development, *water conservation*, energy efficiency, sustainability of materials, indoor environment, and many others. The specific benchmarks a building attains determines which level of LEED certification is awarded. The four levels are certified, silver, gold, and platinum (the highest). Although many view this as a win for the environment, it has been repeatedly proven that it is also a win for those paying the bills for the buildings. The additional costs of building a LEED-certified building are usually recouped within a number of years of operation. [*Energy Star rating; Living Building; energy, ways to save residential; energy-saving apps and programs; light bulb – recent history and legislation*] {usgbc.org/leed}

legumes

These are probably one of the most underappreciated types of plants on our planet, because without them, life would be very different. Almost all organisms require *nitrogen* to live. Nitrogen is the most abundant gas in our *atmosphere*, so you would think it is easy to access. However, most organisms cannot use the nitrogen in the air. Plants and animals can only use nitrogen that has been converted from the "free" nitrogen in the air into an organic form, called "fixed" nitrogen, such as ammonia. This process is called *nitrogen fixation.*

Legumes are one of the few organisms capable of using the free nitrogen in the air and converting (fixing) it into organic nitrogen. When the legumes die and decompose, the fixed nitrogen enters the soil, thus becoming available to all other plants and passed along *food chains* to animals.

Legumes include peas, alfalfa, peanuts, clover, soybeans, and some other plants. When farmers rotate their crops, they typically are planting legumes in alternating years to allow the soil time to be re-nourished with fixed (organic) nitrogen for the other crops. [*conservation agriculture; biogeochemical cycles; biogeochemical cycles, human intervention in; crop rotation; organic fertilizer; nutrient enrichment; safe operating space*]

lemurs

Lemurs are only found in the wilds of Madagascar and have become exceedingly rare. Since a coup in 2009, there has been no government enforcement to protect the animals and their *habitat* as there was in the past. The coup has led to widespread *illegal logging*, resulting in *deforestation* and *habitat loss* and reducing the lemurs' home range. In addition, there has been large scale *poaching* of the lemurs.

Ninety percent of the country's original forest is gone. Ebony and rosewood trees are valuable and need protection or they will be harvested by a poor population that lives on less than $2 per day. Twenty-three species of lemur are on the Critically Endangered *Red List*, 52 on the Endangered List, and 19 on the Vulnerable List, making 91 percent of all species being on a Red List threatened category—the largest of any *mammal* group by far. [*endangered and threatened species lists; biodiversity, loss of; CITES; Convention on Biological Diversity, The United Nations; extinct in the wild; extinction, sixth mass; ivory trade; Lonesome George; poaching; wildlife trade and trafficking*]

lentic

Standing bodies of water such as ponds and lakes. [*aquatic ecosystems*]

Leopold, Aldo

(1886–1948) Aldo Leopold was a conservationist, writer, and philosopher best known for his environmental perspectives in the wonderful essay "*The Land Ethic*," from his book titled <u>A Sand County Almanac and Sketches Here and There</u>. Leopold founded the field of game management. While working for the U.S. *Forest Service* during the 1920s, he helped develop the wilderness policy and advocated regulating hunting to maintain a proper balance of *wildlife* in a *habitat*. His philosophy is summed up by a quote in this book: "A thing is right when it tends to preserve the integrity, stability, and beauty of the *biotic* community. It is wrong when it tends otherwise."

A Sand County Almanac was based on Leopold's life experiences, mostly his later-in-life purchase of an old farm where he wrote of the plants and animals and the changes that occurred. He died fighting a forest fire shortly after his proposal for this book was accepted. It has sold millions of copies and remains today a favorite among anyone interested in nature and our surroundings. [*philosophers and writers, environmental; ethics, environmental; environmental justice; Muir, John; preservation*]

less developed countries (LDC)

LDCs are also called *developing countries* (as opposed to *developed countries*). An LDC is defined by the *United Nations* as a country with low-to-moderate industrialization and low-to-moderate *gross national product*. There are 150 LDCs, with the majority located in Africa, Asia, and Latin America. Most have less favorable climates and *soils* compared with *more developed countries* (MDC).

About 77 percent of the world population lives in LDCs. Living conditions in many of these countries is often at a subsistence level. For example, many of the people in LDCs have unsafe *drinking water* (compared with almost none in the MDCs).

The annual population growth rate for these countries is about 2.1 percent, which is considered very rapid (compared with 0.05 percent for the MDCs, which is considered slow). The average gross national product per capita for those living in an LDC is about $700 per year, compared to about $15,800 per year for MDCs. [*standard of living; hunger, world; water shortage, global; subsistence farming; wood-burning cookstoves, primitive; drinking water*]

levels of biological organization, studying various

Ecological studies often focus on a certain level of biological organization. Studying individuals or a single species in an *ecosystem* is called autecology. Studying one *population* or a few interacting populations is *population ecology*. Studying all the populations within an environment is synecology. *Systems ecology* (or ecosystems science) studies all the relationships between all the organisms and nonliving factors found in an *environment*.

lichens

Organisms that result from a *symbiotic relationship* between *fungi* and a green *algae* or between fungi and *blue-green bacteria*. This is such a close and important relationship that the two organisms in combination are classified as one—something that rarely occurs. The algae or *bacteria* perform *photosynthesis* and pass nutrients along to the fungi. The fungi provide the bacteria or algae with protection, moisture, and minerals.

Lichens are capable of living in the harshest of environments, including bare rock. They produce acids that begin to break down the rock, beginning the process of soil formation. They are often the very first organisms to appear in a *pioneer community* that leads to *ecological succession.*

life-cycle assessment

Previously, people only considered the environmental impact of a product at a single point, during its actual use—for example, an *automobile's* impact on our planet while the car is running. However, in more recent years, we have come to realize that every product has an impact from the time it is created to the time it is disposed of. Some new legislation requires companies to perform *cradle-to-grave* assessments of environmental impact. To understand this impact, companies often perform life-cycle assessments. They determine what polluting substances are created during each stage of the manufacture, use, and disposal of the product or service and what the negative impact might be on the environment. [*cradle-to-cradle; extended product responsibility; corporate social responsibility; corporate sustainability rankings; companies; top ten green global; internalizing costs; integrated bottom line; waste-conscious product development*]

light bulb – recent history and legislation

U.S. consumers might be a bit confused when they go to the store to buy a light bulb. In the recent past, *incandescent* bulbs were the standard, but as far as *light bulb technologies* go, they are terribly inefficient. The United States passed a law mandating that any bulb of 100 watts or higher must use 27 percent less electricity by 2012 and the same for 40 watts and higher by 2014.

In 2012, *compact fluorescent light (CFL)* bulbs were the standard replacement for the *incandescent* bulbs. Unfortunately, CFL bulbs have not turned out well. They did not last nearly as long as advertised and they did not start up at full brightness. Plus, they contain *mercury,* which is *toxic* and should be disposed of as a *hazardous waste* product—but is typically thrown in the regular trash, becoming part of the *municipal solid waste* stream. Many people were not happy with this legislation that forced them to buy a product that appeared less useful than what they had been using for decades.

But the most recent generation of bulbs, called light-emitting diodes bulbs, or *LEDs,* appear they might fit the bill for the future. Even though they are expensive at the moment, they are the most cost effective of bulbs and have the most potential in the digital age.

light bulb technologies

Edison and Tesla revolutionized the way people lit their world—transitioning from kerosene to electricity. Today we have gone through a recent revolution of a sort by going through a rapid succession of light bulb technologies: from standard *incandescent bulbs* to *compact fluorescent bulbs (CFL)* to light-emitting diode bulbs *(LED)*. The driving force behind the change is to reduce energy needs, which is good for you and your wallet, it's good for the country to use less energy overall, and it's good for the environment because it means less *fossil fuel* consumption.

Lights account for 17 percent of U.S. electric demand, so making small improvements to the technology translates to large savings. You can compare the usefulness and efficiency of different types of bulbs in several ways. One is to compare "lumens per watt" meaning how much brightness you get out of each watt. To show the improvements, an incandescent bulb creates 15 lumens per watt, a CFL produces 60 lumens, and an LED 100.

Another way to compare the technologies is how much of the energy consumed is converted into light instead of lost as heat. The standard has been set, of course, by nature, which is about as good as it gets. A firefly produces what is called *cold light* because it converts 99 percent of the energy directly into light and loses only 1 percent as heat. The LED bulb converts 75 percent into light, and the incandescent converts only 5 percent into light, losing 95 percent as heat.

The future will probably result in: the demise of the inefficient incandescent bulbs, the phasing out of CFLs—which never lived up to how good they were supposed to be—and the increased popularity of LEDs, which are believed to be as good as they are hyped up to be. In addition to the energy advantages of LEDs, they are the first lighting technology that is programmable so they can be managed from smartphones and do tricks such as change colors.

light-emitting diode bulbs (LEDs)

The latest in *light bulb technologies* is the LED. These bulbs produce 100 lumens of light per watt (a unit of brightness), compared to 60 lumens per watt for *CFL bulbs* and only 15 for the traditional *incandescent bulbs*. LEDs also convert 75 percent of the energy it uses into light (the remaining 25 percent is lost as heat), compared to the incandescent bulb, which only converts 5 percent into light. This means they are far more energy efficient than both incandescent bulbs and CFLs. New *light bulb legislation* is phasing out the use of incandescent bulbs. Up until recently, CFLs were the only available replacement, but they have not been accepted well by consumers and for good reason.

But now LEDs have become the logical replacement as incandescent bulbs are phased out. Although they are expensive up front, in the long run they are far more cost effective. LEDs last 25 times longer than incandescents and three times longer than CFLs. As time passes, the cost of LEDs will drop and consumers will see their advantages. The end result should be a long-term cost savings to the consumer, reduced energy demand for the country, and less harm to the environment. [*light bulb – recent history and legislation*]

Lighthawk, the Environmental Air Force

Lighthawk is a very unusual environmental *NGO* group. They are dedicated to protecting the *environment* by flying decision-makers, media representatives, and grassroots activists over and into *endangered* areas. This provides these people with the first-hand experience they need to take action.

Lighthawk's goal is "to mobilize enough volunteer pilots, aircraft, and resources to help tip the balance toward sustainability for every major environmental issue within our targeted areas of focus." Some of their accomplishments include helping to cap the largest single source of arsenic *air pollution* in the United States—an aging copper smelter in southern Arizona—and creating the 92,000-acre Bladen National Park/Nature Reserve in Belize, Central America. [*environmental movement; news literacy; environmental literacy; News Service, Environment; Environmental Media Association; social media, environmental*] {http://lighthawk.org/}

light pollution

Seen from space, there is nothing better to determine where the cities and urban life are found around the globe than looking at the lights. Light pollution can occur on a local level, such your neighbor's porch lights. Or on a large scale, where the overall amount of light diffused throughout an entire city brightens the night sky and any thoughts of using a telescope are dimmed.

But light pollution has been linked to more than just making it hard to search the night skies with a telescope. Correlations have been found between light pollution and the health and movements of wildlife and impairment of night driving. Bright lights are needed in many places; the problem arises when these lights are unshielded and brighten surrounding areas. Although reducing the overall number of lights is helpful, shielding those that exist is the bigger issue.

Some examples of light pollution are well known. In some coastal areas, *sea turtles* come to shore to lay their eggs in the sand. When the hatchlings emerge, typically in summer or fall, they are guided by moonlight toward the water to make their escape before predators get them. However, in many coastal cities, the city lights confound their efforts, confusing them and causing them to go the wrong way, usually ending in disaster such as being hit by cars as they cross roads or simply never finding the sea.

Insects are drawn to lights and are killed by the millions. While most people might not think this is bad thing, they play a vital role in all *ecosystems*. If their populations drop, as can happen because of light pollution, there are negative impacts throughout the system. Bird populations typically also drop because insects are their primary food source.

The International Dark-Sky Association follows light pollution issues and has given accolades to Flagstaff, Arizona, and Borrego Springs, California, for passing light pollution laws that limit unnecessary light. In many cases, the reduction correlates with significant cost savings as well. [*visual pollution; malodor pollution; noise pollution; aesthetic pollution; space debris*]

light-water nuclear reactors (LWR)

Almost all existing *nuclear power* plants are of the LWR design. (Light water refers to common water as opposed to *nuclear reactors* that use an isotope of water called heavy water.) The two most common types of LWRs are boiling-water reactors and pressurized-water reactors.

These reactors use about 40,000 *radioactive* fuel rods surrounded by water. The water acts as a coolant and helps control the reaction. A chain reaction of *nuclear fission* is started, and the energy heats the water, which then drives turbines to generate electricity. In a boiling-water reactor, the heat produces steam to drive the turbines, but in pressurized-water reactors, the pressure keeps the water as a liquid throughout the process.

After three to four years, the fuel rods—which contain *radioactive* uranium—are no longer capable of supporting nuclear fission and must be removed from the reactor and disposed. These rods remain highly radioactive and deadly to all forms of life, so they must be transported and stored in a *nuclear waste disposal* facility until they are no longer harmful. This takes about 240,000 years, but a mere 10,000 years if the rods are first processed. At the moment there is not a single long-term storage facility anywhere in the world to take these spent fuel rods, and no country seems capable of coming up with a solution.

An alternative method of disposing of radioactive fuel is to send the fuel to reprocessing plants, which recycle the fuel for use once again. Reprocessing the fuel is more costly than the original processing, so only a few of these plants exist, none of which are in the United States.

limiting factor

That factor in an organism's *environment* that is farthest from the optimum and there-fore limits the organism's chances for survival. Limiting factors might be *biotic*, such as insufficient food (plants to eat or *prey* to be captured), or it might be *abiotic,* such as not enough sunlight, *water*, or *phosphorus* for plants to grow. [*population ecology; ecosystems; tolerance range; ecology; extremophiles*]

Limits to Growth, The

This book attracted the world's attention in 1972 and is still considered an important clarion call that we must change our ways. It was the first research that used com-puter modeling to study the consequences of unlimited growth and consumption on a planet with finite resources. The book sold more than 20 million copies and was translated into 30-some languages. It has been updated every 10 years. The book pre-dicted humans would reach our planet's limits in 100 years (making it 2072).

It calls for a transition from a world based on growth to one that establishes a "condition of ecological and economic stability that is sustainable far into the future." It was the first reference calling for *sustainability,* a word now common-place. It also stated, "The sooner they begin working to attain it, the greater will be their chances of success." (The "they" refers to us.) The book's importance back in 1972 cannot be overstated, and its predictions about 2072 remain pertinent and even poignant today.

The report was commissioned by the *Club of Rome* and the lead author was Donella Meadows (1941–2001), with coauthors Jay Forrester, Dennis Meadows, and Jorgen Randers. [*Sustainability Imperative; Our Common Future; environmental move-ment; decoupling; economy vs. environment; sustainababble*]

limnetic region

The portion of a lake that receives sunlight but is too deep for vegetation to take root. This region usually contains abundant floating *plankton*. [*aquatic ecosystems; ecology*]

limnology

The study of *freshwater* organisms and their *habitats*. [*aquatic ecosystems; ecology*]

lipstick and toxic substances

Recent studies show that *heavy metals*, which are in many cases *toxic* to humans, can be found in lipsticks and lip gloss. A recent test of 32 different types of lipstick found manganese, titanium, chromium, nickel, and aluminum in 100 percent of them; *lead* was found in 75 percent and *cadmium* in 50 percent. The metals are used to get that "perfect" color. These metals can be ingested or absorbed through the skin when the lipsticks are worn.

However, the study showed that the intake was within what is considered acceptable levels for these metals. If you prefer not to use lipsticks with any toxic substances in them, you might consider any of the alternative cosmetic products that are metal free. There are good websites to identify what is in your cosmetics. The *Environmental Working Group* has a website called "*Skin Deep*" to help you. [*cosmetics; hair dyes and PPD; nanoparticles in cosmetics; eco-labels; green consumer; POPs, dirty dozen; lead poisoning; toxic cocktail; body burden; green products; green marketing*] {ewg.org/skindeep}

liquid metal fast breeder reactor (LMFBR)

See *breeder reactor*.

liquefied natural gas (LNG)

Natural gas is difficult to transport, because it can only travel short distances via pipeline. However, it can be liquefied first and then transported. The gas must be cooled to –260 degrees F for liquefaction and then it can be shipped in special tankers (LNG is 600 times as dense as the original gas). Finally, at the destination it must be returned to a gaseous state. [*fracking*]

liquefied petroleum gas (LPG)

See *natural gas extraction*.

lithosphere

The solid Earth portion of the *biosphere* consisting of the Earth's outer crust. [*soil; plate tectonics; supercontinents; biogeochemical cycles*]

litter

A commonly used term for *solid waste* material in places where they don't belong— for example, litter alongside the road. This is different from *garbage*, which is waste material where it "does" belong or at least on its way to where it belongs, for example, in a garbage can or a town dump.

Obviously, these terms, along with rubbish, are used interchangeably and unscientifically. However, one specific aspect about litter is it often ends up causing unintended harm in unintended places. Most commonly, any litter along a road will most likely be washed into a storm drain and usually end up in a stream, a *river*, or the *ocean*. There it often causes harm to *freshwater* or *marine wildlife* when they eat it or become entangled in it, or it decomposes into *toxic* substances. [*municipal solid waste disposal; cigarette butt litter and recycling; plastic shopping bags; plastic pollution; garbage patch, marine*]

littoral zone

The area of shoreline between the high and low tide marks and the immediate area affected by the tides. [*oceans; Ocean Conservancy; ocean pollution; coastal wetlands; coastal flooding; Maraniss, Linda*]

Living Buildings

The ultimate in green building awards, this designation is awarded by the Living Future Institute. It requires zero energy use, extreme energy and water efficiency design, use of locally sourced building materials, and adherence to many other criteria. Throughout the world, only 16 Living Buildings have been certified. [*U.S. Green Building Council; LEED Certification; Energy Star rating; zero-energy building*] {ilbi.org/lbc/certified}

loam

A type of *soil* consisting of 40 percent *sand*, 40 percent silt, and 20 percent clay. Loam is considered the ideal *soil type* for most crops because it has excellent aeration and drainage properties. [*humus*]

locavore

A casual spin-off from scientific terms such as *carnivore* and *herbivore.* Locavores are people who eat foods locally grown and therefore are able to be more carefully scrutinized to ensure the food meets with their approval. This has become a movement of sorts along with the popularity of *organic foods* and the *farm-to-table* concept. For example, most supermarkets carry produce all year long, as if it grew everywhere all the time. Because a locavore only eats local food, they would avoid the vast majority of produce stocked during winter in many parts of the country. [*seasonal foods; food miles; organic farming and food; food's carbon footprint, the best and worst; doggie bags; food eco-labels; seafood eco-ratings; fregan; slow food movement; vegan and vegetarian diets and the environment; vegetarianism; wine, organic and biodynamic*]

lodging, green

Ecotourism has become big business, and green lodging is a logical extension. Some hotels and resorts are *green buildings* and possibly even *LEEDS certified*, but some go beyond this. Some green lodgings have programs to attract the environmentally conscious, especially those in areas with natural wonders. Some even have guests participate in conservation efforts, making conservation education part of their stay. [*ecotravel; resorts, eco-; ecotourism, ten best countries for; hotels, green; lodging, green; voluntourism, environmental; volunteer research, scientific; WWOOF; eco-kibbutz*]

logging, illegal

The *ivory trade* is notorious and well known, but a lesser known form of illegal trade occurs with logging. Many forested areas are protected by local, state, or national regulations. *Illegal logging* occurs in many areas but is most prevalent in the *Amazon Basin*, Central Africa, Southeast Asia, and Russia. Some of this logging is done by local villagers, but the most damage is done by large logging operations. This logging causes *deforestation* and its accompanying ravages. In Cambodia recently, almost 7500 acres of forest was illegally cleared out, mostly for its rosewood that is popular for furniture and guitars.

The *United Nations Convention on International Trade in Endangered Species of Wild Fauna and Flora* and the *Convention on Biological Diversity* are two of many international agreements toward reducing illegal logging. A voluntary certification program called the *Forest Stewardship Council* also works toward this end. [*timber, DNA sequencing of; governments of the forest; forest health, global; Rainforest Alliance; Rainforest Action Network; Statement of Forest Principles*]

London Convention

See *Convention on the Prevention of Marine Pollution.*

Lonesome George

Anyone visiting the Galapagos Islands, (where *Charles Darwin* did his famous research on *natural selection*) over the past few decades probably said hello to Lonesome George, a Pinta Island tortoise. On June 24, 2012, George died at about 100 years old—relatively young for a tortoise—and with him, his subspecies became extinct. He was called Lonesome because for 40 years caretakers tried to get him to mate and have offspring, but they failed (or he failed) and his subspecies came to an end. His notoriety made him the poster child of conservation and the threat of *extinction*. [*biodiversity, loss of; species recently gone extinct; CITES; dodo bird; endangered and threatened species lists; charismatic megafauna*]

longlines, commercial fishing

A large-scale *commercial fishing* method used around the world. Most longlines float or drift just below the surface of the water. They are used to catch yellowfin *tuna*, halibut, and swordfish. The lines are composed of a monofilament and range from a mile to more than 100 miles in length. To remain near the surface, they often use floats made of *Styrofoam*. Baited hooks are attached all along the line.

Each line can hook more than a thousand individuals. The problem is that many of these individuals are not the intended catch—instead they are *bycatch* (unintended species of no commercial value). When these lines are brought in, all of the bycatch is dumped back into the sea, usually dead or dying. These lines kill large numbers of *sea turtles* and especially seabirds. Longline fishing kills an estimated 200,000-plus seabirds every year and is responsible for much of the decline in many bird populations. These nets also ensnare bluefin tuna in the Gulf of Mexico and the Atlantic, a species in steep decline (down more than 60 percent since the 1970s).

Some efforts are just beginning, with hopes of lessening the harm caused by these longlines. The *National Oceanic and Atmospheric Administration* proposed regulations in 2013 that should lessen the bluefin tuna bycatch, if it's passed. Also, new efforts are underway to protect the huge numbers of seabirds killed. New devices weigh down the longlines because keeping it below the surface reduces the likelihood that birds will go for them. Whether fishing fleets bother to use them is unknown. However, many environmentalists believe longlines cause far too much collateral damage to the environment and urge more restrictions on their use, globally. [*fisheries and seabirds; bottom trawling, deep sea; driftnets; fish populations and food; ghost fishing; gillnets, commercial; overfishing; turtle excluder device*]

lotic

Describes *running-water habitats* such as rivers and streams. [*rivers in decline; running water habitats, human impact on; run-of-the-river hydropower*]

Love Canal

Love Canal became a symbol of the problems associated with *hazardous waste disposal*. For about ten years, beginning in the late 1940s, a chemical company dumped about 22,000 tons of *toxic wastes* stored in steel drums into an old canal called Love Canal, named after its builder William Love. In 1953 the company placed *topsoil* over the excavation site containing the drums and turned the property over to the Niagara Falls school district. A school, recreational fields, and almost 1000 homes were built on the site over the next few years.

Beginning in 1976, residents began to notice odd smells and found that children playing around the canal often had chemical burns. The drums were leaking toxic substances into the *sewers*, *lawns*, and even basements of homes near the site. Concerned citizens—led by a concerned mother, *Lois Gibbs*—performed informal health studies that revealed high incidences of many types of disorders. The publicity generated by concerned citizens (including taking some *Environmental Protection Agency* agents hostage for a short while) forced the state to perform formal health studies, and in 1978 the state closed the school and relocated some of the residents living closest to the canal. More than 700 remaining residents finally convinced the federal government to declare the entire neighborhood a disaster area, resulting in almost all the families being relocated.

The site was capped and a drainage system installed to remove the toxic substances as they leak out. The EPA spent about $275 million of taxpayer's money to clean up the site. In 1990 the EPA declared part of the site suitable for inhabitation and the area was renamed Black Creek Village.

In 2008, the 30th anniversary of the Love Canal evacuations, Lois Gibbs and others condemned the final assessment report about what has happened since the site was closed. After reviewing the report, Gibbs is quoted as saying, "Thirty years ago we had to fight to learn why our kids were getting sick. . . . We continue to fight to get the state of New York and every other state to stop allowing schools to be built on toxic and hazardous sites. . . . And now in 2008, we have to fight again to get the New York Department of Health to fully study and present findings about the real results from Love Canal—not just partial data. We don't want a whitewash, we want the truth." [*schools and brownfields; brownfields; cancer alley; cancer villages; hazardous waste; toxic waste; Superfund; toxic cocktail*]

Lovins, Amory B.

(1947–) Lovins is a well-respected and popular spokesperson in the field of sustainable energy strategies. In 1982 he cofounded, with Hunter Lovins, the *Rocky Mountain Institute,* which continues to be a leader in this field. His theme suggests we must follow a "*soft energy path,*" where we get more out of what we already have—instead of constantly extracting, burning, and fouling with *fossil fuels,* the "hard energy path." [*natural capital; negawatts; energy use; energy, A very brief history of U.S.; negawatts; alternative energy sources*]

low-input sustainable agriculture

A method of crop production that shuns *high-impact agriculture* and favors more traditional methods that reduce the use of *synthetic fertilizers* and *pesticides.* It encourages various forms of *conservation tillage* to prevent *soil erosion* and cultural controls such as *crop rotation. Integrated pest management* techniques can be incorporated into this system, also called *conservation agriculture.* [*sustainable agriculture; organic farming and food; regenerative agriculture; permaculture*]

low-till agriculture

A farming practice that reduces *soil erosion* from wind and running water because it reduces the amount of soil exposed to the elements. Instead of tilling all of the cropland, which is the norm, low-till means only portions absolutely in need are tilled for planting. [*conservation agriculture; low-input sustainable agriculture*]

LUCA

See *weird life.*

LULU

Stands for "locally unwanted land uses"—a phrase often used when discussing *environmental justice* issues. [*interim use; brownfields; schools and brownfields; handshake buildings; slum; urbanization and urban growth; zoning, land use*]

Luna Redwood

Luna was the name given to the *coast redwood* tree that *Julia "Butterfly" Hill* climbed up, to prevent *redwood forest destruction* back in 1990. It was previously called the Stafford Giant because it resides in Stafford, California. Believed to be between 600 and 1000 years old, the tree remains standing but was vandalized a year after Butterfly's descent from the tree when someone cut a large swath more than halfway through its base with a chainsaw. Plates have been inserted to try to support the giant, but it is said to be highly vulnerable considering the damage done. [*Save the Redwoods League; ancient forests; old-growth forests; logging, illegal; environmental movement*]

LUSTs (leaking underground storage tanks)

A source of *groundwater pollution.* The *Environmental Protection Agency* (EPA) estimates there to be about 625,000 underground storage tanks that contain fuel or other *hazardous wastes.* The problem and even fear about these tanks and their potential to leak is how readily they could contaminate *groundwater aquifers* that supply most of the *drinking water* in the United States. The concerns are significant enough that in 1985 the EPA created an Office of Underground Storage Tanks to monitor and remediate the situation. Since its inception, the office has cleaned up about 4300 LUST sites and has another 3500 known sites to be cleaned up. [*hazardous waste disposal, historical; hazardous waste, military; remediation of hazardous waste; Superfund; toxic waste*]

Lyme disease and the environment

Lyme disease is caused by *bacteria* transmitted to humans by the bite of infected blacklegged (also called deer) ticks, which are *parasites.* The symptoms can include fever, headache, fatigue, and a telltale skin rash that looks like a ring around the bite. If left untreated, an infection can spread to joints, the heart, and the nervous system.

Environmental issues can contribute to the prevalence of Lyme disease in an area. For example, *fragmented forests* and small forest stands often found in and around cities as recreational areas are known to increase the incidence of Lyme disease.

This occurs because the white-footed mouse, which carries the disease, is far more common in small forested areas than in large ones. Ticks feed on a *host* mouse, pick up the disease, and pass it to humans. However, in large forested areas, ticks have their choice of numerous animals to bite—not just mice—and most of these other animals cannot carry the disease. So in a large wooded area, the odds are far less that a tick will be feeding on an infected mouse as opposed to any other uninfected species. This lessens the overall odds of any tick becoming infected and that lessens the likelihood of a tick spreading the disease to humans. [*bedbugs and remedies; bird flu virus; medical ecology; skin bacteria; spillover effect; superbugs*]

Maathi, Wangari
See *green belt movement.*

macronutrients
See *nutrients, essential.*

Madrid Protocol
See *Antarctic Treaty.*

magazines, eco-
Magazines that cover environmental issues were popular when the *environmental movement* first got off the ground. By the 1990s numerous monthly magazines were published about the environment; however, almost all of those not associated with an environmental *NGO* have since closed down. The one exception that remains is the still excellent *E: The Environmental Magazine.*

Environmental NGOs have good publications such as the *Natural Resources Defense Council's* <u>OnEarth Magazine</u> and the *National Wildlife Federation's* <u>National Wildlife</u> *magazine.* Colorful, informative magazines such as <u>Sierra</u> and <u>Audubon</u> have been popular for years among their members. *Greenpeace* has an excellent magazine with hard-hitting investigative reporting, and *WorldWatch* is always packed with vital information about global issues. In most cases, joining the organization comes with a subscription to the magazine, and these magazines are usually available online for digital subscribers. [*blogs, eco-; microblogs, green; online petition sites and environmentalism; social media, environmental; tweets, eco-; Ranger Rick*]

maggot medicine
Maggots are the larval stage of flies. Granted, they are quite disgusting to look at, but since ancient times they have provided an unusual but valuable medical service.

Some species of fly larvae only feed on dead and decaying flesh, leaving healthy tissue untouched. Because they are small, highly mobile, and adept at finding their meals, they are great at cleaning out wounds.

With the advent of antibiotics, their need dwindled, and with their gross factor, their medicinal use came to an end. Today, however, cleaning out infections is becoming more difficult because of microbes that are resistant to many *antibiotics*, some of which are called *superbugs*.

When new technologies don't work, sometimes the best thing is to go back to the old standards. In this case, maggots are making a comeback but a bit more high tech than the Romans might have used. Medical maggot suppliers provide packages of sterile fly eggs that hatch into sterile little maggots that are placed, by doctors, directly into a wound. The proper terminology for this process is maggot debridement. [*frugal science; biomimetic; bioplastic; robotic bees; synthetic biology*]

Magnuson Fishery Conservation Management Act

This act was created in 1976, with amendments in 1996 and 2006. Its primary purpose is to set U.S. fishing quotas with the goal to stop *overfishing* that has depleted fish stocks. It is managed by the Marine Fisheries Service (an agency within the *National Oceanic and Atmospheric Administration*).

For much of the Act's history, it has been accused of being too weak and ineffective to accomplish its task. And the declining populations of fish in previous years bears this accusation out. Up until just a few years ago, most of these quotas were set by government officials and members of the fishing industry—with no scientists participating. This form of conservation resulted in the fishing industry almost putting themselves out of business with dwindling fish stock. Environmentalists believed these decisions should be made by marine biologists; not politicians and those doing the fishing.

However, more recently, the Marine Fisheries Service, at least in some regions, has turned over to scientists the responsibility of setting quotas, and the results have been positive. A 2010 report by the environmental *NGO* Marine Fish Conservation Network described this change: "To achieve the goal of ending overfishing…Congress strengthened the role of science in the fishery management process and required fishery managers to establish science-based annual catch limits and accountability measures for all U.S. fisheries, with a deadline of 2010 for all stocks subject to overfishing…." The results have been very positive with a significant improvement in U.S. fisheries. [*fishing, commercial; fish populations and food; pirate fishing; illegal, unreported, and unregulated fishing*]

malodor pollution

When people think about *air pollution*, they don't usually think about smell. Malodor refers to a bad odor; a very bad odor, more like a stench. These smells, when bad enough, can make people sick and are considered a public health hazard. So much so that a new science now supports the study—aromachology, about smells that are not toxic and do not make you sick, but are bad enough to trigger a physiological response.

People living near *factory farms* such as hog farms have reported increased anxiety and stress, headaches, nausea, and fatigue. Smell is now considered an environmental stressor the same as *noise pollution*, high heat, or being in a large crowd. [*visual pollution; light pollution; pollution police*]

Malthus, Thomas

(1766–1834) Malthus was an English clergyman and economist, famous for his theory about *population growth*. His theory was put forth in his "Essay on the Principle of Population," first published in 1798. It stated that the human population will increase faster (exponentially) than food supplies (arithmetically). It continued to state that unless fertility is controlled—by late marriage or celibacy—disease and famine would control the growth of populations. Neo-Malthusians believe population growth must be controlled or our population will face famine and a population collapse at some time in the future. [*human population throughout history; Ehrlich, Paul; population explosion, the human; population growth, limits of human; hunger, world; water shortages, global; tipping point; United Nations Population Fund*]

mammals

This large class of warm-blooded *vertebrates*, of which man is included, possess the following characteristics: the offspring are milk-fed by the mother and they have body hair and a diaphragm. It includes three main groups: egg-laying (called monotremes and includes the platypus), those with a pouch for the young to develop in (called marsupials and includes the kangaroo and opossum), and the placental mammals where the young develop within the mother and are born live, such as humans.

There are roughly 4500 species of mammals today, widely distributed around the globe. They have been on Earth for well over 100 million years. Scientists estimate about 19 orders of mammals have gone *extinct* (an order is a large group such as today's hoofed mammals). [*biodiversity; buffalo; elephants; Florida panther; gorillas; greyhound dogs; stereoscopic smell in moles; tagging programs; wolves, endangered; charismatic megafauna; flagship species; extinction, sixth mass; polar bear trade;*

rhinoceroses and their horns; Tasmanian tiger, extinct; United Nations Convention on Biological Diversity; whales, dolphins and porpoises, true]

Man and the Biosphere Programme

A *United Nations* program created in 1970 to link the natural sciences with the social sciences with an endgame to improve the relationship between humans and their environment. The program uses *Biosphere Reserves* as their natural laboratory.

manatee deaths and algae blooms

Florida manatees are an *endangered species* and they are especially hit hard when a toxic red *algae bloom* occurs off the southwest coast of the state. The *algae* produce a nerve toxin, harmful to most animals, that causes respiratory problems in people. However, because manatees typically eat 100+ pounds of the algae a day, it kills them. In 2013 an exceptionally large bloom killed 241 of the manatees. Scientists are studying the causes of the blooms. [*white-nose syndrome; bumblebee decline; colony collapse disorder; coral bleaching; Dutch elm disease; fire blight on apples and pears; Monarch butterfly decline; oysters and climate change; zombie bees*]

mangrove forests

Mangroves are rare and important trees that form dense *forests* in *brackish* waters. They grow in *intertidal* waters—*swamps* and lagoons—in tropical and subtropical regions. Their root systems are mostly underwater, but a portion of their roots are above the water line and therefore capable of absorbing oxygen.

As these plants grow and form forests, they trap silt and mud and gradually build up the environment, making it drier over time. Mangrove forests are important breeding grounds for many fish and shellfish—much like an *estuary*. They are also *habitats* for crocodiles, *alligators*, and numerous fish and small *mammals*.

These unique ecosystems are being destroyed by human activities. They are often cleared for timber harvesting or housing developments along the shoreline. Pollution and *climate change* also contribute to their destruction. Somewhere between 35 and 50 percent of all the world's mangrove forests have already been destroyed. [*habitat loss; urbanization and urban growth; wetlands; coastal wetlands; eco-services*]

Map of Life

An online database created at Yale University with plans to map, in great detail, the range and other information about species. Currently about 25,000 species have been entered, but plans are to include information for the million-plus species on our planet. The goal is to provide an essential tool for scientists to help conserve *biodiversity*. [*Encyclopedia of Life; magazines, eco-; Yale Environment 360 blog; blogs, eco-; microblogs, green; crowd research, scientific; tweets, eco-; social media, environmental*] {mappinglife.org}

Maraniss, Linda

In 1985, Linda Maraniss organized the first of many Texas Coastal Cleanup events, responsible for removing 250 tons of *garbage* annually from almost 200 miles of beaches. This program was the first to foster many others around the country, not only to clean up the trash but to record what they found on data cards, most of which were *plastics*. These records helped persuade Congress to ratify the *Marine Plastic Pollution Research and Control Act* of 1987, as well as an international treaty prohibiting ships from dumping plastic into the sea. By law, these vessels must bring the plastic back to the shore.

Through these types of programs, the beaches are cleaner and marine organisms such as *whales*, birds and *sea turtles* are safer. Coastal and shoreline cleanups have become annual events in numerous regions around the country and the world. [*Ocean Dumping Ban Act; MARPOL; litter; municipal solid waste; plastic pollution; ocean pollution; beach water quality and closings; coastal wetlands; garbage patch, marine; industrial water use; domestic water use*]

Marcellus shale bed

Fracking has made this bed of shale well known. This is a geological formation that contains vast amounts of *natural gas* trapped within the shale and until recently, considered inaccessible. But with the new technologies of fracking and horizontal drilling, the gas is now being extracted at a fever's pitch. The bed covers a 95,000-square-mile area, including portions of West Virginia, New York, Ohio, and Pennsylvania. It is the largest find of natural gas in the United States and believed to contain up to 500 trillion cubic feet. This one bed can supply the equivalent of two years of natural gas for the entire country. Pennsylvania alone already has almost 2000 fracking wells, or what are called *unconventional gas* wells. [*fracking states; fracking process, shale bed; the shale gale*]

mariculture

Also called ocean farming, this is a type of *aquaculture* where marine life—such as shrimp, *oysters*, and seaweed—are cultivated in their *marine habitats* for commercial purposes. Many descriptions of this process state that the organisms are kept "in a natural setting," but mariculture is accomplished by keeping the farmed organisms in enclosures of some sort floating in the ocean—which is not a natural habitat. The high volume of organisms raised in these enclosures results in environmental problems similar to any other form of aquaculture. [*aquaculture producing countries, ten biggest; Aquaculture Stewardship Alliance; fish populations and food; fishing, commercial; overfishing; shrimp farms, aquaculture; Franken-salmon*]

marine ecosystems

Aquatic ecosystems are divided into *freshwater* and marine ecosystems (also called saltwater ecosystems). Ninety-seven percent of all the *surface water* on our planet is in the *oceans*. Marine ecosystems have a high concentration of dissolved salts (salinity). The concentration of salts in the open ocean is about 35 parts per thousand (ppt) but varies greatly elsewhere. For example, the Red Sea, which has no source of fresh water, reaches 46.5 ppt, while the Baltic Sea contains 12 ppt because of a large inflow of freshwater. The study of marine environments is called oceanography.

The various regions of the oceans can be described in many ways, but ecosystems are most easily described by their distance from the continental shores. The region that lies within a few miles of shore, above the continental shelf, is the *neritic zone* (also called the coastal zone), and the water beyond the continental shelf is the *oceanic zone*.

Most of the oceanic zone is relatively nonproductive because few nutrients ever make it out to the open sea. The most productive marine ecosystems are found in the neritic zone, which includes many specialized *habitats*. [*fishing, commercial; fish populations and food; eels and elver fishing; deep sea trenches; gyre; Marine Mammal Protection Act; marine protected areas; marine reserves; MARPOL; ocean acidification; ocean pollution; upwelling; manatee deaths and algal blooms; seismic air guns, whales and; sonar testing, underwater; whale earwax*]

Marine Mammal Protection Act

Established in 1972 and amended in 1988 and 1994, this act protects all *marine* mammals in U.S. waters, including *dolphins, whales,* porpoises, sea otters, seals, sea lions, manatees, dugongs, and *polar bears.* It contains regulations regarding the use of *purse seine nets*, driftnets and *gillnets*.

The Act makes it illegal to harass, feed, hunt, capture, collect, or kill any marine mammal or part of a marine mammal. The responsibilities are divided between the National Marine Fisheries Service (within the *National Oceanic and Atmospheric Administration*), which handles whales, dolphins, and porpoises, and the U.S. *Fish and*

Wildlife Service, which handles all others. The Act also manages regulations for marine mammals in captivity, which is handled by the Animal and Plant Health Inspection Service (part of the *Department of Agriculture*). [*marine ecosystems*]

marine protected areas (MPA)

Most of us are familiar with protected areas such as *national parks* and *forests*, but few are aware of the recent marine equivalents, called marine protected areas. These are areas of the oceans that have been designated as low-impact zones, areas where human activities are limited in some way. Some might allow limited fishing, while others can only be used for recreation such as swimming and snorkeling. There are many possibilities, but what they all have in common is that the limits are imposed to protect the local fish and marine life populations and the overall *biodiversity* of the *habitat* itself. Some of these MPAs are designated as marine reserves, also called *no-take zones* because no fishing is allowed at all.

The purpose for MPAs and marine reserves is to protect dwindling ocean resources and especially protect against *overfishing*, which is of paramount importance. The most successful MPAs and no-take zones are small ones near local economies where fishing is important. When this is the case, the locals help ensure the enforcement of these areas, because their livelihoods depend upon it; people cannot fish for a living if all the fish are gone.

Only 1.2 percent of the world's oceans are MPAs, and a small fraction of that are marine reserves. Environmental groups urge a need to protect at least 10 percent of our oceans by 2020. Because fish and other marine life often migrate long distances in search of food, the exact locations of these MPAs and marine reserves is important to whether they provide any real protection.

Some MPAs are specific to certain species, such as *shark sanctuaries* and swordfish nurseries off the coast of Florida. Others protect all forms of life, such as in *Australia* where plans are to make one-third of all its coastal waters a marine reserve, restricting all fishing and all drilling.

While most of these areas are created by national governments within their territorial waters, some states in the United States are creating state-controlled MPAs. California has approved four such MPAs off their coast, consisting of almost 850 square miles. People can still swim, kayak, and enjoy these waters, but they cannot remove any form of marine life.

Marine Protection, Research, and Sanctuaries Act

See *Ocean Dumping Ban Act.*

marine reserves

See *marine protected areas.*

Marine Stewardship Council Sustainable label

See *food eco-labels.*

MARPOL

The International Convention for the Prevention of Pollution from Ships, adopted in 1973, in combination with the 1978 Protocol that followed, is commonly referred to as MARPOL. This international agreement regulates many forms of pollution from ships at sea, including *oil* pollution, noxious liquids, and other harmful substances being transported, ship's *sewage* and *garbage*, and *air pollution* caused by ships.

These agreements have been signed by 152 countries, representing more than 99 percent of all shipping. [*ship pollution; cruise-ship pollution; plastic pollution; marine ecosystems; Marine Mammal Protection Act; National Marine Sanctuaries Act; Ocean Conservancy; oceans; sea gliders; oil spills*]

marsh

A *wetland* ecosystem characterized by dominant grasses, as opposed to a *swamp* that has trees and shrubs as the dominant vegetation. [*aquatic ecosystems; coastal flooding; wetlands; Convention on Wetlands; dredging; floodplain; nutrient enrichment; estuary and coastal wetlands destruction; National Estuary Program*]

Marsh, George Perkins

(1801–1882) Marsh was one of the original American *environmentalists* with a penchant for *forests* and preventing *deforestation* (before the term existed). His book, Man and Nature, published in 1864, was pivotal to this end. His beliefs are encapsulated in his statement: "The Earth is fast becoming an unfit home for its noblest inhabitants." He was one of the founders of the *Smithsonian Institution,* and his book is considered to have inspired the creation of *Arbor Day* and influenced many, including *Gifford Pinchot* and *President Theodore Roosevelt.* [*dual-use policy; forest management, integrated; preservation*]

mass transit

One of *urbanization's* biggest dilemmas is how to transport large numbers of people to and from work and around a city. The *automobile* chokes many cities to a halt twice each day. Mass transit is an alternative to the automobile and typically includes light *rail*, buses, subways, street cars, or ferries. It is a form of public (shared) *transportation*, available to everyone without reservations. It typically transports passengers within a city and brings them to and from surrounding suburban areas (*urban sprawl*).

Mass transit reduces traffic, conserves energy, saves land (that would be used for roads), and reduces *air pollution*. But even with all these benefits, only about four percent of all Americans use mass transit routinely. New York City is the only U.S. city that has significant mass transit riders, with about 30 percent using it routinely. The next two cities—San Francisco and Washington, DC—only have about 15 percent of all commuters as regular riders.

Other nations, however, embrace the idea. For example, in Russia and *China*, about 50 percent of the population use mass transit regularly. About 30 percent of all Germans, Swedes, Brits, Brazilians, and many other city dwellers routinely use it.

There has been a surge, however, in U.S. efforts to improve mass transit offerings to increase ridership. The availability of good mass transit within cities is now seen as a big plus—and the necessity (and expense) of an *automobile* a big minus. Mass transit use in U.S. cities has steadily increased since 1995. Cities are seeing the light (and the traffic) and beginning to invest more heavily in mass transportation systems and other alternative forms of transportation, such as *bike* lanes, *bike sharing,* and *car sharing* programs.

Some cities are increasing mass transit by reintroducing older technologies, such as in Washington D.C., where they are rolling out a few new streetcar lines to supplement their excellent metro subway. Others are looking to the future for newer technologies and studying vehicles such as the personal rapid transit vehicle (PRT for short). When ready, they will work like small, personal versions of light rail and will work in concert with a *green app* that brings the PRT as close as possible to the commuter's pickup location.

Mather, Steven

(1867–1930) Considered the father of the *National Park Service* and its first administrator. He was a wealthy business man and loved the outdoors. On his trips through *Yosemite*, one of the first parks, he was unhappy with what appeared to be mismanagement. Upon complaining to people in high places, he was invited to come to the capital and improve the situation, which he did. At the time, the parks were under the direct jurisdiction of the Interior Department with no specific agency in charge. He helped create the National Park Service within that department and became its first leader. He fought, along with *John Muir,* to prevent the *Hetch Hetchy Dam*, but lost that battle. [*philosophers and writers, environmental; National Park and Wilderness Preservation System; national parks, global*]

mattresses, recycling

Who is responsible and who will pay for the disposal and harm caused by products when no longer useful? *Cradle-to-grave* and *extended product responsibility* are new ways to think about consumable products and what must be done when their useful life has ended. New laws in Connecticut mandate that mattress makers are responsible for their products, even at the end of their useful life. They must pay for the collection of discarded mattresses and their delivery to *recycling* facilities, where the materials can often be recycled for many other uses. Estimates are that this program will keep 175,000 mattresses out of *landfills* and save the state more than a million dollars per year. [*plastic recycling; recycling metals; paper recycling; motor oil recycling; cigarette butt litter and recycling; downcycling; upcycling; glass recycling; soda can recycling; tires, recycling*]

McKibben, Bill

(1960–) McKibben is one of America's most popular environmental writers and the author of a dozen books. He is probably best known for <u>The End of Nature</u>, considered the first book for the general public that forewarned about the perils of *climate change*.

In 2007, he organized the Step It Up National Day of Climate Action, one of the largest protests about *global warming* to date. More recently, he co-founded 350.org, an international grassroots campaign meant to mobilize a global climate movement with an emphasis toward youth and college campuses. (The 350 refers to parts per million of *carbon dioxide* in the *atmosphere*.) He is a scholar-in-residence at Middlebury College in Vermont. [*philosophers and writers, environmental; Limits to Growth, The; Our Common Future; Spaceship Earth; Orr, David; Quammen, David*]

meat, eating

Eating meat, like owning an *automobile*, is an indication of a nation's wealth. In 2011, the average person in a less developed nation consumed 70 pounds of meat, while the average person in a more developed nation consumed 174 pounds. This number is increasing rapidly in *developing nations* as incomes rise but decreasing in developed nations, probably because of an informed public learning more about the health issues associated with eating meat. There is plenty of evidence showing health problems associated with eating meat including obesity, diabetes, cardiovascular disease, and some cancers. In addition to health concerns, many environmental problems are associated with *meat production*. [*vegetarianism; vegan and vegetarian diets and the environment; food's carbon footprint – the best and worst; food eco-labels; organic food*

labels; organic beef; Meatout Mondays; CAFO; factory farms; farm animal cruelty legis-
lation, egg farms; energy pyramid]

Meatout Mondays

A movement to stop *eating meat* once a week. Several groups are in this movement. Some advance the idea based on *animal rights*, others do so based on health-related issues, and still others on environmental concerns. [*meat production; food eco-labels; organic beef*] {livevegan.org/meatout-Mondays and meatout.org and meatlessmon-day.com}

meat production

Eating meat comes with all sorts of costs—some personal and some environmental. It starts with the fact that raising livestock is a very inefficient process. Almost eight pounds of feed is needed to produce one pound of beef and about six pounds of feed to produce one pound of pork. The majority of U.S. crops go—not to feed people—but to feed livestock. So for every pound of meat you eat, eight pounds of grain could have been eaten directly by people.

Then there is the *water* consumption issue. About 1000 tons of water are needed to produce 1 ton of grain. Most of this water is used to grow the grain to feed the live-stock—instead of using the water directly for human consumption. About 60 pounds of water is needed to produce one pound of wheat, but about 5000 pounds of water is needed to produce one pound of meat.

Another factor is the *fuel* needed to get the water, harvest and transport the grain, and all the other processes involved. All of this might not make you stop eating meat, but you can see how it is not the most efficient use of our natural resources.

Many other environmental factors should be considered. *Methane* is a *green-house gas*—one that makes *carbon dioxide* look good. Cattle produce methane gas during digestion and release it as belches and flatulence—a lot of it. Estimates have about 20 percent of all methane gas released into the *atmosphere* coming from live-stock, with one cow emitting roughly 100 gallons of methane a day.

Next comes *nitrous oxide,* which is produced from *synthetic fertilizers* used to grow the crops. It too is far worse than carbon dioxide as a greenhouse gas.

Then there is the *deforestation* caused by clearing forests to make pastureland. Large regions around the globe are cleared to make room for grazing livestock. Some of this land is not cultivatable, but most could be used to grow crops for people—far more people than can be fed by the meat.

And the final problem is the resulting solid waste. You can only imagine the amount of waste produced when you have thousands of cattle or pigs in concentrated areas such as in *CAFO*. A single open pit can hold three million gallons of urine and feces in a typical hog operation. In many cases, instead of being properly composted, the contents of these pits are bulldozed into the ground. Much of this organic matter then decomposes *anaerobically,* producing even more methane gas.

U.S. *factory farms* generate well over a billion tons of animal waste, which is a major cause of *water pollution*. Waste spills kill fish and contaminate water supplies. The New River hog waste spill in North Carolina spilled 25 million gallons of excrement and urine into water in 1995, killing more than 10 million fish and closing more than 360,000 acres of coastal shell fishing.

Still other concerns about meat production include the use of *antibiotics for livestock*. Factory farmed animals are the most medicated creatures on Earth. And finally, many people consider the treatment of the livestock to be inhumane and unethical.

medical ecology

A new science that studies how the environment directly affects human health. It links environmental degradation to various states of poor health. The term was first coined by the famous microbiologist, *René Dubos*.

One fascinating aspect of this new science pertains to the medical and health fields—specifically, the study of the human body with emphasis on the microbes that live in and on us, called the *human microbiome*, and the impact it has on our health. [*puberty and the environment, early; autism studies and environmental factors; breast milk and toxic substances; body burden; toxic cocktail; nanoparticles and human health; superbugs; maggot medicine*]

megacities

Cities that become home to 10 million people or more are often called megacities. The three biggest megacities are Tokyo, Japan; Delhi, *India*; and Mexico City, Mexico. Megacities are one of the outcomes of *urbanization*. In the United States, one city does not typically grow to such a large size; instead cities and their *urban sprawl* tend to grow together into one vast populated area often referred to as a *megalopolis*. [*handshake buildings; smart cities; mass transit; urbanization and urban growth; population explosion, the human; noise pollution; light pollution*]

megafloods

Geological studies show enormous floods, caused by rainfall, have hit the California coast roughly every 200 years and have done so over the past 2000 years. The last such flood occurred in 1862. That storm started on Christmas Eve of the previous year and continued for more than 40 days straight. Thousands of people died and almost a million cattle drowned. These megafloods have occurred in many regions across the globe and might be caused by *atmospheric rivers.* [*extreme weather and climate change; Bangladesh and sea rise; climate change; sea level rise*]

megalopolis

Large, continuous, densely populated regions, where *urban sprawl* has connected many cities and formed one large city complex called a megalopolis. Some examples are the Boston to Washington D.C. corridor, the eastern seaboard of Florida, and the southern coastline of California. Also, the region from London to Dover in England and the Toronto Mississauga region of Canada are considered megalopolises. [*mega-cities; urbanization and urban growth; population explosion, human; air pollution in Beijing, China; ground-level ozone, ten U.S. cities with the worst*]

megawatt (MW)

Equal to one thousand *kilowatts* (one million watts)—enough power to meet the typical demands of about 720 homes daily. [*watt and kilowatt hours*]

meltdown

The coolant (usually water) in a *nuclear reactor* is responsible for keeping the radioactive fuel rods and the entire system from overheating. If the coolant isn't working properly because of leaks or other malfunctions, the fuel rods and surrounding structures overheat and physically melt, releasing radioactivity. Meltdowns are supposed to be prevented automatically by sophisticated monitoring equipment, by plant operators, or by both, but as history shows, these don't always work. [*Chernobyl; Three Mile Island; Fukushima*]

melt ponds

There has always been a belief that the *Arctic* region blooms with life only after the spring thaw; a very limited timeframe. However, recent studies show that green blooms of *phytoplankton* can be found under remaining spring ice still over three feet thick. This means the Arctic is far more biologically productive than previously thought because life can exist for a longer period of time each year.

Light was thought to be the *limiting factor*—the thick ice blocked light from reaching the water below, preventing *photosynthesis.* However, recent studies show that melt ponds form below the thick ice, allowing 50 percent of the light to pass through to the water below and resulting in under-ice blooms of phytoplankton and the *food chain* that follows. [*Arctic ice, changes in; arctic drilling; climate change*]

Mendes, "Chico"

(1944–1988) Mendes was a rubber tapper in the Brazilian *Amazon*. He became a successful activist, defending the land where he lived and protecting his livelihood against *deforestation* caused by the *logging* industry and cattle ranching. He believed in nonviolent organizational tactics and garnered the support of his fellow workers during the late 1970s and early 1980s after founding the National Council of Rubber Tappers. His activism cost him his life when he was assassinated in 1988 by a cattle rancher—an act that brought his cause to a worldwide audience and made him an environmental martyr to many. In response to the public outcry, *Brazil* set aside a forest reserve specifically for the sake of rubber-tapping sustainability. [*governments in the forest; logging, illegal; Rainforest Action Network; Rainforest Alliance; rainforest destruction; slash-and-burn cultivation; Statement of Forest Principles; Dian Fossey; Silkwood, Karen*]

Merchant, Carolyn

(1936–) Merchant is a philosopher, environmental historian, and leading *eco-feminist* writer. One of her best known works is <u>The Death of Nature</u> (1980), which states that the scientific revolution and the philosophy of the founding fathers caused the subjugation of both women and the natural world, and this intersection unites *ecology* and the women's movement. [*Women's Earth Alliance; women farmers; eco-psychology; environmental justice; ethics, environmental; philosophers and writers, environmental*]

Mercury and Air Toxics Standards (MATS)

Passed in 2012, this U.S. law mandates *coal*-burning *electric power plants* to reduce the amount of *mercury*, *heavy metals,* and other *toxic substances* they release and to do so by 2017. The health benefits should be huge, including preventing as many as 4700 heart attacks and 11,000 early deaths. This law forces these power plants to pay for the costs of these improvements.

This law will force many of the older coal plants to close down or be converted into gas-burning plants, because the cost of *natural gas* in the United States has plunged. Of course, this has no impact in other nations where coal-fired electric power plants are the norm, because they are not in the midst of a natural gas price drop. [*air pollution; Clean Air Act; atmospheric brown cloud; shale gale, the*]

mercury poisoning

Mercury poisoning can result in *acute* illness or disabilities such as numbness and garbled speech. *Chronic* exposure affects the nervous system and in the womb can impede brain development.

Mercury poisoning can occur by the direct intake of mercury, but it usually happens when someone eats contaminated organisms such as fish with high concentrations of mercury in their bodies and it gradually accumulates in the person's body over time—a process known as *bioaccumulation.*

Mercury is also used in the manufacture of some products such as LCD computer screens, but most notably, small amounts are found in *compact fluorescent bulbs* (CFLs). As is the case with so many products, they are safe and cause no problem when they are being used. But once disposed of, these small amounts become huge amounts and they become a problem. CFLs are supposed to go to *hazardous waste sites,* but many people simply throw them in the trash so they become part of the *municipal solid waste.* They end up in *landfills* or *incineration* facilities and thereby make their way into the environment and then into our bodies through *food chains.* Today, *CFL bulbs* are phasing out in favor of *LED bulbs,* which do not use mercury. [*heavy metal; hazardous waste; toxic waste; body burden; breast milk and toxic substances; medical ecology; nanoparticles and human health; puberty and the environment, early; autism studies and environmental factors*]

metal organic frameworks (MOF)

Carbon sequestration storage is a hot topic these days because it offers a possible solution to *carbon emissions.* Scientists are looking for compounds that can absorb vast quantities of CO_2 so it can be removed from our *atmosphere* and stored in something and placed somewhere. MOFs, organic molecules containing a metal atom, are one possibility. They have the unique property of having an almost unbelievable amount of internal surface area compared to the size of the molecule, a feature that enables them to absorb lots of gas such as CO_2. One single gram of this molecule, if unfolded, would theoretically cover an entire football field of surface. MOFs might be a way to make carbon sequestration storage a reality. [*climate change*]

metamorphosis

The process of obvious change in form during the development of an organism. For example, a caterpillar building a cocoon and emerging as a butterfly. Also used in geology to refer to changes in rock. [*insects; eclosion; plate tectonics*]

meteorology

The study of the *atmosphere*, especially as it pertains to weather. [*extreme weather and climate change; drought, the latest; global warming; greenhouse effect; Keeling Curve*]

methane digester

A device that converts *biomass* (plants and animal wastes) into *methane gas* to be used for fuel. The process uses *anaerobic* organisms (*bacteria*) that feed on the biomass and produce the gas. The gas is separated, stored, and then used for heating and cooking. Millions of these digesters are used in *China* and *India*. [*biomass energy; alternative fuels; biofuels; corn to biofuel; sugarcane to biofuel; sunflower oil as a biodiesel fuel*]

methane gas

Methane is a potent *greenhouse gas* that contributes to *climate change*. Comparing equal amounts of methane gas to *carbon dioxide* over a 100-year period would result in methane causing roughly 20 times more global warming. About 60 percent of all the methane gas released into the *atmosphere* is caused by human activities, as opposed to natural processes. Roughly 12 percent of global warming is now attributed to increased levels of methane in the atmosphere.

Methane gas is created naturally by *anaerobic bacteria* (living with little or no oxygen). These bacteria are naturally found in waterlogged soils and *wetlands* where methane gas is produced without any assistance from humans. The microbes in the guts of termites, which digest wood, also produce methane.

But methane is also produced in artificial environments such as *rice paddies* and *landfills*. By far the biggest methane producers are ruminant domesticated animals such as cattle and sheep, which contain these bacteria in their digestive systems. A single cow belches 100 gallons of methane gas a day. Livestock contribute about 28 percent of all the methane emitted into the atmosphere. Most of this comes from *factory farms* and *CAFO* with their vast numbers of livestock.

The dramatic rise in *natural gas* wells using *fracking*, (natural gas is methane gas) in the United States has raised concerns about leaking wells and the potential for increased levels being released and increasing global warming.

In some parts of the world, especially in developing nations, methane gas is generated as a *biofuel* and burned in *methane digesters*. [*rice paddies and methane; methane gas eruption; methane hydrates; permafrost, Siberian*]

methane gas eruption

Billions of tons of *methane gas* are trapped just beneath the East Siberian *Arctic ice* shelf. As the ice sheets melt due at least in part to *climate change*, this gas is gradually being released. Estimates have about 10 million tons of methane being released each year. This was probably occurring long before human-caused climate change, but the process is now accelerating. The fear is that pressure from the gas can soon burst through the ice, releasing about 50 billion tons all at once. Scientists studying the region say such a release is "highly possible at any time." Because methane is a far more potent *greenhouse gas* than *carbon dioxide*, this would have an immediate and unfortunate impact on the process of climate change. [*methane hydrates; permafrost, Siberian*]

methane hydrates

In some regions of the ocean floor are regions containing large amounts of a crystal-like substance that contains trapped *methane gas*—and lots of it. They are called methane hydrate deposits and can be mined as a possible energy supply. Methane gas is the primary component of *natural gas,* so mining these regions is similar to other *natural gas extraction* methods. The Gulf of Mexico has large areas with this potential. [*methane gas eruption; permafrost, Siberian*]

microbiome and microbiota, human

The collection of organisms—mostly *bacteria* but also *viruses* and *fungi*—living in and on us is called our microbiota. Microbiome is often used interchangeably with microbiota, but this refers to the genetic makeup of our microbiota. Whichever term is used, this topic is a huge scientific development you will be hearing about in the near future.

Most of these microbes live in your gut (the digestive system from top to bottom), but some live on your skin and elsewhere. Scientists have always known that people harbor a vast array of bacteria and other microbes, but only recently have they discovered the extent and importance of this relationship thanks to *DNA sequencing technology*. This research is conducted in labs in many countries, but most notably in the United States with the *Human Microbiome Project*.

About 100 years ago, scientists learned that killing all of the bacteria or "germs" in the gut of a monkey made the monkey sick—not healthier. But they didn't know why. It was impossible to study the bacteria while it was living in animal, and the bacteria behaved differently in a petri dish. With recent advances in DNA sequencing, where organisms can be studied by just looking at their genes, scientists were able to learn more. And what they learned is fascinating.

The human body is composed of about 10 trillion cells, but we now know that our bodies also harbor more than 100 trillion bacteria cells and other microbes. You can think of your microbiome as a rather unusual organ in your body, because, in total, it weighs about the same as your other organs—between two and four pounds. Your solid waste illustrates this point. Most people think solid waste is mostly leftover food. However, about half is actually bacteria. Because bacteria reproduce quickly, the biomass lost in this waste is almost immediately replaced in the gut.

Medical research has always looked at our cells to figure out how our bodies remain healthy and why we get sick. Now they have learned they must also look at our microbiome. In this *mutualistic relationship*, the bacteria have a place to live and gain nourishment from their hosts—us. But we get much in return. The relationship begins at birth, as a baby passes through the birth canal picking up a collection of bacteria from the mother. (Infants born via Caesarian section are known to begin life with a different microbiome.)

Our microbiome is important to many of our functions, but possibly most to our digestion. About ten percent of our calories come from plant carbohydrates we cannot digest ourselves and must rely on our bacteria to do so. Also, surprisingly, babies cannot digest their mother's milk because it contains a carbohydrate our cells cannot digest—only our bacteria can do so.

Our microbiome also provides some of our vitamins. Some bacteria produce vitamins B2, B12, and folic acid. This is an example of good bacteria helping keep harmful bacteria—what we commonly call germs—from overwhelming our bodies. Another example is the microbes that cause diarrhea, which are typically controlled because they are attacked by our good bacteria.

Recent research shows that our microbiome's importance goes even further. It is believed to be at least associated with the prevention of diseases as different as obesity and malnutrition, both types of diabetes, heart disease and hardening of the arteries, and the list goes on and on. Scientists believe our bacteria produce substances that help regulate our human cells, making them function properly. If certain bacteria are not present, the human cells go awry and illness can follow. Many of these illnesses are inherited, but the belief is that the microbiome is also inherited.

The applications from this new science are immense. Today you hear yogurt companies telling you the virtues of *probiotics* because they contain a few bacteria species believed to calm a malady called irritable bowel syndrome. However, with what is now being learned, probiotics might take on a much more important role in the future. Creating prescribed probiotics for certain conditions or illnesses might be the wave of the future. Some examples of this research are striking. For example, the microbiome of thin people appears different when compared to that of heavy people. And if a thin person gets heavier or vice versa, the person's microbiome changes.

Some of this research is being applied today. Doctors have recently used *stool transplants* to replace the microbiome of a sick person with the microbiome of a well person. This has been used successfully for a disease that occurs after a person has had a regimen of antibiotics that wiped out their normal microbiome. This disease gives a certain species of bacteria an opportunity to dramatically increase—often resulting in a distended gut that can kill the person (14,000 people die in the United

States each year from this disease). Doctors take a microbiome culture from a healthy person and implant it in the bowel of the sick person. As gross as it sounds, it works, and typically, the person is cured.

Some scientists now think of our bodies as an *ecosystem* instead of that of a single organism. A new field called *medical ecology* is finding inroads into journals and has become mainstream. Others say our human sense of self has been totally changed, because the term self must now include a hundred trillion microbes. [*antibiotics and the human microbiome; poop pills; self-concept and human microbiome*]

microblogs, green

A microblog is similar to a traditional *blog* but contains less content and is therefore typically more interactive. Because the intent is to be brief, Twitter is the most common platform used and a microblog can be simply a twitterfeed. Microblogs, like traditional blogs, have become an important tool for environmental activism and the *environmental movement*.

The most dramatic example of a microblog advancing environmental issues is that of Chinese billionaire Pan Shiyi using Sina Weibo, which is *China's* equivalent to Twitter. Mr. Pan microblogged about the horrible air pollution in Beijing in 2013 by posting $PM_{2.5}$ readings and explaining the meaning of these readings. Most Chinese never heard about this form of air pollution and the harm it causes to people's lungs.

Mr. Pan's following became huge and vocal and demanded something be done. Quite surprisingly, the government responded to this flurry of activity and stated their intent to improve the air quality. What comes of it is anyone's guess; however, there is no question that the citizenry learned a valuable environmental science lesson and many became environmental activists, thanks to microblogging. [*apps, eco-; cloud computing and the environment; crowd funding, green; crowd research, scientific; crowd sourcing, eco-; data centers, energy consumption of; online petition sites and environmentalism; social media, environmental; tweets, eco-; tweets, red panda on the loose; tweetstorm, Rio+20; energy-saving apps and programs*]

microfinance, green

See *microloans*.

microgreens and urban gardening

Tiny versions of our regular veggies. They can be grown in tiny gardens in containers anywhere they can find some sun provided by nature and some soil and water provided by you. Microgreen varieties of kale, broccoli, arugula, peas, brown mustard, radish, basil, and cabbage are available, to name a few. They grow to their diminutive maturity in less than two weeks, reaching about 1 to 2 inches high. At that time they are ready for your small harvest and can be used immediately in your salads or other dishes.

Some research has been done that shows microgreens are healthier than standard veggies because it appears the same nutrients are packed into smaller packages. Vitamins C and E were found to be six times greater than their larger versions. Also, they lose less of the nutritional value because there is no lag time during transportation—they can be grown right in your kitchen. [*urban gardens; gardens, rain; straw-bale gardening; community gardens; guerrilla gardening; nutrients, essential; teaching farms*]

microgrids, power

The *power-grid* is the network of power transmission lines that spread across the country supplying electric power to most people's homes. The problem with this hundred-year-old technology is the distance between the power source—usually a *coal-* or gas-burning *electric power plant*—and the people needing the power. Because the transmission lines are usually high above ground, they are vulnerable to wind and weather and often come down, leaving large segments of the population with no power. During Superstorm Sandy in 2012, some states in the northeast were totally without power for many days or, in some cases, weeks.

A power microgrid is an alternative to the normal power grid, where the power source is distributed instead of centralized. The idea that is being circulated in a few states is to have many small-scale power plants that can supply power to localized microgrids. In normal situations, this microgrid would simply contribute to and be part of the large grid; however, when large-scale power outages occur to the major grid, these microgrids could still supply power to their smaller constituencies. They could be large enough to keep entire communities up and running or small enough to at least keep hospitals and other emergency facilities in operation. [*states, ten most energy-efficient; time shifting; baseload power; smart energy; energy, ways to save residential; energy-saving apps and programs*]

microlending

In 1974, Muhammad Yunus, a lecturer at a university in Bangladesh, lent the equivalent of $27 to a few impoverished villagers to start a small business enterprise. This was the official beginning of microfinancing with microloans; he was awarded the Nobel Peace Prize in 2004 for this idea.

Microlending is a system that provides credit or loans to the very poor, who would never be able to borrow money through the regular banking system. It has become popular in many nations and is especially popular for efforts to support *sustainable development.* Microlending institutions provide small loans—usually $100 or less—to worthy entrepreneurs to start a small business. Today, millions of dollars are on loan through this type of enterprise.

Almost all of these loans are paid back because all of the debtors share the burden of repaying loans that remain unpaid. The process has empowered those who could not typically advance themselves; it creates self-esteem and helps build an economy where one would typically not exist. Because many of the loans pertain to a business that enables sustainability, it is a win-win-win situation—for the individual, the economy, and the environment. [*nurture capital; slow money movement; money market funds, environmental; shareholders, corporate*]

micronukes

See *nuclear reactors.*

micronutrients

See *nutrients, essential.*

migration

The movement of populations from one area to another. (When speaking about *human populations* crossing international borders, it is called emigration and immigration.) Animal populations often migrate annually as part of their normal behavior patterns. Some migrate for breeding purposes, others simply to avoid harsh conditions. Migratory birds use established *flyways* across continents. [*Migratory Bird Hunting and Conservation Stamp Act; migratory bird protection; land acquisition, wilderness; Migratory Bird Act; Wild Bird Conservation Act*]

Migratory Bird Hunting and Conservation Stamp Act

In 1934, the United States passed this act requiring waterfowl hunters to purchase a duck stamp to help fund the purchase of easements and other lands that benefit waterfowl. This is a win-win-win situation: for hunters, waterfowl, and philatelists (stamp collectors). [*migratory bird protection; Migratory Bird Act; Wild Bird Conservation Act; land acquisition, wilderness*]

migratory bird protection

The United States has many laws to protect migratory birds. Some of the earliest are the Lacey Act of 1900 and the *Migratory Bird Act* of 1918. Other United States acts include the Bald Eagle Protection Act, the Waterfowl Depredations Act, the Fish and Conservation Act, the Wild Bird Conservation Act, and the *Endangered Species Act.*

In addition, many international agreements protect birds, such as *The Antarctic Treaty* that protects native birds of the region, *CITES,* and the Bonn Convention, which helps protect migratory birds around the world. [*Migratory Bird Hunting and Conservation Stamp Act; land acquisition, wilderness; Migratory Bird Act; Wild Bird Conservation Act*]

Migratory Bird Act

The Lacey Act of 1900 was one of the original acts of Congress attempting to protect birds from commercial hunting and shipping. The Migratory Bird Act of 1918 expanded this to protect most *migratory birds* that pass through our borders, en route elsewhere. This act makes it illegal to kill, capture, collect, buy, sell, ship, import, or export any migratory bird or its eggs or its parts, including feathers.

These acts came about primarily to protect birds from the millinery industry, which used birds' exotic feathers for hats—quite the fashion statement of this period. *Theodore Roosevelt* and *John Burroughs* were opponents of this trade. Burroughs wrote about the slaughter of birds just for their feathers to help convince the public, and Roosevelt used a big stick to help get these acts passed.

Some states are allowed to manage their own resident game birds and can make exceptions for some common birds. [*migratory bird protection; Migratory Bird Hunting and Conservation Stamp Act; Wild Bird Conservation Act; land acquisition, wilderness*]

Millennium Development Goals (MDG)

In 2000 a set of eight Goals to reduce global poverty and disease was agreed to at the United Nations Millennium Summit. The intent was to fulfill these goals by 2015. All 189 United Nations countries signed the document. The eight Goals are 1) eradicate extreme poverty and hunger, 2) achieve universal primary education, 3) promote gender equality and empower women, 4) reduce child mortality, 5) improve maternal health, 6) combat HIV/AIDS and malaria, 7) ensure environmental sustainability, and 8) develop a global partnership for development.

Each of the Goals is divided into Targets. For example, Goal 5—to improve maternal health—is divided into Target 5A: Reduce by three quarters, between 1990 and 2015, the maternal mortality ratio; and Target 5B: Achieve, by 2015, universal access to reproductive health. Each of these Targets has specific criteria to determine the degree of success. For example, Target 5A uses a maternal mortality ratio and the proportion of births attended by skilled health personnel.

These goals can be viewed as averages across all countries in need. Some were reached ahead of schedule, many are well behind schedule, and some certainly cannot be attained. For example, reducing extreme poverty and hunger as well as promoting gender equality in education have made great progress, but improving maternal health will not be met. The results so far have been a mixed bag, but progress cannot be denied.

These goals and their results can also be viewed specifically by country. For example, some of those countries that have made the most progress meeting these goals include Nigeria, Turkey, and Bulgaria, while those with the least progress include Nigeria, Jordan, and Kyrgyzstan.

As the MDGs draw to their target date of 2015, there is hope that a new set of goals will be established, called the *Sustainable Development Goals*. During the *United Nations Conference on Sustainable Development*, also known as Rio+20, the plan was to have SDGs in place to pick up where the MDGs leave off when they expire in 2015. However, the SDGs were never developed at that summit. Instead, they agreed that the SDG goals will become part of a "post-2015 Development Agenda" to be established in 2014. Time will tell if this comes to be. [*United Nations – global environmental issues and the; United Nations Development Program; hunger, world; water shortages, global*]

mimicry

One organism resembling another organism, usually as a method of protection from *predators*. [*aggressive mimicry; aposematic coloration; Batesian Mimicry; directive coloration; Müllerian mimicry*]

mineral exploitation (mining)

Minerals such as iron, nickel, copper ore, and many others are a necessity to our lives, but mining these and other resources almost always has a negative impact on the *environment*. *Recycling* manufactured products reduces the demand for mineral exploitation and helps minimize this damage.

When mineral deposits are located, mining for the ore often destroys the area by using *strip-mining* operations or other *surface mining* techniques. Crushing the rock to extract the minerals produces large quantities of leftover rock called *mining tailings*, which are unsightly, often cause *erosion*, and may release harmful substances such as *asbestos*, *lead*, or *radiation* into the area. The water some mining operations use during processing becomes contaminated before it is released back into the environment, causing *water pollution*. Many manufacturing processes also release contaminants into the air, causing *air pollution*, and waste products into water or soil. Exploration for new mineral deposits sometimes occurs in *national parks* and other nature reserves, damaging these natural settings and reducing public access.

Recycling materials such as *aluminum*, *glass*, and many other products, plus *source reduction*—to minimize the quantity of minerals needed in the first place—reduces the harm produced by mineral exploitation.

Most of the developed nations have mining regulations to reduce environmental damages to some degree. Most environmentalists believe the U.S. regulations are far too weak. Most of the developing nations have few if any regulations to protect the environment from the ravages of mineral exploitation. [*acid mine drainage; gold mining; Mining Law of 1872, General; mountain top removal mining; overburden; Pebble Mine; rare earth metal environmental problems; rare earth metals; room and pillar method; spoil banks*]

mini-mill industry
See *automobile recycling.*

mining
See *mineral exploitation (mining).*

Mining Law of 1872, General
The General Mining Law was signed into law by President Grant in 1872 and remains in effect—basically in its original form—even today. The fact that it has not been revised for more than 140 years is testimony to how the federal government can be pressured into protecting the needs of special interest groups—in this case, the *mining* lobby—instead of protecting the American public and the environment.

The law was passed in 1872 to encourage the settlement and development of western public lands—something of great value then but hardly needed now. To think that this law has remained untouched for all this time should boggle the minds of sensible citizens. The law allows any person who discovers a valuable mineral on public lands to lay claim to that land. The claim holder pays no rent to the federal government. In addition, the claim holder can rent or lease the land to others and charge them rent or royalties. Even more absurd, the claim holder can then purchase the land and its contents from the government for $5.00 per acre! Yes, that amount has not been increased for more than 140 years. More than 1.2 million of these claims have been filed for more than 25 million acres of what was previously *public land*; 2000 of these acres are in *national parks*.

The environmental damage left by these mines is immense. Tens of thousands of mines are left abandoned and the *toxic wastes* must be cleaned up at the taxpayers' expense. In many cases, the claimed lands aren't mined but instead are developed and built on at huge profits for the owners—from the original investment of $5 per acre on what had been public land.

Efforts to amend or rewrite this law have failed for decades. On occasion, changes appeared imminent but were then reversed. That a law so antiquated remains unchanged at the expense of U.S. taxpayers and the environment is hard for many people to believe. But until enough people become vocal, this law will never change. The Mineral Policy Center (now part of *Earthworks*) has been fighting along with other *NGOs* to change the law but to no avail. [*mineral exploitation (mining)*]

Monarch butterfly decline

Monarch butterflies have the longest *migration* route of any *insect*. Those that live west of the Rocky Mountains during the summer travel to southern California and live in eucalyptus trees; however, those that live in the eastern United States during the summer travel thousands of miles south to Mexico to feed on oyamel fir trees.

The most amazing aspect of this journey—besides its distance—is that the butterflies tend to return to the same trees each year, even though they are not the same individuals. By the time a full year passes, four generations will have reproduced. So it is the great-great-great grandchildren that actually return to the same tree its ancestors visited. It is not known how they do this.

Unfortunately, this migration is disappearing. Because it is impossible to count the butterflies, the size of their winter grounds is used to indicate their numbers. Their Mexican wintering grounds once covered more than 50 acres of land but are now reduced to less than three, meaning only a fraction of individuals arrive compared to what was once there. The decline was rapid, dropping from 44 acres between 1994 and 2003, to only three today.

The cause is probably twofold. First, in recent years an ongoing drought and excessive heat spells have reduced their numbers. But more long-term harm has possibly come from *habitat loss* in the United States, where they spend the warmer months. Their main food source used to be milkweed plants that grew between rows of corn or soybeans. However, with the advent of a *genetically modified organism* crop that uses, *herbicide* resistant corn and soy crops, the milkweed cannot grow and the butterfly's food source is significantly diminished. [*biodiversity, loss of; migratory bird protection; ash tree demise; bumblebee decline; Dutch Elm Disease; fire blight on apples and pears; honeybees, pollination and agriculture; manatee deaths and algae blooms; oysters and climate change; white-nose syndrome*]

money market funds, environmental

Many money market funds specialize in investing monies only in institutions they deem socially and environmentally responsible. Companies with histories of environmental violations, for example, are excluded from their portfolio, while those that adhere to the tenets of *corporate social responsibility*, do well in *corporate sustainability ratings*, or manufacture products with *extended product responsibility* as part of their business model would be included.

A few of the many funds available are Green Century Funds, Working Assets Money Fund, Calvert Social Investment Fund, Freedom Environmental Fund, Global Environment Fund, Parnassus Fund, Pax World Fund, Domini Portfolio 21, and Winslow Green Mutual Fund. [*companies, top ten green global; corporate acquisitions of small green companies; internalizing costs; integrated bottom line; investing, green; nurture capital; polluter pays principle; shareholders, corporate; triple bottom line; waste conscious product development*]

monkeywrenching

A process used by some radical *environmentalists* of performing destructive activities in the name of environmental protection; also known as sabotage or *ecotage*. The idea was advanced in books such as The Monkey Wrench Gang, by *Edward Abbey*, and Ecodefense: A Field Guide to Monkeywrenching, by *David Foreman*. [*EarthFirst!; environmental movement*]

monoculture

When a single species is planted over a large area of land, it is called a monoculture. Farmland covered with thousands of acres of corn or wheat is an example of monoculture. Monocultures are ultra-simplified *ecosystems* and required for *high-impact agriculture*. A single species of plant exists where once was a *grassland* or *forest* ecosystem with hundreds or even thousands of species. These simplistic ecosystems are a very efficient method of growing crops but are highly vulnerable to change and must continually be protected and provided for by artificial means. Monocultures need extraordinary amounts of synthetic *pesticides* to protect them and *synthetic fertilizers* to provide *nutrients*.

For example, unwanted *weeds* naturally try to invade monocultures, so *herbicides* are used. Insect pests that feed on a particular crop have an unending supply of food in a monoculture, so *insecticides* are used.

Insects reproduce rapidly and develop resistance to these insecticides through the process of *natural selection*, so greater dosages of insecticides must be used just to maintain the same level of control. Monocultures also invite plant diseases to establish themselves, so *fungicides* are also used to defend the crop.

Crops grown in monocultures are constantly harvested, so there is little if any part of the plants that remain to nourish the soil for the next crop, so the soil must constantly be enriched with fertilizers. Many *organic farming* and *conservation agricultural* practices reduce the need for synthetic pesticides and fertilizers and provide alternatives to the monocultures of high-impact agriculture.

Montreal Protocol

This international agreement was originally signed in 1987 by 24 countries, including the United States, Canada, Japan, and the European nations and went into effect in 1989. Since then, 192 more parties have signed it and it has been revised many times. The signatories pledged to phase out the use of all *CFCs*—used as a refrigerant—by 1999. CFCs are potent, *ozone-depleting gases* responsible for the so-called hole in the *stratospheric ozone* layer that had formed at that time. They also pledged to phase out other ozone-depleting gases such as *halons*.

CFC concentrations in the *atmosphere* grew rapidly from the 1960s and continued after the ban, until about 2003, at which time they have begun to taper off. In 2006, scientists declared that the hole in the ozone was improving because of the success of the Montreal Protocol. Many consider this one of the greatest achievements in global environmentalism because it turned the tide of what could have been a global disaster.

As new substitutes for CFCs were created, the protocol was amended to mandate switching to those that caused the least environmental harm. HFCs are the latest improvement and cause no harm to the ozone layer.

However, the Montreal Protocol never anticipated another issue, that of *greenhouse gases* causing *climate change*. Unfortunately, although HFCs—the newest refrigerants—don't harm the ozone, they are powerful greenhouse gases. HFCs are a 2100 times more potent greenhouse gas than *carbon dioxide*.

So while the Montreal Protocol did a great job of protecting the Earth's ozone layer, it is unwittingly contributing to the destruction of its climate. Efforts to amend the protocol to mandate the use of a new category of refrigerants that are neither greenhouse gases nor ozone-depleting gases have been to no avail. [*ozone-depleting gases and greenhouse gases*]

moon power

See *tidal power*.

moo-shine

Unpasteurized milk. Because it is illegal to sell in many states, it is sometimes referred to as moo-shine. [*organic farming and food; apples, antibiotics in organic; night soil – biosolid; green manure; sugar, sugar alternatives, and the environment; farm-to-table; organic agriculture; wine, organic and biodynamic; guerilla labelling*]

Moran, Thomas

(1837–1926) A member of the *Hudson River School art movement*, Moran specialized in western landscapes instead of the eastern landscapes painted by most members of this movement. As with other members, he helped instill a sense of pride in the natural wonders of the United States during its formative years. Moran was commissioned by the United States government to motivate citizens to move west and to appreciate the wide open spaces. His art is credited with supporting the establishment of *Yellowstone*—the first *national park* in the United States.

morbidity

The frequency and distribution of disease within a *population* and its effect on that population. Morbidity data is used by public health officials to control the spread of disease and determine its cause. [*medical ecology; biosensors; nanoparticles and human health; pathogens; demography; population dynamics; mortality*]

more developed countries (MDC)

Also called industrialized countries, *developed countries*, or advanced countries. Defined by the *United Nations* as a country with high industrialization and a high *gross national product*, (compared with *less developed countries*). There are 33 MDCs, including the United States, Canada, Japan, *Australia*, New Zealand, and all the countries in Western Europe. Most of these countries have favorable climate and fertile soils.

About 23 percent of the world population live in MDCs, but they use about 80 percent of the world's energy resources. The annual population growth rate of these countries is about 0.5 percent, which is considered slow (compared with 2.1 percent for the LDCs, which is considered rapid). The average *GNP* per capita for those living in an MDC is about $15,800 per year, compared with someone in an LDC who averages about $700 per year. [*standard of living; demographic transition, theory of*]

mortality

The probability of dying within a *population*. Mortality can be used for any type of organism but is most often associated with people.

The probability of people dying is related to age, sex, race, occupation, social class, and other many other factors. The incidence of death usually reveals a great deal about a population's *standard of living*.

Mortality for any species is usually measured as the crude death rate. This is calculated by taking the number of deaths at the midpoint of a specific period of time, such as a calendar year, dividing it by the population at the beginning of that period, and then multiplying by 1000. The resulting figure is expressed as deaths per 1000 for a certain period of time.

For example, if there were ten deaths in a population of 3000 individuals in one year, the death rate for that population is roughly 3.3 per thousand, per year (10 divided by 3000, multiplied by 1000). [*less developed countries; human population throughout history; immigration; overpopulation; population dynamics; population stability*]

mortgages and fracking

Some mortgage lenders have become concerned about the environmental problems associated with fracking and are adding mortgage clauses stating the homeowners cannot lease their land for *fracking*. Some companies are adding riders to existing policies, stating the policy does not cover damage caused by fracking. [*fracking, ongoing environmental concerns; cemeteries and fracking; frackademia; Gasland (the movie); campus drilling for fossil fuels*]

motor oil recycling

Do-it-yourself *automobile* motor *oil* changes produce more than 200 million gallons of used oil each year in the United States, with much of it illegally dumped and not recycled. *Recycling* a single gallon of drained motor oil produces the same amount of fresh clean oil as extracting 42 gallons of crude oil from a well. Most towns and cities have their own motor oil collection sites, facilities, or regulations. [*barrel of oil; oil recovery or extraction; oil spills in U.S.; glass recycling; paper recycling; grass recycling; composting; recycling metals; soda can recycling; tires, recycling; upcycling; pipeline spills*]

mountain top removal mining

A form of *strip* or *surface mining* where mountain ridges are dynamited to expose *coal* seams. The resulting blasted rock, soil, and debris are dumped in the valleys below, often disrupting or blocking streams. This form of *mining* is commonly used in parts of Appalachia. It has all of the negatives of traditional surface mining plus the loss of mountain ridges and accompanying damage to the rivers below. [*acid mine drainage; mineral exploitation; overburden; Pebble mine; room and pillar method; spoil banks*]

movies, environmental

Even with the advent of the Internet, *social media, blogging, microblogging,* and television networks dedicated to environmental topics, movies remain a powerful force in the environmental movement. The *Environmental Media Association* gives out Environmental Media Awards each year, and the *Earth Communications Office* helps promote movies with an environmental message. Many excellent movies, both non-fiction and fiction, have immersed audiences in issues of environmental importance. Many websites list what they believe to be the top ten best environmental movies of all time, and educators have a long list of fine movies that turn a textbook topic into a full-length story.

Mowat, Farley

(1921–) A prolific and popular writer of *Arctic* cultures, primarily in Canada, Mowat writes compassionately about both the original residents and the animals of the regions. He writes about the differences between traditional culture that is united with the land in a harmonious way as opposed to modern societies where land and animals are disregarded and exploited. His best known book is Never Cry Wolf (1963). [*Native American environmentalism; primitivism; ethics, environmental*]

MTBE

This stands for methyl tertiary butyl ether, a gasoline additive that replaced *lead* when it was phased out in the 1970s. MTBE was added to gas to improve mileage and reduce *air pollution*. However, although it was lessening air pollution, it was beginning to contaminate *groundwater*. The contamination is believed to most likely be coming from *LUSTs*. California and New York were the first to ban MTBE in gasoline. It was later banned in numerous states and finally banned by federal legislation in 2003, but not before it began appearing in our *drinking water*. [*leaded gasoline; groundwater pollution; precautionary principle*]

muck

A fine-grain *soil*, saturated with water and of a thick consistency, containing a high concentration of decomposed *organic matter*. But it often refers to any soil considered offensive, such as in livestock pens where lots of excrement is mixed in with the soil. [*night soil –biosolids; industrial livestock industry*]

Muir, John

(1838–1914) John Muir is considered by many to be America's greatest naturalist and conservationist and is called the father of the *preservation* movement. Through his writings and arm-twisting of then President *Theodore Roosevelt*, he was in part responsible for the establishment of many *national parks*, including Sequoia, *Yosemite*, Rainier, and the *Grand Canyon*.

After attending the University of Wisconsin, Muir worked on mechanical inventions, but he abandoned that career to devote himself to nature. He walked from the Midwest to the Gulf of Mexico and kept a journal published in 1916 as <u>A Thousand Mile Walk to the Gulf</u>.

In 1868 he went to the Yosemite Valley and took numerous trips to Nevada, Utah, Oregon, Washington, and Alaska. Through his travels, he theorized that the formations in Yosemite were *glacial* erosions, now accepted as fact. In 1876, he urged the federal government to adopt a *forest* conservation policy. In 1892, Muir—along with many of his friends—founded the *Sierra Club,* which continues to this day to preserve our national parks.

His books include intensely detailed narratives about the flora and fauna of the *habitats* he roamed. He went into nature as few did at that time. To experience a winter snowstorm in the Sierra mountains, Muir climbed high into a 100-foot Douglas spruce, hung on through the entire dangerous storm, and later wrote about it in <u>The Mountains of California</u> (1894).

His belief in the preservation of nature was a religious conviction that followed the transcendentalist philosophy of *Henry David Thoreau* and *Ralph Waldo Emerson.* He was a good friend of *John Burroughs* and spent time with *President Theodore Roosevelt.* Muir spent two days camping with Roosevelt in Yosemite; soon after, it was declared a national park.

One of his most famous quotes is often used in introductory ecology courses: "When we try to pick anything out by itself, we find it hitched to everything else in the universe." [*philosophers and writers, environmental; Hetch Hetchy; Pinchot, Gifford*]

mulch

This is material placed on the ground to protect the *soil* and plants. It lessens the chance of *soil erosion*, maintains water supply, and reduces *weed* growth. *Inorganic* mulch can be composed of small rock, vermiculite, a fine soil called dust mulch, and many other materials. *Organic* mulch can include *leaf litter*, sawdust, grass and leaf clippings, peat moss, straw, animal manures, among others. As organic mulch decomposes, it has the added benefit of adding *nutrients* to the soil. [*cover crops; green manure; loam; biogeochemical cycles; carbon cycle*]

Müllerian mimicry

A form of *mimicry* where many species of distasteful or poisonous organisms all look similar, making them simple for predators to avoid. For example, many species of poisonous butterflies all have similar coloration to warn birds to stay away. [*Batesian Mimicry; directive coloration; aposematic coloration; cryptic coloration*]

multiple use policy

Much of our *public lands* follow a multiple use policy (sometimes called a *dual-use policy* when specifically used for *forests*), where the land is used for *outdoor recreation* or for *preservation* of natural resources in their natural state, as well as for resource extraction such as *logging* and *mining*. These competing uses have been hotly debated since the inception of the *U.S. National Forest Service*, and other public lands. [*wise-use movement; dual-use policy; conservation; utilitarian environmental ethic*]

municipal solid waste (MSW)

Simply put, MSW is *garbage*. But it specifically refers to a municipality's solid waste, which is usually collected by the town or a private company and delivered to a collection site, or what is often called the town dump. MSW does not include the other forms of solid waste, such as *mining* wastes, *agricultural* wastes, *sewage*, *industrial waste* products, and, obviously, any waste created outside of urban areas (the municipality).

MSW includes waste that gets *recycled* and *composted,* because, in most cases, waste destined to be recycled or composted is not separated until after it has entered the *waste stream* and is delivered to a central location. About 55 percent of the U.S. waste stream ends up in a *landfill,* and about 34 percent gets recycled or composted. In addition to landfills, recycling, and composting, MSW can also be *incinerated*, but only about 15 percent of the U.S. waste stream is incinerated.

Globally, about 1.3 billion tons of MSW are generated each year. This amount increases every year because of increased prosperity in many parts of the world and an overall increase in *urbanization*. The United States is the world leader in MSW volume. In the United States, more than 620,000 tons of solid waste are produced every day; this is more than five and a half pounds of garbage per person, per day—the highest of any nation. This is up from just three pounds per person, per day 20 years ago. In comparison, people in *India* produce less than three-quarters of one pound per day.

The largest component of U.S. MSW is paper products, making up about 28 percent. Then comes *food waste* at about 15 percent; yard waste (grass clippings and the like) about 14 percent; *plastic* almost 13 percent; *metals* almost nine percent; rubber, leather, and textiles combined at a little more than eight percent; wood about six percent; glass about five percent; and the remaining miscellaneous round out the last few percent.

Municipal solid waste disposal is one of the biggest environmental challenges we face. What we do with these growing piles of waste and how we prevent them from damaging our environment is a challenge every country faces.

In the United States, local governments historically were responsible for MSW management, but the federal government became involved in 1965 with the Solid

Waste Disposal Act. Later federal legislation cleaned up many of the worst offending aspects of landfills. The 1976 *Resource Conservation and Recovery Act* (RCRA) and follow-up amendments eliminated open dumping, promoted solid waste management programs, established standards for landfills and air *emissions*, and regulated *hazardous wastes*.

Today some of the MSW is converted to energy in *waste-to-energy facilities* and offers a promising alternative to letting the waste just sit in a landfill for hundreds of years. [*municipal solid waste disposal*]

municipal solid waste disposal

Municipal solid waste can be disposed of in three ways: 1) dumping it in *landfills*, 2) burning it in *incinerators*, and 3) *recycling* it, which can include *composting*. In some cases, the energy created by incineration is also converted into a power source in *waste-to-energy* facilities.

Landfills are the most common disposal method, because they are economical and were originally thought to be safe. About 55 percent of U.S. MSW goes into landfills, whereas about 90 percent of the United Kingdom's MSW goes into landfills. About 15 percent of U.S. solid waste is incinerated, and this number has been increasing with improved technologies.

Much emphasis has been placed on *recycling* the waste before it ever gets into landfills or incinerators. At present, however, about 25 percent of MSW globally is recycled, with wide differences between countries. The United States recycles about 34 percent, but South Korea, Singapore, and Hong Kong are close to 50 percent.

Finally, *source reduction* is designed to reduce the amount of garbage that must be disposed of by simply producing less to begin with.

music, Earth

See *jazz, Earth*.

mutagenic

Any substance that can cause genetic defects. [*teratogens; REM; radiation; Chernobyl; nuclear reactor safety and problems*]

mutualism

A *symbiotic relationship* in which both organisms benefit. Insect-plant relationships of this type are common. For example, the yucca plant is pollinated by the yucca moth, enabling the continued survival of both. The moth lays its eggs in the plant, and the young depend on the plant's seeds for nourishment. Both plant and *insect* survives because of this relationship. *Nitrogen fixation* is another important mutualistic relationship between *bacteria* and *legumes*.

A recently discovered and unusual example of mutualism involves some *corals* and their "bodyguard fish," gobies, a small, colorful, *tropical fish* often seen in aquariums. It works like this: Corals and seaweed compete for the same sunlit locations in the shallow waters. Some species of seaweed produce a deadly chemical in an effort to kill coral and secure the best locations. When coral is damaged by the seaweed poison, the coral produces a chemical that signals the gobies for help. The gobies arrive within minutes and begin eating the threatening seaweed. The coral gets protection, and the gobies use the coral for protection and a good location to find food.

mycowood

Violin-makers typically use healthy, stiff, solid wood to produce high-quality sound in their instruments. However, recent research in Switzerland has shown that violins made of wood that has been attacked by a specific type of *fungus* produces a superior sound—comparable to the famous violins made by Stradivarius. They call this wood mycowood, myco meaning fungus.

nanoceuticals
See *nanoparticles in food*.

nanoparticles
Particles so small they are measured in nanometers—one-billionth of a meter. This is smaller than a red blood cell and thousands of times smaller than the width of a human hair. These particles are so small, they possess unusual properties because of the relatively high surface area of the particles, compared to larger objects.

They are typically classified as hard or soft nanoparticles and can be natural or synthetically produced. They are used in numerous products for a wide range of purposes, including foods, medicines, and many processes such as environmental remediation.

Natural nanoparticles include terpenes, which are produced and released by volcanoes and fires; synthetic examples include carbon nanotubes and numerous food additives. *Nanotechnology* has made this a bourgeoning business with numerous applications. While the value of nanoparticles is obvious, there are also risks involved. [*nanoparticles and human health; nanoparticles in cosmetics and personal care products; nanoparticles in food; nanopesticides*]

nanoparticles and human health
Nanoparticles can enter our bodies both intentionally in our food and unintentionally from our environment by ingestion, contact, or inhalation. The FDA does not believe there is any need for the regulation of nanoparticles in food, cosmetics, or other consumer products; however, many *NGOs* and some government agencies in other countries are looking into the subject and disagree.

Although research remains limited, evidence of potential problems exists. For example, research has shown that some *nanoparticles in foods* can be absorbed in the intestines, pass into the body, and accumulate within certain tissues. The effects of this are unknown. Other scientists are concerned that nanoparticles are small enough to pass through the blood-brain barrier, something few substances can do. The effects of this are also unknown.

Carbon nanotubes are used in computer chips. These nanoparticles resemble asbestos in some ways and have been suspected of damaging lung tissue. (This would be an occupational hazard to those involved in the manufacture process, as opposed to users of computers.)

Nanoparticles are an example of a technology that got ahead of appropriate policies. No regulations of any sort exist pertaining to their use. Foods and many consumer products are all regulated in efforts to protect consumers from harm, but nanoparticles are not even routinely tested to see if they are dangerous. At the moment the only effort to document products containing nanoparticles is the Woodrow Wilson International Center for Scholars Initiative, which has a limited listing.

Probably the most common consumer nanoparticle is titanium dioxide, which is used for many purposes, including a whitening agent in toothpastes. You can assume you are ingesting this nanoparticle on a daily basis. [*nanotechnology; nanoparticles in cosmetics and personal care products; nanopesticides*]

nanoparticles in cosmetics and personal care products

Nanoparticles are finding their way into numerous products, many of which are personal care products and cosmetics—so much so that this new field even has a name: nanocosmetics. It has taken off so quickly because of the unique properties that can be created by these tiny particles, such as intense color and improved transparency and solubility.

Nanoparticles' first use in care products was as ingredients in sunscreens. Titanium oxide protects against dangerous UV light, but when it is applied to the skin, it appears white. When titanium oxide is packaged on nanoparticles, the creams don't turn an unsightly white, because their tiny size keeps the cream transparent. And that was the start of something big. It was found to be useful in numerous personal care products for many reasons.

The problem is the classic case of the cart getting in front of the horse. Nanoparticles are already found in numerous cosmetic products, but little serious safety testing has been done, either for humans who use these products or for the environment, where they typically end up as waste products.

Cosmetics are not tested by the *Food and Drug Administration*. The research that has begun indicates there can be serious concerns, because these particles, unlike almost any other ingredient, are small enough to enter individual cells. (This is being used to advantage in some products, such as in *nanopesticides*—they freely enter the plants cells and reside there to kill off pests.) Some research is raising red flags. Nanoparticles are not even required to be listed as an ingredient on products containing them. In many other nations, the *precautionary principle* is the accepted standard, but not so in the United States, where people have often become the research lab mice. [*nanotechnology; nanoparticles and human health; nanoparticles in food*]

nanoparticles in food

Most people are not even aware *nanoparticles* exist, let alone used in many products—such as our food—and have been for years. These tiny particles have unusual properties that lend themselves to act as food additives. They are used to: extend shelf life, act as a thickening agent or anti-caking agent, protect against UV light in packaging, prevent mold growth, enhance flavor, and provide nutrients in food supplements (called nanoceuticals).

The *Food and Drug Administration* (FDA) is charged with ensuring we have a safe food supply. However, they do not perform any testing on foods or *cosmetics* containing nanoparticles; they don't even know which foods and cosmetics contain them. Many people are concerned that they have made their way into our food system but have never been tested to see if they are harmful when ingested or placed in contact with our bodies. This is a case where technology has gotten ahead of the safety process, and these situations often end badly.

We also ingest nanoparticles that were never intended to be eaten. For example, zinc oxide is used in many sunscreens. When the sunscreen makes its way into wastewater—after a shower—the zinc oxide nanoparticles are so small they pass through the normal *sewage treatment facility* filters and end up in the solid waste collected at the end of the treatment process. This solid waste, called *biosolids*, is then collected and often sold as *fertilizer*. The nanoparticles from the sunscreen then makes its way into plants such as soybeans, and you end up with zinc oxide not only on your skin but on your dinner plate.

Nanoparticles also enter the food supply from food packaging when they migrate into the food or drinks. Examples include: nano-clay used to prevent gases from exiting plastic bottles, nano-aluminum used to keep aluminum foil from sticking to food, and nano-silver used to prevent microbe growth on some containers.

At the moment, the FDA has left the labeling of foods and other products containing nanoparticles up to the manufacturers. Needless to say, very few of them place it on their labels when they are present. [*nanotechnology; nanoparticles and human health; nanoparticles in cosmetics and personal care products; nanopesticides*]

nanopesticides

Numerous *nanoparticles* are in use, and one of the newer applications is nanopesticides. These are *pesticides* encapsulated in tiny containers small enough to enter the cells of an organism. For example, certain nanopesticides can be absorbed by a plant to kill a pest that eats the plant. The pesticide can target specific pests, and it won't wash off in the rain. The possibilities are great, but so too are the risks, because so little is known about how nanoparticles affect our health and the health of *ecosystems*. [*nanotechnology; nanoparticles and human health; nanoparticles in cosmetics and personal care products; nanoparticles in food*]

nanotechnology

One of the latest technologies to revolutionize the manufacturing process. As the name implies, it pertains to the size of the products the technology produces, which are called nanomaterials. These tiny synthetic substances are typically less than 100 nanometers in size and are created and used for a variety of purposes. (A human hair is 100,000 nanometers.) Compounds of various metals have been created, using nanomaterials to make the metals stronger or more long-lasting. Hundreds of nanomaterials are already in consumer products, including packaging, cosmetics, clothing, health and beauty products—even food. They include *nanopesticides*, nano-carbon tubes, nano-food additives, nanoceuticals, and others.

Health and environmental critics are concerned that we are producing new substances, that never before existed, without proper testing for long-term harm to people and the planet. Concerns are that as these nanomaterials break down, they will enter food chains or—by contact or absorption—enter organisms and cause unknown harm. [*nanoparticles; nanoparticles and human health; nanoparticles in cosmetics and personal care products; nanoparticles in food*]

natality

The birth rate of a *population*. More specifically, it refers to the "crude birth rate." This is calculated by taking the number of births in a specific period of time, such as one calendar year, dividing it by the estimated population at the midpoint of that period, and multiplying by 1000. The resulting figure is expressed as births per 1000 for a certain period of time.

For example, if a population of 2000 individuals has ten births in one year, natality for that population is 5 per thousand per year (10 divided by 2000 and multiplied by 1000). Natality figures can be generated for all types of organisms, including humans. [*population dynamics; population growth, limits of human; population explosion, human*]

National Audubon Society

Originally started by a group of ornithologists, in 1886 in New York. It was originally created to stop *poaching* and large-scale hunting of birds and other wildlife. One of its first projects was a crusade to stop the wildly popular fashion of using wild bird feathers on women's hats, a style that caused the destruction of millions of birds each year. The society was joined by nature writers such as *John Burroughs* and by President *Theodore Roosevelt,* both of whom abhorred the practice.

It was incorporated and officially became an *NGO* in 1905. It works to protect *wildlife* and its *habitats*, with special emphasis placed on birds. It is also active in *land management* and pollution issues. It does so primarily through education, research, and political action. Its namesake is *John James Audubon*. The Society has more than 550,000 members in almost 500 chapters nationwide, who are involved in numerous local conservation issues. It maintains more than 100 sanctuaries and *nature centers* in the United States. Its <u>Audubon Magazine</u> is world renowned for its nature photography. [*NGO; environmental education; environmentalist; migration; flyways; Wild Bird Conservation Act; Migratory Bird Hunting and Conservation Stamp Act*] {audubon.org}

National Environmental Policy Act (NEPA)

Passed in 1969, NEPA's primary purpose is "to promote efforts which will prevent or eliminate damage to the environment and *biosphere* and stimulate the health and welfare of man." In a nutshell, the act mandates all federal agencies to assess the potential environmental impact before beginning a project. This is typically done by submitting an *environmental impact statement* (EIS). As part of this process, citizens get to voice their opinions, including opposition to any federal project on environmental grounds. The federal agencies involved must consider alternatives and improvements to the proposal based on public participation.

NEPA and the requirement for an EIS has many critics, including those who state it slows progress. Many attempts have occurred in Congress to water down this act, but it remains intact for now. [*cost/benefit analysis; eco-services; economics and sustainable development*]

National Environmental Scorecard

See the *League of Conservation Voters*.

National Estuary Program

This program was created by an amendment to the *Clean Water Act*. It recognizes the inherent value of these important *ecosystems*—something not realized in earlier years. The program's goals are to identify, restore, and protect important *estuaries* in the United States. It is administered by the *Environmental Protection Agency*. State governors submit nominations to the Environmental Protection Agency (EPA), which administers the program and determines who is covered. [*estuary and coastal wetlands destruction; eco-services; wetlands; coastal wetlands; coastal flooding; Everglades; marsh*]

National Marine Fisheries Service

This service supports the conservation and management of living *marine* resources, including protecting *habitats* and the populations within. This service has been around a long time; it was originally created by President Ulysses S. Grant in 1871. It oversaw the building of the first federal fish research lab—the Woods Hole Oceanographic Institute of Massachusetts. The service is now under the auspices of the *National Oceanographic and Atmospheric Administration*. [*marine ecosystems; Marine Mammal Protection Act; marine protected areas; marine reserves; overfishing; fishing, commercial; oceanic zone ecosystems; bycatch; fish populations and food; fisheries and seabirds; Magnuson Fishery Conservation Management Act*]

National Marine Sanctuaries Act

This act created the National Marine Sanctuary System, which is designed to protect *oceanic zone ecosystems* in the same fashion that *national parks* and *forests* were protected in the late 1800s and early 1900s. This act is administered by the *National Oceanic and Atmospheric Administration (NOAA)* and protects many *marine ecosystems*, such as the *coral reefs* off the Florida coast, from *water pollution* and *urban development*. Fourteen sanctuaries are now protected.

This act has been joined by international efforts to create a series of worldwide *marine protected areas (MPA)* and *marine reserves*. NOAA works with many international agencies to advance these efforts. [*National Marine Fisheries Service*]

National Oceanographic and Atmospheric Administration (NOAA)

NOAA was created in 1970 by President *Richard Nixon* and resides within the Department of Commerce. Its purpose is to coordinate United States policy dealing with *oceans* and the *atmosphere*, with the exception of *air pollution,* which falls under the *Environmental Protection Agency*. Most people know this office for its National Weather Service, but it is also responsible for the *Coastal Zone Management Act*, the *Marine Mammal Protection Act*, and parts of the *Endangered Species Act* and the *Ocean Dumping Act*. NOAA has an outstanding educational program and does a wonderful job of communicating with the public on matters of environmental and scientific importance. Its website is a wealth of information and easily searched and navigated. [*marine ecosystems; oceanic zone ecosystems; ocean pollution*]

National Park and Wilderness Preservation System (U.S.)

When President Johnson signed the *Wilderness Act* in 1964, it established about nine million acres of federally protected lands. Today the more than 110 million acres of protected wilderness are in more than 500 areas, with the vast majority found in Alaska (more than 50 million acres). These areas are designated as *wilderness*, contain no roadways, and were protected to remain as such by law. However, these protections have been challenged in the courts, and some are not as wild as they once were.

Wilderness land includes *national forests*, managed by the *U.S. Forest Service*; *national parks*, managed by the *U.S. National Park Service*; wildlife refuges, managed primarily by the *U.S. Fish and Wildlife Service*; and other lands, managed by the *Bureau of Land Management*. Wilderness areas make up only a small portion of all *public lands*. [*National Wildlife Refuge System*]

national parks (global)

Globally, this is a vague concept, but generally speaking, these are areas set aside and protected in some manner. The *International Union for the Conservation of Nature* (IUCN) defines a protected area as follows: "A protected area is a clearly defined geographical space, recognized, dedicated and managed, through legal or other effective means, to achieve the long term *conservation* of nature with associated *eco-services* and cultural values."

In 2009, the *United Nations* estimated that roughly 14 percent of the land on our planet was set aside as protected—called parks, wildlife refuges; wilderness areas, or other such names. Because many of these so-called protected areas have *dual-use*, only about seven percent are truly protected. (This does not take into account illegal *logging*, *mining*, and *poaching*, which are common, making the number significantly lower.) Many *conservation* biologists believe that at least 20 percent (and some think much more) of our planet needs to be protected to ensure the continuity of global *biodiversity*.

These numbers do not include *marine protected areas* and *marine reserves*. [*National Park Service; National Park and Wilderness Preservation System; National Wildlife Refuge System*]

National Park Service (U.S.)

A bureau within the *U.S. Department of Interior*, the National Park Service was established in 1916 to protect natural areas from development and provide Americans with areas for *outdoor recreation*. Educating their 275 million visitors per year has always been an important part of their role and they do it well. In recent years and throughout their 100-year existence, they have often faced budget restraints, depending on who is running the country at the time and their thoughts about nature. They manage 27,000 historic structures, 2461 historic landmarks, 582 natural landmarks, 401 *national parks*, and 49 national heritage areas.

The service was first created after the *Hetch Hetchy* valley was dammed within *Yosemite National Park* and the public demanded better protection of these new parks. At that time, no one agency was responsible for protecting the parks that had been established.

In its early years and up to the early 1960s, the service emphasized recreation—getting people to the parks. But as the parks became more and more popular, the service started placing more emphasis on keeping visitors happy while also preserving the integrity of the parks. [*National Park and Wilderness Preservation System; National Wildlife Refuge System; Blueways, National; national parks, global*]

National Petroleum Reserve – Alaska (NPR–A)

Previously called the Alaskan Western Reserve, this is a 23.5 million acre area of *public land* considered the largest tract of undeveloped land in the nation. It consists of mountains and coastline that hold the largest *wetland* in Alaska, numerous *migratory bird* species, and the largest U.S. caribou herds, among other natural wonders.

But as the name implies, the most significant factor about this land is that it holds over 600 million barrels of *oil* and estimates of over 17 trillion cubic feet of *natural gas*. Unlike its neighbor the *Arctic National Wildlife Reserve*, the NPR-A is managed by the *Bureau of Land Management,* which means at least some of this reserve can be used for oil and gas extraction.

Recently, dozens of hunting groups, the *National Wildlife Federation*, the *National Resources Defense Council,* and other environmental groups won court decisions to protect 11 million acres of the reserve from exploration. Oil companies are still allowed to drill in other areas of the reserve, but this decision protects some of the most important *habitats* for *wildlife*, including caribou, migratory birds, *polar bears*, seals, and *whales*. [national parks, global; *National Park Service; National Park and Wilderness Preservation System; National Wildlife Refuge System; national parks, global*]

National Priorities List

A list of the worst hazardous waste sites in the United States that are now abandoned and scheduled for cleanup by the *Superfund*.

National Wildlife Federation (NWF)

With 5.8 million members, this is the largest *conservation NGO*. Founded in 1936, it focuses on conservation education, promoting the appropriate use of our *natural resources*. The group believes the welfare of *wildlife* and that of humans are inseparable and that wildlife is the best indicator of the environmental quality of our planet. In addition to education, they work to influence conservation policy by legislative means.

Many children grew up reading *Ranger Rick*, one of their children's publications, and now as adults, maintain their own Certified Wildlife Habitats—a program that lets people establish their own wildlife sanctuaries. So far individuals have created more than 300,000 acres of these mini-reserves. The NWF publishes National Wildlife and International Wildlife magazines, which are well written, informative, and always a good read. {nwf.org}

National Wildlife Refuge System

President Theodore Roosevelt created the first *wildlife refuge* in 1903 to protect Florida's Pelican Island, where herons, egrets, pelicans, and many other bird populations were being decimated by hunting; mostly for their feathers. Today, however, it is a national network of federally owned sanctuaries for *wildlife*. It includes over 550 refuges, plus almost 40 *wetland* areas, across some 150 million acres located in every state and a few U.S. territories. Its current primary role is to protect *habitats* of *endangered species* and protect *migratory* waterfowl such as ducks and geese.

One of refuge system's first high-profile projects was protecting the whooping crane in the 1930s. Estimates at that time had the total number of these cranes at about 20. The Arkansas National Wildlife Refuge was created (in Texas) to protect the bird, which has since made a comeback although still on the *endangered species list* with fewer than 600 known to exist. [*National Park Service; National Park and Wilderness Preservation System; Arctic National Wildlife Reserve; national parks, global; flagship species*]

Native American environmentalism

Many people credit Native Americans as being the first *environmentalists* on the North American continent. Today they would be considered *biocentrists*—believing humans are members of life on Earth, not rulers of life on Earth. Their ability to live off the land foreshadowed what we now call *sustainability* and living within a small *ecological footprint*. [*"Land Ethic", The; primitivism; Biophilia hypothesis; Gaia Hypothesis; ethics, environmental; eco-psychology; ethnobotany; governments of the forest; utilitarian environmental ethic*]

native species

A species that naturally evolved in a specific area is considered native to that area. By contrast, species introduced by people into an area are considered *alien species* and possibly *invasive species*. [*indigenous*]

natural capital

A term coined and defined in <u>Natural Capitalism: Creating the Next Industrial Revolution</u>, by Paul Hawken, *Amory Lovins*, and L. Hunter Lovins (1999). This was the first book to fully articulate the problems with the currently accepted "industrial capitalism" and present an alternative new model. The book states that the current system "liquidates its capital and calls it income. It neglects to assign any value to the largest stocks of capital it employs—the natural resources and living system, as well as the social and cultural system that are the basis of human capital."

Natural capital has fostered significant improvements, such as the use of *eco-services* to assign value to natural resources, and new measures, such as the *Genuine Progress Indicator*, that include natural capital in their calculations.

The endgame of natural capital was nicely summed up by President Clinton when he said, "Natural capitalism basically proves beyond any argument that there are presently available technologies, and those just on the horizon, which will permit us to get richer by cleaning, not spoiling, the environment."

The concept has been advanced in numerous books and articles. <u>Climate Capitalism: Capitalism in the Age of Climate Change</u>, by L. Hunter Lovins and Boyd Cohen (2011) focuses on capitalism and the most pressing issue of the day, *climate change*. [*Beyond GDP; CERES; climate capitalism; companies, top ten green global; corporate social responsibility; corporate sustainability rankings; Sustainability Imperative; economics and sustainable development; green net national production; integrated bottom line*]

natural food coloring

Artificial food dyes are used to make foods more colorful. Research has led many to believe these artificial dyes pose health risks—so much so that artificial dyes have been banned in much of Europe, where natural food colorings are used as alternatives to artificial dyes. Natural food colorings come from plants, animals, and minerals. For example, the *Food and Drug Administration* has approved natural food colorings from beets, carrots, grapes, tomatoes, red cabbage, turmeric, paprika, *chlorophyll*, iron, and an *insect* called Coccus Cacti L. Many foods in the United States are sold with artificial food colorings, while the same products are sold in Europe with natural food coloring. [*organic farming and food; food eco-labels;*

organic food labels; sugar, sugar alternatives and the environment; apples, antibiotics in organic; organic food supermarkets; genetically modified organisms you routinely eat; Franken-salmon]

natural gas

Natural gas is a *fossil fuel*. It is believed to have been formed under conditions similar to *oil formation* but then continued to change into a lighter *hydrocarbon—methane gas* or what we commonly call natural gas. It can be found by itself or often together with crude oil deposits and sometimes accompanied by butane and propane. [*energy sources, historical; natural gas as an energy source; natural gas extraction; natural gas extraction, conventional; natural gas extraction, unconventional; natural gas loophole; fracking*]

natural gas as an energy source

Natural gas, a *fossil fuel*, supplies about 21 percent of the world's energy supply. In just the past few years—because of new *unconventional natural gas extraction* techniques—its importance on the world stage has increased. Estimates are that natural gas could provide 25 percent of the world energy mix by 2035.

Natural gas supplies are now expected to outlast oil. Only a few years ago, the known *reserves* of natural gas were expected to last about 60 years at current consumption levels. However, with these new unconventional extraction techniques, the supply could last roughly 240 years if the pace of use remains the same.

About 32 percent of U.S. electricity comes from natural gas, putting it on equal footing with *coal*. Only five years ago the United States was expected to become a natural gas importer—but today plans are being made to export *liquefied natural gas*. It is possible it will surpass *oil* as the most important fuel in the United States within 20 years. This sudden rush of U.S. natural gas has been called the *shale gale* because of vast new *reserves* found in shale formations such as the *Bakken* and *Marcellus shale beds*.

Currently, almost all of the U.S. demand for natural gas is for home heating but with its new abundance, it is finding uses in industry, such as being turned into *synthetic fertilizers* and *plastic* products. The possibility of a significant number of *natural gas vehicles* might be around the corner, as well. Because of recent legislation regarding *air pollution* and *climate change*, many coal-burning *electric power plants* are converting to natural gas.

Natural gas is cleaner than other fossil fuels and has higher heat content. It produces fewer pollutants and about half as much *carbon dioxide* (a *greenhouse gas*) compared with *coal*.

But there are concerns about its growing popularity. *Methane*—a potent greenhouse gas far more damaging than carbon dioxide—is the main component of natural gas. There are fears that methane is escaping during its extraction and that will contribute to climate change. This is currently being researched to determine if it happens and if so, how often and how much.

In the past, when natural gas was found during conventional oil exploration, it was simply released into the atmosphere. Because the gas is a powerful greenhouse gas, it was preferable to burn it off, called flaring, and that became the norm. Recent U.S. legislation makes it illegal to flare the gas and it must now be captured at the source.

Other issues associated with natural gas include the problem of transport. It can only be piped a short distance. When natural gas is too far from where it is needed, it is called "stranded gas" and must be liquefied for transport. Some people are concerned that shipping the gas is dangerous because it is highly flammable.

Even though phasing out coal use and replacing it with natural gas will reduce air pollution and carbon emissions, as it is already doing, many environmentalists are concerned that this increased use of a fossil fuel will slow the transition to *renewable energy sources* such as *solar power* and *wind power*. [*energy sources, historical; natural gas as an energy source; natural gas extraction; natural gas extraction, conventional; natural gas loophole; fracking*]

natural gas extraction

Conventional natural gas extraction methods have been used since the early 1800s, starting in both *China* and the United States. However, beginning in the late 1980s and early 1990s, new advanced technologies allowed companies to perform what is now called *unconventional natural gas extraction* methods, creating a boom in gas exploration and recovery. [*energy sources, historical; natural gas as an energy source; natural gas extraction, conventional; natural gas extraction, unconventional; natural gas loophole; fracking*]

natural gas extraction, conventional

Similar to crude *oil*, *natural gas* is brought to the surface by wells. Conventional extraction methods include primary recovery, which involves tapping and removing the gases by natural forces. Then, secondary recovery involves pumping air or water into the well to force the remaining gas from the deposit.

Natural gas is composed primarily of *methane* gas but also contains small amounts of butane and propane. The methane is separated, dried of moisture, cleaned of impurities, and pumped into pipelines for distribution short distances or stored. It can then be converted into *liquefied natural gas* (LNG) at very low temperatures, so it can be shipped longer distances or by tanker for export to other countries.

The butane and propane are usually separated and liquefied to form *liquefied petroleum gas* (LPG), which is stored in pressurized tanks and shipped to rural areas that don't have natural gas pipelines. Newer methods of natural gas extraction are considered to be unconventional methods. [*unconventional oil and gas; energy sources, historical; natural gas as an energy source; natural gas extraction; natural gas extraction, unconventional; natural gas loophole; fracking*]

natural gas extraction, unconventional

Unconventional gas extraction provides access to gas that could not be extracted by using *conventional gas extraction methods*. This includes shale gas, extracted by *fracking; coal-bed gas,* found in coal seams; and tight gas, which is found trapped in rock formations.

natural gas loophole

Some people are under the false impression that companies using the process called *fracking* to extract *natural gas* must adhere to the *Clean Water Act*—the important regulations that do a good job of protecting our water supply. If this were true, we could be far less concerned about fracking polluting our *groundwater*. However, this is not the case. Companies extracting natural gas, even with unconventional methods such as fracking, are exempted from any regulations put forth by the Clean Water Act.

This legal loophole exempts these companies from adhering not only to the Clean Water Act but also the *Safe Drinking Water Act*, the *Comprehensive Environmental Response Compensation and Liabilities Act*, and others. Many environmentalists view this loophole as a travesty of justice. [*natural gas as an energy source*]

natural gas vehicles (NGVs)

With the bonanza of *natural gas* in the United States has come renewed interest in producing and using vehicles that run on natural gas. Natural gas burns cleaner than regular gasoline and produces far less *carbon dioxide*, a *greenhouse gas*.

Today about 130,000 vehicles in the United States run on natural gas; but that is only about 1 percent of the total number of vehicles on the road. These vehicles are mostly in fleets, such as taxi companies, city busses, garbage trucks, and delivery services. Although few places exist to fill up with natural gas, fleet owners have their own filling stations. Some U.S. car makers are producing natural gas vehicles, but they are targeted toward these fleet sales. For example, AT&T plans to become the biggest owner of NGVs by ordering 8000 of them over the next five years.

Because natural gas costs about 20 percent less than gasoline at the moment, there is little doubt that many people would buy these vehicles if infrastructure supported it, such as natural gas stations. More than 115,000 gas stations exist in the United States, but only about 500 are natural gas stations. If these natural gas stations do begin to appear, they will have to provide two types of natural gas as most NGV *automobiles* use *compressed natural gas* (CNG), but large NGV trucks use *liquefied natural gas.*

Natural gas is inexpensive in the United States but costly elsewhere, so a move toward NGVs would likely only occur in the United States—if the prices stay low, which is not a sure thing.

One of the few other countries with abundant natural gas is Russia, where they are showing an interest in skipping the *hybrid vehicles* popular in the United States and going directly to NGVs. [*ammonia as a car fuel; automobile propulsion systems, alternative; autonomous automobiles; electric vehicles; greenest cars, top ten; vehicles, hydrogen fuel-cell; biofuels*]

natural history

The study of nature, including all natural objects, living and nonliving, and natural processes. It is typically used to express the natural history of a particular organism. For example, you might want to study the natural history of the *bald eagle*, potato *beetle*, or crabgrass. [*ecology; environment; environmental science*]

naturalist

To some, a naturalist is someone who studies nature. However, today it is used informally for anyone who simply enjoys nature; particularly being in nature, such as *walking*, hiking, or *bicycling* in the woods or fields and observing the flora and fauna of the area. Some find interest in particular types of *wildlife* or *habitats*, such as birders or spelunkers, (cavers). It can also include those who use their interests to capture nature in some way, such as artists or photographers.

Professional naturalists include those who educate others about nature. They are employed by the government as Park Rangers in *national parks* and in public and private *zoos*, museums, *aquariums*, gardens, and *nature centers*. [*environmentalist; environmental movement; environmental education; environmental literacy; primitivism; nature-deficit disorder*]

natural pesticides

Although most of the *pesticides* applied each year are synthetic chemical pesticides, a few are found in nature and do not need to be created in the lab. These are botanicals, which are derived from plants; mineral pesticides, which come directly from the *soil*; and microbial pesticides, which are organisms (or toxins they produce) that can control pests. (The botanical and microbial pesticides combined are called *biological pesticides*.)

Pyrethrin, a botanical extracted from chrysanthemums, was the original model for many synthetic *pyrethroid* pesticides. Nicotine, from the tobacco plant, and rotenone, from *legumes*, are also botanicals.

Boric acid, which contains boron, is a mineral pesticide. Sulfur and copper have been used as natural *fungicides*. Other natural methods include the use of smoke to repel insects and light oil sprayed on crops.

But the most commonly used natural pesticides now are microbes such as *Bt*, which is a type of *bacteria*. Bt was originally used as a *biological control methods* on many crops. The advantage of Bt and other biological control methods is that they target a specific pest and do not kill or harm other forms of life as do most synthetic pesticides. Bt is now used to create many varieties of *genetically modified organisms* and is routinely used for many of our crops. [*nanopesticides; pesticide dangers; pesticide regulations*]

natural resources

Substances, structures, or processes that are used by people but cannot be created by them. For example, the sun, land, and oceans are natural resources and their uses are obvious. Iron ore is a natural resource because we use it to make steel, and the *Grand Canyon* is a natural resource because it is a natural wonder and a popular tourist attraction.

Natural resources can be either *renewable* or *nonrenewable*. Renewable resources include the sun, *soil*, plants, and animal life, because they all naturally perpetuate themselves. Some of these renewable resources, such as the sun, are used as *renewable energy* sources. Nonrenewable resources are those that do not perpetuate themselves; if continually used by humans, they will someday be "used up." For example, the supply of a mineral such as iron ore is finite and will become exhausted someday. Most of the world's energy supply comes from *fossil fuels,* which are *nonrenewable energy* sources and will be gone at some point in the future.

All natural resources can be categorized into four major components of the *biosphere:* the *lithosphere* (the solid Earth), the *atmosphere* (the air), the *hydrosphere* (all of the waters), and the *biota* (all forms of life).

Using natural resources takes an environmental toll on our planet by causing some form of *pollution* or damage. *Strip-mining* for *coal* destroys the land, and burning it for electricity causes *air pollution*. Using our natural resources must be balanced between our needs and the impact on our *environment*. [*environmental movement; environmentalist; economy vs. environment; Spaceship Earth*]

Natural Resources Conservation Service

This is an agency within the *Department of Agriculture* whose mission is to help private landholders conserve, improve, and sustain *natural resources* and the *environment*. It was formed (under a different name) after the *Dust Bowl* to prevent such catastrophes from happening again. It is primarily concerned about the *soil—topsoil,* specifically—and *soil erosion*.

Natural Resources Defense Council (NRDC)

In 1970, a few Yale University law students created the *NGO*, NRDC to provide legal advice to *environmentalists* and to litigate environmental cases, if needed. They have grown into one of the preeminent environmental *NGOs* with almost 1.5 million members and 350 lawyers, scientists, and various other professionals. The New York Times calls them "one of the nation's most powerful environmental groups" and The National Journal says of them, "A credible and forceful advocate for stringent environmental protection."

The NRDC accomplishes its goals primarily through research and legislation. In the past, they focused on *acid rain* and fought for proper enforcement of the *Clean Air Act*. Their "Mothers and Others for *Pesticide* Limits" campaign helped eliminate the use of *alar* on apples. They were one of the first to push for *forest management* reform and fought federal *coal*-leasing and *oil*-drilling policies that threatened wildlife.

Today they focus on *climate change*, environmental health, *toxic* substances, *marine pollution,* and *China's* impact on our planet and publish many informative books on environmental topics. [*law and the environment; environmental justice; SLAPP; OnEarth magazine*] {nrdc.org}

natural selection

The process—originally proposed by *Charles Darwin* in his classic book On the Origin of Species—where a species gradually adapts to its *environment*. Natural selection occurs because individuals of a particular species, although all similar, have some genetic variability due to mutations. This genetic variability sometimes makes these individuals better suited to their environment—so they are more likely to survive. And sometimes it makes them less suited—so they are more likely to die.

As generations pass, those with the best genetic makeup are naturally selected for survival, carrying these genes to future generations. Over long periods of time, the species evolves, becoming better suited for its environment. If some of the individuals become separated from others, entirely new species can form—a process called *speciation.*

Comprehending processes that normally take thousands, millions, or even hundreds of millions of years to occur is difficult, but here are two short-term examples of natural selection. When you hear stories about *super bugs* that refer to disease-causing *bacteria* that have become resistant to *antibiotics*, you are hearing about natural selection. That's how these "bugs" become resistant to the antibiotics—they evolve via natural selection. When you hear about *insect* pests becoming resistant to *pesticides*, you are hearing about natural selection. These bacteria and insects have evolved and are now capable of surviving substances that previously killed individuals of the same species. [*genetically modified organisms—resistance in; Bt; glyphosate, (GMO and super weeds); geological time scales*]

nature centers

Many urban centers have so little of the natural *environment* left that people, especially children, have little or no knowledge about the wilds of nature. Perhaps for this reason, nature centers have become popular in or near many big cities. A cross between a park and a *zoo*, these educational facilities teach people about natural environments. Nature centers are often funded by state or local governments or nonprofit organizations. [*zoos, captive breeding programs in; Zoo, Frozen; zoos, future; touch-pools;* aquariums; *National Park Service; urbanization and urban growth; straw-bale gardening; gardens, rain; urban gardens; microgreens and urban gardening; urban open space; environmental education*]

Nature Conservancy

The Nature Conservancy is the leader in the purest form of the *environmentalism:* preserving and protecting unspoiled lands. Established in 1951, they identify, acquire, and manage unique natural environments so they cannot be destroyed by human intervention. They have helped protect more than 119 million acres and thousands of miles of rivers in 50 U.S. states and 35 other countries. They have purchased and now manage more than 1600 sanctuaries, most of which are home to *endangered species.* They have a nationwide network of local and state chapters, with more than a million members. Many people prefer this group's straightforward approach: the best way to save land is to purchase and protect it. Unlike some environmental groups, the conservancy gets the job done in a nonconfrontational manner. [*land acquisition, wilderness*] {nature.org}

nature-deficit disorder

A nonmedical condition—or possibly what might be better called a situation—where children are rarely out of doors and therefore have little understanding and little interest in nature and the world around them. Many might call this nothing more than unfortunate and some might consider it normal. Woody Allen is believed to have once said "The outdoors is something that must be passed through to get from one building to the next."

However, the basic premise about this disorder is a serious concern. People who have little or no interest in nature and their planet are probably not going to be concerned about environmental degradation and issues that are affecting them and their children.

The more people show interest in the environment, the more likely they will become involved in resolving these issues. As the old adage goes, "If you are not part of the solution, you are part of the problem." Those with little interest about their environment are more likely to be part of the problem. [*Biophilia Hypothesis; gaia hypothesis; primitivism; outdoor recreation; zoos; green fatigue; environmental literacy; environmental education; news literacy; Student Conservation Association*]

NBA arenas, green

Five NBA arenas now have *solar panels* generating electricity for the facility. The Atlanta Hawks arena has even been awarded *LEED* certification. [*baseball stadiums, green; NFL stadiums, green; universities, top ten green; cities, ten best American; green gym movement, green; bicmaqiunas*]

near earth objects (NEOs)

Asteroids and comets that pass our planet and have the possibility of becoming a threat are considered near earth objects. A few programs attempt to identify and track NEOs, the intent being to forewarn citizens of *Spaceship Earth* of impending threats with the hope that we could do something about them—although exactly what could be done is unknown at the moment.

The following facilities track NEOs. The Catalina Sky Survey based in Arizona and *Australia* has identified about 600 NEOs since the mid-2000s. The first of four new Panoramic Survey Telescope and Rapid Response System telescopes in Hawaii is up and running in search of NEOs. The Asteroid Terrestrial-Impact Last Alert System will open in 2015 with the intent of finding 50-meter-across "city-killers" and warning everyone a week ahead of time. [*extinction and extraterrestrial impact; Blue Marble photographs, original and updated; outer space; space debris; space-debris cleanup*]

Negative Population Growth

A *NGO* that, "educates the American public and political leaders about the devastating effects of *overpopulation* on our environment, resources and standard of living. We believe that our nation is already vastly overpopulated in terms of the long-range *carrying capacity* of its resources and *environment.*" [*population explosion, human; Population Reference Bureau; Population Connection*] {npg.org}

negawatts

Short for "negative watts," meaning electricity that has been saved as opposed to used or wasted. Coined at the *Rocky Mountain Institute*, the word helps define and encourage conserving electricity. For example, how much money can a company save by finding negawatts in their buildings and factories? [*energy, ways to save residential; energy-saving apps and programs; energy, A very brief history of U.S.; ecological footprint; soft energy path; Lovins, Amory*]

nematodes

A large group of organisms that few people know about, think about, or care about—even though they are of utmost importance in most *ecosystems*. Nematodes—also called roundworms—are microscopic unsegmented worms of the phylum Nemata. More than 20,000 species exist, and they are one of the most numerous multicellular organism on Earth. Many are *parasites* of plants, animals, and especially *insects*. Others are free living, feeding on *bacteria* or *fungi* or are *predators* of other nematodes.

Nematodes are close to indestructible; those onboard the ill-fated space shuttle Columbia survived the 2003 crash even though the ship disintegrated. They have even been found a mile deep in bedrock, making some of them *extremophiles*.

neonicotinoid pesticide and honeybees

This *pesticide* is considered by recent research to be the probable cause of *colony collapse disorder,* which has been decimating *honeybee* populations. It attacks the insect's nervous system, paralyzing it and usually resulting in death. This *persistent pesticide* remains in the plant and in the environment for long periods of time. The toxic substance has been banned in some countries; others are considering doing so but are waiting for more conclusive research. At the moment, it remains legal in the United States. [*honeybees, pollination and agriculture*]

neritic zone (coastal zone)

Marine ecosystems can be divided into those found in the open ocean (*oceanic zone*) and those found along the coast, called the neritic zone. The neritic zone includes all the waters above the *continental shelf*, beginning at the shore and extending out to sea several hundred miles in some areas. The immediate coastal area is continually affected by the tides and contains some of the most productive (*net primary productivity*) *environments* on Earth.

These coastal waters are home to many *fish*, clams, *oysters*, crabs, sponges, anemones, and *jellyfish*, among others. In much of this water, the sunlight penetrates to the bottom, allowing plants to attach to the substrate and providing shelter for many other organisms. The types of ecosystems found in these areas vary widely because of the variations in the types of shorelines. The ecosystems of rocky shores, for example, are very different from the ecosystems of sandy shores.

The coastal zone has many unique, highly productive *habitats,* including *estuaries*, coastal *wetlands*, and *coral reefs*. Some areas in this zone exhibit *upwelling,* which provides a habitat for unique ecosystems. [*estuary and coastal wetlands destruction*]

net primary production (NPP)

NPP is used to compare the plant productivity of different *ecosystems*. NPP is the rate at which plants build *organic* molecules containing usable energy (via *photosynthesis*), minus the amount of energy used up by the plants themselves in *cellular respiration,* for a given area, at a given time. It is usually measured in dry grams per square meter per year ($g/m^2/yr$). *Wetlands* and *tropical rainforests* have the highest NPP levels, ranging from 1000 to 3500, while the lowest are found in *tundra* (10 to 400) and *deserts* (0 to 250). [*biomes; anthromes*]

news literacy

A new term of utmost importance to everyone, but especially those interested in *environmental literacy*. News literacy is the "use of critical thinking skills to judge the reliability and credibility of news reports and news sources." Today, finding information is as easy as pressing a few keystrokes in a search engine. The hard part is determining what is valuable information and what is not. And most of it is not. Literate individuals must know the difference between fact and *opinion*, between *science* and *pseudoscience*, between green news and *greenwash*.

The concept was established, in 2009 by Howard Schneider, who left a career at "Newsday" where he was editor, to become Dean of the Stony Brook State University School of Journalism in New York. News literacy started out as one course and evolved into an entire program. In addition to a news literacy curriculum, the school hosts annual conferences to teach others how to teach news literacy. Their mission is to train the next generation of "citizen news consumers"—and that includes everyone.

The only question is why this form of literacy is not taught everywhere. Until it is, the hucksters, conspiracy theorists, deniers of all sorts, and pseudoscientists will have too strong a voice with too large an audience. [*magazines, eco-; blogs, green; climate change deniers; Agenda 21 conspiracy*] {ens-newswire.com and environlink.org and sej.org}

News Service, Environment (ENS)

The Environment News Service is an international news agency dedicated to gathering and disseminating worldwide environmental news. Their news is delivered to online services, magazines, newspapers, and organizations. They tap a resource pool of thousands of experts in environmental fields to help find, decipher, and explain the news. It was founded in 1990 and is privately owned. You can use their website to locate environmental news categorized by topics such as air/climate, energy, wildlife, government/politics, oceans, and others. [*news literacy; blogs, green; magazines, eco-; PIRG, U.S.*] {ens-newswire.com}

NFL stadiums, green

The Philadelphia Eagles plan to generate 100 percent of the electric needs at their stadium by using a combination of 11,000 *solar panels* and 14 *wind turbines*. [*baseball stadiums, green; NBA stadiums, green; universities, top ten green; cities, ten best American; green gym movement, green; Facebook's carbon footprint; bicmaqiunas*]

NGOs (nongovernmental organizations)

NGOs are organizations that are neither governmental nor for profit. They include a vast array of organizations, including numerous environmental advocacy groups. Some are small grassroots groups working at the community level, others are regional or national, and many of the best known are international. Most are membership based and stress volunteerism. The *United Nations* works closely with many NGOs to help establish communications on a global scale.

Historically, NGOs have played an important role in environmentalism. During the 1960s and 1970s, NGOs in the United States and parts of Europe helped stoke the fires of the *environmental movement* that spread worldwide. NGOs are common in democratic societies but scarce in more controlled societies such as *China* because they are limited in their capabilities, and they are nonexistent in communist countries.

NGO is the common acronym used, but recently people have started using ENGOs specifically for environmental groups. [*Arbor Day Foundation; Audubon Society; Boone and Crockett Club; Center for Health, Environment, and Justice; Center for Science in the Public Interest; Charity Navigator; Citizens Clearing House for Hazardous Waste; Coastal*

Society, The; Desert Protective Council Inc.; Ducks Unlimited; Earth Communications Office; Earth First!; Earth Island Institute; EarthWorks; Environmental Defense Fund; Environmental Working Group; Friends of Earth; Greenpeace; HEAL; Izaak Walton League of America; League of Conservation Voters; Lighthawk, the Environmental Air Force; National Audubon Society; National Wildflower Research Center; National Wildlife Federation; Natural Resources Defense Council; Nature Conservancy; Outward Bound USA; PIRG, U.S.; Rocky Mountain Institute; Sea Shepherd Conservation Society; Sierra Club; Student Conservation Association; Student Environmental Action Coalition; Wilderness Society; World Wildlife Fund; Worldwatch Institute; Xerces Society; green parties, political; greens, the; Group of 10]

niche, ecological

Every *species* is unique from every other species, each placing distinctive demands on an *environment* and contributing in its own way to that environment. This specific role in the environment is called an organism's ecological niche. A niche is different from the organism's *habitat*, because niche is concerned with the impact of the organism on the environment instead of the environment's impact on the organism, as with habitat. You can think of a habitat as where a species lives and its niche as its job within that habitat. [*acoustic niche; apex consumers and ecosystems; biosphere, shadow; high-altitude ecosystems; hydrothermal ocean vent ecosystems; coral reef; cold water; forams; keystone species; hummingbirds and flight; oysters and climate change*]

night soil – biosolid

Night soil refers to using human manure as an *organic fertilizer*. It has been used for thousands of years and is still used in many *less developed countries*. Today it is making a comeback in developed nations as well, including the United States. New *sewage treatment facilities* are converting waste materials, including human wastes, into safe *compost* called biosolids for use as a *organic fertilizer*. [*biosolids and wastewater; toilets-to-tap; composting, large-scale; San Francisco and compostable waste*]

NIMBY syndrome

NIMBY is an acronym for "not in my backyard." It refers to local opposition to new projects people often don't like, such as *landfills*, *incineration* plants, and *hazardous waste disposal* sites. When people speak about running out of room for waste, such as landfills, they really mean running out of places where people will allow them to be built. Now that people realize the dangers inherent in many of these facilities, they take action to prevent them. Not all of the projects opposed by NIMBY activists are universally disliked by *environmentalists*. For example, a great deal of NIMBY activity is against *wind farms*, but most people consider *wind power* to be environmentally friendly. Another great example of NIMBY along with its opposite, YIMBY FAP (yes in my backyard, for a price) can be seen with *fracking*.

NIMBY is also viewed as an aspect of environmental injustice, because people with the most power tend to be far more successful in their NIMBY demands than those without power. [*environmental justice; ethics, environmental; environmental literacy; environmental education; environmentalist; precautionary principle*]

nitrogen compounds

One of the five primary pollutants that contribute to *air pollution*. The most common of these compounds are *nitrogen oxide* and nitrogen dioxide. As with other primary pollutants, nitrogen compounds come primarily from *automobiles* and *electric power plants*. These compounds play a major role in the production of *secondary air pollutants* that create *photochemical smog*. They also contribute to the development of *acid rain*. (Do not confuse nitrogen oxide with *nitrous oxide*.)

nitrogen cycle

Nitrogen is important to all organisms. It is a primary component of proteins and DNA. Even though 78 percent of the *atmosphere* is nitrogen, plants are incapable of using it in this "free" form. Most nitrogen becomes available to green plants (and therefore to all life) thanks to the process of *nitrogen fixation* that takes free nitrogen from the air and converts (fixes) it into *organic* nitrogen that green plants can absorb through their roots and incorporate into their tissues. This nitrogen is then passed through *food chains* to animals.

When plants and animals die, the organic nitrogen found in their bodies also becomes available to green plants, but first it must be broken down into a usable form. Certain types of *bacteria* break down organic molecules containing nitrogen into a form that can be absorbed by the roots of green plants.

For this cycle to be complete, organic nitrogen must have a way to return to the atmosphere as free nitrogen. Other types of bacteria, called denitrifying bacteria, can convert organic nitrogen in the soil directly into free nitrogen to begin the cycle over again.

Finally, lightning also plays a minor role by combining nitrogen in the atmosphere with oxygen to form a usable form of nitrogen, which falls to the ground with precipitation and becomes available to plants. [*biogeochemical cycles; biogeochemical cycles, human intervention in; planetary boundaries*]

nitrogen fixation

Even though 78 percent of the air we breathe is nitrogen, organisms cannot use it in this form. Nitrogen in the *atmosphere* is called free nitrogen, but plants require fixed nitrogen, meaning it has been incorporated into organic molecules such as ammonia or nitrates. This occurs during the process of nitrogen fixation.

This process occurs because of a *symbiotic relationship* that exists between *bacteria* that live in the roots of plants such as beans, peas, and clover (*legumes*). The bacteria produce swellings or nodules in the plant's root. The bacteria in the nodule takes nitrogen from the soil (actually air that permeates the soil) and incorporates it into its own cells. When the bacteria in the nodules die, the organic nitrogen becomes available to the plants. When these plants are eaten, the fixed nitrogen is passed along *food chains* to animals.

There are also nitrogen-fixing bacteria that live freely in the soil or water, and some trees and grasses have a symbiotic relationship with nitrogen-fixing *fungi*. [*nitrogen cycle*]

nitrous oxides

Nitrous oxide gas is produced from the breakdown of *fertilizers* and *livestock wastes* and from burning *fossil fuels* and other forms of *biomass* such as wood. (Do not confused this with the *nitrogen compound*, nitrogen oxide.) Nitrous oxide is a *greenhouse gas* and contributes to *global warming*. About six percent of global warming is attributed to increased concentrations of this gas in the *atmosphere*.

Nixon, Richard M.

(1913–1995) Most people do not think of our 37th president (1969–1974) as an environmentalist. However, many of the actions President Nixon took in office during his presidency were monumental to the *environmental movement*: The *Environmental Protection Agency* was established, the *National Environmental Policy Act* was signed, the *Clean Air Act* was passed, and in 1971 Nixon designated April 18⁻24 as Earth Week, providing the backdrop for the first *Earth Day*.

No Dirty Gold campaign

See *gold mining*.

noise pollution and control

Any noise that annoys or harms individuals is considered noise pollution. At certain levels, noise can pose a health risk. Noise pollution is measured in decibels (db). To take into account the pitch of the sound, dbA is used to indicate the A scale.

The following examples indicate when noise pollution becomes a problem: a whisper is about 20 dbA; normal conversation, 50 dbA; a vacuum cleaner, 70 dbA; a garbage disposal or nearby train can cause slight hearing damage at 80 dbA; a diesel truck or food blender for extended periods of time can cause permanent hearing damage at 90 dbA; a chain saw or thunderclap causes pain at 120 dbA; and a jet taking off less than 100 feet away can rupture an eardrum at 150 dbA.

In addition to immediate ear damage, noise pollution has been blamed for headaches and many other ailments. The Noise Control Act of 1972 was the first federal attempt to control noise pollution in the United States. In 1987 this act was amended with the Quiet Communities Act, which allows states to control noise at the local level. States and local governments have enacted many of their own laws governing noise pollution. [*visual pollution; malodor pollution; aesthetic pollution; water pollution; air pollution*]

no net loss

This phrase is used by local, state, or federal agencies regarding the loss of environmentally sensitive lands. It is often used in the following context: If a human activity—such as building a mall—destroys a natural resource such as a *wetland*, an artificial replacement will be created elsewhere to replace it. The idea then is that there is no net loss of environmental productivity. The thought that artificially created environments are equal to natural ones is viewed with great skepticism by most environmentalists. [*eco-services; habitat loss; biodiversity, loss of; Hetch Hetchy; Biosphere 2; wilderness; Wilderness Act; Nature Conservancy; Rigs-to-Reefs Program*]

Non-GMO Project Verified label

See *food eco-labels* and *GMO*.

non-point source pollution

See *pointless pollution*.

nonrenewable energy

For energy, *more developed countries* depend primarily on *fossil fuels*, including *coal*, *oil*, and *natural gas*. These fuels are considered nonrenewable because they are not being replenished and will become depleted over time. *Renewable energy* sources, however, are considered to be available for an infinite length of time, because the supply is continually replenished.

nonstick cookware

The material found on nonstick cookware typically contains small amounts of PTFE (polytetrafluoroethylene, commonly known as Teflon). When heated, it can degrade, releasing a gas called PFOA (perfluorooctanoic acid), which is considered a likely human carcinogen. The people most at risk are workers where this cookware is manufactured. The United States is phasing out all use of PTFE by 2015. There are other nonstick alternatives labeled to be PFOA free. (Alternatively, it might say PTFE free.) [*BPA; BPAF; chemical body cleanse; hormone disruptors, endocrine; pesticide residues on food; beryllium in computers; POPs, dirty dozen; fruit waxing; outgassing; PERC; Safe Chemicals Act; silent killer; THM; VOCs; toxic cocktail; body burden; couches and indoor pollutants; HEAL*]

nontarget organism

Most people believe *pesticides* kill a specific organism—the pest. Any organism affected by a pesticide that is not the targeted pest is called a nontarget organism. Most pesticides kill more nontargeted organisms than pests, so some people prefer to use the term *biocide* instead of pesticide, because this refers to killing many forms of life. [*pesticide dangers*]

noosphere

That portion of the *biosphere* that is affected by human influence. [*environmental science; extremophiles*]

northern spotted owl

The northern spotted owl became a symbol of the battle between the *environment* and the economy. In June of 1990, the northern spotted owl was placed on the *threatened species* list under the *Endangered Species Act*. The bird lives in the *old-growth forests* of the Pacific Northwest, and—like most *predatory* birds—requires large expanses of land to hunt for food. But less than 10 percent of these forests remain and most of it is on *Forest Service* managed lands where *logging* is allowed.

The only way to protect the owl is to protect these forest *habitats* from logging. Saving this bird from possible *extinction* would require protecting and thereby preventing further logging in these forests. The logging industry believes that protecting these lands would cost thousands of logging jobs. Opponents believe this number simply represents a continuation of an already ongoing decline in the number of logging jobs. The number of acres of old-growth forest set aside to protect the owl's habitat is constantly challenged in the courts. Whether the owl survives or not remains to be seen. [*economy vs. environment; economics and sustainable development; Beyond GDP*]

no-take zones, marine

Properly called marine reserves, these are areas where no marine life of any sort can be removed. The primary goal is to allow populations of fish and other marine life to rebound from *overfishing*. No-take zones are a subset of *marine protected areas*.

no-till agriculture

Planting crops without turning over the soil as is normally done in *conventional agriculture*. This dramatically reduces *soil erosion*. [*conservation agriculture*]

nuclear fission

The process of splitting an atom to release energy. This energy can be released for destruction, as with nuclear bombs, or it can be controlled and harnessed as an energy source called *nuclear power*.

The nucleus of an atom is composed of neutron and proton particles bound together by energy. In most elements, the nucleus remains stable, meaning the particles and the energy contained within stay bound in the nucleus infinitely. Some elements, such as uranium and plutonium, are *radioactive*, meaning the nucleus is unstable and particles are released—a process called decomposition. During decomposition, the energy that binds the particles together is released along with the particles. In its simplest sense, this is nuclear power. The decomposition of nuclei can be harnessed to produce useful power in *nuclear reactors* to generate electricity.

nuclear fuel cycle

A *nuclear reactor* uses *radioactive* materials (usually uranium) that must be mined, milled, enriched, fabricated, used, and finally, disposed of. Low-grade uranium ore is first mined and then goes through a milling process that involves crushing and treating it to concentrate the uranium. Once milled, the mixture is called *yellowcake*. The crushed uranium ore that remains is called *mining tailings*.

Yellowcake then goes through an enrichment process to increase the amount of radioactivity so it can sustain a chain reaction. Once enriched, yellowcake goes through a fabrication process and turned into pellets to fill fuel rods about 4 meters (13 feet) in length. The fuel rods are placed within the reactor core. The energy released by *nuclear fission* heats water that turns turbines to generate electricity in a nuclear power plant. After three to four years, the fuel rods can no longer support the chain reaction (they are considered spent) and must be removed from the reactor core.

One of two things happens with the spent fuel rods. Some countries, such as France, reprocess these rods at great expense to make them usable again; others dispose of them in some way. *Nuclear waste disposal* has become a problem confronting every nation using nuclear power. [*nuclear power; nuclear waste dilemma, the global*]

nuclear fuel cycle hazards

Each step in the *nuclear fuel cycle* has inherent dangers. Uranium miners are exposed to low-level radiation from the ore. *Chronic* exposure to this radiation is believed to increase the miners' likelihood of getting certain diseases, such as lung cancer. During the milling process the ore is crushed, leaving low-level radioactive tailings that can harm organisms and *leach* radioactivity into the groundwater.

During enrichment and fabrication, individuals must be protected from exposure to the high-level radiation released by the fuel. Once in the reactor, the intense levels of radiation must be contained. Radiation leaks can be catastrophic as seen in the *Chernobyl* and *Fukushima* disasters.

The biggest problem associated with nuclear power is disposing of the used but still highly radioactive fuel rods. *Nuclear waste disposal* technology is just beginning to be developed. In addition, every step of this cycle involves transporting the radioactive material from one location to another, with the danger of an accidental release of radioactivity. [*nuclear power*]

nuclear fusion

Conventional *nuclear power* plants use *nuclear fission*—the splitting of atoms—to release energy. However, when nuclei are combined (fused), energy is released as well. The sun is powered by nuclear fusion. The process involves combining two hydrogen nuclei to form helium. Controlling this process has been elusive to scientists who have been studying it for decades. Nuclear fusion might someday supply the world with all the energy it needs, but this will not happen for decades to come, if ever.

nuclear power

Nuclear power is an *alternative energy* source that became popular in the 1970s and proliferated in the early 1980s. Today the United States has 104 nuclear power plants providing about 20 percent of our energy needs. Predictions were that by 2010, 40 percent of the U.S. energy supply would come from nuclear plants. However, after the *Fukushima nuclear disaster*, the only two reactors on the drawing board were scrapped when financing dried up. Most people believe this to be a short-term response to this disaster and that more will be built in the States over the coming decade.

Nuclear power in the United States is on hold not only for safety concerns but also because of the abundance of cheap *natural gas*, making the capital expenditure of nuclear power plants not economically feasible. At the moment, nine proposals in the United States have been withdrawn.

Globally, only a little more than 5 percent of all power generation and about 13 percent of all electricity comes from nuclear power. In 2011, 433 nuclear reactors were in operation globally, which was about a dozen fewer than the year before. The global trend downward might also be temporary, and 65 reactors are being built today, spread among 14 countries. *Germany*, Switzerland, and Italy announced plans to phase out all nuclear power, but about 50 countries are moving ahead with plans to keep existing plants open and build more. *China* plans to build 50 plants by 2020 and *India* will invest $100 billion in nuclear by 2030. France has stated its intention to remain a nuclear power proponent as well.

Nuclear power offers many advantages, the biggest being it does not produce *greenhouse gases,* which cause *climate change*. It also does not cause *air pollution* and produces minimal *water pollution*. (It does, however, cause *thermal pollution*.) The two major problems with nuclear power are safety and cost.

The Fukushima nuclear reactor meltdown was not the first wake-up call. The *Three Mile Island* accident in 1979 was the first to raise public awareness of *nuclear safety* problems, and the *Chernobyl disaster* renewed and reinforced these fears in 1986.

Additional safety issues include inherent building flaws and associated containment problems that may allow the release of radioactive gases. Some plants are showing signs of premature aging, which may force them to close before their licenses expire. Besides accidents, *nuclear waste disposal* of the spent radioactive materials causes serious safety concerns. Some believe this to be the biggest problem with nuclear power and reason alone to discontinue its use.

New designs for smaller, safer, and less expensive reactors—informally called micronukes—are being studied. These new technologies will generate only 100 to 300 MW of power, compared with today's reactors that generate 1000 MW. These units could be built quickly with less capital expenditure. Another major advantage is that they will be more flexible in changing power output, so they can be used in concert with *solar* and *wind power* generation. This is important because power utilities must deal with *baseload* aspects of power.

These new designs use passive safety features, meaning they use simple natural forces such as gravity to shut down the reactor in an emergency instead of human operations and complex electronic computer systems that have failed in the past.

nuclear reactor

A nuclear reactor controls *nuclear fission* and uses the energy released to generate electricity. Nuclear reactors allow a chain reaction of nuclear fission to occur in which one atom is split, then the particles released are used to split other atoms, and so on.

A supply of a radioactive element such as uranium-235 is placed in a reactor in the form of fuel rods. A slow-moving neutron from another element strikes the *uranium*, initiating the chain reaction. The energy released from the chain reaction superheats surrounding water that is removed from the reactor and used to drive turbines and generate electricity. (The water also acts as a coolant in the system, keeping it from having a *meltdown* in which the fuel rods melt from overheating.) The speed of the chain reaction is controlled by control rods containing *cadmium* and boron, which absorb the nuclei.

Four types of nuclear reactors are currently in use: *Light-water nuclear reactors* (the most common), *heavy-water nuclear reactors*, *high-temperature gas-cooled nuclear reactors,* and *breeder reactors.*

However, new types of reactors are being developed. The traveling wave reactor (TWR) and the waste-annihilating molten salt reactor are on drawing boards. The beauty of these designs is that they use the spent fuel from existing reactors, which has become a *nuclear waste disposal* nightmare for most of the world. Bill Gates is an investor in the development of the TWR technology.

One of the biggest problems with the existing reactor technologies is cost. To overcome the staggering expense and time needed to build a reactor, future designs might include smaller reactors called micronukes. [*nuclear power; nuclear reactor safety and problems; nuclear waste dilemma, the global*]

nuclear reactor safety and problems

The accidental release of radiation from a *nuclear reactor* has had catastrophic results. Accidents at *Chernobyl*, *Three Mile Island*, and *Fukushima* have made the world aware of these dangers.

The primary danger is a reactor core meltdown in which the coolant no longer keeps the nuclear reactor within safe temperature limits and the reactor overheats. As a result, it either explodes or it becomes breeched, releasing radioactivity.

Nuclear power plants use many safety features to prevent accidents. These include systems that automatically shut down the reactor in case of an emergency such as a loss of coolant, concrete and steel walls surrounding the core of the reactor to prevent the release of radiation, concrete and steel containment walls around the entire reactor to contain radiation if the core ruptures, emergency electrical systems to assure the cooling system doesn't fail, and emergency cooling systems in case it does fail. However, history has shown that even with all these safety features, major disasters occurred.

Besides a core *meltdown*, other problems exist. For example, about one-third of the heat generated by nuclear power plants is used to generate electricity, while the other two-thirds is carried away by the coolant, causing *thermal water pollution*. Also, at the end of a nuclear facility's life, it must go through *decommissioning* (be taken out of service). Decommissioning is difficult and extremely costly because the facility is contaminated with radioactivity. However, many people believe the biggest safety problem is not with the reactor itself but with *nuclear waste disposal* of spent fuel rods.

nuclear reactors, decommissioning

Most *nuclear reactors* have a life expectancy of less than 40 years, but many are retired from service earlier. These facilities are contaminated with radioactivity, so they cannot simply be knocked down and bulldozed under like other buildings. Three options are currently available. The plant can be dismantled and all the radioactive components shipped to *nuclear waste disposal* sites. This produces large amounts of *radioactive waste* materials and puts workers at risk of exposure to the radiation.

Another option is to mothball the plant until the radiation levels drop (although still highly dangerous) before dismantling the plant. This means the plant must be secured for roughly 10 to 100 years before taking it apart and disposing of the radioactive materials.

The third option is to create a solid concrete tomb over the entire reactor to prevent radiation from escaping until the radiation level is insignificant. The tomb would have to last thousands of years before becoming safe or until new technology could be developed to decontaminate it. Even if the tomb prevented radiation from leaking into the air, there is still the danger that it could *leach* into the soil and the water supply.

Japan, France, and Canada plan to mothball their nuclear plants before dismantling them, even though the United States plans to dismantle most of its plants. The costs for decommissioning these plants are estimated to be between 50 million dollars and a few billion dollars. Dismantling a small U.S. *nuclear power* plant, decommissioned by the *Department of Energy* back in 1974, took three years and cost more than six million dollars. More than 3000 cubic meters of dismantled materials from the plant were buried.

Decommissioning costs are so high that about 20 plants worldwide have been shut down but are waiting to be decommissioned. Many more are approaching retirement age. [*nuclear waste dilemma, the global*]

Nuclear Regulatory Commission (NRC)

This commission's mission is to protect the public's health and safety as it pertains to *nuclear power*. It regulates *nuclear reactors* and the transport, storage, and disposal of *nuclear wastes*.

nuclear waste dilemma, the global

Today, about 270,000 metric tons of high-level *radioactive* wastes—mostly spent fuel rods from *nuclear reactors*—are in temporary storage facilities around the globe, 65,000 tons of them in the United States alone. Another 10,000 tons is added globally every year.

Most of these *radioactive wastes* are stored in temporary above-ground *cooling ponds*, awaiting a final destination. This high-level waste does not have a permanent storage facility anywhere in the world. Every country using *nuclear power* is grappling with the same dilemma: what to do with some of the most dangerous materials humans have ever created.

The problem is not just about protecting us from this waste; it is about protecting people—whoever they might be—10,000 or 200,000 years from now, because that is how long these wastes remain deadly. Besides the technological challenge of figuring out how to do this, there is the cost involved. We must plan to protect people thousands of years from now, but today's generation is paying the bill to create such a facility.

The *Environmental Protection Agency*, the agency responsible for setting these safety standards, has told the *Department of Energy*, who is responsible for building a permanent storage facility in the United States, that the facility must be able to contain the radioactivity for 10,000 years.

Once the fuel rods used in a nuclear reactor have lost much of their radioactivity (they are "spent"), they are removed and usually placed in pools of water to cool for at least five years. Then they are placed into dry concrete and steel casks about the size of a small truck. Both the pools and the casks are stored outdoors near the reactor, which many believe is a serious security risk, especially in light of the 9/11 attacks and the *Fukushima* disaster. These casks can theoretically retain the remaining radioactivity for a few hundred years, if needed, although many are skeptical of this estimate.

The United States, the United Kingdom, *Germany*, Canada, and Scandinavia have been searching for ways to dispose of spent fuel rods for decades and all have failed. Numerous ways to permanently store this waste have been and continue to be suggested, including dropping it into a volcano, shooting it into space, and burying it in a deep seabed. None of these have been deemed worthy, capable, or safe.

The best idea put forth so far is to find a deep, stable underground location to bury the nuclear waste. Some countries have found that the problem is not only finding the proper geological location, but also finding one where people do not oppose its presence with *NIMBY* protests.

In the 1980s, Congress designated two locations to become permanent repositories for nuclear waste materials. The first is the *Waste Isolation Pilot Plant (WIPP)* near Carlsbad, New Mexico, which opened in 1999 and became the first permanent nuclear waste storage site anywhere in the world. However, it only accepts wastes from military sites and does not accept spent nuclear reactor rods. The second, in the *Yucca Mountains* of Nevada, was scheduled to start accepting spent fuel rods decades ago; however, the planned site was closed down in 2010 before it ever actually opened.

Some scientists now believe a better alternative might be borehole disposal— that is, drilling holes many miles down and packing the spent rods into them.

Nine U.S. states have banned construction of new nuclear reactors until the government can figure out what to do with the spent rods. Other countries, such as Sweden and France, are also using temporary sites while they study permanent possibilities.

After Yucca Mountain was closed, a Blue Ribbon Commission on America's Nuclear Future suggested that to buy time until a permanent storage facility is created, the United States can create intermediate sites that would centrally locate and store all of the temporary casks stored around the country. However, many see this as simply turning a temporary solution into an intermediate solution until someone can figure out a permanent solution. [*nuclear waste disposal*]

nuclear waste disposal

Even if a totally safe *nuclear power* plant could be constructed, disposal of the leftover *radioactive waste* would still be a major problem. Until 1983, most radioactive waste—an estimated 90,000 metric tons—was dumped into the *ocean*. Since the early 1980s, alternative methods of disposal have been suggested, studied, and tested, but none are in full operation.

Radioactive waste can be divided into high-level waste, which includes spent fuel rods from nuclear power plants as well as *nuclear weapons*, and low-level waste, which includes any other radioactive materials from nuclear power plants, such as the coolant, and waste from medical facilities that routinely use radiation for treatments and diagnostics.

High-level radioactive waste—including tens of thousands of tons of spent fuel rods—are stored in temporary facilities around the globe, awaiting a final destination. The most recent plans were to place this waste in deep underground geological formations. One such facility was going to be *Yucca Mountain*, although at the moment, that project has been shut down.

An alternative to burial of the high-level radioactive waste is to reprocess the material for reuse. Reprocessing is more expensive than storing but eliminates the need for disposal. Reprocessing plants exist in France and the United Kingdom but not in the United States.

Low-level radioactive waste is usually placed into steel drums and shipped to special facilities. Many of these sites have been closed because of radioactive contamination of the *groundwater*. However, one permanent low-level radioactive waste facility, called the *Waste Isolation Pilot Program*, has opened and is deemed a success. There are concerns that much of the low-level radioactive waste has probably been illegally disposed of. [*nuclear waste dilemma, the global*]

nuclear weapons

Any weapon designed to release nuclear energy is deemed a nuclear weapon. The first of these weapons was developed by the Manhattan Project to create the first nuclear (atomic) bomb. These bombs were later dropped on Hiroshima and Nagasaki, Japan, ending World War II. Nuclear bombs not only kill virtually all forms of life within a certain radius of the detonation, but they cause radioactive fallout well beyond that range that remains in the environment, continuing to cause harm long after the detonation. As nuclear weapons are dismantled, they create problems similar to *nuclear reactors,* including what to do with the leftover *nuclear waste*. However, the danger from nuclear weapon waste materials far exceeds that caused by spent nuclear fuel rods, the biggest problem associated with nuclear power. [*hazardous waste, military*]

nurdles

If you manufacture products composed of *plastic*, this is what you start with. These tiny pieces of plastic, about the size of a lentil bean (less than one-fifth of an inch across) are the preproduction form of plastic that gets melted down and processed into whatever form is needed—be it a toy, a container, or just about anything else. They are shipped in vast quantities all around the world. Because of their size and weight, they are easily dropped, spilled, and otherwise lost in our environment, contributing to *plastic pollution*.

nurture capital

This is a spin-off of the term venture capital. Venture capital is the funding of start-up businesses with an emphasis placed on opportunities for large returns on an investment in new, usually high-risk, startup companies. Nurture capital, however, places the emphasis on funding local businesses that support sustainable practices; the emphasis is on building local economic and environmental health, not just wealth. This can be accomplished in many ways, from large-scale investments to small *microloans*. [*Beyond GDP; CERES; climate capitalism; companies, top ten green global; corporate social responsibility; corporate sustainability rankings; Sustainability Imperative; economy vs. environment; economics and sustainable development; Genuine Progress Indicator; green net national production; integrated bottom line; natural capital*]

nutrient enrichment

The unintentional input of excessive *nutrients* into bodies of water, often resulting in regions within the water with *environmental hypoxia* or even *dead zones*, where no life can exist.

These nutrients come in *fertilizer runoff* from agricultural farms or in livestock wastes from *factory farms*. Vast quantities of these nutrients enter rivers, streams, ponds, and lakes every year. Even though these bodies of water are being "enriched" with nutrients, the end result is harmful.

Natural *ecosystems* cannot handle excessive amounts of nutrients—especially *nitrogen* and *phosphorus*. When fertilizer or animal excrement enters the water, nutrient enrichment results in population explosions of *algae*, resulting in *algal blooms*.

As the algae dies in large numbers, it floats to the bottom, where it is decomposed by *bacteria*. This large amount of dead algae results in an explosion of the bacteria population. Bacteria perform *respiration*, taking in oxygen and releasing *carbon dioxide*, just as we do. These large numbers of bacteria use up most, if not all of the oxygen in the water, leaving none for other aquatic life. The resulting deficiency of oxygen is called hypoxia. In some cases, the end result can be a complete dead zone, where virtually nothing can live.

When this occurs, it is often an ecological, commercial, and recreational disaster. Many of these bodies of water are important *fisheries*, used for *drinking water* or for *outdoor recreation*.

Many environmental groups have tried to get the *Environmental Protection Agency* to regulate the amount of nitrogen and phosphorus that can be released by farms, but this has not yet happened. New *conservation agriculture* techniques have been developed that would lessen the amount of fertilizer runoff, but they are costly and comprise only a small percentage of total agriculture.

Other ideas include burning animal wastes to convert it into usable energy with *methane digesters*, extracting phosphorus from the *soil* with special crops that can then be harvested and sold—a form of *bioremediation*, incentives such as tax credits for farmers if they reduce their nitrogen and phosphorus use, and even nutrient-credit trading schemes similar to *carbon trading*.

nutrients, essential

Substances required for life that we must obtain from the environment because they cannot be created within our bodies. Organisms need about 40 different essential nutrients, which can be divided into three categories.

Those required in large quantities every day are called bulk elements and include hydrogen, *carbon*, and *nitrogen*. The next group of nutrients is needed to a lesser degree and collectively called macronutrients. They include *phosphorus*, potassium, calcium, magnesium, and sodium. Finally, the ones called micronutrients or trace elements are needed in very small amounts. These include iron, copper, zinc, and iodine. [*biogeochemical cycles; synthetic fertilizers; rice, contaminated; Golden Rice*]

ocean acidification

The amount of *carbon dioxide* (CO_2) found in the *atmosphere* is rising. More CO_2 in the *atmosphere* results in more CO_2 being absorbed by *oceans*. When CO_2 is absorbed by ocean waters, it creates carbonic acid. As the name implies, this substance makes oceans more acidic and this overall process is called ocean acidification.

Ocean acidification is occurring faster than at any time in the history of our planet. The process occurs mostly near the surface of the water. Surface waters today are 30 percent more acidic than they were prior to the *Industrial Revolution*. Estimates have the ocean's acidity doubling before the end of this century compared to pre-industrial times—a global change of epic proportions.

The change has a dramatic effect on all forms of marine life, but especially on organisms whose shells or exoskeletons are made from calcium carbonate, because acids dissolve their shells. This includes *corals*, mollusks, and many *plankton*. For example, more than 80 percent of *oyster* larvae in the U.S. Pacific Northwest commercial hatcheries do not survive, probably because of ocean acidification.

The best way to reduce the process of ocean acidification is to control the amount of CO_2, a *greenhouse gas*, entering the atmosphere, which means controlling *climate change*. [*carbon sink; oysters and climate change; coral reef sperm bank; indicator species; whale earwax*]

Ocean Conservancy

Formerly called the Center for Marine Conservation, the Ocean Conservancy is an *NGO* that "educates and empowers citizens to take action on behalf of the *ocean*." Their home page says it all: "From the *Arctic* to the Gulf of Mexico to the halls of Congress, Ocean Conservancy brings people together to find solutions for our water planet. Informed by science, our work guides policy and engages people in protecting the ocean and its *wildlife* for future generations."

And then they provide some great facts: 144,606,491 = the number of pounds of trash removed from *beaches* during the International Coastal Cleanup over the past

25 years. 2,300,000 = the estimated number of jobs created by America's ocean and coastal economy. 1972 = the year the Ocean Conservancy was founded. 1 = the number of people it takes to make a difference for our ocean.

The Ocean Conservancy protects *marine* wildlife and conserves *coastal waters* and *ocean* resources. Some of the projects they are involved with include reducing the use of *driftnets*, *gillnets*, and *purse seine nets* and helping to enforce the worldwide moratorium on commercial *whaling*. [*marine ecosystems; oceanic zone ecosystems; ocean pollution*] {oceanconservancy.org}

ocean-current power

A form of *nontraditional hydropower*. *Ocean currents* are being tested to see if they can commercially be used to generate electricity. Instead of using moving water—either as it passes through a man-made dam or as it runs downstream, as with *traditional hydropower*—this technology uses *ocean currents*.

A pilot program will attempt to use the Gulf Stream of the Atlantic Ocean off the Florida coast. An ocean-current generator will be installed off the coast of Miami that will use the underwater movement of the Gulf Stream to turn a turbine to generate electricity.

These generators would float between 100 and 200 feet below the surface. They are about 65 feet in diameter, with a large opening where the water enters. Because the output from one device is not sufficient to be economically feasible, generating electricity will require a series of these devices working in concert. [*tidal power; wave power; run-of-river hydropower; pumped-storage hydropower*]

ocean currents

The wind drives ocean surface waters in huge circular patterns across the globe, called *gyres*. These surface currents are important for many reasons, but none as important as modifying the planet's climate. These gyres move warm waters north and south from the equatorial regions. As the warm water travels, heat is lost to the atmosphere, impacting—or actually creating—weather and overall climate. These ocean currents might be capable of generating *ocean-current power* in the future.

In addition to these surface currents, "density currents" occur when cold water sinks and warm water rises. Other factors such as salt density and pressure also play a role in these currents. Density currents move far more slowly than surface currents. And finally, a third type of ocean current exists, called *upwelling*. [*hydropower, nontraditional; hydropower, traditional; extreme weather and climate change; ocean thermal energy conversion*]

Ocean Dumping Ban Act

Formally called the Marine Protection, Research, and Sanctuaries Act (MPRSA) and passed in 1972, this act prohibits dumping of material into the *ocean* that would unreasonably degrade or endanger the health of humans, marine life, or *marine ecosystems*. It was amended in 1988 to ban dumping *industrial wastes* and *sewage sludge* into oceans. All ocean dumping is now illegal unless a permit is obtained from the *Environmental Protection Agency*. [*marine protected areas; MARPOL; ocean pollution; oceanic zone ecosystems; sea gliders*]

ocean farming

See *mariculture*.

oceanic zone ecosystems

Marine ecosystems are divided into the coastal region (*neritic zone*) and the open sea (oceanic zone). The oceanic zone is found beyond the *continental shelf,* which can protrude hundreds of miles from the shore. Light penetrates only the surface layer of the open sea down to about 600 feet. This is called the *euphotic zone,* where *autotrophs* perform *photosynthesis* and most marine life exists. Beneath this layer is the *aphotic zone* where little or no light penetrates.

The ocean receives a constant flow of *nutrients* from rivers and water *runoff* from the surrounding land, but most of these nutrients remain near the coastal region, so the oceanic zone is surprisingly one of the least productive environments (*net primary productivity*) on the planet.

In regions where *ocean currents* can carry nutrients out to the open ocean, *food chains* exist. *Phytoplankton*, found in the euphotic zone, perform most of the *photosynthesis* of the sea. *Zooplankton* consumes the phytoplankton, which are in turn eaten by many types of small fish or crustaceans, such as *shrimp*. *Tuna* and *shark* eat these fish and complete the chain. Many of the larger fish live below the euphotic zone but visit the surface in search of food.

Organisms that live beneath the euphotic zone and don't venture up to it must depend on the constant downward flow of dead organisms (mainly plankton) as their source of food, meaning they are *scavengers* or *decomposers*.

Oceanic circulation supplies all these deep-sea organisms with oxygen to survive. A few deep-sea *predators* have been discovered in the lower portion of the aphotic zone, and some organisms have been found living on the floor of the oceans, even at depths of 3.5 miles to 5 miles down. These bottom dwellers, called *benthic organisms*, include unusual species of sea urchins, starfish, and *tube-dwelling worms*. Although they don't live in the water, many birds also play a role in these ecosystems because they feed on animals near the surface. [*ocean pollution; whale earwax*]

oceanography

The study of *marine* organisms and their *habitats*. [*marine ecosystems; oceanic zone ecosystems; ocean pollution*]

ocean pollution

Oceans have long been a favorite dumping ground for humans. Its vastness results in quick dispersal and dilution. A sad old saying states, "The solution to *pollution* is dilution." This works deceivingly well until the volume of pollution overwhelms the volume of water— something that is visibly happening along coastlines where half the people of the world live. About 80 percent of all ocean pollution comes from land-based sources. These sources can be divided into two categories.

    Point source pollution can be traced to a single source. This includes *industrial wastes* and *sewage treatment facilities* where you can actually see the pipes (the point) where the waste pours out. Point source water pollution has been significantly improved since the *Clean Water Act* and subsequent amendments helped reduce much of the industrial *effluent* dumped into rivers and bodies of water. *Combined sewage systems* remain a problem because they continue to dump sewage wastes into waters during heavy rain events. But overall, the *Environmental Protection Agency* (EPA) believes point source pollution is no longer the primary problem and has shifted their focus to non-point sources.

    Pointless pollution, also called non-point pollution, is pollution from many sources that cannot be traced to a single source such as an effluent pipe. It includes water runoff containing *fertilizers* and *pesticides* from *high-impact agriculture*, along with oils and solvents from cars and industry, which all sooner or later make their way to the ocean, often collecting along shorelines and causing high pollution levels that damage these ecosystems. We cannot see the harm done, but know things are not as they should be when *beaches* are closed because of these contaminates.

    Two relatively new indications of ocean pollution are *dead zones* and *marine garbage patches*. *Nutrient enrichment* caused by fertilizers from high-impact agriculture has created numerous dead zones in ocean waters. And *plastic pollution* collects in the seas in the form of vast marine garbage patches that threaten *marine ecosystems*.

    Most recently, *noise pollution* has resulted in a new form of ocean pollution. *Underwater sonar testing* used for oil exploration and military exercises has become a threat to marine life.

The Federal Water Pollution Control Act of 1956 (amended to include the Clean Water Act of 1977) and the *Ocean Dumping Act* of 1972 authorize the EPA to regulate ocean pollution. The *United Nations Environment Programme* works on an international scale at controlling ocean pollution. [*Ocean Conservancy; whale earwax*]

oceans

Our planet is often called the water planet, because more than 70 percent of the surface is covered with water and 97 percent of that water is found in the oceans. All water from streams and rivers ultimately end up in the oceans.

Oceans play a key role in regulating the planet's climate by storing heat and distributing it through worldwide currents. Oceans help regulate the *water* (hydrologic) *cycle* and play an important role in many other *biogeochemical cycles*. The oceans also store vast quantities of *carbon dioxide,* acting as a *carbon sink* that helps regulate the overall temperature of the planet. Finally, oceans are home to roughly 250,000 species.

Human activities are taking their toll on oceans, including *ocean pollution*, *dead zones*, vast *garbage patches*, *ocean acidification,* and *coral bleaching*. [*ocean currents; whale earwax*]

ocean thermal energy conversion (OTEC)

OTEC is a *nontraditional hydropower* method of generating power that uses differences in temperature between the ocean's colder, deeper water and the warmer, more shallow water. OTEC power plants are anchored to the *ocean* bottom and float at the surface. Japan has one up and running, and the United States has an experimental OTEC power plant in operation. These types of facilities produce no pollutants, similar to *traditional hydropower*. [*ocean currents*]

Odum, Eugene

(1913–2002) Odum is known as the father of *ecosystem* ecology. He made the term *ecology* a household word and is considered one of the most influential people in the field of *environmental science*. He pioneered the idea that the *environment* must be studied as an integrated system—everything interacting with everything else. The first generation of ecologists all studied his landmark textbook, <u>Fundamentals of Ecology</u> (1953), coauthored with his brother, Howard Odum. To say the book was ahead of its time is an understatement, because it was the only environmental textbook available for almost 10 years. [*ecological studies; computer modeling, ecological; applied ecology; community ecology; deep ecology; industrial ecology; zoogeography*]

OECD (Organization for Economic Cooperation & Development)

An organization consisting of 34 wealthy nations, created in 1961 and headquartered in Paris. They state their mission to be "to promote policies that will improve the economic and social well-being of people around the world." They provide excellent statistics and generate reports about member nations. The information they provide is routinely cited by the press. [*CERES; economy vs. environment; countries, ten greenest; countries, ten most energy-efficient; hunger, world; water shortages, global*] {oecd.org}

Office of Technology Assessment (OTA), former

The OTA was created in 1974 and had a board of six Senators and six House Representatives with an Advisory Council of 10 eminent scientists who were called upon regularly for input.

The OTA would report to Congress on the scientific and technological impact of government policies and proposed legislation. It studied science and technology issues for Congress, attempting to resolve conflicting claims, beliefs, and data. The goal was to provide those in power with unbiased, scientific knowledge to advance government policy. The office helped ensure that the strategic direction the United States took regarding environmental and energy issues was based on *science*—not opinions and *pseudoscience*.

After 23 years of providing this service, Congress voted in 1994 to defund the OTA, and the agency ceased to exist in 1995. It appears that those in Congress no longer felt they needed scientific information and advice from the most eminent scientists the country had to offer. The distinction between science and pseudoscience has become blurred for many people, which is unfortunate—but none more so than those who lead the country.

Many people believe the OTA should and could be reinstated. The Technology Assessment Act of 1972 remains on the books and was never formally abolished. All that is needed to restart the OTA—to educate those in Congress about science—would be an appropriation of funding. This is unlikely to happen in the current Congress but not out of the question sometime in the future when sounder minds might prevail. [*opinions and science; climate change deniers; argumentum ad ignorantiam*] {princeton.edu/~ota}

Ogallala aquifer

The world's largest freshwater *aquifer*, the Ogallala lies beneath portions of eight U.S. states from South Dakota to Texas. It was probably formed about 15,000 years ago by receding *glaciers* from the last *Ice Age*. This one aquifer has been, in part, responsible for the amazing agricultural productivity in an otherwise dry region. Prior to becoming the primary irrigation source for farms, it is estimated to have contained 16 times more *freshwater* than all of the surface freshwaters, such as lakes and rivers, in the country.

Scientists estimate that 3 percent of the aquifer's *water* was used up back in 1960 and 30 percent was gone by 2010. And if current trends continue, almost 70 percent of the aquifer will likely be drained by 2060. It would probably take somewhere between 500 to 1300 years to completely refill the entire aquifer if left unused.

In the past, water—even in an aquifer—has been considered a renewable resource, because the water *recharges* naturally. But with excessive withdraws due to *irrigation agriculture*, it can no longer be considered a renewable resource. [*water shortages, global; water conservation; water pollution*]

oil

Oil began to replace *coal* during the early 1900s, primarily because of the *automobile*. Oil has a high heat content and burns far more cleanly than coal. Crude oil (also called petroleum) is a thick, black liquid substance composed mainly of *hydrocarbons* with small amounts of other chemicals such as *sulfur*. It is found deep in the Earth's crust or beneath the *ocean* floor.

Today more than 30 percent of the world's primary energy supply comes from oil, but this percentage has been dropping for 12 consecutive years. In the United States, *natural gas* has been replacing both oil and *coal*; however, the amount of oil extracted each year continues to increase. Currently, more than 80 percent of all *oil reserves* belong to *OPEC* nations. The largest producer of oil is Saudi Arabia, recently surpassing Russia. However, in 2013, quite surprisingly, the U.S. surpassed Russia and has become the second largest oil producer in the world.

Only a couple of years ago, there was a great deal of talk about *peak oil* and the likelihood of the end of oil as our primary energy source. But now the world's oil reserves are constantly being upgraded as new sources—such as the *Bakken* formation in the United States—are found and new technologies such as *fracking* and horizontal drilling are developed. Projections estimate 100 times more oil might be still undetected within the Earth.

To many, these new reserves are good news; however, most *environmentalists* fear it will slow the necessary transition away from *fossil fuels* and toward *renewable energy sources* to control *climate change*. Burning oil produces *carbon emissions,* a major component of *greenhouse gases* and the biggest contributor to climate change. Also, *oil spills* such as from the *Exxon Valdez* and the *BP Deep Horizon* explosion cause environmental destruction. [*oil recovery or extraction; oil formation; oil spills*]

oil formation

Oil, like other *fossil fuels,* was created from accumulations of dead organisms that were exposed to intense heat and pressure for hundreds of millions of years. While *coal* was created from vegetation, oil is believed to have formed from microscopic *marine* organisms that died and accumulated in the sediment at the bottom of the oceans. As these organisms died, they released small quantities of oil. The sediment gradually became *shale* that contained the accumulated oil. Oil deposits were created when geological conditions allowed the oil in the shale to be forced into surrounding layers of sandstone that soaked up the oil like a sponge. [*geological time scales*]

Oil Pollution Act

Signed by President George H. Bush in 1990 in response to the *Exxon Valdez* oil spill, this act expanded oil spill prevention measures and created new regulations for oil transportation, cleanups, and response capabilities of the federal government as well as the industry. It also improved the nation's ability to prevent and respond to *oil spills* by providing funds and resources necessary. It accomplished this by creating the national Oil Spill Liability Trust Fund, which makes available up to one billion dollars per spill incident. [*oil pollution in the Persian Gulf; oil recovery or extraction*]

oil pollution in the Persian Gulf

Oil tankers in the Persian Gulf (also called the Arabian Gulf) usually wash out their holds and dump about two million barrels of *oil* each year, directly into the gulf. This large volume of oil pollution was dwarfed, however, during the Persian Gulf War of 1991, which produced the largest *oil spill* in history with four to eight million barrels of oil spilling directly into the gulf waters within a few weeks. Seven hundred wells were blown up by retreating Iraqi forces, most of which were set afire and burned for almost one year, blackening the skies.

Almost all the oil remained within 250 miles of the spill along the Kuwait and Saudi Arabian coastlines, killing *wildlife* and causing respiratory illness in people living near the shore. It is also believed to have contaminated some of the *groundwater* drinking supply. The cleanup, including the cost of putting out all the well fires, ran about 20 billion dollars.

oil recovery or extraction

Oil is extracted in one of three ways. Primary oil recovery involves drilling a well and pumping out the oil that accumulates at the bottom. Secondary recovery attempts to get the thicker, heavier oil that does not accumulate by the natural force of gravity. In this method, water is injected into an adjacent well in an effort to force remaining oil into the central well. Both primary and secondary recovery extracts only about one-third of the total oil content.

Tertiary (enhanced) recovery is expensive and involves pumping steam into the well in an attempt to reduce the viscosity of the heavy oil and get it to flow into the well. Newer methods of oil and gas extraction are called unconventional recovery methods. *Unconventional oil and gas* extraction methods have become important, especially in the United States.

Once removed from the well, the crude oil is usually sent through a pipeline to a refinery, where it is separated by heat into *gasoline*, heating oil, asphalt, and many other substances. [*energy, A very brief history of U.S.; barrel of oil; oil spills; tar sands oil extraction; fracking*]

oil spills in history, ten worst

The ten worst oil spills (from any source) in history are, in descending order (rough estimate of millions of gallons): 1) the *Persian Gulf War*, 1991 (350), 2) *BP Deepwater Horizon* offshore oil platform, 2010 (210), 3) the Ixtoc 1 oil well in Mexico, 1979 (140), 4) the Atlantic Empress *oil tanker* in the West Indies, 1979 (88), 5) Fergana Valley oil well in Uzbekistan, 1992 (88), 6) the Nowruz oil field in the Persian Gulf, 1983 (80), 7) the ABT Summer oil tanker off the coast of Angola, 1991 (80), 8) the Castillo de Bellver oil tanker off the coast of South Africa, 1983 (79), 9) the *Amoco Cadiz oil* tanker, 1978 (69), and 10) the Odyssey oil tanker off the coast of Nova Scotia, Canada, 1988 (43). [*oil spills in U.S.; ship pollution; pipeline spills*]

oil spills in U.S.

The worst oil spills (from any type of source) to have occurred in the United States are as follows (oil spilled in millions of gallons) in descending order: 1) *BP Deepwater Horizon* offshore oil platform in 2010 (210), 2) *Exxon Valdez* oil tanker in 1989 (11), 3) Argo Merchant *oil tanker* off Nantucket in Massachusetts in 1979 (8), 4) various sources from Hurricane Katrina, New Orleans, Louisiana, in 2005 (7), and 5) Mega Borg oil tanker off the coast of Galveston, Texas, in 1990 (5).

Even though these oil spills are dramatic and get lots of news coverage, they are not the biggest cause of environmental harm caused by oil spills in bodies of water. Offshore drilling operations and spills or leaks from tankers contribute only about eight percent or less of the total.

Most of the oil spilled into the oceans (700 million gallons per year, globally) comes from other sources. About 50 percent comes from land-based sources such as the improper disposal of used *motor oil*. The rest comes from the routine maintenance of ships (about 20 percent), *hydrocarbon* particles from onshore *air pollution* (about 13 percent), and natural seepage from the seafloor (about 8 percent). [*oil spills in history, ten worst; ocean pollution; Oil Pollution Act; ship pollution; pipeline spills*]

oil tankers

Huge ocean-going ships that transport *oil* across the globe. They are considered the largest moving human-made objects. The first oil tankers were built in the late 1800s. Today, oil tankers carry around 2,000,000,000 metric tons of oil each year. Most are large, but the size varies based on the location of their travels. Those that travel inland routes are small compared to those making transcontinental trips. Because of the volume of oil they can carry, transportation costs come out to only a few cents per gallon.

The size of some of these vessels is truly amazing. The largest are called super-tankers—or very large crude carriers (VLCCs) in the trade—that carry between about 175,000 and 350,000 tons of oil and some of the newest, even more.

The largest tanker afloat today, is longer (taller) than the Empire State Building in New York City.

In the past, tankers were single-hulled. After the *Exxon Valdez* oil spill, all newly built tankers are double-hulled to lessen the possibility of spills; however, it does not prevent *oil spills* from occurring.

old-growth forests

To a forester, old growth means trees that are at their maximum harvest size. To *environmentalists*, it refers to a stand of large, old, living trees that have never been harvested (called primary forests). A classic old-growth forest, according to the *Wilderness Society,* contains at least eight big trees per acre, each exceeding 300 years in age or measuring more than 40 inches in diameter. [*Save the Redwoods League; old-growth forests; logging, illegal; Luna Redwood; deforestation; forest health, global; environmental movement*]

oligotrophic lakes

Deep, clear, cold lakes containing few *nutrients* to sustain life. [*aquatic ecosystems; eutrophication, natural; eutrophication, cultural; nutrient enrichment*]

omnivore

Animals that eat both plants and animals are omnivores. Bears, raccoons, foxes, and most humans are omnivores. [*carnivores; herbivores; vegetarianism; insectivore; food chains; trophic levels; chemoautotrophs; energy pyramid*]

OnEarth magazine

An excellent quarterly magazine published by the *Natural Resources Defense Council*. It contains well-written full-length articles reporting on environmental topics. Joining the NRDC automatically provides a subscription to this informative magazine. [*magazines, eco-; blogs, green; microblogs, green; tweets, eco-; social media, environmental*] {onearth.org}

online petition sites and environmentalism

Many online petition sites exist for people who want to campaign for a cause. They can be for virtually anything or anyone but are often used to advance health or environmental issues. Some examples include a petition that helped obtain health care for victims of water poisoning at a U.S. army base and another that helped get an investigation started to stop animal cruelty in Belarus. These sites are usually for profit, as opposed to *NGOs*, which are not. [*environmental movement; coffee, environmental impact of; environmentalist; voluntourism, environmental; volunteer research, scientific; social media, environmental; tweetstorm, Rio+20*] {actionnetwork.org and change.org}

oology

The study of eggs. [*applied ecology; autecology; synecology; caliology; cryptozoology; entomology; terrestrial ecology; egg farms*]

ooze

Fine-grain sediment found at the bottom of deep waters, containing the remains of *marine* organisms.[*sedimentation; oceanic zone ecosystems; benthic organisms*]

OPEC (Organization of Petroleum Exporting Countries)

Created in 1960 when five of the world's major *oil*-exporting countries got together and formed an oil *cartel* to control the global price of oil. Thirteen countries are in OPEC, including seven Arab countries—Saudi Arabia, Kuwait, Libya, Algeria, Iraq, Qatar, and the United Arab Emirates—and six non-Arab countries—Iran, *Indonesia*, Nigeria, Gabon, Ecuador, and Venezuela. Saudi Arabia is the largest producer in the world. These countries control more than 80 percent of the entire world's *oil reserves* and about 40 percent of the current oil market. [*energy sources, historical; energy, A very brief history of U.S.; peak oil*]

opinions and science

An opinion is simply someone's belief about something, or a conclusion someone has reached. Everyone is entitled to their opinions, but that doesn't mean it is a fact. *Science* is fact based. It uses the scientific method to determine cause-and-effect relationships that, once verified, are considered facts and become part of the body of knowledge.

Although it might be fine to make personal decisions based on one's opinion, to do so about important issues, obviously, is not wise. You would not want to go to a medical doctor who read magazines based on people's opinions and guesswork instead of a medical journal based on science. The same is true of environmental issues. Distinguishing between opinion and fact can be difficult, especially when people go beyond opinions and try to actually deceive you with *pseudoscience*. [*environmental literacy; environmental education; NGOs; National Oceanographic and Atmospheric Administration; blogs, environmental; magazines; eco-; climate change deniers; Agenda 21 conspiracy; law and the environment; argumentum ad ignorantiam*]

organic

Organic has many meanings, depending on the circumstance and use. From a chemical perspective, it is a compound that contains the element carbon in combination with other elements, often in long chains, and usually produced by or derived from an organism. It can be used to mean naturally made—as opposed to human-made or synthetic such as *organic fertilizer*. It can also mean pertaining to or derived from an organ in the body.

Finally, it is often used to designate foods or materials that have been grown without *pesticides*, *synthetic fertilizers,* or containing artificial additives such as *food dyes*. Organic foods and even *organic beef* have specific *organic food labels* or *food eco-labels* that ensure they are what they say they are. [*organic farming and food; organic clothes; Organic Consumers Association; organic fertilizer; organic food supermarkets; apples, antibiotics in organic; farmers' markets; sugar, sugar alternatives and the environment; wine, organic and biodynamic*]

organic beef

Beef that has been raised without *antibiotics*, growth hormones, or synthetic chemicals of any kind. There are specific credible *organic food labels* to ensure the beef is raised as claimed. [*organic; organic farming and food; food eco-labels*]

organic clothes

Although most of us are familiar with *organic farming and foods*, a new industry is also popping up providing *organic* clothes. Still a loosely used term, it refers to clothes that are manufactured in a manner that causes less harm to the environment and to the wearer than traditional clothes. A few boutique companies, such as bgreen and Uranus, have been in this line for many years. More recently, big names such as Patagonia and American Apparel have joined the movement.

The emphasis for organic clothes is on the materials used and how they are produced. They are, of course, natural materials as opposed to synthetic—such as 100 percent organically grown cotton or wools that come from organically raised sheep. In some cases, materials such as a soy fabric might be used. Also, recycled materials might be used as well as natural dyes for color or material simply left as its natural color. Whether this becomes popular or not remains to be seen, but indications are we will be hearing more about it in the future. [*recycling; upcycling; consumers, green*]

Organic Consumers Association (OCA)

This *NGO* represents thousands of *organic* farmers and their consumers. Some of their goals are to convert at least 30 percent of U.S. agriculture to become organic by 2015, to establish a global moratorium on *genetically modified organism* crops, to phase out the most dangerous aspects of *factory farms* and *high-impact agriculture,* and for the United States to become energy independent by transitioning to *renewable energy* sources. [*organic farming and food*] {organicconsumers.org}

organic farming and food

Organic farming is a new term used to define old farming practices. The International Federation of Organic Agriculture Movements provides the following definition: "Organic agriculture is a production system that relies on ecological processes rather than the use of synthetic inputs, such as chemical *fertilizers* and *pesticides*." The goal is to use s*ustainable farming* practices or *conservation agriculture*. The movement began in the early 1960s as a response to the new, at that time, use of synthetic pesticides on crops. As research began to show the health and environmental problems associated with these chemicals (as exposed by *Rachel Carson* and others) used on *high- impact agricultural* farms, people looked for alternatives.

The amount of energy needed to run an organic farm is less than that used on high-impact agriculture with *monocultures*, but the yields are generally lower. However, studies confirm that organic agriculture provides better yields during times of drought and the *topsoil* remains fertile longer without the need for synthetic fertilizers. Studies also show that organic farms have about 30 percent greater *biodiversity* of species than *conventional agriculture* farms.

Efforts to ensure that people get what they think they are getting has resulted in more than 80 nations establishing organic food regulations. There are now many reliable *organic food labels and food eco-labels* attesting to the veracity of these foods.

Many countries have seen dramatic increases in the popularity of organic farming. The regions with the largest share of organic farming are *Australia*, New Zealand, and Pacific Island nations, followed by Europe and then Latin America. Next come Asia and North America with significant numbers. Japan alone has more than 20,000 organic farms in operation.

Organic farming existed long ago, before any synthetic chemicals were available, so it is not surprising that organic farming is common today in poor nations. People in these nations don't think of it as a modern method but as the way it has always been done. Today, the three countries with the largest number of people working on organic farms are *India*, Uganda, and Mexico.

In 2010, sales of organic foods and beverages equaled more than $26 billion annually, an eight percent increase over the previous year. However, even though organic agriculture is a booming business, it remains a miniscule portion of the total—less than one percent of all agricultural land is organically farmed. [*tomatoes, tasteless but red; farm-to-table movement; farm-to-school movement; fruit waxing; imported food dangers; pesticide residue on food; campus quad farms; teaching farms; WWOOF; community gardens; community-supported agriculture; heirloom plants; Organic Consumers Association; organic fertilizer; organic food supermarkets; apples, antibiotics in organic; farmers' markets; sugar, sugar alternatives and the environment; seasonal foods; slow food movement; food miles; locavores; food dyes; microgreens and urban gardening; wine, organic and biodynamic; fruit waxing*]

organic fertilizer

Organic fertilizer, like *synthetic fertilizer*, enriches the soil by adding nutrients required for plant growth. However, organic fertilizer comes from *organic* or natural matter as opposed to being synthetically produced from petroleum products.

There are three types of organic fertilizer. 1) *Animal manure*, composed of dung and urine usually from livestock—although some *less developed countries* use human dung called *night soil* as an organic fertilizer. 2) *Green manure* consists of any plant materials that are plowed into the *soil* or *grass clippings* left on a lawn, and 3) *compost*. With the exception of organic farms, virtually all U.S. crops are fed synthetic fertilizers instead of organic fertilizer.

organic food labels

Foods grown on organic farms are becoming increasingly popular and have resulted in a flurry of marketing and advertising claims, many of which have little to do with reality. Recent surveys show that many Americans believe most organic labels are nothing more than *greenwash*, meaning they do not believe what the labels say.

Savvy consumers know not to believe something is truly environmentally sound just because a label uses the word *green* or has *eco-* before the product's name. However, some *eco-certification* programs, *eco-labels* and organic food labels are accurate and can help consumers know what they are buying.

Two groups that have been around since organic farming got its start are the Organic Crop Improvement Association (OCI), which is international, and the Northeast Organic Farmers Association (NOFA), which is regional. Both have established reliable certification programs assuring products have been organically grown. The U.S. *Department of Agriculture* has its own food eco-labels for organic products as well. [*food eco-labels; Organic Consumers Association*] {ocia.org and nofa.org}

organic food supermarkets

Many people concerned about their health and that of the environment prefer to eat foods that haven't been grown with *synthetic fertilizers* and *pesticides* and meats that haven't been raised on synthetic hormones and shot up with preventative *antibiotics*. For that reason, organic food companies and outlets are experiencing dramatic growth. Twenty years ago, only about 50 supermarkets specialized in *organic farming and foods*, but now they are common. Organically grown produce is sold in almost all supermarkets, and *organic beef* and other livestock are also readily available. U.S. organic food sales grew by about 20 percent per year in the early 2000s and remains at about 10 percent annual growth today. Bread & Circus, Whole Foods Market, and Alfalfa's are just a few of the organic food supermarket chains. [*organic food labels*]

organism, most abundant

Have you ever wondered which species has more individuals than any other on our planet? Until this year, that accolade went to a *bacteria* called P. ubiques, found in *oceans*. This makes sense because most of our planet is covered by oceans. However, in 2013, scientists discovered a *parasite* that lives in this bacteria.

In most cases, multiple parasites live in a single *host*, and this appears to be the case here. So if the most abundant species has a parasite, that makes the parasite of the host, the most abundant organism on Earth. But...

There are *viruses* that live in these bacteria. (A virus that lives in a bacteria is called a bacteriophage, or a *hyperparasite*.) So the most abundant organism in the world is a bacteriophage called HTVC010P. [*biodiversity*]

organophosphates (OPs)

One of four major categories of synthetic *insecticides*. OPs, along with *carbamates,* are considered *soft pesticides* because they break down into harmless chemicals after application, usually 1 to 12 weeks. OPs kill by affecting the normal functioning of nerve cells. Even though they are less persistent than *chlorinated hydrocarbons*, they are usually far more *toxic* to humans, meaning they are especially dangerous to people who apply the chemicals or are in the vicinity of the application. Because OPs pose a significant health risk, many have been replaced with carbamates.

Two examples of OPs are malathion, often used to control mosquitoes, and diazinon, commonly used for pest control in household gardens. Many of the less expensive flea and tick collars for pets use OPs, while the more expensive collars use safer alternatives. [*pesticides; pesticide dangers; natural pesticides; biological pesticides*]

Orr, David

Orr is one of our preeminent environmental educators. He is a professor and Chair at Oberlin College and author of many excellent books, including <u>Ecological Literacy</u> (1992). His focus has always been on students and their relationship with our planet's resources, and he is a leading proponent of sustainable design on college campuses. He spearheaded the effort to design and build an exemplary Environmental Studies Center at Oberlin College. It was described by the <u>New York Times</u> as "the most remarkable" of a new generation of college buildings. A citation presented to Orr by the Connecticut General Assembly says it well. It was awarded for Orr's "vision, dedication, and personal passion" in promoting the principles of sustainability. [*philosophers and writers, environmental; environmental literacy; environmental education; buildings, green*]

Our Common Future

In 1983 the *United Nations* convened the *World Commission on Environment and Development* with the intent to develop "a global agenda for change." This agenda was published in book form in 1987 and called <u>Our Common Future</u>. The book was the agenda used for the 1992 *Earth Summit*.

Today it remains an iconic book, warning of a need for action to prevent a future of environmental decay, poverty, and limited resources (much as *<u>The Limits to Growth</u>* did 15 years earlier).

The report, more commonly known as the Brundtland Report, coined the now ubiquitous phrase *sustainable development* and provided what is still considered its best and most eloquent definition: "Development that meets the needs of the present without compromising the ability of future generations to meet their own needs." It called for new multilateral solutions to resolve environmental issues and helped engrain the idea that environmental issues cannot be separated from social issues.

Today, when the environment is discussed, poverty and the human condition are always part of the discussion.

Perhaps this book, more than any other, clearly states that decisions we make today could harm those who come after us, as seen in these quotes directed toward students and educators: "Most of today's decision makers will be dead before the planet feels the heavier effects of *acid precipitation*, *global warming*, *ozone depletion*, or widespread *desertification* and species loss. Most of the young voters of today will still be alive."

It continued, "First and foremost our message is directed towards people, whose well-being is the ultimate goal of all environment and development policies. In particular, the Commission is addressing the young. The world's teachers will have a crucial role to play in bringing this report to them."

Even though the book is more than 25 years old, the underlying premise remains valid and even more in need of being taught and finding a young audience who are, we hope, willing to listen. [*hunger, world; water shortages, global; land grabs; decoupling; economy vs. environment; safe operating space; economics and sustainable development*]

outdoor recreation

One of the stated goals of *national parks* and other *public lands* is to provide opportunities for outdoor recreation. Pinpointing exactly what is meant by outdoor recreation is difficult and numbers are hard to determine because many activities are often lumped together or defined in many ways. However, it appears that seven of the most popular outdoor recreational activities in the United States—all with more than 50 million participants—are (in alphabetical order): *bicycling*, camping, fishing, gardening, hiking, *walking* (most reports separate these two), and running/jogging (most reports group these two). [*greenways; multiple use policy; nature-deficit disorder; urban open space; outdoor recreation and motorized vehicles; urban gardens*]

outdoor recreation and motorized vehicles

To some people, *outdoor recreation* means motorized vehicles such as all-terrain vehicles and skimobiles. Nonvehicular outdoor recreation causes some damage to the *environment*, but motorized vehicles inflict a great deal of damage. For example, all-terrain vehicles kill the vegetation along trails, usually resulting in *erosion* and substantial damage to the local *ecosystem*. Because U.S. *public lands* are owned and paid for by everyone, there is a great deal of controversy over this type of land use.

[*greenways; multiple use policy; National Park Service (U.S.); nature-deficit disorder; urban open space*]

Outdoor Writers Association of America (OWAA)

The OWAA is an *NGO* that represents professional communicators who write and report about the outdoor experience. Members are writers, journalists, editors, broadcasters, producers, photographers, artists, publishers, bloggers, PR professionals, and others. Many members also have degrees in *wildlife management, environmental science*, forestry, and allied fields. They provide and share information through conferences, workshops, newsletters, environmental committees, and other venues. [*magazines, eco-; blogs, green; microblogs, green; environmental literacy*] {owaa.org}

outer space

Outer space has helped humanity understand the concept of our planet as our spaceship—floating in emptiness. This first became obvious when astronauts onboard Apollo 17 took the *Blue Marble photo* that gave the world an "oh my!" moment; we really are on *Spaceship Earth*.

There is no specific definition about where the *atmosphere* ends and outer space begins. You can think of it as above the *noosphere*— that part of our planet where humans have an effect, where planes fly, satellites orbit, and *space debris* proliferates.

Beyond that altitude—outer space—might not seem of any immediate importance to us. But it is only a matter of time before outer space becomes "just around the corner." Only a few decades ago, the *Arctic* was untouched, but today it is fast becoming the next *tragedy of the commons*. In only a few tomorrows, outer space will become the next commons.

A couple of global treaties are meant to manage or protect outer space. They include the Treaty on Principles Governing the Activities of States in the Exploration and Use of Outer Space (1967) and the Convention on International Liability for Damage Caused by Space Objects (1972). [*space-debris cleanup; near earth objects; panspermia; extinction and extraterrestrial impact; exobiology; terraforming*]

outgassing

A form of evaporation that occurs when unstable molecules are released into the air. A good example is the release of *formaldehyde* from building materials and products such as particleboard, paneling, carpeting, and furniture. Outgassing occurs when formaldehyde escapes from the glue found in these products. *Volatile organic compounds* and many other substances perform outgassing. Intense outgassing usually occurs the first few weeks of a product's life and then levels off to low-level outgassing that can occur for years. Outgassing is an important aspect of *indoor air pollution*. [*couches and indoor pollutants*]

Outward Bound USA

An adventure-based educational *NGO* program designed to teach leadership skills through challenging activities in a *wilderness* setting. Founded in 1941, they now serve more than 70,000 students and educators every year. Their focus is to help individuals learn about themselves. The instructors are professional outdoor educators. The adventures include whitewater rafting, canoeing, sailing, sea kayaking, backpacking, horsetrailing, and trekking. Students must be fourteen years and older. [*outdoor recreation; environmental education; nature-deficit disorder; voluntourism, environmental; WWOOF; primitivism*] {outwardbound.org}

overburden

The material found above a vein of *coal* that must be removed during *strip-mining*. [*surface mining; fossil fuel; gold mining; mineral exploitation (mining); mountain top removal; Pebble Mine; spoil banks*]

overfishing

No one owns the open *ocean*. Modern high-tech *commercial fishing* fleets use an array of methods to cast a wide net and harvest vast quantities of fish and other marine life quickly. *Factory ships* are virtual manufacturing plants on the seas. The result has been overfishing, where more fish are caught than are naturally replaced, resulting in the depletion of many species of fish around the globe. The most recent estimates have many commercial *fish populations* down by 75 percent and some close to 99 percent.

Few people disagree that overfishing is a worldwide threat, but there has not been a great deal of international agreement about what to do about it. In response to the declining available seafood, *aquaculture* now supplies more than half of all seafood eaten globally.

Humans have once again managed to use technology to come to the rescue of an environmental crisis and to still put (fish) food on everyone's plate, even after it has been seriously depleted. However, the environmental aspect of this crisis—the serious decline in most of our wild fish species—continues to exist. Overfishing might be our ultimate *tragedy of the commons*. [*hunger, world*]

overgrazing

Animals feeding (*grazing*) on vegetation at a rate that exceeds the vegetation's ability to recover, resulting in damage to the *ecosystem*. [*desertification; Amazon River basin; buffalo; cacti; deforestation; ecological restoration; grassland, temperate; habitat loss; herbivore; meat production; savanna*]

overpopulation

In general terms, overpopulation means there are more individuals of any one species than the resources can provide for. Too many rabbits without enough food, water, or shelter will result in overpopulation, death, and a decline in the number of rabbits. More specifically, the term overpopulation is used when the *carrying capacity* of an area is exceeded, resulting in a degradation of the environment and a decline in the population by natural forces such as starvation or disease.

When speaking about human populations, we include not just survival, but also the health and welfare of the individuals in the population when determining whether overpopulation exists. The *human population* is a unique situation, and over-population is often referred to as a *population explosion*. [*baby boom; doubling time in human populations; population growth, limits to human; hunger, world; water shortages, global*]

overstory

The uppermost layer of a *forest* community, formed by the highest trees. The highest *canopy* forms the overstory. [*niche, ecological; habitat*]

oviparous

Animals that produce eggs that develop and hatch outside of the maternal parent. Includes birds, *amphibians*, and most *insects*. [*oology; ovoviparous*]

ovoviviparous

Animals that produce eggs that develop within the maternal parent but hatch either inside or immediately upon leaving the parent's body. *Sharks*, lizards, and some *insects* and snakes are ovoviviparous. [*oology; oviparous*]

oysters and climate change

Coral reefs and shellfish including oysters are threatened by *climate change*. As humans emit *carbon dioxide* into the *atmosphere*, the *oceans* absorb much of it. This occurs because oceans act like a *carbon sink*, storing the carbon. Oceans absorb more than 20 million tons of carbon dioxide every day.

As carbon levels increase in the seas, the water becomes more acidic—a process called *ocean acidification*—and that is the problem. Oysters, corals, and other organisms use calcium carbonate to make shells or other structures, but acidic water slows the process or even destroys what has already formed. The next time you go to an oyster bar, the price of oysters might be an indication of climate change. [*climate change and marine life; coral bleaching; Petermann Glacier ice detachment; iron hypothesis; assisted migration; vineyards and assisted migration*]

ozone

From an environmental perspective, there are two types of ozone. The first is naturally occurring ozone located in the upper *atmosphere*, called *stratospheric ozone*, or the ozone layer. *Ozone depletion* of this layer was one of the most pressing environmental problems 20 years ago, but much has changed since then.

The second type of ozone is produced by human activities, called *ground-level ozone*. It is the main component of *photochemical smog*. It irritates eyes and can cause serious respiratory problems. It also damages *chlorophyll* in plants.

ozone, ground-level

Nitrogen compounds are primary pollutants that react with oxygen, in the presence of sunlight, to create ground-level ozone, a *secondary pollutant*, and a form of *air pollution*. Ground-level ozone damages *chlorophyll* in plants and lung tissues in animals. In people, it can cause many respiratory problems and has recently also been linked to heart problems.

After the ground-level ozone has formed, it can combine with other primary pollutants (such as *hydrocarbons*) to form new substances called PANs (short for peroxyacyl nitrates) that cause severe eye irritation. Ground-level ozone and PANs form *photochemical smog*.

ozone-depleting gases

Stratospheric ozone depletion—commonly called the hole in the *ozone* layer—is caused by gases introduced into the atmosphere almost exclusively by human activities. The primary cause of ozone depletion was a gas called *CFCs* (chlorofluorocarbons), previously used in refrigerants and aerosol sprays. Other contributors include *halons* and a few solvents, such as carbon tetrachloride. As these substances were released, they accumulated in the *atmosphere,* interfering with the ozone layer.

Under normal conditions, ozone (O_3) in the upper atmosphere absorbs ultraviolet light and breaks down into a molecule of oxygen (O_2) and a free oxygen atom (O).

These two parts then recombine to form another molecule of ozone and perpetuate the existence of the ozone layer.

Chlorine from CFCs, however, interferes with this process by reacting with ozone, resulting in the production of an oxygen molecule but no free oxygen atoms because they are tied up with the chlorine. This means CFCs prevent ozone from continually re-forming and thereby deplete the ozone layer.

CFCs were phased out of use on a global basis, thanks to the *Montreal Protocol,* and replaced with *HCFCs*. This substance still caused ozone depletion but to a far lesser degree. As technologies advanced, HCFCs were replaced with *HFCs* (hydroflourocarbons) that do not cause ozone depletion at all.

Unfortunately, there appears to be a relationship between *ozone-depleting gases and greenhouse gases.*

ozone-depleting gases and greenhouse gases

Unfortunately, *ozone-depleting gases* are also *greenhouse gases.*

As countries transitioned from one refrigerant to another under the requirements of the *Montreal Protocol,* two environmental problems benefitted: *stratospheric ozone depletion* and *global warming*. This occurred because *CFCs*, the first substance banned by the Montreal Protocol, is both an ozone-depleting gas and a greenhouse gas—a very powerful greenhouse gas about 15,000 times worse than *carbon dioxide*.

The refrigerant called CFCs was replaced with a substance called *HCFCs*, which—although a greenhouse gas—didn't deplete the ozone layer nearly as badly. And now *HFCs* are replacing the HCFCs because they cause no harm to the ozone layer, but they are still a greenhouse gas—more than 2000 times worse than carbon dioxide.

In total, all of these substances (CFCs, HCFCs, and HFCs) cause about 20 percent of global warming. So even though stratospheric ozone depletion is on the mend—thanks to banning the original CFCs—their replacements still cause global warming.

Many environmentalists are urging nations and those individuals who manage the Montreal Protocol to phase out the currently approved refrigerants (the HFCs) and replace them with new refrigerants that not only are safe for the ozone layer but also do not cause global warming. New hydrocarbon refrigerants are commercially available that do not deplete the ozone layer and are not greenhouse gases—the best of both worlds.

ozone depletion, stratospheric

Depletion of the *stratospheric ozone* layer—the so-called hole in the ozone—became one of the most important and pressing environmental concerns of the recent past—and one of the greatest environmental success stories. The problem was studied, a solution was found, and the world—through international agreement—acted on it, resulting in what appears to be a remedy.

Ozone absorbs the sun's ultraviolet (UV) light, so the ozone layer is a vital protective layer for life on Earth. UV light is harmful to most organisms, including humans. It disrupts DNA and causes genetic defects in both plants and animals. It causes skin cancers and eye cataracts in humans and reduces yields in food crops such as corn and rice. A one percent loss of the ozone layer is expected to result in a five percent rise in the incidence of skin cancer.

During the 1980s, a hole containing 50 percent less ozone than typically expected was found in the ozone layer over the *Antarctic* during September and October of each year. On some occasions the hole was the size of the entire continental United States. The ultraviolet radiation reaching the Earth's surface increased about 20 percent during these periods. Scientists later discovered a hole over the *Arctic* as well, missing 20 percent of its usual complement of ozone. After a few months, this hole broke up and dispersed over much of Europe and North America. Estimates in 1988 had the overall annual loss of ozone over most of North America, Europe, and Asia at about three percent when compared with 1969.

Ozone depletion is caused by *ozone-depleting gases* that accumulate in some regions of the stratosphere during certain times of year. These gases take a long time to work their way into the atmosphere and remain active for many years, so resolutions will take a long time to fix the problem.

Scientists stated 20 years ago that if we stop putting these ozone-depleting gases into the air (which we have done), it will take decades if not centuries to recover. It has been a few decades, and recent readings appear to show a definitive trend—ozone depletion or the hole-in-the-ozone is getting smaller.

ozone satellite

This is a common name for a U.S. space satellite properly called the Upper Atmosphere Research Satellite (UARS). It was in a unique situation to both help identify a global environmental problem and to help confirm that the problem is being resolved, something never done before.

UARS was launched in 1991 to study the relationship between *CFCs* and the formation of a hole in the *stratospheric ozone* layer. UARS confirmed the link between the two and that led the way for the *Montreal Protocol*, an international agreement to ban the use of CFCs and other *ozone-depleting gases*.

In 2002 UARS then helped confirm that the Montreal Protocol was working and the ozone hole was diminishing—probably the most notable international environmental success story of our era.

The data UARS collected has also helped with the study of *climate change*, even though at the time of its launch few people understood climate change and fewer were worried about it. The satellite was decommissioned in 2005, and it fell to Earth in 2011. [*space debris*]

ozone, stratospheric

Stratospheric ozone forms the ozone layer, which protects life on Earth from harmful ultraviolet light (UV) produced by the sun. This layer is found roughly 10 to 18 miles above the Earth's surface.

Ozone is composed of three oxygen atoms bound together that are capable of absorbing UV light from the sun. As ozone absorbs the UV radiation, it breaks into smaller pieces. These smaller pieces consist of one oxygen molecule (composed of two oxygen atoms) and one free oxygen atom. The oxygen molecule and free oxygen atoms readily recombine to re-form ozone; thus the process perpetuates itself, continuing to protect us from UV radiation; unless *ozone depletion* occurs.

palm oil

Palm oil is found in numerous items, including packaged foods, body care products, cleaning products, and many others. It is sometimes labeled as palm oil but also by the names palmitate, sodium lauryl sulphate, and stearic acid.

Most of this oil comes from *Indonesia* at great environmental cost. Since 1950, when palm oil became popular, over 40 percent of the Indonesian *rainforests* have disappeared to make room for palm oil farms. These rainforests are *habitats* for many *endangered species*, thereby accelerating their demise. Because of the increased demand for this oil, these *forests* are now disappearing at a rate in excess of 3.5 million acres per year, and the country plans to dramatically increase production over the next few years.

In addition to clearing rainforests, *peatlands* are commonly destroyed to make room for palm oil farmland, also with devastating ecological impact. Destruction of peatland releases far more *greenhouse gases* than the destruction of rainforest.

Eco-certification programs were recently established to let consumers know what they are buying and the environmental destruction it causes. Major doughnut chains have recently agreed to produce their doughnuts using only palm oil from properly certified plantations deemed as *sustainable agriculture*. [*rainforest destruction; forest health, global; forest, sustainable; logging illegal*]

palynology

The study of *pollen* and spores. [*pollination; aerobiology; allergens, natural; asthma and the environment; bumblebee senses; climate change and human health*]

Pangaea

Three hundred million years ago, there was only one giant *supercontinent*, which we call Pangaea. It began to break up into the existing continents about 200 million years ago. Pangaea rested on the equator near where Africa sits today. The earliest dinosaurs were not roaming the planet as we know it; instead, they roamed this one giant supercontinent. [*plate tectonics; geological time scales*]

panspermia

A theory that life on Earth was initiated by the introduction of life from somewhere else in the Universe, meaning it was extraterrestrial. [*extinction and extraterrestrial impact; outer space; near earth objects; space debris; Blue Marble photograph, original and update; Earthrise photograph*]

pantheism

A religious belief where nature and God are considered one and the same. [*primitivism; Biophilia Hypothesis; Gaia Hypothesis; nature-deficit disorder*]

paper recycling

Landfills are not "filling up" with *disposable diapers* or *plastics* nearly as much as with paper—the largest component of *municipal solid waste*. *Recycled* paper products help reduce the solid waste dilemma, but there is a great deal of confusion about what is and isn't recycled paper. Saying a paper product is recycled tells only a portion of the story. More important, is the percentage of the product that is recycled from *post-consumer waste*. In other words, how much of the product was already used in another consumer product, and therefore has truly been recycled for a second use.

The terms post-industrial and *pre-consumer waste* mean the material was never in the consumer's hands but recycled from some industrial process. For example, wood chips or sawdust from a timber mill could be collected and called recycled pre-consumer, or post-industrial waste. This is better than simply disposing of the material, but not truly recycled.

Whenever you purchase recycled paper products of any kind, look for the amount of post-consumer waste, post industrial waste (or pre-consumer waste) and "virgin" materials included. For example, some environmentally conscious companies are producing paper products marked as recycled from 80 percent post-consumer waste, 10 percent pre-consumer waste, and 10 percent virgin materials. If the paper is not clearly marked with this information or it is not readily available on the packaging, assume the worst.

paper shopping bags

Plastic shopping bags are considered more harmful than paper bags. The Los Angeles, California, city council recently passed an ordinance banning the use of plastic grocery shopping bags completely and charging 10 cents for paper bags.

However, the question of which is better—paper or plastic—is misleading, because the real answer is neither. While *recycling* both plastic and paper is good, reusable bags are the only ones that won't continue to be incinerated or dumped into our *landfills*. [*municipal solid waste disposal*]

parasite

When one organism lives on or in another organism for at least a portion of its lifetime to gain nourishment, it is called a parasite and the relationship is parasitism. The organism gaining nourishment is the parasite and the organism providing nourishment is the *host*. The parasite obviously gains from this relationship, and the host is usually harmed to some degree but not killed, because killing the host eliminates the food supply for the parasite.

Numerous *insects*, such as lice and fleas, are *ectoparasites* that use humans as well as other animals as hosts. The tapeworm, some protozoans, and many *bacteria* are *endoparasites* in humans and other animals. The mistletoe is an example of a flowering plant that parasitizes trees.

Scientists do not understand well the importance of parasites in our *biosphere*. By sheer numbers alone, their impact on *ecosystems* must be significant. Some scientists believe parasites are not appreciated and therefore not studied as they should be; some go so far as to believe parasites are looked at with disdain as opposed to appreciation for their importance.

As examples of this attitude, only one species of parasite is on the IUCN's *Red List of Endangered Species,* and the majority of *conservation* textbooks surveyed between 1970 and 2009 do not mention parasites at all or speak of them only negatively. [*symbiotic relationships; predator / prey relationships; flagship species; charismatic megafauna*]

parasitoidism

A *symbiotic relationship* that is a cross between *parasitism* (with its *parasite/host* relationship) and a *predator/prey relationship*. Parasitoids lay eggs in a host. When the young hatch, they feed on the host, either internally or externally. Once mature, the adult parasitoid exits the body of the host. The host usually dies as a result of the invasion. Most parasitoids are *insects* from the Hymenoptera (wasps) or Diptera (flies) families. Parasitoids are sometimes used as an alternative to *insecticides* and can be a component of an *integrated pest management program*. [*biological control methods*]

parent rock

Soil is formed when solid rock breaks down into small particles. The type of solid rock (along with the climate and animal life that exists in the region) dictates the type of soil that will form. The rock from which a soil is formed is called the parent rock.

The parent rock is broken down by different types of *weathering*. Physical weathering involves the effect of temperature changes (freezing and thawing); mechanical abrasion is caused by rock rubbing against rock, pressure exerted by plant roots, or a *glacier* moving over the rock. Chemical weathering is caused by chemical contact from exposure to air, water, or products of plants and animals. Soil forms as the particles become smaller and biological processes become involved. [*soil horizons; soil particles; soil profile; soil erosion; soil organisms; topsoil*]

parthenogenesis

The creation of an individual from an unfertilized egg; a form of *asexual reproduction*. Some *insects,* such as aphids, are parthenogenic. [*hermaphrodites; sexual reproduction*]

particulate matter

One of five types of primary pollutants contributing to *air pollution*. It is any tiny solid particle, such as dust, which is fine soil particles; soot, which is fine carbon particles; and any other fine particles dispersed from *pesticides*, *asbestos*, or thousands of other products. Many of these are simply an annoyance, but others are *toxic*. Even harmless particulate matter often attracts and carries toxic chemicals through the air, such as dust carrying *sulfuric acid*. Particulate matter is the most noticeable type of air pollution because it is readily visible.

Particulate matter can be reduced by pollution control devices in industrial and *electric power plants*. Scrubbers, filters, separators, and precipitators are either built into new construction or often retrofitted into old facilities. *Automobiles* also have emission control devices such as *catalytic converters* to minimize particulate emissions.

In developing countries, primitive *wood-burning cookstoves* burn fuel wood inefficiently and release a great deal of particulate matter causing serious respiratory health issues. In developed nations, many municipalities regulate the number and efficiency of wood-burning stoves to reduce these emissions. Large numbers of wood-burning stoves and fireplaces release enough particulates to produce a symptom called *brown cloud*.

There are two categories of particulate matter based on the size of the particles. They are generally called big particulate matter, or PM_{10}, about 10 microns in thickness—the size of a human hair strand, visible to the naked eye. PM_{10} is emitted from woodstoves, fireplaces, incinerators, and wildfires and are carried by the wind. Fine particulate matter, $PM_{2.5}$, is 2.5 microns or less (30 times smaller than a human hair strand). It is not visible to the naked eye. They consist of tiny bits of smoke and soot from brush fires or *heavy metals* and *toxic* fumes. Industrial *fly ash* emits these fine particulates, causing serious air pollution problems in some cities.

PM$_{2.5}$ is a far greater health risk because these tiny particles can enter the bronchial tubes and get into the lungs. *Green micro-blogs* in *China* made PM$_{2.5}$ a household word because of some of the worst air pollution ever recorded.

parts per million (PPM)

PPM is the unit of measurement used when quantifying the minute amounts of a gas in the *atmosphere*. For example, concentrations of 0.25 ppm of *formaldehyde* in the air (given off by office furniture and carpeting) is not irritating to most adults, but 0.5 ppm can cause burning of the eyes and respiratory irritation in many individuals. [*climate sensitivity; air pollution; aerosols*]

passenger pigeon

Some species become icons of *extinction,* and the passenger pigeon is such an organism. In the late 1800s, the skies of North America could become so full of passing flocks of these birds that day seemed like night. Many nature writers have written about these extraordinary events. It was considered the most common bird in the eastern United States. Hunters would simply aim a shotgun toward the sky and shoot, bringing down dozens. It is hard to believe that only a few decades ago, in 1914, the last of these birds died in the Cincinnati Zoo. Today they are extinct and only found stuffed, in museums. [*biodiversity, loss of; extinction, sixth mass; dodo bird; species recently gone extinct; Red List of Threatened Species; migratory bird protection; Migratory Bird Act*]

passive smoke

See *environmental tobacco smoke.*

passive solar heating systems

Passive solar heating systems simply absorb the sun's energy in the form of heat for immediate use. Insulated windows and greenhouses, for example, allow the radiant energy from the sun in without letting the resulting heat out. Specially designed and oriented walls made of concrete, adobe, brick, or stone absorb the sun's energy and gradually release the heat into the building after the sun sets.

Homes and other buildings designed to use solar heat dramatically reduce the need for other energy sources. Hundreds of thousands of homes and tens of thousands of commercial buildings in the United States use passive solar heating. These buildings can get about 20 to 35 percent of their energy needs directly from the sun, although most require backup conventional heating systems. The technology currently exists, however, to supply about 80 percent of a new building's heating needs in some regions. Solar heat can also provide hot water for a home with the use of passive solar water heating systems that pass water through panels heated by the sun, producing hot water.

A good quality passive solar heating system in a *super-insulated* home is believed to be the least expensive way to heat a home in regions where there is enough sunlight. Even though it raises the cost of a new home slightly, it reduces the overall annual operating costs of the home significantly. [*active solar-heating systems; solar power; concentrated solar technology; solar ponds; solar thermal power plants; homes, green; LEED Certification*]

patch clear-cutting
See *clear-cutting*.

pathogens
Disease-producing organisms such as *viruses, bacteria, fungi*, and *parasites*. [*bird flu virus; bubonic plague; Lyme disease; maggot medicine; medical ecology; skin bacteria; spillover effect; superbugs; vector; antibiotics for livestock*]

payroll deduction, environmental
Payroll deductions are often used as a vehicle to donate money to charitable organizations of all sorts. Earth Share of Washington, D.C., offers a payroll deduction plan that allows people to make charitable donations exclusively toward environmental organizations. [*land trusts; grant-making foundations; NGOs*] {earthshare.org}

PBDE (polybrominated diphenyl ethers)
PBDEs (for short) have been used commonly since the 1970s as a *flame retardant* in consumer goods such as pillows, *mattresses*, and furniture as well as electronic devices such as televisions. (The flame retardant substance that produces PBDEs is called BFR for brominated flame retardants.) This is a persistent substance that does not break down for long periods of time.

In the 1990s, research showed that PBDE appeared in *breast milk* and was becoming a public health risk. It does not enter our bodies via *food chains* and our food; instead, it is in the air, attached to dust we breathe in. Research shows this substance can cause hyperactivity, attention deficit, low sperm count, and possibly damaged brain development in the very young. Scientists believe it might be acting as a *hormone disruptor*, disturbing normal body processes.

U.S. citizens have the highest concentrations in their bodies—10 to 40 times more than people in Europe. It is believed almost all Americans have at least some PBDE in our blood. The amount found in breast milk has risen sharply since originally detected in the 1990s.

There have been attempts to remove the substances from consumer goods. Many companies are voluntarily removing them in advance of regulations and in response to public outcry. [*body burden; toxic pollution; toxic cocktail; hazardous air pollutants; Center for Health, Environment and Justice; Safe Chemicals Act ; bioamplification*]

PCB (polychlorinated biphenyls)

PCBs are a type of *volatile organic* compound commercially produced since 1966 and responsible for widespread contamination. They were used in plasticizers, lubricants, paints, adhesives, inks, and many other products, but primarily in electronic components such as capacitors and transformers.

They are persistent substances, meaning they remain in the environment for long periods of time before breaking down, because they are not water soluble. They *bioaccumulate* in individual organisms and are passed along *food chains* due to *biological amplification,* so they harm most forms of life. Humans get started on a diet of PCBs from birth, because traces exist in most human *breast milk*. It is a known *carcinogen* and *mutagen*.

PCBs were banned in consumer products in the late 1970s in the United States, but remained in electrical equipment until 1990. Just as *DDT* was the environmentalist's icon of contamination in the 1960s, PCBs were considered the same in the 1970s and 1980s. A legacy we continue to live with today; beginning the day a person is born.

Although they are no longer being manufactured, they still enter our environment from *landfills* and *hazardous waste sites*. [*persistent organic pollutants; POPs, dirty dozen; body burden, toxic waste; toxic cocktail*]

PCV valve

The positive crankcase ventilation (PCV) valve is required on *automobiles* sold in the United States. It reduces *hydrocarbon emissions,* a major component of *air pollution*. [*catalytic converter; leaded gasoline; diesel engine fumes*]

peak oil

This term was originally used to describe that time when the amount of *oil* extracted from our planet peaks and begins to decline. The idea being that at that time we must prepare for a non-oil world. Originally proposed in 1949, the timing has been reviewed, revised, redacted, and recalculated many times since and continues to be pushed back because of new extraction technologies. For the record, the *International Energy Agency's* 2010 report states crude oil production had already peaked and the scientific consensus as of 2012 was that it actually happened in 2006.

The point appears irrelevant to many, because we must plan for a non-oil (and non-fossil-fuel) world in advance of its depletion. *Climate change, pollution,* and related environmental problems caused by *fossil fuels* necessitate a transition to *alternative energy sources* before we run out of fossil fuels. Possibly, it was wishful thinking that we would be forced to shift to clean energy sources because we will run out of fossil fuel.

People are now talking about the peak oil pertaining to consumption, as opposed to extraction of oil. As extracting oil becomes more and more difficult, the economics are changing. Also, the move to *natural gas* in the United States because of the *fracking* revolution is reducing our need for oil. The term peak oil will continue to be used, but exactly what it means will change over time. [*bridge fuel; renewable energy*]

peat

A rich, fertile *soil* composed of at least 50 percent *organic* matter. Often found in large areas called *peatlands*.

peatlands

When vegetation dies in most *ecosystems*, it breaks down rapidly, thanks to *decomposers* that break down the *organic* matter. These decomposers are *aerobic*, meaning they require oxygen to live. Peatlands, however, are found in regions that remain waterlogged much of the time and therefore contain little oxygen. (Most peatlands are found in *wetlands*.) Because the usual complement of decomposing organisms cannot live in these environments, the vegetation breaks down slowly. This causes the dead vegetation to accumulate, building up large deposits of partially decomposed organic matter called peat and forming peatlands. The formation of peat is a very slow process. It takes roughly 20 years for about one inch of peat to form. But this accumulation can occur over thousands of years. Most peatland deposits were created at the close of the last ice age almost 10,000 years ago.

Almost all peatlands are found in northern latitudes, such as in Ireland, Scotland, Germany, and Scandinavia. In North America they are located in Canada and Michigan but can occur in the south, as in the Florida *Everglades*. They do not all form in the same way. Depending on certain factors, the peat produced can vary and the peatland is named accordingly. For example, peatlands can be called mires, fens, moors, or *bogs*.

Peatlands have many environmental issues associated with them. They are vast *carbon sinks*, storing carbon and keeping it out of the *atmosphere*. Scientists believe there might be more carbon stored within peatlands than in all of the *tropical rainforests*. If large amounts of this carbon were released, it would significantly increase *global warming*. This occurs when people in *developing countries* burn peat as an inexpensive fuel, releasing *carbon dioxide*. Another concern, also found in *developing countries*, is the draining of these peatlands for land development. Once dry, the aerobic organisms begin to decompose the organic matter they could not decompose while waterlogged. This decomposition also results in carbon dioxide production.

However, the biggest environmental concern about peatlands is *climate change*. The world's largest peatland is found in Western Siberia in the form of *permafrost*, where the peat has been frozen for more than 10,000 years. In recent studies, climate change appears to be causing the permafrost to begin to thaw. This massive peatland—the size of France and Germany combined—could release billions of tons of *methane gas* into the atmosphere if it were to melt. The potential for large releases of methane gas from melting and carbon dioxide from drying has many environmentalists and scientists concerned about peatlands' potential contribution to climate change. [*permafrost, Siberian; palm oil; carbon emissions*]

Pebble Mine

This is a proposed gold and copper open pit mine proposed to be created in the pristine Bristol Bay area of Alaska. If it is built, it will be the biggest such mine in North America. These large mines have long track records of being environmentally devastating to the local *ecosystems*—in this case the local ecosystem is the headwaters of the largest *salmon* fishery in the world and an extensive wild *watershed*. Nine indigenous tribes are trying to stop the mine because it will destroy the salmon runs and their way of life. [*gold mining; mineral exploration (mining); Mining Law of 1872; rare earth metals*]

pelagic marine ecosystems

Marine ecosystems can be divided into two major types: *benthic* marine ecosystems are found at the bottom of the oceans and seas, and pelagic marine ecosystems are found in open bodies of water. Pelagic ecosystems are then divided into two zones: neritic (the coast) and oceanic (the open seas).

The *neritic zone* is found above the *continental shelf*. It begins at the continental shores and extends out to sea several miles, reaching a depth of about 600 feet. This zone takes up only 10 percent of all the oceans but contains 90 percent of all the marine plants and animals. *Oceanic zone ecosystems* are found beyond the continental shelf and are considered relatively nonproductive (*net primary productivity*) when compared with the coastal regions and most terrestrial ecosystems (*biomes*).

penguins

Most people are not surprised when they are told *alligators* have been around since dinosaurs roamed the land, but they are amazed to learn that so have penguins. Over the past 70 million years (dinosaurs became *extinct* 65 million years ago), penguins evolved into 17 different species living on four continents including *Antarctica*, where they originated. At least five of these 17 species are under threat of *extinction* because of *commercial fishing*; vanishing populations of krill, an important food source; and *habitat loss*, *invasive species*, and other unique situations. Populations of Humboldt Penguins in Peru and Chile are in decline because villagers are harvesting their *guano* to be sold as *fertilizer*. (People harvest the guano for fertilizer, but the birds need it to build their nests.)

A 2011 study estimated that a few species of penguins have about 150 years left before they will be gone. However, all is not bad news. Penguin diversity overall should protect them from mass extinctions, and some species populations are even increasing. Conservationists must concentrate on the specific species in trouble. Most people only see penguins in *zoos*, where these few individuals are ambassadors for their wild brethren.

PERC

See *dry cleaning*.

percolation

The downward flow of water through *soil* or *solid waste*. For example, rainwater percolates through a *landfill,* producing *leachate* that enters the soil beneath and might contaminate the *groundwater*. Also, rainwater percolates down into recharge zones to refill *groundwater aquifers*.

peregrine falcon

Believed to be the world's fastest animal, this falcon can attain speeds of more than 200 miles per hour in an aerial dive. This bird almost became *extinct* during the late 1960s and early 1970s because of the *insecticide DDT*. During the mid-1970s, the birds made a recovery, thanks to restricted use of DDT and efforts from organizations such as the Peregrine Fund and the *National Park Service*. [*endangered and threatened species lists; dodo bird; passenger pigeon; flagship species; Lazarus taxon; Lonesome George; species recently gone extinct, ten; penguins*]

perennial plants

Plants that live more than three years and have specific growth periods each year. Vines, shrubs, and trees are perennials. [*annual plants; succession, primary terrestrial*]

periodic table of elements

If you haven't looked at the periodic table for a while, it has changed. In 2012, two elements were added, Flerovium (FL) and Livermorium (LV), and in 2013 one more was added that has not been named yet. All of the elements that exist in nature have been identified, so all of these new elements are manufactured and only exist for brief moments, in laboratories. [*nutrients, essential*]

permaculture

A term that combines the words permanent and culture, or permanent and *agriculture*. The philosophy was developed about 30 years ago in *Australia* by Bill Mollison. In 1978, Mollison, a wildlife biologist, and a student of his, David Holmgren, published a book called <u>Permaculture One</u>, introducing the idea of a "design system for creating sustainable human environments, based on observations of natural systems."

These ideas are similar to what is often now called *sustainable agriculture* and *organic farming*. Permaculture emphasizes the fact that nature doesn't have to work hard at creating sustainable habitats and humans can do the same—if they work with nature instead of against it. Permaculture emphasizes education, so it is a good way to get more people knowledgeable about and involved in sustainable agriculture. [*conservation agriculture; low-input agriculture; regenerative agriculture; agroforestry*]

permafrost, Siberian

Permafrost is any part of the Earth's solid surface that remains frozen for two or more years. The largest single area covered in permafrost is in Siberia. It is so large it covers almost a quarter of the northern-hemisphere landmass of the entire planet. This particular region of permafrost is composed of *peat* and considered a *peatland* region.

This permanent frost is believed to hold about 1 trillion tons of *methane* and *carbon dioxide*, both *greenhouse gas*es. Estimates are that 1/6 of all the *methane* found naturally in land is found in this Siberian permafrost. As the permafrost melts from *climate change*, it releases this methane and carbon dioxide, which in turn causes more warming, creating a positive feedback loop. At the moment, an area the size of France and Germany combined is melting. [*carbon sink; carbon emissions*]

Persian Gulf War pollution

Although brief, the Persian Gulf War of 1991 produced an environmental disaster. Almost 700 *oil* well fires were set in Kuwait by the retreating Iraqi forces, and another 100 wells produced a multimillion-gallon *oil spill* in the gulf. Severe *air pollution* in the skies over Kuwait and Saudi Arabia, caused by the fires, forced about ten million people to breathe the polluted air during this time. Much of this oil contained sulfur, which, when burned, produces *sulfur dioxide* that contributes to *acid rain*. The oil that spilled into the gulf caused immediate damage by killing thousands of birds and other *wildlife*, but the long-term effects on the entire *ecosystem* are still being assessed. [*oil spills in U.S.; oil spills in history, ten worst; oil tankers; pipeline spills*]

persistence (populations)

See *population stability.*

persistent organic pollutants (POPs)

POPs are synthetically produced compounds used as *pesticides*, or in some industrial processes, created as a result of the combustion of certain substances. The worst twelve are often called the *dirty dozen POPs.* Just a few are *DDT, PCBs, dioxins,* and furans.

These pollutants are called persistent because once applied or released, they remain in the *environment* for long periods of time as they do not break down into harmless substances. They collect in individual organisms through the process of *bioaccumulation* and can reach very high concentrations in organisms high up food chains through the process of *biological amplification.* Almost all of these substances are known to cause cancer and damage the liver, nervous system, and reproductive system in humans and most wildlife.

POPs can travel on air currents and have been found in almost all locations around the globe. This universal threat prompted an international agreement called the *Stockholm Convention on Persistent Organic Pollutants* that took effect in 2004. Countries signing this treaty agreed to reduce and eliminate these substances over time. It originally only addressed the dirty dozen, but additional substances were added in 2009.

The widespread long-term use of these substances, without previous sufficient (or any) testing to study their safety is a lesson about what happens when countries do not use the *precautionary principle* or even simple logic about deadly chemicals. [*pesticide dangers; persistent pesticides; Toxic Substances Control Act; Federal Insecticide, Fungicide, and Rodenticide Act; toxic cocktail; body burden*]

persistent pesticides

Once applied, persistent *pesticides* last for many years, so they need not be applied often. These long-lasting, synthetically manufactured pesticides are also called "hard pesticides." Because they remain in the environment long after application, their *toxicity* remains as well. Eight of the most notorious persistent pesticides have been designated to be phased out globally by the *United Nations Environment Programme* via the *Stockholm Convention on Persistent Organic Pollutants*. (This *convention* also controls substances other than pesticides.)

These eight are *DDT*, toxaphene, mirex, heptachlor, *chlordane*, *aldrin*, endrin, and dieldrin. They have been banned in the United States (with few exceptions) but are still common in our environment, posing health risks. [*body burden, toxic cocktail*]

personal portable air-pollution sensing device

If you have never heard of it, it is because it doesn't exist yet; but it will soon. The term refers to a small, wearable device that would measure the air you breathe for pollutants as well as your physiological response in real time, such as heartbeat and breathing rate. Because it would read both factors at the same time, it could provide cause-and-effect relationships for you and also for specific demographic groups such as elderly people, children, expecting mothers, and those with some form of preexisting condition. The *Environmental Protection Agency* has a competition called "My Air, My Health Challenge" to help advance this type of research. [*biosensors; biotechnology; geomedicine; geoengineering; nanotechnology*]

pest

Any organism found in a location where humans prefer it not to be. Although the term can be used for both animals and plants, *weed* is often used for plant pests.

But, the term is very subjective. One person's pest might be another person's pet and one person's weed might be found in another person's vase. [*pesticide; biocide*]

pesticide

A generic name for poisons that kill unwanted organisms. They can be categorized as either synthetic pesticides or *natural pesticides*. Pesticides have historically saved lives by protecting people from insect-transmitted diseases as well as protecting our food supply. Pesticides were at first thought to be a panacea, but many *pesticide dangers* have become evident over the past few decades.

About 5.2 billion pounds of pesticides are used each year, globally. Almost all of these are synthetically produced as opposed to natural pesticides. Pesticides can be designated by the type of pest targeted for destruction. *Insecticides* kill insects, *herbicides* kill weeds, *fungicides* kill fungus, rodenticides kill rodents, and so on. In the United States, about two-thirds of all pesticides used are herbicides, about 23 percent insecticides, and 11 percent fungicides. The vast majority of these pesticides are used on only four crops: corn, cotton, wheat, and soybeans. Of all pesticides used in the United States, 23 percent are not applied to crops but to *lawns*, gardens, *golf courses*, and other nonfarming lands.

pesticide dangers

Each year, more than one billion pounds of *pesticides* are used in the United States, and they come with numerous environmental and public health problems. Environmentally, the biggest issue is the fact that pesticides do not just kill pests. Most are harmful to all organisms, including humans. For this reason, many people prefer to use the term *biocides*—coined by *Rachel Carson*—because they can kill all forms of life.

Numerous studies have found birds, *mammals*, and fish to be poisoned by pesticides, either by direct contact or from *biological amplification*. Pesticides have reduced many species populations and driven some to the brink of *extinction*. Many pesticides *leach* through the *soil* and contaminate *groundwater,* which supplies 50 percent of U.S. *drinking water*. Pesticides have been found in drinking water supplies throughout the United States. In addition to our water, pesticides are in and on our foods in the form of *pesticide residues*.

Studies have shown many of these pesticides increase the risk of cancer, birth defects, and many chronic illnesses. Besides long-term concerns, roughly 50,000 cases of direct pesticide poisoning are reported in the United States annually. In most cases, these are people that apply these chemicals on a professional basis.

In addition to harming the environment and the public health, there are other issues. Many pesticides have become less effective because the pests become resistant to the poison through the process of *natural selection*. The more resistant they become, the more chemicals must be used. The quantity of pesticide required to kill often increases by a factor of 100 after just a few years. In the past 50 years, the quantity of pesticides used has dramatically increased, but the quantity of crops lost to the pests also continues to rise.

Pesticides can also turn nonpests into pests. Some insect pests that are normally controlled by *parasites* and *predators* increase in numbers dramatically when pesticides kill their parasites and predators. Increased numbers often turns an otherwise harmless insect into a pest.

Alternatives do exist. Organic foods avoid the use of pesticides and have skyrocketed in popularity. Many *eco-labels* and *eco-certification programs* ensure you are getting pesticide-free products. On a larger scale, farmers have alternative methods of protecting their crops without extensive use of pesticides, including *integrated pest management* (IPM) and *biological control methods. Genetically modified organisms* are also used to control pests.

pesticide regulations

The threat posed by pesticides was first brought to the public's attention by *Rachel Carson* in her classic 1962 book <u>Silent Spring</u>. The dangers have been well founded and still exist today. The *Federal Insecticide, Fungicide, and Rodenticide Act* of 1972, called FIFRA for short, is the primary regulation controlling pesticides. The *Environmental Protection Agency* (EPA) is responsible for enforcing this act. (But the *Federal Food, Drug, and Cosmetic Act* controls pesticides when used on foods.) Approximately 20,000 pesticides are in use and regulated by FIFRA.

Originally considered a weak regulation, FIFRA was beefed up in the 1990s. Now chemical manufacturers must prove their products won't cause "unreasonable harm" to the environment or human health.

The Federal Food, Drug, and Cosmetic Act, however, sets a maximum tolerance level for each pesticide and *pesticide residues on food* must be below the acceptable tolerance level when harvested. [*pesticide dangers*]

pesticide residue on food

The *Environmental Protection Agency* regulates *pesticides* as established by the *Federal Insecticide, Fungicide, and Rodenticide Act*. However, the safety of our food supply is established by a separate law called the *Federal Food, Drug, and Cosmetic Act*. This act creates tolerance levels for pesticides on foods for interstate sale in the United States and for exportation.

Pesticides can only be registered by the *Environmental Protection Agency* (EPA) for use on a food if a tolerance is first established and granted. The EPA has approved about 300 pesticides for food use.

Almost everyone agrees that some level of pesticide residues exist on many foods, including fruits, vegetables, and grains, but the question is, are they safe? The EPA believes they are safe, but many environmental groups say they are not.

The FDA tests about one percent of the food supply for pesticide residues. More than half of the foods that get tested have pesticide residues. But, as long as the amount of pesticide found on the tested food doesn't exceed the accepted tolerance levels, these residues are legal. And this is where the main argument rests.

Pesticides banned in the United States are still exported to other countries. Many of these banned pesticides are used in other countries on foods that are exported back into the United States. Therefore, pesticides banned by the United States may make their way back to our tables on imported foods. This has been called the *circle of poison*. FDA testing consistently shows more pesticide residues on imported fruits and vegetables when compared to domestic-grown produce—but still, in most cases, below the required tolerance levels.

United States regulations also allow pesticides to contain impurities. These impurities can include the active ingredients of banned pesticides. For example, some pesticides sold today contain *DDT* as an impurity—a pesticide banned decades ago. [*fruit waxing*]

PET

PET is an acronym for polyethylene terephthalate, a *plastic* used to produce rigid containers such as carbonated soda bottles. PET comprises about 25 percent of all the plastic produced in the United States but is one of the most successfully recycled plastics. In the United States, almost 30 percent of all PET plastics are recycled.

Consumer interest in *plastic recycling* programs has boosted the recycled product market for PET and forced companies to use more and more recycled PET. Today, many major manufacturers make their plastic bottles out of 50- to 99-percent recycled PET plastic. These products are stamped with the number 1 surrounded by a triangle made of arrows. [*plastic pollution; nurdles; bottled water*]

Petermann Glacier ice detachment

The Petermann *Glacier* is located in the Arctic, in northwest Greenland. In 2010, a section four times the size of Manhattan broke off (calved), creating an iceberg 100 square miles in size with walls 3,000 feet high. It is the largest such iceberg from Greenland. Another piece appears poised to calve from the same shelf. The *Arctic ice* sheets overall are melting, and this is seen as another indication of *climate change*.

Other such events have occurred on the Larson B ice shelf in the Antarctic. The biggest recent iceberg calved from the northern rim of a Canadian island in the Arctic in 1962.

petrochemicals

Petrochemicals are substances obtained from crude *oil* or *natural gas*. The oil or gas is refined in some way to create hundreds of these petrochemicals, such as acetylene, benzene, ethane, ethylene, *methane*, and propane. These petrochemicals are in turn used to create thousands of other substances used as fibers, *solvents*, paints, *pesticides*, *fertilizers, feedstock,* and countless other products including *plastic*. Many people do not realize just how many of the products we use today come from these *fossil fuels*.

petroleum

See *oil* and *petrochemicals*.

pH

The scale used to determine the acidity or alkalinity of a solution. All water, regardless of what it contains (the solution), falls somewhere between being highly acidic and highly basic (alkaline). Because all life, both plants and animals, require water to survive, the pH of the water in or around an organism is critical. Changes to the water's pH have a direct impact on the organism and whether it can survive or not.

pH is measured on a scale of 0 (most acidic) to 14 (most basic), with 7 being neutral (plain water). The scale is logarithmic, meaning it increases by 10 for each number; that is, four is ten times more acidic than five. Anyone who has ever had tropical fish knows they can only survive if the water is within the correct pH range. In nature, all organisms have optimum pH levels and can only survive within a specific pH *tolerance range. [limiting factors; natural selection]*

phantom load

See *vampire energy*.

pheromones

Pheromones are a subset of *semiochemicals*. Pheromones are substances produced by an organism to communicate information to other members of the same species. For example, individuals of one species of ants, if attacked by a *predator*, secrete an alarm pheromone warning other members in the colony of danger. Other pheromones are *sex attractants* that have been successfully used as nontoxic *natural pesticides,* as in the case of the tomato pinworm, a serious pest. When synthetically produced sex pheromones of the pinworm are applied to tomato vines, the male pinworm finds it impossible to find a mate amongst what appear to be females everywhere. They end up not mating, and the pest population plummets.

Contrary to popular belief—and what many perfume manufacturers might tell you—no sex pheromone has yet been discovered for humans. *[biological control; biological control methods; biological pesticides]*

philosophers and writers, environmental

There have always been men and women who helped us develop an awareness about our relationship with nature. In the early 1800s, *Ralph Waldo Emerson's* essay <u>Nature</u> and, later that century, *Henry David Thoreau's* <u>Walden Pond</u> raised our environmental conscience. *John Muir's* writings and his establishment of the *Sierra Club* in 1890 helped move beyond awareness to create environmental activism.

In the early 1900s, *Aldo Leopold* combined philosophy and solid science in his writings and studies. He philosophized in <u>A Sand County Almanac</u> and proposed scientific hypotheses in the <u>Bulletin of the American Game Association</u>. *Rachel Carson* melded the two into one with the classic, <u>Silent Spring</u> in the early 1960s. Many believe this book, about *pesticides*, was the catalyst for today's *environmental movement*.

Eco-theologians are people who link spirituality with *ecology*. One such, *Thomas Berry*, brings a new approach to our worldview with his book <u>The Dream of the Earth</u>. [*Abby, Edward; Audubon, John James; Bartram, William; Berry, Wendell; Brower, David; Brown, R. Lester; Burroughs, John; Commoner, Barry; Darling, J.N. "Ding"; Darwin, Charles; Dillard, Annie; Dubos, Rene; Ehrlich, Paul; Eiseley, Loren; Fuller, Buckminster; Goodall, Jane; Gore, Al; Hardin, Garret; Hayes, Denis; Lovins, Amory B.; Marsh, George Perkins; McKibben, Bill; Merchant, Carolyn; Mowat, Farley; Orr, David; Quammen, David; Roosevelt, Theodore; Ward, Barbara; White, Gilbert; Whitman, Walt; Wilson, E.O.*]

phosphorus cycle

Phosphorus is an essential nutrient used by organisms to build DNA and cell membranes. Inorganic phosphorus is found in phosphate rocks. *Weathering* and *erosion* breaks down the rock, allowing phosphorus to dissolve in water and become available for green plants to absorb through their roots. The phosphorus is passed along *food chains* to animals. When plants and animals die, the organic molecules containing phosphorus are decomposed back into simple inorganic phosphorus to be used by other green plants. [*biogeochemical cycles; biogeochemical cycles, human intervention in; safe operating space*]

photochemical smog

Burning a *fossil fuel* releases five primary air pollutants that cause *air pollution*. Many of these primary pollutants react with each other in the presence of sunlight to create *secondary air pollutants*. Some of these secondary pollutants produce a form of air pollution called photochemical smog.

The process begins when *nitrogen compounds* (a primary pollutant) react in the presence of sunlight to create *ground-level ozone* (a secondary pollutant). Ozone damages *chlorophyll* in plants and lung tissues in animals. After ground-level ozone has formed, it can combine with other primary pollutants (such as *hydrocarbons*) to form new substances called *PANs* (peroxyacyl nitrates), which cause severe eye irritation. The combination of ground-level ozone with these other substances is called photochemical smog. Certain climate and geographical factors can intensify the smog in a process called *thermal inversion*.

photosynthesis

Radiant energy from the sun is captured by green plants and converted into chemical energy during the process of photosynthesis. The chemical energy is in the form of sugar (glucose) molecules. Photosynthesis uses *carbon dioxide* and *water* to form this sugar and releases oxygen in the process. This energy is then released during the process of *cellular respiration* and used by plants and animals (by way of *food chains*) to survive.

photosynthetically active radiation (PAR)

Those wavelengths of light capable of driving the process of *photosynthesis*.

photovoltaic cell (PV)

Photovoltaic cells, also called solar cells and used in solar panels, convert solar energy (light) directly into electricity. They were first created in 1954 by Bell Laboratories and became well known years later in solar calculators. Unlike *fossil fuels, solar power* from photovoltaic cells produces little *air pollution* or *water pollution* and no *carbon dioxide* to contribute to *climate change*.

Today, PV cells provide electricity to 13 million homes globally. *Germany* has almost half of the world's installed capacity, with Italy second at about 17 percent, and the United States has about 5 percent. The strong European interest in PV is a result of feed-in tariffs, policies that ensure a financial incentive for anyone producing solar energy and feeding it into the *power grid*.

PV cells contain purified silicon (from sand) and trace substances such as cadmium sulfide that conduct small amounts of electricity when struck by sunlight. Because each cell produces a small amount of power, they must be linked together to produce the power needs of a home, which is why you see arrays of numerous PV panels on roofs.

Large-scale applications, containing many square miles of these cells, can act as *solar power plants*. It is estimated 25 percent of the world's energy needs and 50 percent of U.S. needs could be supplied by solar power plants within 50 years.

At present, photovoltaic cells are expensive and not cost competitive, but this is expected to change within the next few years thanks to new technologies and increased use. Now only about 15 percent of light striking the cells is converted into power, but newer PV technologies will increase this percentage and make it more competitive.

A new cone-shaped solar cell is being tested that spins the cells away from the sun when they begin to overheat, which reduces efficiency. This type of cell is installed on a pole-like device that eliminates the need for rooftop arrays. The first commercial installation is scheduled in 2014.

Even though PV power is considered one of the best bets for the world's future energy needs, U.S. research dropped by 75 percent during the 10-year period beginning in the early 1980s, causing the United States to lose 50 percent of its market share in the PV world market. [*chlorophyll-based solar cells; photovoltaic cell film; solar charging of electronic devices; baseball stadiums, green; crowd funding, green; eco-cities; solar cookstoves; solar glitter*]

photovoltaic cell film

A recent invention allows *photovoltaic cell* functionality in a thin film that can be applied to almost any flat surface as opposed to the standard panels. The film uses infrared light, which means it can be placed on windows and still allow visible light to penetrate. This type of photovoltaic cell only has an efficiency of 10 percent (converting 10 percent of the sunlight into power), compared to 15 percent in typical photovoltaic solar panels. However, the ability to cover almost any flat surface and turn it into a solar cell is enticing in spite of its lower efficiency.

phthalates

Pronounced THAL-ates, this is a substance used to soften up *PVC*—a common *plastic*—used for things such as old vinyl records, many toys, shower curtains, computer cables, and numerous others; they are even found in some medicines. Like so many *toxic substances*, they have entered our *waste stream*, then entered our *environment,* and now are found in our bodies. Minute quantities of phthalates are found in most of us.

Some research has linked them to birth defects and asthma and they are probably endocrine *hormone disruptors* and possibly *carcinogens*. They have been banned in Europe and will likely be banned in the United States sometime in the near future. [*precautionary principle; body burden, toxic cocktail*]

phytoplankton

Phytoplankton, also called microalgae, are free-floating microscopic plants consisting mostly of green *algae*, *diatoms*, and desmids. (It can include dinoflagellates, but they move by their own power.) Phytoplankton are the most important *autotrophs* (producers) in *aquatic ecosystems*, both *freshwater* and *marine*. They use more *carbon dioxide* than all plants on land combined. They live near the water's surface because they need sunlight to perform *photosynthesis*. Research shows that phytoplankton populations have dropped about 40 percent since 1950, probably from *climate change*. Because they are at the bottom of these *food chains*, changes in phytoplankton impact entire *ecosystems*.

phytoremediation
The use of plants to detoxify a contaminated *environment*—for example, *fungi* and *algae* have been used to clean up the radioactive soil and water around a *nuclear power plant meltdown* such as *Chernobyl*. [*bio-ore; bioremediation; hyperaccumulators; chledophyte; remediation of hazardous waste*]

pig-gate
Social media has many benefits. However, national news networks and anyone serious about delivering news to audiences should know better than to assume something is factual because it was seen on the Internet. In February of 2013, "NBC Nightly News" showed a YouTube video of an pig allegedly saving a baby goat from a muddy pond in a petting *zoo*. Even though the news anchor stated they did not know if it depicted a real event or one staged, they showed it to their national audience as news. The next day they admitted they had been fooled—the video was an elaborately staged hoax. Social media has labeled the event "pig-gate." [*news literacy; science, pseudoscience; opinions and science; blogs, green; microblogs, green; tweets, eco-; tweetstorm, Rio+20; social media, environmental*]

Pinchot, Gifford
(1865–1946) Pinchot graduated Yale, was trained in *forestry* in France, and then returned to the United States. He played a major role in the creation of the *U.S. Forest Service* (called the Bureau of Forestry at that time) and was its first chief officer. He is considered the father of modern forestry. Pinchot worked with (and sometimes against) *President Theodore Roosevelt*. Pinchot believed in *dual-use* (also called *wise-use* or the *conservation* ethic), meaning forests should be used as a resource for humans to exploit but also used in a sustainable manner so they will be here for future generations. His beliefs were opposed by *John Muir* and even his boss, Roosevelt. Muir and Pinchot had a well-publicized feud over the creation of the *Hetch Hetchy* dam in the *Grand Canyon*—a battle Pinchot won, much to Muir's disdain. [*forest health, global; forests*]

pioneer community

The first stage (or sere) in the process of *succession*. *Lichens* are often the first organisms to establish themselves in a pioneer community. These pioneer species prepare the *environment* for the latter stages of succession.

pipeline spills

While *oil tankers* and *oil* rig spills get most of the attention (and for good reason), many smaller but environmentally important spills occur from normal *port operations* and pipeline leaks. In 2012, an oil pipeline burst in Laurel, Montana, near the Yellowstone River, losing more than 40,000 gallons of oil into the water before it could be sealed. The oil traveled 240 miles downriver, resulted in an evacuation, and caused serious harm to the local *ecosystem*. The pipeline carried especially *toxic* oil from the *Canadian tar sands*. The Governor of Montana used *crowdsourcing* to track the pollution as it flowed downstream.

A larger pipeline spill occurred in the Kalamazoo River in Michigan in 2010. Many environmentalists are calling for stricter federal pipeline regulations to ensure better maintenance and oversight so these spills happen less often. [*oil spills in U.S.; Keystone XL pipeline*]

pirate fishing

Any form of illegal fishing. There is an official category of fishing called IUU, standing for *illegal, unreported, and unregulated fishing*. It is considered pirate fishing if they are using fishing gear such as *longlines* or *purse seine nets* that are banned by the country owning the waters, fishing for protected species, or fishing in *marine preserves* that are no-take zones. For example, bluefin *tuna* cannot be caught with purse seine nets during certain seasons, and *driftnets* are banned in all Italian waters. [*fishing, commercial*]

PIRG, U.S. (Public Interest Research Group)

U.S. PIRG (United States Public Interest Research Group) is a national lobbying office that represents state PIRGs around the country. Their purpose is to perform research and lobby for national environmental and consumer protection legislation. They focus on strengthening the *Clean Water Act*, supporting *recycling* policies, and alerting consumers regarding unsafe products, and they support new product safety laws. [*NGOs; Office of Technology Assessment, former; League of Conservation Voters; environmental literacy; environmental education*] {uspirg.org}

planetary boundaries

See *safe operating space*.

plankton

Plankton are microscopic organisms found—usually in large numbers—in slow-moving or standing bodies of water such as the *ocean*, *estuaries*, lakes, and ponds. Plankton is divided into plant plankton called *phytoplankton* (which include floating green *algae*, desmids, *diatoms*, and *cyanobacteria*) and animal plankton called *zooplankton* (which include *crustaceans*, rotifers, and protozoans, most of which can move about in the water).

Food chains of many aquatic ecosystems begin with phytoplankton that acts as the *autotrophs* (producers). They are eaten by the zooplankton—the primary consumers (*heterotrophs*)—and provide a food source for higher organisms up the food chain, such as fish. [*bacteria; algal bloom; Arctic ice, changes in; forage fish; melt ponds; oceanic zone ecosystems*]

planned obsolescence

It is hard to believe, but products used to be built to last a long time and people kept them for as long as they lasted. During the late 1920s and early 1930s a few people came up with the idea of planned obsolescence: make things that only lasted a short while or became outdated quickly. Then people would have to (or want to) buy replacements. This was a great idea for economic growth and it has worked that way ever since. Some credit the *automobile* industry with starting this idea with new models every year. Others give credit to Bernard London who wrote a booklet called "Ending the Depression Through Planned Obsolescence" and still others to Vance Packard's 1960 book titled The Waste Makers.

Today we have taken planned obsolescence to the next phase; we don't even wait until it breaks because something better becomes available. This is especially true of electronic gadgetry. When was the last time your phone or other device actually stopped working before you bought a new one? From a personal perspective, it is everyone's right to buy whatever they want. But planned obsolescence comes with environmental degradation that we must live with.

All of what we make comes from somewhere—it must use resources from the Earth and those resources are finite. And all of what we make—most of which is not *biodegradable*—must go somewhere when it becomes waste, and those places are also finite. [*municipal solid waste disposal; recycling; product packaging; cradle-to-grave;*

deadtime; waste minimization; plastic pollution; e-waste and you; e-waste; conspicuous consumption; Spaceship Earth; ecological footprint; sustainable development]

plant-oil fuels

One of three sources of *biomass energy*, where plants and animals are used to produce energy. Some plants—such as sunflowers, soybeans, and rapeseed—produce natural oils within their seeds. These oils have been used as a replacement for *diesel* fuel in the past and hold some promise for the future. Plant oils must be refined so they don't foul the machinery they power. (This should not be confused with using plants to create *biofuels*.) Rapeseed can be grown off season so it doesn't compete with food crops. Some forms of *algae* also produce natural oils and are being researched for their potential use. Plant-oil fuels are not currently cost competitive with other fuels, but there is a great deal of interest and research on the subject.

plastic

The plastic products we all live with on a daily basis are synthetically manufactured products made from petroleum (*oil* or *natural gas*). Plastic can be used to make so many products because of its ability to flow into any shape when heat or pressure is applied and then harden into a solid. Its uses are legendary and the environmental problems they now pose are legendary, as well.

Plastic is composed of molecules consisting of long polymers—a linkage of hundreds or thousands of smaller units called monomers. These long chains result in large molecules that are responsible for the biggest environmental problem—they do not decompose; they are too big for most *bacteria* and *fungi*, the most common *decomposers*, to feed on them. Their disposal results in *plastic pollution*. The few alternatives we have include *plastic recycling* and possibly, in the future, *bioplastics*. [*nurdles; plastic shopping bags; bottled water; styropeanuts; HDPE; LDPE; PET; PP; PS; PVC; garbage patch, marine; Plastic Pollution, Research, and Control Act, Marine*]

plastic pollution

Between 100 and 200 million metric tons of *plastic* are manufactured globally each year. *Plastic recycling* is not as popular as people think. Only about 15 percent is recycled, so most of it finds its way into *landfills* or *incinerators*. Plastics make up about 25 percent of U.S. *municipal solid waste*, but much of it never gets into the *waste stream*, instead becoming *litter* on land or in bodies of water. With few exceptions, plastics take so long to decompose, they are considered non-*biodegradable*.

If you consider that only a small percentage of the plastics manufactured are recycled (about eight percent in the United States), an even smaller amount is incinerated, and they do not decompose, we can logically assume that almost all of the plastic that has ever been manufactured still exists.

We know that plastic just sits in landfills, but we are now learning about some of the problems that occur when it makes its way into bodies of water. Plastic is especially insidious, because it remains intact, acting like booby traps for many organisms. *Sea turtles* mistake plastic baggies for *jellyfish*, eat them and die from blocked intestines. Birds swallow bits of plastic foam and choke. Animals often become entangled in plastic debris and either drown or starve. The extent of this damage has only recently been discovered as scientists learned that a great deal of it has made its way into the *oceans* and formed large *marine garbage patches* where it wreaks havoc on *marine ecosystems*.

Human health issues are also of concern. *Toxic* chemicals *leach* out of plastic waste as it resides in landfills or is released into the air during incineration. For example, the incineration of *PVC* products produces *dioxin* emissions, and *phthalates* and *BPA* are now found in *food chains* of which we are part.

Plastic Pollution Research and Control Act, Marine

This act was signed into law in 1988 and prohibits ocean dumping of *plastic* waste. It was signed by then President Ronald Reagan. There have been attempts to update and upgrade the act but they have lingered in committees with no action.

plastic recycling

Only a small percentage of *plastic* products are recycled and that results in a lot of *plastic pollution*. This is unfortunate, because some of the most common types of plastic are readily recyclable. If people were more aware of the problems posed by plastic when it enters *landfills*, causes *litter* and *ocean pollution,* and creates large *garbage patches* of plastic in the oceans, they might recycle more often. Anyone concerned enough to recycle their plastic should make an effort to help the cause further by looking for new products made from recycled plastic. *Recycling* programs only work if a market exists for these products.

The potential for recycling depends on the type of plastic, identified by a number stamped on the product, surrounded by a triangle composed of arrows. These numbers, contrary to common belief, simply identify the resin used to make the plastic. They are not recycling numbers and often mislead people into thinking they are recycled when, in fact, many of them are not. Recycling rates for each of the resins follow.

PET (1) used for *bottled water* and other drinks can be recycled into many products. It is the most common plastic recycled, at about 20 percent, and typically collected at roadside pickups. It is unfortunate that more are not returned, because a strong demand exists for this used material.

HDPE (2) used in laundry and milk jugs and *grocery shopping bags*, can be recycled into many products and often collected at roadside pickups. About 28 percent of the jugs and only 10 percent of the bags are recycled.

PVC (3) used in pipes, plumbing parts, siding, and some bottles. It is rarely recycled or picked up.

LDPE (4) used in squeeze bottles, garbage bags, carpets, and furniture. It can be made into new products, but only about one percent is recycled. Some cities are beginning to include this resin in their pickup programs, so possibly this percentage will improve.

PP (5) used in yogurt containers, ketchup bottles, caps and straws, and medicine vials. It can be used to make new products, but only about five percent is recycled now. Some cities are beginning to collect PP in efforts to improve this number.

PS (6) used in disposable cups, plates, and egg cartons. It is difficult to recycle, likely to cause litter, and often the target of plastic bans. Less than one percent is recycled and it is not typically picked up.

Most environmentalists will tell you that although recycling is good, the best thing you can do for the environment is to use less plastic to begin with.

plastic shopping bags

Paper or *plastic*? You've heard it before at the supermarket and elsewhere. But the answer is actually neither. The environmentally correct thing to use is a reusable bag made of a natural product. However, to answer the question, plastic loses. Even with *plastic recycling* programs, *plastic pollution* is a huge problem. Possibly a sign of things to come, Los Angeles recently passed a city ordinance banning plastic supermarket bags. They are not the first, but are by far the largest city to have done so. But they also assess a 10-cent charge for each *paper shopping bag*. [*three Rs*]

plate tectonics

The accepted theory that states the solid Earth (*lithosphere*) is comprised of huge plates that slowly move. This movement is called continental drift and is responsible for the formation and location of the continents at any given time in the Earth's history. There are eight major plates (and many smaller plates) roughly 60 miles thick, moving a few inches each year. They move because the lithosphere rests on top of another layer called the asthenosphere, which consists of molten (liquid) lava, allowing the plates to seem to float over it.

The boundaries between these plates are unstable regions, resulting in volcanoes and earthquakes. The largest single boundary of these plates beneath the oceans is called the "ring of fire" for this reason. When two plates, moving toward one another, meet at a boundary, they can be responsible for the creation of mountains. The Himalayan Mountains are the highest peaks in the world because they are at the boundary of two plates banging into one another and forcing the Earth upward.

The most recent plate movement is believed to have begun about 200 million years ago when Earth consisted of one giant *supercontinent* called *Pangaea*; that since has broken up into today's continents.

plutonium pit disposal

Now that the Cold War is over, the United States must dispose of tens of thousands of *nuclear weapons*. A plutonium pit is that part of a nuclear bomb that ignites the thermonuclear blast. Plutonium remains deadly for more than 25,000 years and is considered the most deadly substance on Earth. Grains much smaller than *sand* can and probably will cause cancer.

These plutonium pits have been temporarily stored in concrete and steel bunkers in the *Department of Energy's* Pantex Plant in Amarillo, Texas, for decades. This facility can store about 20,000 of these pits and is close to capacity now.

The United States and other countries have the same problem with nuclear weapons as they do with *nuclear waste* from power plants; they cannot determine what to do with it, now that they don't need it. At the moment, scientists believe dismantling these pits to prepare them for use as fuel in a *nuclear power* reactor might be a solution. But currently there is no ongoing process to permanently dispose of these pits. [*military hazardous waste; Hanford Nuclear Reservation; nuclear waste disposal; nuclear waste dilemma, the global*]

PM2.5 and PM10

See *particulate matter*.

poaching

The illegal capture of protected *wildlife*—such as those classified as *threatened species* or *endangered species*. International attempts to regulate poaching and the illegal *ivory trade* are carried out by *CITES* and its member nations. Local poaching might be handled by federal, state, or local officials when and if available. Poaching occurs to species of every size and shape—from magnificent African *elephants* to tiny *elvers* (baby eels).

The ivory trade is a good example of how poaching affects wildlife populations—drastically reducing the remaining numbers of the African elephant. The black *rhinoceros* population, also in Africa, dwindled from 65,000 in 1976 to only 10,000 a few years later as poachers slaughtered them for their two horns, which are worth tens of thousands of dollars on the black market as a reputed aphrodisiac.

Demand for African ivory dramatically increased in the past few years, primarily because of the rising wealth in *China* and their fascination with anything made of ivory. This Chinese demand is probably responsible for a doubling of poaching kills in *Kenya* alone.

Efforts to stop the slaughter had previously been at the enforcement end, which is expensive and difficult because of the vastness of the regions that must be guarded. New campaigns are targeting consumers—educating them about the inhuman and useless killing of these animals. Many consumers are not aware of how the trinkets they buy wreak havoc in the wild.

The use of *conservation drones* is another new method being used to reduce the slaughter. One *drone* can survey an area that would require dozens of guards. [*wildlife trade and trafficking; elephant poaching; biodiversity, loss of; charismatic megafauna; endangered and threatened species lists; extinct in the wild; extinction, sixth mass; polar bear trade; seafood eco-ratings; doomsday seed vault; species recently gone extinct; United Nations Convention on Biological Diversity; whaling; Wildlife Conservation Society; habitat loss; animal law; Lighthawk, the Environmental Air Force; conservation drones*]

podzol

See *soil types*.

pointless pollution

Pointless pollution (also called non-point source pollution) does not originate from any one source, but from many unrelated sources. This conglomeration of pollutants often occurs when rainwater collects residues of *automobile gasoline* and *oil* on roads, *fertilizers* and *pesticides* from farms and *lawns*, and *animal wastes* from feed lots. The water containing pointless pollution is collected and concentrated in drainpipes and drainage ditches and often runs directly into bodies of water. It may also be diverted into bodies of water when municipal *sewage treatment facilities* overflow.

Coastal areas are most significantly affected, often resulting in closed *beaches* and contaminated *fishing* grounds and shellfish beds. [*point source pollution; water pollution; domestic water pollution; Clean Water Act; combined sewer systems; detergents, phosphates in; nutrient enrichment; rivers in decline; running water habitats, human impact on; runoff; algal bloom; Great Lakes Environmental Assessment and Mapping Project*]

point source pollution

Pollution that comes from a single, identifiable source, such as a manufacturing plant's *effluent* pipe that dumps its *industrial waste*water into a stream. Point source pollution is more easily controlled because it is easily identifiable, compared to *pointless pollution*, whose source is often difficult to determine or control. [*water pollution; domestic water pollution; Clean Water Act; combined sewer systems; detergents, phosphates in; nutrient enrichment; rivers in decline; running water habitats, human impact on; runoff; algal bloom; Great Lakes Environmental Assessment and Mapping Project*]

polar bear trade

While the *ivory trade* is better known, there is also a polar bear trade where these animals are slaughtered for their skins (pelts), teeth, and claws. At the moment, the only country where it is legal to kill polar bears for sport and then for international sale is Canada. The Canadian *Arctic* is one of the best remaining *habitats* on Earth where these animals can survive; however, the country has allowed an unsustainable kill to quench the global market for these pelts and other parts. Russia and the United States are urging a trade ban be passed by *CITES* to stop the Canadian kills, but more countries must back the proposal before it will be approved. [*poaching; elephant poaching; rhinoceroses and their horns; wildlife trade and trafficking; biodiversity, loss of; endangered and threatened species lists; extinction, sixth mass; species recently gone extinct; United Nations Convention on Biological Diversity; whaling; Wildlife Conservation Society; habitat loss*]

political ecologist

A person who studies the politics of environmental issues—especially as they relate to how the issues impact people. More specifically, it is a person who deals with the relationships that exist between environmental issues and the political, economic, and social factors associated with these issues. This includes topics such as *environmental justice* (or injustice); environmental degradation as it pertains to the human condition, the *environment*, and conflict; and environmental identities and social movements, to name a few. [*water shortages, global; water diversion; blue-jean pollution; Colorado River; water grabs; energy poverty; land grabs; fracking, ongoing environmental concerns; golf courses and the environment; irrigation, agricultural; Millennium Development Goals; Our Common Future; hunger, world; food desert; food waste; tragedy of the commons*]

pollen

The male reproductive cells of either flowering plants (angiosperms) or naked seed plants (gymnosperms). Most angiosperm pollen is transported to the female reproductive cells via *insects*, while gymnosperm pollen typically floats through the air. [*pollination; pollination, wild*]

pollination

The fertilization of plants that produce seeds. [*pollination, wild; sexual reproduction; seeds and climate change, Project Baseline; bats; black finger of death fungus; doomsday seed vault; robotic bees; heirloom plants; jactitation; plant-oil fuels; quiescence; vegetable seeds, heirloom*]

pollination, wild

Colony collapse disorder is a major concern for most farmers and many scientists, and it should concern us all because it is decimating honeybees that we all depend upon to put food on the table. This rapid decline has prompted a great deal of scientific research looking for ways to stop the disorder.

However, farmers cannot wait and are looking for alternatives to honeybee *pollination* right now. About 30 percent of all the food we eat is pollinated by millions of commercially rented honeybee hives. New research finds that encouraging wild pollination instead of using commercial hives is probably a better method anyway.

Research performed on more than 40 crops in 19 countries found that wild insect pollination, meaning by a variety of insects instead of just honeybees, resulted in far greater numbers of flowers forming seed or fruit (the purpose of pollination). Wild pollination is performed by many types of bees, flies, *beetles*, butterflies, and others. Possibly the solution is to let nature take the lead. [*honeybees, pollination, and agriculture; honeybee sperm bank; robotic bees; bumblebee decline*]

pollination crisis

See *colony collapse disorder*.

polluter pays principle (PPP)

See *extended product responsibility*.

pollution

A negative change in the quality of some aspect of our *biosphere*. These changes, if left unchecked, can cause annoyance, illness, death, or the *extinction* of a species.

Pollution is often identified by the part of the planet harmed, such as with *air pollution*, *water pollution*, or contaminated *soil*. Pollution can also be categorized by its source, as with *pointless pollution* and *point source* pollution or—more specifically—as with *Persian Gulf war pollution* and *cruise-ship pollution*. Finally, pollution can be identified by the actual product or cause of the pollution, as with *plastic pollution* and *noise pollution*.

Pollution problems were documented as far back as the early Roman empire, but the magnitude of these problems has catapulted in modern times for two reasons: *human population growth* and the proliferation of manufactured products. The number of people and the number of products we manufacture, both of which typically occur in dense *urbanized* areas, have increased and concentrated our wastes, including *sewage*, *municipal solid waste*, manufacturing byproducts, and *automobile emissions*. More recently, not only has the number of people accelerated pollution problems, but the number of those living in industrialized nations with high standards of living and a larger *ecological footprint* has also increased pollution problems globally. [*toxic pollution; malodor pollution; light pollution; aesthetic pollution; pollution police; visual pollution; hazardous air pollutants; airplane pollution; lawn-mower pollution; seaport-related air pollution; cancer alley; cancer villages; unintentional pollutants; deep-well injection sites; hazardous waste; Love Canal; Superfund; persistent organic pollutants; pesticides; POPS, dirty dozen; schools and brownfields; e-waste; Citizen's Clearinghouse of Hazardous Waste; Student Environmental Action Coalition*]

pollution police

Almost every state attorney general's office has some type of environmental offense unit. Some states have specially trained environmental crime squads who identify and investigate environmental crimes and enforce environmental laws, much as any felony would be handled. These units are not formally called pollution police, but they fulfill that role. California and New Jersey, among others, have squads that perform this function. In New Jersey, these squads consist of local police officers, sheriffs, and environmental protection officials who aggressively enforce environmental regulations. [*Environmental Protection Agency, U.S.; Environmental Defense Fund; Lighthawk, the Environmental Air Force; NGOs; United Nations, global environmental issues and the*]

polychlorinated biphenyl (PCB)

See *PCB*.

poop pills

The *human microbiome* has been found to play a vital role in our health—so much so that replacing the microbiome of a sick person with one from a healthy person can cure at least one deadly illness. This was accomplished by using *stool transplants* to transfer the microbiome from one person to another. However, just recently, these procedures have been replaced with poop pills that contain the needed stool matter, harboring the "healthy" bacteria. The patient takes about 25 pills. It has worked in nine out of every ten patients.

POPs, dirty dozen

POPs are *persistent organic pollutants*. The dirty dozen were targeted by the *Stockholm Convention on Persistent Organic Pollutants* for control globally. They are *aldrin, chlordane, DDT*, dieldrin, endrin, heptachlor, hexachlorobenzene, mirex, toxaphene, *PCBs, dioxins*, and furans.

population

A group of individuals of the same *species* living within a specified area to be studied. All of the populations within this area make up a *community*. The community of organisms, along with the nonliving parts of the *environment* such as the soil and water and factors such as climate, make up an *ecosystem*. [*population ecology; population dynamics; population stability*]

Population Connection

Originally called ZPG (Zero Population Growth), this organization advocates stabilizing the human *population* in the United States and the world. ZPG publishes several reports and provides teaching kits that describe how to stop population growth and explains its importance as it pertains to our environment. [*population explosion, human; population growth, the limits of; human population throughout history*] {populationconnection.org}

population dynamics

Populations are always under some form of environmental stress that threatens to disturb the stability of the population. For example, the stress might be natural—such as extreme cold temperatures—or caused by humans—such as using *pesticides* or *clear-cutting* a forest. Changes to the size of a population and the factors creating these changes are called population dynamics.

Populations respond to stress in four ways. a) the most obvious is an increase in the death rate (*mortality*) or a decrease in birth rate (*natality*); b) organisms often move (*migrate*) to other areas to avoid stress; c) organisms that reproduce rapidly and have many offspring might be capable of adapting to the stress through *natural selection*; d) any population that cannot successfully respond to stress may perish from that ecosystem or even become *extinct*. [*population ecology; population dynamics*]

population ecology

Population ecology is the study of *population* statistics—the factors that affect a population, such as birth and death rates, and the causes for changes in these rates. The size of a population in a specific area is called population density and is central to understanding population *ecology*.

A population's density is controlled by two types of *limiting factors*. Density-independent factors, as the name implies, have no correlation to the existing density. For example, extreme weather conditions can reduce populations, but this has nothing to do with the population density. Density-dependent factors occur as a result of the existing population density. For example, the food supply dwindles as population grows, resulting in less food for more individuals and causing the population to decline. [*J-curve; S-curve*]

population explosion, human

To give you some idea of the historical speed of *human population growth*, consider the following: It took all of human history—roughly 10,000 years—to reach the first billion people, around 1800. By 1930, there were two billion; by 1960, three billion; by 1974, four billion; by 1987, five billion; by 1999, six billion; by 2011, seven billion. And projections for 2024 estimate eight billion and for 2050, ten billion. But projections are just that—projections—and they could be wrong, because factors can change.

The primary factor that appears to be changing is a slowing of the population growth rate, which can greatly reduce the latter year's estimates. The *theory of demographic transition* states that as nations become more advanced and their people more educated, fertility rates drop. Some believe this theory no longer remains valid, but numerous studies clearly show that fertility rates rise with increased poverty and drop with improved education.

Further, many recent surveys appear to confirm the demographic transition theory. For example, a survey taken from 2006 to 2008 shows the differences in attitudes between women of different nations. When asked how many children they believe is ideal, the answers ranged from a low of 1.6 in *Australia* to a high of 9.1 in Nigeria.

But even if we assume fertility rates will continue to drop as the world becomes more developed—poverty declines and education improves—the consensus remains that the global human population will reach roughly ten billion people. And the question then remains—can our planet feed and otherwise support that many people— not only more people but more affluent people, meaning a larger *ecological footprint* as well.

The world has trouble feeding those who are already here. In 2007, estimates have almost six million children under five dying from hunger-related diseases. In 2009, some 24,000 children died each day from diseases caused by a lack of clean *drinking water* or water for sanitation.

These problems will only be exacerbated by the fact that most of these additional billions of people will be in the very nations that cannot care for themselves now. Projections show that almost all of the forthcoming growth will be in less developed nations, mainly in sub-Sahara Africa. Famine and conflict are now common there, so what might the future hold?

Today, the top five most populous countries are 1) *China*, 2) *India*, 3) United States, 4) Russia, and 5) Japan. Projections for 2050 show it will change to 1) India, 2) China, 3) Nigeria, 4) United States, and 5) *Indonesia*. The only region expected to lose population is Europe. [*population growth, limits of human; baby boom; age distribution in human populations; hunger, world; water shortages, global; China's One-Child Program; human population throughout history; Title X of the Public Health Service; United Nations Population Fund*]

population growth, limits of human

People have been debating how many people planet Earth can sustain ever since *Thomas Malthus* wrote <u>An Essay on the Principle of Human Population</u> in 1798. He stated that human populations would grow geometrically but the food supply would grow arithmetically—and people would starve.

In the early 1960s, the *Green Revolution* and its accompanying advances in agricultural technologies appeared to be capable of feeding everyone—no matter how many of us there were. But by the end of that decade, our numbers had tripled since 1800, and human population growth was at the forefront of conversation, with books such as *Paul Ehrlich's* <u>The Population Explosion</u>. A debate raged about human population, but the global catastrophes never came, and by the 1990s human population concerns were put on the back burner and have remained there ever since.

But regardless of whether the issue is in the news or not, or whether there is an existing catastrophe, the human population continues to grow at a staggering rate. Many people appear content to sit and wait for the *demographic transition* to slow growth rates down and hope the *human population explosion* levels off before we hit ten billion people. Others believe the issue is far more urgent and press for family planning on a global scale to ensure planet Earth can provide enough resources for all.

At the moment, three perspectives exist about the relationship between human population growth and economic growth. To oversimplify, some people—called doomsters or neo-Malthusians— believe we are doomed because we will have far more people living on the planet than can be supported and economies will collapse. Others—the boomsters—believe technology will always come over the hill to save us, and in doing so will boost the economy. And more recently, a third perspective, called neutralism, sees no connection between population growth and economic growth, and states that the first two groups are wasting their time debating the issue. [*population explosion, human; age distribution in human populations; baby boom; doubling time in human populations; human population throughout history; hunger, world; water shortages, global*]

Population Reference Bureau (PRB)

A private, scientific, and educational organization that collects, interprets, and disseminates information on human population trends. Established in 1929, the PRB serves as a link between scientific research and the public and those who affect policy worldwide. They offer many excellent publications and teaching materials. They have an information service and one of the oldest and largest population libraries in the world. If you ever want to know (almost) exactly what the world human population is at any given moment, just go to their website to see their World Population Clock. [*population explosion, human; Population Connection; Negative Population Growth*] {prb.org}

population stability

Populations within a *community* are always under some form of environmental stress. Natural stress factors include extremes in temperature, lack of moisture, and floods, while human-induced stress includes *clear-cutting* the land for crops, using *pesticides*, and *deforestation*. Populations attempt to maintain a stable population. The three aspects to this stability are 1) persistence—the ability of a population to resist change, 2) constancy—the ability of a population to maintain a certain size, and 3) resilience—the ability of a population to bounce back to its original condition after being exposed to some form of stress. [*population dynamics; population ecology*]

Porter, Eliot

(1901–1990) Although photography was first a hobby, Porter became one of the leading nature photographers of our time. He was inspired by *Ansel Adams* and did a great deal of his early works in black and white, as did Adams. As he established himself, he later used color photography and specialized in birds. He is said to have photographed birds with the same art and expertise as *Audubon* had painted them. In 1962 Porter gained worldwide recognition when the *Sierra Club* published <u>In Wildness Is the Preservation of the World</u>, a color photography book of the New England woods combined with excerpts from the writings of *Henry David Thoreau*. He later expanded his focus to include places as diverse as the Galápagos, East Africa, and *Antarctica*. His photos were credited with helping people understand the ecological diversity and the environmental stresses these locations faced. [*Hudson River School art movement*]

postconsumer waste

When most people think of waste, they are thinking about postconsumer waste—waste that has been used for its intended purpose by the consumer and then thrown away. This waste ends up in *landfills*, being *incinerated*, or possibly *recycled* or even reused. It seems so obvious that the descriptor "post" doesn't seem necessary; however, it must be differentiated from *preconsumer waste*. [*municipal solid waste disposal; three Rs; recycling, moving toward zero-waste; gold fingers*]

potable water

Water that is fit for human consumption or commonly called *drinking water*. It can come from *surface waters* or *groundwater*. It must contain very low levels of salts and little or no animal wastes or bacterial contamination.

The nonavailability of potable water is a life-threatening problem in many regions of the world. About 800,000 people worldwide do not have regular access to potable water. However, this number is down from more than 1.5 billion people just a few years ago. One of the *Millennium Development Goals* (MDG) was to halve by 2015 the number of people without continual access to safe potable water and basic sanitation. This is one of the few MDG goals that has been met and even exceeded. However, about three million people still die each year from waterborne diseases. [*water shortages, global; domestic water use; Safe Drinking Water Act; toilet-to-tap drinking water*]

power grid

This is the entire transmission system that supplies the nation with electricity. It consists of a network of power providers (utility companies) and the consumers who use the power. They are all connected by transmission and distribution lines. The continental United States does not actually have one national grid but three regional grids:

the Eastern Interconnect, the Western Interconnect, and the Texas Interconnect. Alaska and Hawaii have many smaller systems.

Most people in the United States are content with the power grid. However, with what appear to be more *extreme weather* events and a slew of storms during the past few years, people in some parts of the nation feel the power grid cannot be trusted and are buying generators to power their homes when the grid goes down. Many homes in the upper price range now come with built-in standby power generators under the assumption they will be needed. [*microgrids, power; light bulb – recent history and legislation; light bulb technologies, domestic; smart energy; watt and kilowatt; baseload power; soft energy path*]

power islands

Because the wind doesn't always blow and the sun doesn't always shine, both *wind power* and *solar power* facilities have a problem generating electricity constantly. The most common method to supplement these facilities to make energy available 24/7 is a process called *pumped-storage hydropower (PSH)*. However, because this can only be used in some regions, companies are searching for new ways to provide power when the wind and sun are not available. Power islands provide one of the more promising possibilities as an alternative to PSH.

A power island is a small artificial island with a large central reservoir below water level, with pipes connecting to the sea. When the pipes are open, water from the sea rushes in, filling the reservoir and turning a turbine that generates power. Then, when the wind or solar facilities are generating power, the water is pumped up and out of the reservoir. The next time there is a need to supplement the wind or sun, the water is again allowed in, repeating the process and generating energy. A large power island is up and running in Denmark in concert with an *offshore wind farm*. [*energy (fuels); energy use*]

PP (polypropylene)

PP is an acronym for polypropylene; a *plastic* used in such diverse products as snack food packaging, *disposable diapers*, bottle caps, and encasements for car batteries. It makes up about ten percent of all the plastics produced in the United States and is rarely recycled. These products are stamped with the number "5" surrounded by a triangle made of arrows. [*plastic pollution; plastic recycling*]

prairie

See *grassland*.

precautionary principle

When a new product or process is developed and money to be made in the market-place, there is often a rush to cash in before taking time to test it out to be sure it is safe—both to people's and the planet's health. Many people believe a product is safe until it has been proven harmful. To many others, this is a backwards logic, because you want to find out if something is harmful before releasing it to consumers. The precautionary principle is a *risk assessment* strategy that assumes something should be considered harmful until it is proven safe.

The world is full of examples where this principle was not followed and problems ensued: *lead* in gasoline, *DDT,* and *PCBs* just to name a few. Many examples exist today as well—*genetically modified organisms (GMOs)* and *nanoparticles.*

The principle is mentioned in international agreements such as the *Rio Declaration on Environment and Development,* and it is implied in some United States legislation, such as the *Clean Air Act.* The principle is more accepted in Europe than in the United States. [*Toxic Substances Control Act; toxic pollution; hormone disruptors, endocrine; unintentional pollutants*]

precision farming

Also called smart farming, this is farming gone hi-tech. Using technology such as mapping equipment, sensors, GPS, and specialized software, farmers can reduce the amount of water, *synthetic fertilizers,* and *pesticides* they use; only applying exactly what is needed, when it's needed, and where it's needed. This use of technology can increase yields and profits while also reducing environmental harm.

Some examples of precision farming programs include the On-Farm Network of the Iowa Soybean Association, which helped farmers reduce fertilizer use by 33 pounds per acre, saving them $16 per acre with no loss of yield. Another is called variable-rate irrigation (VRI), currently being tested in Georgia with encouraging results. The upfront costs are high, but the results show significant cost savings after continued use.

Although promising, precision farming methods are rarely used. In the United States, less than one percent of cropland applies these smart farm advantages—even though, after the initial expense, they provide very real financial benefits. [*conservation agriculture; drip irrigation; integrated pest management; low-input sustainable agriculture; organic farming and food; permaculture; regenerative agriculture; sustainable agriculture; vertical farms*]

preconsumer waste

When most of us think about waste, we are probably thinking about *postconsumer waste*—a product we have used and no longer want that becomes postconsumer waste when we throw it out. But another type of waste—preconsumer waste—never made it into the consumer's hands. This type of waste is most often a problem faced by manufacturers. Whenever a product is made, plenty of materials never get used and become waste. Because it never made it into a consumer's hands, it is considered preconsumer waste.

An example would be on the assembly line where cars are made. You can imagine a lot of materials never make it into the car and must be disposed of, such as sheet metal for the body or leather for the interior that is cut off and falls on the floor. Also, the sheet of latex used to stamp out a dozen soles of *flip-flops* will have a lot of leftover material. These all become preconsumer waste.

This type of waste used to be ignored and simply thrown out. However, many companies now want to reduce their *ecological footprint*; minimizing this preconsumer waste is one way to do so. Some companies are even attempting to attain *net-zero waste*. To do so requires managing both preconsumer and postconsumer wastes.

Preconsumer waste, just like postconsumer waste, can be *recycled* or reused. Companies can also use *source reduction* by trying to use less to begin with. In the examples given, starting with smaller sheets of metal, leather, or latex means less waste at the end. [*automobile recycling; zero waste; recycling, moving toward zero waste; waste minimization*]

predator
See *predator-prey relationship.*

predator/prey relationship
When one animal kills and eats another animal, the killer is the predator and the animal killed is the prey. The interaction between the two is a predator-prey relationship—for example, a bird eating a worm, a cat eating a mouse, and a lion eating a zebra. This type of interaction is a common and important aspect of most *food webs*. On rare occasions, plants can be predatory, such as a Venus fly trap. [*parasite; host; parasitoidism; tropic levels; symbiotic relationships*]

prescribe burn
A *forest management* practice where fires are purposely set in wooded areas to remove forest litter such as broken branches and dead leaves. These controlled fires make it less likely there will be a wildfire that cannot be controlled. Some species of trees and other plants require fire for propagation and would not reproduce without a fire. [*forest health, global; forest, forest, sustainable; governments of the forest; logging, illegal*]

preservation

Although often used synonymously with *conservation*, they are quite different. Preservation refers to saving something, such as nature, for its own sake; its own intrinsic value makes it worth saving for future generations to see. Conservation pertains to sustaining nature while it is being used or consumed; its economic value makes it worth sustaining for future generations to use.

Preservation became popular in the United States in the late 1800s by a triumvirate of factors: *John Muir* and *John Burroughs* were great nature writers with vast audiences. The *Hudson River School art movement*, with artists such as *Thomas Cole* and *Frederic Church*, were painting magnificent works of nature, and Currier and Ives were producing beautiful nature prints. And charismatic, popular, and pro-environment *President Theodore Roosevelt* completed the trio. Americans learned to love their natural surroundings by reading great writers, appreciating great art, and listening to a president defending nature.

Many *NGOs* advocate preservation. The *Sierra Club* was founded by John Muir and continues his preservationist beliefs. More recently, preservationists have worked to save the giant redwoods of northern California so future generations might someday see them. [*redwood forest destruction; philosophers and writers, environmental; ethics, environmental*]

prey

See *predator-prey relationship*.

primary air pollutants

See *air pollution*.

primate research

Thousands of primates, such as monkeys and tamarins, are used for scientific research. They are valuable for medical research; however, many animal rights activists insist they need not be used and the research could be performed by other methods. Possibly the science community is beginning to listen.

In recent years, many research labs have decided to stop using primates for research. For example, the Harvard Medical School Research Center is sending their 2000 primates, most of them macaques, to other facilities. The center states the reason for the change is economic. Others believe it might be a sign of things to come and science is finding other ways to perform research. [*Goodall, Jane; Fossey, Dian; vivisection; animal rights movement and animal welfare*]

primitive technologies

These are aboriginal survival tools and techniques, used to demonstrate how humans survived without the use of modern-day "high-tech" equipment. These include weapons such as atlatls, rabbit sticks, bolas, and slings. More domesticated tools and techniques are using primitive hand drills, making fire from scratch, and preparing foods and medicines. These and other primitive technologies are taught in survival schools and camps. *[ethnobotany; governments of the forest; utilitarian environmental ethic]*

primitivism

A romantic philosophical belief that happiness and overall well-being are inversely proportionate to the degree of civilization. The more we live with nature and wilderness, the happier we are. The less we live with it, meaning in a more civilized society, the less happy we are. Proponents of this belief included Jean-Jacques Rousseau and *William Bartram*. [*Biophilia Hypothesis; Gaia Hypothesis; ecofeminism; ethics, environmental; ethnobotany; nature-deficit disorder; self-concept and the human microbiome*]

probiotics

Microorganisms that, when ingested, provide some benefit to their *host*. For example, some *bacteria* in yogurt have been found to calm a gastrointestinal condition called irritable bowel syndrome and are therefore considered probiotics. Many people—about nine million—take probiotic supplements. The research, however, is very sketchy as to how much help these probiotics provide.

But the future of probiotics, in the hands of scientists and medical doctors instead of marketing firms, looks to be very important. Recent research is leading scientists to consider using probiotics in other more advanced ways to help manage the human microbiome. For example, *stool transplants*, which can be considered a type of probiotic, have cured many people of a life-threatening disease. [*microbiome and microbiota, human; antibiotics and the human microbiome*]

producers

See *autotrophs*.

Production Tax Credit (PTC)

A tax credit provided by the federal government to any company that provides energy to the *power grid* by using *renewable energy* sources such as *wind, solar, biomass,* or any of the other renewable sources. For each *kilowatt-hour* generated, the utility gets a few cents of credit for the first 10 years of the facility's production. This incentive makes the investment worthwhile and encourages renewable energy development.

These incentives are needed while new technologies are getting started. Some forms of renewable energy have already been shown to be competitive but will advance even more with some form of incentive. This credit has been on and off for many years and its future is always in doubt. [*fossil fuel subsidies; tweetstorm, Rio+20*]

product packaging

Product packaging is one of the biggest components of *municipal solid waste*. Many efforts have been made to manage and reduce this type of waste. Consumers are primarily responsible for two of the *three Rs*: reuse and *recycle*. However, the third R, reduce, is up to manufacturers when it comes to packaging.

In the United States, there is little legislation and no incentive for companies to use less packaging materials. However, Europe and other parts of the world have regulations that hold manufacturers responsible for the cost of waste disposal of their products. These are called *extended product responsibility (EPR)* laws. If you were to compare the same product's packaging in the United States with that in many European countries, the difference EPR makes is quite amazing. [*waste minimization; source reduction*]

profundal region

The region in a lake where sunlight cannot penetrate. [*freshwater ecosystems*]

PS (polystyrene)

PS stands for polystyrene, a *plastic* used to manufacture foam, paper cups, and food containers. Commonly called Styrofoam, it makes up about 11 percent of the total plastics manufactured in the United States. This type of plastic does not recycle well, even though programs do exist. These products are stamped with the number 6 surrounded by a triangle formed by arrows.

It also is so light that it is often seen flying around *garbage* cans and *landfills*, landing everywhere as *litter* and often in bodies of water, where it is harmful to aquatic life. Many large international companies that use these types of coffee cups have already switched or are planning to switch to other materials. [*plastic pollution; plastic recycling; styropeanuts; nurdles*]

pseudoscience

Science is based on facts and facts are established by the scientific method. People can usually tell the difference between an *opinion* and a fact. That's because most people do not try to hide the fact that an opinion is just that—someone's personal thoughts about something. Pseudoscience, however, is the attempt to mislead the public by making information sound like science when in fact it is not.

Pseudoscience is rampant, especially on the Internet, where groups with scientific sounding names and "scientists" with advanced degrees in some arcane field produce so-called scientific experiments and publish in so-called scientific journals. The idea is to create the appearance of science when, in fact, it has nothing to do with real science. Instead, it usually has to do with an agenda someone or some group wants to advance. This is accomplished by making the science appear to have come to the conclusions and facts that were desired from the start.

The best way to avoid pseudoscience is to believe facts when they come from reputable, established sources. There are reputable scientific magazines both online and in print, websites and blogs on just about every subject. Many that pertain to the environment are listed throughout this book. Find resources you know to be reliable and use those as your source of information when you need the facts. [*opinions and science; argumentum ad ignorantiam; magazines, eco-; blogs, green; microblogs, green; news literacy; environmental literacy; environmental education; News Service, Environment; Group of 10; NGOs*]

puberty and the environment, early

American children are reaching puberty earlier than in the past. Although the exact cause is unknown, some research indicates it might be due in part to the toxic mix of synthetic chemicals (called a *toxic cocktail*) in our environment that have made their way into our bodies—also called our *body burden*. Recent studies at the U.S. Centers for Disease Control showed that girls exposed to high levels of some common household chemicals had their periods seven months earlier than those who had no exposure. Some believe it might be because of *hormone-disrupting* substances found in many products. These substances would only be partially to blame, because increased obesity in children also contributes to the change. [*hormone disruptors, endocrine; Toxic Substances Control Act; POPs, dirty dozen; hazardous air pollutants; persistent organic pesticides; toxic pollution; autism studies and environmental factors; antibiotics and the human microbiome; blue baby syndrome; breast milk and toxic substances; nanoparticles and human health*]

public lands, U.S.

Most of the public lands in the United States are managed by the *Bureau of Land Management*, the *National Park Service*, the *Forest Service*, the *Fish and Wildlife Service*, and the Bureau of Reclamation, as well as the *National Oceanic and Atmospheric Administration* (because public "lands" actually includes public waters).

The U.S. Congress passes the creation of new public lands such as *national parks.* The only exception is national monuments, which the President can create by executive order, using the 1906 *Antiquities Act.* [*National Parks and Wilderness Preservation System*]

pumice

A porous volcanic rock that used to be heavily mined for one reason: to give jeans a "stonewashed" look. To meet this need, regions were *strip-mined,* with all the accompanying destruction. Environmental groups pressured jean companies to use less environmentally damaging methods, and alternatives are now routinely used. Using enzymes to chemically "stonewash" the jeans is the standard. It does not require pumice mining, requires less water, and produces less waste. So stonewashed jeans today can be viewed as an environmental success story. But with that problem resolved, another has taken its place—*blue-jean pollution.*

pumped-storage hydropower (PSH)

A method of storing the power generated from *wind power* and *solar power* facilities so energy can be available even when the wind is not blowing or the sun is not shining. This process of saving some energy to be used at a later time is called *time-shifting.*

PSH involves pumping water from a low-lying area such as a river up to a storage tank at higher elevation, such as up on a high bank of that river. Then, when the wind is not blowing or the sun is not shining, and therefore no power is being generated, the water is released from the tank, flows downhill back into the river, turns turbines, and generates electricity.

At this time, 99 percent of worldwide energy storage for renewable power is in this form. This process has serious limitations because the topography of the land must be suited for such a process—and such an area is not always available. For this reason, newer technologies are being studied for other forms of time-shifting, such as *power islands.*

purse seine nets

A purse seine is a weighted net that hangs down from a boat into the water. It has floats on the top to keep it at the surface and rings on the bottom with a rope running through the rings. The fishing vessel encircles a school of fish and the rope is pulled tight to prevent the fish from escaping. This *commercial fishing* method is used to capture fish that travel in schools close to the surface, such as certain species of tuna, sardines, herring, and salmon.

During the 1950s, tuna fisherman discovered that yellowfin tuna are often found swimming below *dolphin* herds. (It's believed the tuna follow dolphins because of their superior ability to locate food.) Tuna boats routinely use a technique of encircling dolphin herds with purse seine nets and hauling in the tuna catch below, along with the herd of dolphins above. The dolphin, as any unintended catch, is called *bycatch*. And as any other bycatch, the dolphins usually die before being released. An estimated 100,000 or more dolphins die in these nets each year, globally.

The tuna industry in the eastern tropical region of the Pacific is alone believed responsible for more than seven million dolphin deaths since the advent of purse seine fishing there in 1959. The U.S. government has tried to ban the importation of tuna caught via purse seines that harm dolphins but has met with international resistance. However, public pressure within the United States has resulted in many tuna canners selling *dolphin-safe tuna* or other *eco-labels* such as the *Flipper Seal of Approval*, that offer the consumer some level of assurance that the tuna in those cans were not caught in *driftnets* or purse seine nets. [*overfishing; fish populations and food; fisheries and seabirds; bottom trawling, deep sea; longlines; gill nets; pirate fishing; illegal, unreported, and unregulated fishing; ghost fishing; Magnuson Fishery Conservation Management Act*]

push-pull hypothesis

A theory stating that conditions within one country, such as poverty or unemployment, push people out (*emigration*), while conditions within other countries, such as a high standard of living and plenty of jobs, pull people into that country (*immigration*). [*standard of living; demography; human population throughout history; overpopulation; S-curve; population explosion, human*]

PVC

An acronym for polyvinyl chloride—a common *plastic* used for construction and plumbing. It is the material surrounding most computer cables, the power brick that comes with your computers and other electronics. It is also used in other products from pipes to clothes. Although its use is declining in some products, it can still be found in *water bottles* and things needing water proofing, such as baby-changing mats and *mattress* covers. PVC contains *phthalates,* a substance with many of its own health and environmental problems.

Environmentally speaking, PVC is the worst of the plastics because it is not usually recycled, and when disposed of, it causes a variety of human health and ecological problems.

The only proper disposal method is for PVC to go into a *hazardous waste* site. However, many people simply throw this material away, so it usually goes into regular *landfills,* where there is always the potential for it to *leach* out and contaminate *groundwater.* If not put in landfills, it is disposed of via *incineration,* where it produces *toxic, carcinogenic* gases such as *dioxin.* The *Environmental Protection Agency* links it to serious respiratory problems, immune suppression, and cancer.

Because of these issues, many companies have stopped using PVC in their products. For example, Apple no longer uses PVC in any of their hardware and associated products. As alternatives, look for plastics or waterproof items that state they are PVC-free. [*toxic pollution; toxic waste; hazardous waste disposal; toxicology; toxic cocktail; body burden*]

pyrethroids

One of the four categories of synthetic *insecticides.* Pyrethroids are synthetically produced versions of a natural insecticide substance found in wild chrysanthemums, called pyrethrin. They are considered a *soft pesticide,* because they break down into harmless chemicals shortly after application—usually a few days to a few weeks. Some examples of synthetic pyrethrin are permethrin and tralomethrin, which are sold under many brand names. [*pesticides; pesticide dangers*]

qualitative studies

Studies that concentrate on descriptions as opposed to numerical analyses. [*descriptive ecology; ecological study methods; taxonomy*]

Quammen, David

(1948–) Quammen is one of America's most popular nature writers. His entertaining book, Natural Acts: A Sidelong View of Science and Nature (1985), is a collection of 15 years-worth of articles published in "Outside Magazine." His most recent book is about the *spillover effect*, Spillover: Animal Infections and the Next Human Pandemic (2012). He has won numerous awards, including the *John Burroughs* Medal for Nature Writing (1997).

quantitative studies

Studies that are based on counts, measurements, and other numerical analyses. [*ecological study methods; experimental ecology; theoretical ecology*]

queenright

A honeybee colony that contains a queen. [*honeybees, pollination, and agriculture; colony collapse disorder; honeybee sperm bank*]

quiescence

A temporary resting stage (dormancy) that usually occurs because of unfavorable environmental factors such as a lack of moisture for seeds to germinate or cold temperatures for *insects*. [*torpor; hibernation; estivation; cold resistant; natural selection*]

.

Rachel Carson Council, Inc.

This *NGO* was founded in 1965, according to *Rachel Carson's* wishes to establish a nonprofit scientific and educational corporation that focuses on *pesticide* contamination of our environment and its effects on human health. They inform and advise the public about these chemicals and promote the use of alternatives. The council responds to inquiries from individuals and organizations around the world and provides information through publications, meetings, and government programs. [*insecticides; DDT; pesticide dangers; pesticide residues on food; Toxic Substances Control Act; Federal Insecticide; Fungicide, and Rodenticide Act*] {rachelcarsoncouncil.org}

radiation

Some elements have unstable nuclei and are constantly throwing off particles, releasing the energy that once bound these particles together. The release of these particles produces radiation, of which three types exist. Alpha radiation particles are composed of two neutrons and two protons. These particles move rapidly but only short distances and can be stopped by material as thin as paper. Beta radiation consists of electrons that also move short distances and can be stopped relatively easily. Gamma radiation is a form of electromagnetic radiation, such as X-rays, and travels greater distances and can pass through thick concrete walls. When something is struck by radiation, it is said to be irradiated. *Food irradiation* is sometimes used as a method of preserving food. X-rays—another controlled use of irradiation—are used for diagnostic purposes.

Nuclear power and nuclear weapons use radioactive elements such as *uranium* and plutonium. The radiation emitted by these substances, both during use and after disposal, poses a significant threat to all forms of life. *Nuclear waste disposal* is one of the great unsolved environmental problems of our time.

[*nuclear waste disposal; nuclear waste dilemma, the global*]

radioactive waste disposal

See *nuclear waste disposal*.

radon

A naturally occurring radioactive gas that is odorless, tasteless, and invisible. It is released during the natural breakdown of uranium, which is found in many types of rocks and *soils*. When released into the air, radon becomes diluted and harmless. If, however, the gas becomes trapped in an enclosure such as a basement, it accumulates and the concentration can become dangerous. The amount of radon that enters a building is based on the amount of radon present in that area and the construction of the building.

The problem was first discovered in the 1960s when high concentrations of radon were found in homes built with contaminated materials from uranium mines. This led to greater awareness, and many contaminated homes were found across the country, because of natural causes such as seepage into foundation cracks.

Exposure to radon is believed to increase an individual's risk of developing lung cancer. It is estimated that between 5000 and 20,000 deaths from lung cancer may be blamed on radon exposure. People exposed to roughly ten times the normal indoor concentration of radon gas have about the same likelihood of developing lung cancer as someone who smokes half a pack of cigarettes a day. Long-term exposure to low levels of radon is thought to be more dangerous than short-term exposure to high levels of the gas.

Although only a small percentage of buildings have the problem, a simple test can determine if danger exists. Towns and local agencies often provide test kits, or professional services can be contracted. Homes found to be contaminated can usually have the problem corrected by structural changes. [*indoor air pollution; indoor ecology; radiation*]

radura

An internationally accepted symbol required on all foods that have been irradiated. The symbol was originally created in the Netherlands back in the 1970s. Bearing a striking resemblance to the *Environmental Protection Agency's* logo, it appears as a broken circle that contains highly stylized waves and the sun. [*food irradiation*]

rail, high-speed (HSR)

There is more than one definition for HSR, but a good consensus might be trains running at over 150 miles per hour, using special tracks, or 125 miles per hour if running on existing tracks. Although the number of miles of this type of rail is growing, it still is only about four percent of all rail lines when including passenger and freight. Countries with the most HSR miles of these tracks are *China*, Japan, Spain, France, and *Germany*.

In countries that invested in HSR, it appears to have paid off. France uses HSR for over 60 percent of all passenger rail service. It has connected isolated regions of the country and improved economic development in these areas. In Japan, for routes that have both HSR and air travel available, HSR routinely gets 75 percent of the market.

HSR has the least environmental impact per person when compared with *automobiles*, planes, and standard *rail transportation*. In Japan, 18 percent of all intercity travel is done by HSR. In European countries it is between five percent and eight percent, and in the United States it is less than one percent.

Rails-to-Trails Conservancy

The Rails-to-Trails Conservancy was established in 1986 and has created a national network of *bicycle* trails that provide exercise, alternate ways to commute to work, and a lot of fun for young and old riders. This *NGO* converts abandoned railroad tracks into trails for cyclists, hikers, and naturalists to enjoy. Many of these trails (also called *greenways*) connect metropolitan areas, rural communities, and local parks. Some offer commuters alternatives to congested highways. The Conservancy works with municipalities to acquire and convert these trails. They went from having established about 2000 miles of trails 20 years ago to more than 20,000 miles today. The Rails-to-Trails Conservancy is one of our best advocates for biking and hiking. [*bicycle-sharing programs; bicycling cities, ten best*] {railstotrails.org}

rail transportation and mass transit

The biggest advantage of rail is the number of passengers one train can transport. It would take a fifteen-lane highway to carry as many people as one train track. The three forms of rail *transportation* used for *mass transit* are light, commuter, and heavy rail lines. (*High-speed rail* is not used for mass transit.)

Light rail is the modern day version of the old streetcars. They are electric powered, usually above ground, consisting of only a few attached cars. Light rail routes are typically less than 15 miles. Examples of light rail include the San Diego Trolley, MetroLink in St. Louis, the Metropolitan Area Express (MAX) in Portland, and the Sacramento Regional Transit District system.

Commuter rail lines can be electric or diesel and can take passengers longer distances, usually 20 miles and higher. They consist of many cars, carrying large numbers of passengers. They are only found in and around large cities. Commuter rail systems are found in Boston, Chicago, Montreal, New York, Philadelphia, San Francisco, and a few other cities, with limited service.

Heavy rail are electric, run on dedicated tracks, with runs usually between 5 to 15 miles. Heavy rail includes many of the largest city metros such as the New York City subway system, the Washington DC Metro, Atlanta's MARTA, San Francisco's BART, and the El in Chicago.

Rainbow Warrior

This ship became an icon to many environmental activists in the 1980s. Rainbow Warrior was the flagship of *Greenpeace* and was in preparation to disrupt French *nuclear weapons* testing in the South Pacific in 1985. The ship was in dock in Auckland, New Zealand, when explosions sunk the ship, killing one person on board. The French government was believed to have sunk it to prevent its interference with the nuclear tests. Greenpeace continued its historic use of ships at sea with the Rainbow Warrior II, retired from service in 2011, and the christening of the Rainbow Warrior III in the same year. [*Group of 10; NGOs; environmental movement*]

Rainforest Action Network

If *tropical rainforests* interest you and you want to get involved, this is the *NGO* for you. Founded in 1985, its first big initiative was boycotting a restaurant chain for using "rainforest beef," which is cattle raised on lands cleared of their original rainforests. Their campaign worked and the chain discontinued the practice. They also fought to save *old-growth* or *ancient forests* from *logging* by getting large retailers to commit to no longer buying old-growth timber or products. They have thousands of volunteers and a large annual budget to get things done.

They believe in aggressive action and are considered (by the Wall Street Journal) "some of the most savvy environmental agitators in the business." [*rainforest destruction; forest health, global; governments of the forest; logging, illegal; Rainforest Alliance*] {ran.org}

Rainforest Alliance

The Rainforest Alliance works to preserve the *biodiversity* of *tropical rainforests* with an emphasis on *forest sustainability*—and they do so by using the power of the marketplace. The following statement from their website explains how they accomplish this, "We believe that the best way to keep forests standing is by ensuring that it is profitable for businesses and communities to do so. That means helping farmers, forest managers and tourism businesses realize greater economic benefits by ensuring *ecosystems* within and around their operations are protected. . . . Once businesses meet certain environmental and social standards, we link them up to the global marketplace where demand for sustainable goods and services is on the rise." It is important to note that their efforts only work when consumers are *environmentally literate* and what some would call *green consumers.*

The alliance has an *eco-certification* program called the Rainforest Alliance Certified seal and an *eco-label* called the Rainforest Alliance Verified mark. Their work has helped reduce *deforestation,* unsustainable *logging* practices, and the *loss of biodiversity* in general. [*rainforest destruction; forest health, global; Rainforest Action Network*] {rainforest-alliance.org}

rainforest destruction

Tropical rainforests are some of the richest, most productive *ecosystems* on the planet, playing vital roles in *biogeochemical cycles* and containing the majority of the world's *biodiversity.* While covering only seven percent of the Earth's surface, they are home to over half of all life on the planet. Rainforests are being destroyed at a staggering rate; well over 30,000 square miles are converted to nonforest use every year.

Forests are destroyed during *slash-and-burn cultivation* to make room for crops. The vegetation is burned to add *nutrients* to the *soil* for crops. This type of *organic fertilizer* only enriches the soil for a few years before becoming useless. Much of this *deforestation* is caused by exploding populations trying to produce more food for more people. Many forests in Central America have been cleared to make room for pastures to raise cattle for beef. Forests are also destroyed by *clear-cutting* for timber.

The effects of this destruction are numerous, including the impact on *global warming*, *soil erosion,* and *biodiversity loss.*

In addition to human impact, *climate change* is also affecting rainforests. Dry spells and what were once rare droughts have hit places such as the western *Amazon* more often. It is a vicious cycle, because fewer trees mean less rain and less rain means fewer trees. This occurs because rains in most rainforests are generated by the evaporation and transpiration of the trees in that forest. [*climate change and biodiversity; deforestation and climate change*]

Ramsar Convention

See *Convention on Wetlands.*

range

The geographical distribution of a *species* is called its range. [*Map of Life; zoogeography; ecosystem; ecological study methods*]

range anxiety

According to many surveys, people who own electric vehicles (EV) are some of the most loyal car owners. However, probably the number one concern many people have about buying an electric vehicle is running out of power before getting to their destination—called range anxiety. Most of these vehicles have a relatively short driving range before needing to be recharged. This is gradually being resolved as electric charging infrastructure is being built around the country and the world. *Electric vehicle highways* are gradually forming, with charging stations all along the way.

Those loyal EV owners have obviously already overcome this anxiety. And range anxiety will lessen as batteries improve and infrastructure builds up. [*automobile propulsion systems, alternative; electric vehicles, charging; tire, high mileage*]

Ranger Rick

Depending on your age, you might remember going to a doctor's office and reading a children's magazine, Ranger Rick, which is produced by the *National Wildlife Federation*. This magazine has introduced many budding *environmentalists* to the world around them. The magazine targets 7- to 12-year-olds. In 2013 a new magazine targeting 4- to 7-year-olds appeared, called Ranger Rick, Jr., to get kids hooked on *ecology* even younger. [*environmental education; magazines, eco-*] {nwf.org/Kids/Ranger-Rick.aspx}

rare earth metals

Also called rare earth minerals, these are elements most of us never heard about—unless you enjoyed studying the periodic table. These substances have become of huge importance, because they are used in the manufacture of our electronic gadgets such as computers, televisions, and smartphones. They are also used in *electric cars*, *wind turbines*, and *CFL* bulbs. With names like cerium, europium, promethium, thulium, and yttrium, they sound like they are rare.

But their name is somewhat of a misnomer because they are not really all that rare. For example, cerium is the 25th most common element on the planet and yttrium is twice as common as *lead*. The "rare" refers to the inaccessibility of these elements. They occur in hard rocks such as granite, making them difficult to extract. Since they became economically important quickly, only a few places are mining and processing them at the moment, with most of it coming from Mongolia in *China*.

China holds about a third of the total supply, but the United States and other nations—*India*, for example—have plenty as well. Until these other nations get their mining operations up and running, China will continue to produce almost all of it. It is only a matter of time before other countries invest in the infrastructure to extract and process their own rare earth supplies. A few such mines are planned in the United States (California and elsewhere), but it might be years before they come online.

These metals are used in electronic devices for many reasons but primarily for their fluorescence. Their atomic structure makes most of them glow and produce brilliant colors; hence they are found in televisions and smartphone screens. Also, many of them are strong magnets, so they are used in generators and electric motors and therefore in *electric vehicles*. The Toyota Prius contains eight different rare earth metals, about 25 pounds of the car's total weight. [*rare earth metal environmental problems, recycling metals; beryllium in computers; e-waste; mineral exploitation (mining); Pebble Mine; gold fingers; Basel Convention; rare earth metals environmental problems*]

rare earth metal environmental problems

Rare earth metals are used in our electronic gear, but they pose a major public health concern. Most of these metals are *toxic*, a few are *carcinogenic*, one is *radioactive,* and many cause a variety of other ill effects. These are not of concern while being used in your products but become a concern during the product's life cycle: from how they are mined from the Earth, to how they are manufactured, to how they are disposed of.

These metals are one of the main reasons *e-waste* has become a huge global problem. Reports state that the major source of rare earth metals—mines in Baotou, *China*—is an environmental disaster. So much so that China released an unusually harsh assessments of its own mines and the public health hazards they pose, promising to begin cleaning them up.

Because of the limited supply of these elements, scientists are working on finding alternatives. Toyota, for example, has announced plans for a "rare-earth-free" car in the future. [*recycling metals; beryllium in computers; mineral exploitation (mining); Pebble Mine; gold fingers; Basel Convention; e-waste and you*]

ratified vs. signed

Many international agreements can be signed and/or ratified and to many, the difference is confusing. For example, the United States signed the *Kyoto Protocol* but has never ratified it. Signing an international agreement typically means the nation agrees in principle but cannot yet adhere to or be bound by it. The logic is that each nation's representatives must go back to their country for approval. For example, the U.S. Congress must pass legislation approving such an agreement before it can become binding. Then representatives can actually ratify the agreement, meaning they will abide by the laws, rules, conventions, or whatever the agreement entails.

However, as is often the case, some countries use signature as an appearance of approval when, in fact, they have no intention of ever ratifying an agreement. This was the case with the United States and the Kyoto Protocol. [*United Nations, global environmental issues and the; convention vs. treaty*]

realms

A method of describing animal distribution within major regions of the planet. [*zoogeography; ecological study methods; bioregions; synecology*]

recombinant DNA

Many years ago, a new *biotechnology* hit the news, called recombinant DNA. Scientists learned how to recombine an organism's genetic makeup—the DNA. By doing so, they are able to create new forms of life. When first discovered, the technology was viewed as a great science breakthrough for mankind. Today, recombinant DNA has become commonplace and is used to create what are called *genetically modified organisms (GMOs)*. GMOs, the end results of recombinant DNA, are not as well accepted as the recombinant DNA—the science that creates them. [*DNA sequencing technology*]

recycling

Recycling can be as simple as leaving your lawn mower clippings on the ground instead of bagging them (called *grass cycling*) or as complex as recovering materials from the *waste stream* to remanufacture into new products such as *plastic recycling* where water bottles might be made into wastebaskets.

Recycling reduces the volume of *municipal solid waste,* which means less is put into *landfills* and *incineration* plants. Recycling programs have been mandated in many cities and states. Even with these programs and a new public awareness, only about 11 percent of U.S. solid waste is recycled, compared with countries such as Japan that recycle almost half of their solid waste.

In addition to reducing what must be disposed of, recycling also reduces the need for *mineral exploitation* and the resulting depletion of *natural resources*. Recycling has improved in the United States, but still has a long way to go. For example, the United States imports almost all of the *aluminum* it uses in manufacturing, but throws away about $400 million worth of aluminum each year, most of which could be recycled and be back in another aluminum can within six weeks. Also, one Sunday issue of the New York Times uses between 60,000 to 75,000 trees, but we recycle less than 20 percent of all our paper waste.

A successful recycling program has many aspects. The logistics involve establishing methods of collection, separation, and transportation. Technology must be available to process the used materials, prepare it for reuse, and then remanufacture it into new products. Markets for the new product must be identified and available. Collecting and processing accomplishes nothing if these raw materials don't have a market. Finally, incentives, enforcement, and educational policies must be established to complete the process.

Recycling programs have been initiated with everything from appliances to computer printer toner cartridges. Here are some examples: Corrugated cardboard boxes are recycled into cereal and shoe boxes. This week's newspaper is recycled into next month's newspaper. Computer printout paper is recycled into toilet paper. Glass food and beverage containers are recycled into new bottles, glass wool, and highway reflectors. Metal food containers are turned into new cans and household appliance parts. *PET plastics* can be turned into fiberfill for jackets, carpeting, and other new plastics. *HDPE* plastics can become auto bumpers and drainage pipes, and scrap metal can become new car parts and *refrigerators*. [*automobile recycling; appliance recycling; refrigerator recycling; refrigerator doors; energy, ways to save residential; automobile recycling; mattresses, recycling; cigarette butt litter and recycling; downcycling; upcycling; motor oil recycling; paper recycling; glass recycling; recycling metals; plastic recycling; e-waste recycling; single-stream recycling; composting, large-scale; recycling logos; recycling, moving toward zero-waste; single-stream recycling; soda can recycling, aluminum; tires, recycling/reprocessing; wine bottle corks, recycling*]

recycling, moving toward zero-waste

Recycling of *municipal solid waste* is growing steadily. The less waste that must be placed in *landfills* or *incinerated* means less costs and less environmental harm.

Some municipalities are better than others, but none have reached the ultimate goal of *zero-waste*, meaning everything is recycled. The national U.S. average of waste entering into a city's *waste stream* but diverted for recycling is about 34 percent. (This number includes waste that will be *composted*—another form of recycling.) Some large cities, though, do much better than that. The best three are Portland, Seattle, and *San Francisco*. San Francisco recycles an amazing 78 percent of their *waste stream* and hopes to reach 85 percent.

Recycling started back in the 1970s with newspapers and other paper or cardboard products. Then glass bottles and aluminum cans were added. Many years later some types of *plastic* were added. More recently, since *large-scale composting* facilities have come online, *food wastes* and yard debris can be collected. You cannot produce a lot of *compost*, however, unless you can do something with it. In many cases, either local markets are willing to pay for it or it is simply given away to residents as *fertilizer* because it saves the communities money.

The most difficult types of waste to recycle include *disposable diapers* and pet wastes. Seattle plans to begin picking up these wastes in the near future in separate bins so the feces can be processed for *methane digesters* to produce energy.

recycling logos

Three twisting arrows are often marked on packaged goods to symbolize the product has been recycled or is recyclable. Three arrows indicate the product is supposedly recyclable. Three arrows within a circle supposedly mean the product has been recycled. The logo was originally designed in 1973 and has become a universal symbol of *recycling*. However, these logos are not trademarked. They are not regulated and are therefore worthless as purchasing guidelines. There are far better ways to know the environmental impact of a product by using more meaningful *eco-labels*, *environmental rating programs*, and *eco-certification* programs.

Another so-called recycling logo was never meant to indicate recycling. The logos commonly seen for each of the different types of *plastic*, such as a number 6 with three arrows chasing one another (for *PS* plastic products) are sometimes thought to mean the item is recycled. They do not. They are only meant to convey the type of plastic the product consists of.

recycling metals

Metals are a valuable natural resource and extracting them from the Earth almost always causes environmental harm. Recycling metals is often economically feasible and always environmentally favorable; however, most of the metals used in industry end up in *landfills*. A good indicator of metal recycling is to look at what percentage of a metal is recycled at the end of its product life. Today, only 18 out of 60 commonly used metals are recycled more than 50 percent of the time. The majority are recycled either not at all or less than one percent of the time.

Some countries have made recycling of metals, and recycling in general, a much higher priority than others. *Germany* and Japan are creating what can be called a "circular economy," where policy dictates that everything that can be reused or recycled should be.

REDD+ (Reducing Emissions from Deforestation and Degradation)

Tropical *deforestation* and burning the wood is responsible for about 15 percent of all *greenhouse gas* emissions. The only way the world can ever hope to keep *climate change* under control will be to reduce dramatically the destruction of *tropical rainforests*. (Not to mention the *biodiversity* lost and other forms of environmental degradation that occur.) Many of these *forests* are located in *developing countries* that do not have the money or wherewithal to make significant progress preventing this destruction without some form of financial assistance.

REDD is an international agreement, originally created at the *United Nations Framework Conference on Climate Change* in 2007, that provides financial incentives to nations to prevent deforestation. It changes the standard model where forests are more economically valuable cut down than left standing. REDD is designed to reverse this by assigning an *eco-services* value to a country's forests. This value includes many services provided by forests, the most important being *carbon sequestration* (keeping carbon in the trees and out of the *atmosphere* acting as a greenhouse gas). REDD was expanded in 2009 and became REDD+, placing more emphasis on *conservation* and management measures.

Exactly how REDD+ will work once fully implemented on a global scale is uncertain. However, it is based primarily on developed nations (or private entities within these nations), providing funds for conservation projects in developing nations that would reduce deforestation. The developing nations would usually get these funds directly to replace the revenues lost (by not cutting down the forests). The developed nations (or entities) providing the funds would receive carbon credits to be used in *cap-and-trade carbon markets.* These carbon credits could be used to offset their own carbon requirements if needed or could be sold for profits.

REDD+ has shown some progress, but remains a big question mark at the moment on exactly how it will work and if it will work at the global level.

Red List of Threatened Ecosystems (IUCN)

The *International Union for the Conservation of Nature* is best known for their *Red List of Threatened Species*, and they are in the process of developing a Red List of Threatened Ecosystems as well. Some scientists feel too much effort is spent on single species when entire *ecosystems* are of more concern to our planet's health. For this reason, the well-respected Red Lists will soon include a list of endangered ecosystems such as the Long Island (NY) Pine Barrens and the Cape Flats Sand Fynbos of South Africa. {iucnredlistofecosystems.org}

Red List of Threatened Species (IUCN)

The *International Union for the Conservation of Nature* (IUCN) compiles and maintains the Red Lists—species at risk of extinction. These highly respected lists are used to make the world aware of species that won't be with us much longer if something is not done, and they explain what must be done to prevent extinction of these species. They have separate lists for plants and for animals, which they update every few years.

Species are classified into categories such as *extinct, extinct in the wild*, critically endangered, endangered, vulnerable, near threatened, and least concern. Extinct and extinct in the wild are obvious; critically *endangered species* are at a high risk of extinction in the near future; vulnerable means at risk of becoming endangered in the near future, and near threatened means likely to be endangered in the near future.

The numbers of species endangered for extinction are truly staggering. The IUCN Red List shows that almost 30 percent of all *amphibians* are threatened (meaning listed as vulnerable or higher), 20 percent of all *mammals*, and more than 10 percent of all birds.

In addition to these concerns, we cannot forget that only about 4 percent of all known species have been well studied. What potential good the remaining 96 percent hold will never be realized if they become extinct before we ever get to study them.

The IUCN has no authority to enforce any type of protection for species on these lists. They can only provide this information and work with nations with the hope that the nations will impose protective measures—and many do just that. [*endangered and threatened species lists; Endangered Species Act, U.S.; extinction, newly identified species and; climate change and biodiversity; CITES; de-extinction; seed banks and vaults; species recently gone extinct; Lazarus Taxon; extinction and extraterrestrial impact; extinction, mistaken*] {iucnredlist.org}

red tides

A sudden and dramatic population explosion of microscopic *algae* (*phytoplankton*) along the coast and often in harbors. This may be caused naturally but is usually attributed to human activities that cause imbalances in nutrient cycles. Raw or partially treated *sewage* dumped into the water or *fertilizer* carried by rainwater *runoff* adds excessive nutrients, causing the increase in the *plankton*; a process called *nutrient enrichment*.

Although called red, they may actually be brown, yellow, or green. The slimy, thick mats that form can block the sunlight from reaching the waters below, affecting the *aquatic ecosystem*. The harm really begins, however, when these masses of algae die and the *bacteria* that feeds on them has a population explosion of their own. They reduce the amount of oxygen in the water, which can kill large numbers of fish and other organisms. This is called *hypoxia*.

In a few instances, the algae produce toxins that poison shellfish and in turn can poison whoever eats the shellfish. [*algal blooms; manatee deaths and algae blooms*]

redwood forest destruction

Probably the most famous trees in the world, the *coast redwood* and the giant sequoia, are both commonly called redwoods. They typically live for hundreds of years but have been recorded more than 2500 years old. They can be higher than a 30-story building with a trunk diameter more than 25 feet across. These trees have been photographed by *Ansel Adams* and written about by *John Muir* and others. Everyone should have seeing these trees on their bucket list, but you must hurry. In the past 150 years, all but 4 percent have been cut down. It seems amazing to many that a private logging company can come in and simply cut down an *ancient forest*. But that is what happened during the late 1980s, when the company that owned a huge swath of land decided to liquidate their inventory of these *old-growth* trees.

It didn't happen without a battle in 1990—to be known later as the Redwood Summer. Environmentalists, including a group spurred on by *EarthFirst!,* organized resistance to the logging. *Julia "Butterfly" Hill* became front-page news for climbing a redwood and remaining there for two years. The campaign resulted in a few concessions from the logging company, but overall, environmentalists lost and 360,000 redwoods came down. Today this land remains privately owned but is now under the supervision of the *Forest Stewardship Council.* If you decide to add it to your bucket list, today redwoods can only be found on one strip of land about 450-miles long that runs from the northern coast of California into Oregon. [*Save the Redwood League*]

reforestation

Replanting stands of trees in areas that have been previously *logged* or otherwise degraded by human use. The amount of time required between harvesting these trees is called the logging cycle. Most logging cycles today are about 60 years but could be as long as 150 years. *Ancient forests* and *old-growth forests* that contain trees hundreds or even thousands of years old, of course, can never be replaced.

The *biodiversity* of a *forest* is believed to never return to its original richness once a virgin forest has been logged.

The use of reforestation, along with *afforestation,* has resulted in a net increase in the total forest acreage of some developed nations. [*forest health, global; forest management, integrated; forest, sustainable; logging, illegal; deforestation and climate change; dual-use policy; Redwood forest destruction; tree farms*]

refrigerator doors

Conserve energy by checking the seal on your refrigerator door. Close the door on a piece of paper that is half in and half out of the refrigerator. If you can easily pull the paper out, you're wasting energy and your money. The seal can be replaced or the door adjusted. [*energy, ways to save on residential; energy-saving apps and programs; bathroom water-saving techniques; tub bath vs. shower; Christmas tree lighting; Energy Star rating; smart energy; negawatts; states, ten most energy-efficient*]

refrigerator recycling

Americans dispose of nine million refrigerators each year. The good news is this removes the old, energy-inefficient machines. The bad news is this becomes a lot of waste; much of it releasing Freon, the cooling agent—an *ozone depleting gas*—and other *greenhouse gases* into the *atmosphere*. Most of these defunct appliances end up in *landfills*, but in some places much of this material is recycled. The Appliance Recycling Centers of America (ARCA) can recycle 100,000 refrigerators a year, resulting in over 5 million pounds of material that did not have to go into landfills. The shredding devices that tear them to bits also capture the harmful gasses before it is released into the *atmosphere*. [*refrigerator doors; recycling; CFCs*]

regenerative agriculture

In many ways, regenerative agriculture is similar to *conservation agriculture*. The main difference is that regenerative agriculture emphasizes making poor *soil* productive again (whereas conservation agriculture tries to keep the soil healthy to begin with). It is used not only to improve farmlands but also *rangeland* for livestock.

Regenerative agriculture uses *conservation tillage* methods, *composting,* and increased plant *biodiversity,* among other techniques. It also uses *perennial* grasses for rangeland. All these actions enrich the soil, making it *friable* (crumbly), which means it is rich in *organic matter* and can withhold moisture for plant growth.

Regenerative agriculture has an added benefit that gives it some recent panache. Rich, healthy soils are *carbon sinks*, storing vast amounts of carbon in *organic matter* and keeping it out of the *atmosphere* so it cannot become a *greenhouse gas*. If cropland and rangelands are managed properly, they not only produce food and feed livestock but can also soak up a significant percentage of *carbon dioxide* and help reduce *climate change.* Using soil as a carbon sink can be thought of as a natural form of *carbon sequestration storage*.

Regional Greenhouse Gas Initiative (RGGI)

This is the first mandatory *cap-and-trade program* for *carbon dioxide* in the United States. RGGI sets a cap on emissions of carbon dioxide from *electric power plants* and provides a mechanism to trade emissions credits. The program began by capping emissions at current levels (in 2009) and aims to reduce these emissions by 10 percent by 2018.

RGGI was created in 2005 by the governors of seven states: Connecticut, Delaware, Maine, New Hampshire, New Jersey, New York, and Vermont. Massachusetts, Rhode Island, and Maryland joined in 2007. New Jersey has since opted out and no longer participates. [*carbon markets; carbon tax; REDD+*] {rggi.org}

REM

Short for roentgen equivalent man, a unit that measures the exposure of an individual to *radiation*. A typical X-ray of the intestines releases 1 REM. Exposure to low levels of radiation (less than 10 REMs/year) is believed to be safe, although that remains debatable.

Moderate levels of radiation (10 to 1000 REMs/year) are believed to increase the likelihood of disease, such as some cancers. High dosages (above 1000 REMs/year) will cause ill effects and the likelihood of death. [*nuclear power; nuclear weapons*]

remediation of hazardous waste

The process and business of cleaning up *hazardous waste* sites. Remediation may be as simple as manually mopping up an area or involve high-tech equipment and sophisticated chemical processes. New methods of remediation include the use of microbes, called *bioremediation*, that eat hazardous wastes. *Phytoremediation* uses plants to absorb hazardous wastes such as *heavy metals* or radiation. Hazardous waste remediation is a rapidly growing sector of the U.S. economy. [hyperaccumulators]

remote-control car starter

A device that starts an *automobile* remotely, much like a remote control used for a television. The idea is you can warm up the car on a cold day before venturing outside. This option is commonly found on new cars. Trucks often leave the engines on (idling) to remain warm in winter months. Both idling trucks and using remote starters to turn on and idle car engines use gas and, of course, contribute to *air pollution*. Some environmentalists urge people not to use this option to both save gas and reduce emissions. Of course, someone sitting in a cold car on a freezing morning might not agree. [*idling your car; hotel load; autonomous automobiles*]

rendering

The act of processing livestock into a protein meal substance that is used for things such as food for pigs, poultry, and pets. The animals used for rendering are often dead or dying at time of slaughter. This includes "downers"—animals that collapse before they can be slaughtered. Millions of dead, dying, or collapsed animals are used for rendering each year. [*CAFO; factory farms; farm animal cruelty legislation; egg farms; animal rights movement and animal welfare*]

renewable energy

Renewable energy sources are considered to be inexhaustible, even if people continuously use them. Renewable energy and *nuclear power* are the two major *alternative energy sources*; alternatives to *fossil fuels*.

Renewable energy includes *solar power*, *wind power*, *hydropower*, *biomass energy* sources, *geothermal energy*, and *ocean thermal* energy. Renewable sources are usually nonpolluting and produce none of the *hazardous wastes* nuclear does. Renewable sources can produce energy in the form of electricity, heat, and *transportation* fuel as opposed to nuclear, which only produces electric power. Many renewable sources are already cost competitive when compared to fossil fuels and are becoming less expensive as their use expands.

A problem facing many renewables is energy storage. Many of these renewable sources are variable in nature, meaning they cannot provide *baseload* power. Solar availability varies because of day and night cycles and short-term weather conditions, wind because of short-term weather conditions, and hydro because of long-term weather conditions such as drought. Energy storage technologies—which can smooth out the supply—are improving rapidly and should resolve this issue in the near future. This process is called *time shifting*. Many existing renewable power plants use fossil fuels for backup purposes during down time and peak usage periods.

Some countries are much further along with using renewable energy sources than the United States. For example, Iceland gets close to 100 percent of its energy from renewables; 78 percent comes from geothermal and the rest from hydropower. Denmark has more than 5000 wind turbines, which in this small country comes to about one wind turbine for every 1000 residents. And Canada gets 68 percent of all its electricity from hydropower.

Renewable energy sources supply over half of Japan and Israel's energy needs, but in the United States, all forms of renewable energy fulfill only 13 percent of our electric power needs. Hydropower is the most popular, generating about 60 percent of all U.S. renewable energy power. In 2011, a milestone was reached when renewable energy sources surpassed nuclear power for the first time.

renewable resources

See *natural resources*.

resorts, eco-

See *ecotourism*.

Resource Conservation and Recovery Act (RCRA)

An important piece of legislation, originally enacted in 1976, that empowered the *Environmental Protection Agency* to protect the country's water and air from contamination. Specifically, it defines exactly what are considered *hazardous wastes* and how they can be transported, stored, treated, and disposed of. It initiated the use of *cradle-to-grave* accountability of a product. In 1984 an amendment expanded its role to include *LUSTs* and many other issues. [*Superfund; water pollution; air pollution*]

respirable suspended particulates (RSP)

RSPs are airborne particles small enough to be inhaled deep into the lungs but large enough to get stuck and remain there indefinitely. The nose, throat, and bronchial tubes filter out particles larger than 1.5 microns (a micron is one millionth of a meter), and particles smaller than 0.1 micron are usually exhaled normally, so RSPs range in size between 0.1 and 1.5 microns.

Asbestos and *lead* are two examples of RSPs. Cigarette-smoke particles carry an RSP that is probably responsible, at least in part, for the increased risk of lung cancer among cigarette smokers. [*toxic pollution; air pollution; hazardous air pollutants; particulate matter*]

respiration, cellular

Green plants perform *photosynthesis* to create sugar. These sugar molecules act as fuel for plants and animals. Cellular respiration is the process in which the molecules of sugar are burned (broken down) to release the energy stored within. This released chemical energy is used by organisms to survive (grow, move, reproduce, etc.). Cellular respiration occurs when the sugar (glucose) molecule combines with oxygen to form *carbon dioxide* and *water* and to release energy. [*biogeochemical cycles; biogeochemical cycles, human intervention in*]

rhinoceroses and their horns

Many species of this magnificent animal are in danger of *extinction* and much is being done to save them. In November of 2013, the western black rhino was officially considered *extinct in the wild*. Southern black rhinos are not much better off, but a few thousand still remain wild—mostly in Africa but a few in South America. Efforts to save the rhinos include the *World Wildlife Federation's* Black Rhino Range Expansion Project, where sedated rhinos, safely harnessed by their feet, have been airlifted (upside-down) out of areas with no roads into new safe *habitats*.

The rhino's horns are not *ivory* but similar to our fingernail material, keratin. It is as valuable as—if not more than—ivory, fetching in recent years $30,000 a pound. A *CITES* ban on selling rhino parts was established in 1977, but horn *poaching* is decimating the few remaining populations of rhinos. Both species, the black and the white rhinos, are believed to be destined for extinction within a couple of decades if poaching continues as it is today.

The most unfortunate aspect of this slaughter is that if their horns are shaved, they grow back, making it a sustainable harvest. However, poachers don't want a few scrapings and don't care about creating a sustainable harvest; they just want the money.

Many people believe the best way to protect rhinos is to legalize the selling of their horns. The price would plummet, reducing the poaching and letting people protect rhinos so they could harvest the horns in a sustainable manner. A South African entrepreneur, John Hume, owns and cares for 800 rhinos and has over 2,000 pounds of sustainable horns. If he was allowed to sell them, it would help drive the price down, reduce the poaching, and protect the rhinos. This, however, is a very controversial topic among environmentalists. [*poaching; elephant poaching; wildlife trade and trafficking; biodiversity, loss of; endangered and threatened species lists; extinction, sixth mass; species recently gone extinct; United Nations Convention on Biological Diversity; whaling; Wildlife Conservation Society*]

ribbon sprawl

A type of *urban sprawl* in which commercial and industrial growth occurs along *transportation* routes such as major roads or commuter rail lines. [*urbanization and urban growth*]

rice, contaminated

Globally, humans consume more rice than any other food. It is often the first solid food given to young because it is bland and readily digestible. In some regions of the world, rice is typically served three times a day. It has plenty of vitamins and minerals and is full of nutrients. So what is not to like about rice? Some consumer groups are concerned that *arsenic* is often found in rice and are asking that maximum levels of arsenic be regulated for this food, just as is already done for water.

Part of this problem is the abundance of arsenic in our overall environment. It is actually found in many places, with rice just one more way people are exposed to it. When people eat a lot of rice in addition to their other exposures to arsenic, the total intake can become a health issue.

Agriculture and in-house *pesticides* often use synthetic arsenic, which makes it common in some *soils*. It can also seep into *groundwater*. Because rice is grown in soil saturated with water up to the roots, it absorbs arsenic from the soil more readily than other crops. Also, rice is a common crop in the U.S. south, where cotton fields were typically treated with arsenic pesticides. There are arsenic tests for water, but none for foods. [*Dirty Dozen – fruits and vegetables; food dye; food eco-labels; nanoparticles in food; GMOs in your food, the five most common; trace elements; apples, antibiotics in*

rice paddies and methane

Even though livestock is one of the biggest emitters of *methane*—a potent *greenhouse gas*—into the *atmosphere*, rice paddies are as well. Rice paddies are believed to contribute possibly 20 percent of all methane emissions. The methane is produced because rice paddies are continually soaked in water, allowing *anaerobic microbes* to continually *decompose* the *organic matter* and produce the gas. Rice is the primary food for almost half of all people on the globe, so the problem is projected to get worse, especially with *global warming* contributing to the problem. [*climate change; food's carbon footprint – the best and worst; assisted migration; climate change and biodiversity*]

Richter scale

A scale used to measure the intensity of geological disturbances, ranging from 1.5 for barely detectable tremors to 8.5 for catastrophic earthquakes. The scale is logarithmic. A reading of 7 is 30 times greater than a reading of 6. [*plate tectonics; supercontinents; Pangaea*]

ride-sharing

See *car-pooling*.

Rigs-to-Reefs Program

About 650 old, dormant *oil* rigs, mostly in the Gulf of Mexico, have become man-made reefs full of underwater life. Over time these structures become inhabited with *biodiversity*, similar in many ways to natural *coral reefs*. *Eco-tour* guides use them as well as natural reefs to view sea life. Scientists estimate that after a few years a typical oil platform turns into a two-to-three-acre reef *habitat*. Each platform can support more than 10,000 fish and other life. Considering the assault natural coral reefs are under, these artificial reefs could be considered a welcome addition.

However, a law requires oil companies to destroy old nonperforming platforms and remove them within a few years. The intent was to prevent oil companies from simply abandoning these facilities and leaving them to decay. The law is now viewed as an environmental dilemma because within a few years they will become a useful artificial habitat, loaded with life. Destroying a rig takes large amounts of explosives, kills hundreds or even thousands of fish and other marine life, and destroys their habitat.

Now efforts are being made to allow oil companies to leave many of these rigs in place as artificial reefs. The Interior Department has a program called Rigs-to-Reefs that allows companies to leave most of the rig and cut off the rest, so no explosives are needed and the habitat is left intact. Companies must apply to this program and be approved. A *NGO* was formed to help insure and maintain these platforms. [*ship-sinking program*] {infosources.org/what_is/Rigs-to-Reefs.html}

right-to-dry laws

As recently as the 1950s, the standard way to dry clothes was on a clothesline. Some people still like to use them, saying the clothes smell fresher. Regardless of the smell, using an electric or gas drier adds a great deal to the size of a person's or city's *ecological footprint*. Some people trying to get back to basics and use clotheslines to reduce their footprint cannot do so because many homeowner associations (HOAs) forbid them as being unattractive. Some states are fighting back by passing legislation banning the bans on clotheslines. These same laws typically also encompass the right to put up solar panels for *solar power*. The states with right-to-dry laws are Florida, Utah, Hawaii, Colorado, Maine, Vermont, and Maryland. [*laundry and the environment; noise pollution and control; outdoor recreation and motorized vehicles; zoning, land use*]

rill erosion

A form of *soil erosion* caused by rain and overflowing *surface waters* that produce a series of grooves or channels. [*conservation tillage*]

Rio+20 Earth Summit of 2012

See *United Nations Conference on Sustainable Development.*

Rio Declaration on Environment and Development

See *United Nations Conference on Environment and Development.*

Rio Earth Summit of 1992

See *United Nations Conference on Environment and Development.*

risk assessment, environmental

The process of gathering facts and making assumptions about potential dangers to human health or to the environment caused by some project, product, or technology. These dangers could include various forms of *pollution, radiation, pesticide* poisoning, and other negative effects on *ecosystems* or *biogeochemical cycles.*

This assessment attempts to differentiate between *true and perceived risks*. Proper risk assessment should lead to *risk communication* to educate the public, and *risk management* in which the knowledge gained is used to eliminate or reduce dangers, look for alternatives, and balance the economic advantages with the risks. [*cost/ benefit analysis*]

risk communication

Risk assessment distinguishes fact from fiction about the risks involved in proceeding with a particular project, product, or technology. Risk communication can then educate the public about these risks. Public hearings or other forums give the public an opportunity to voice their opinions and influence final decisions about any forthcoming environmental risk. [*risk management, environmental; risks, true vs. perceived; science; pseudoscience; opinions and science; environmental justice*]

risk management, environmental

Risk assessment clarifies facts so the decision-making process, called risk management, can proceed. *Risk communication* ensures public participation in the process.

Environmental risk management balances advantages and disadvantages of a new project, product, or technology. Managing environmental risk means weighing negative environmental impact against economic gains, and includes public interest and legal issues. Risk management is difficult because much of the information gleaned from studies cannot be based on fact, but on assumption. In other cases, what's good for the economy may be harmful to the environment, and the correct decision is highly subjective. [*cost/benefit analysis; risks, true vs. perceived*]

risks, true vs. perceived

The public often has great difficulty determining the true risks of a new project, product, or technology when they are bombarded with information, some factual, some not. This information comes from industry, government, and environmental groups. An informed public is the best way to assure that risk management is based on the best interests of the citizens. The difficulty involved in differentiating true vs. perceived risks is underscored by comparing what the *Environmental Protection Agency* and the public believes to be an environmental risk.

For example, the EPA believes that *indoor air pollution*, safe *drinking water*, and workers' exposure to on-the-job chemicals rate as "very high" risks to our health, but the public considers all three to be "low" risks. The public, however, thinks chemical manufacturing plants are a "very high" risk, but the EPA considers this category only a "mid-level" risk. [*risk assessment, environmental; science; pseudoscience; opinions and science*]

rivers in decline

Rivers around the world are degrading in many ways. Some of the problems include *nutrient enrichment* from *fertilizers*, *water pollution* from *pesticides* and *urban run-off*, *erosion* from *urbanization*, and *dam* construction and associated flood control impediments.

Many people get their water supply from rivers. Degradation of rivers doesn't just threaten *aquatic habitats* and *agriculture* but also the health of those that drink the water. The *NGO*, American Rivers, releases an annual ten most endangered rivers and many of them are highly polluted, yet still provide *drinking water* to local communities. In 2012 the Potomac River, which runs through our nation's capital and provides drinking water to about five million people, was designated one of the most endangered rivers in the United States.

The most endangered rivers 2013 list (and the primary cause) is as follows: 1) Colorado River (outdated water management), 2) Flint River (outdated water management), 3) San Saba River (outdated water management), 4) Little Plover River (outdated water management), 5) Catawba River (coal ash pollution), 6) Boundary Waters (copper and nickel mining), 7) Black Warrior River (coal mining), 8) Rough & Ready and Baldface Creeks (nickel mining), 9) Kootenai River (open-pit coal mining), 10) Niobrara River (improper sediment). [*blue-jean pollution; Clean Water Act; industrial water pollution; running water habits; running water habitats, human intervention on; domestic water pollution; in-stream water use*]

robotic bees

Picture a robot that is about the size of a bee, looks a little like an *insect*, and is capable of controlled flight—that's the robo-bee or robotic bee. And they do exist. Researchers at the Wyss Institute for Biologically Inspired Engineering at Harvard have produced a miniscule, motor-powered, wing-flapping robot capable of flight. The research began with the thought of producing entire colonies of these "insects" to *pollinate* crops in light of the advance of the *colony collapse disorder* that is decimating many natural bee colonies.

They are two-thirds of an inch long with a 1-inch wingspan and weigh in at 2/100 of an ounce. The hope is they will be able not only to pollinate but also to be used for search and rescue, surveillance, and traffic monitoring.

At the moment, they are not ready for prime time, but they can hover and perform simple maneuvers. The problem is now they still must be wired to a computer, so going wireless is the next logical step and supposed to be coming soon. Another stumbling block is the battery to provide power, which needs improvement because one small enough to fit into these robots can only keep it in flight for a few minutes. [*biomimetic; biosensors; biotechnology; nanotechnology; geoengineering*]

Rocky Mountain Arsenal National Wildlife Refuge

This is a 15,000 acre *wildlife refuge* northeast of Denver, Colorado. At first blush, it seems like any small *wildlife refuge* that has about 35,000 visitors each year who enjoy trails to hike and bike, go on tours to see reintroduced *buffalo,* and learn about environmental issues. However, it is very different than most other refuges because it was at one time one of the most contaminated sites in the United States. During World War II, it was used to produce nerve gas, mustard gas, and napalm. After the war, it was used as a weapons stockpile. Later some of the land was leased to Shell Oil Company to produce *DDT. Hazardous wastes* were dumped by both the U.S. Army and Shell into unlined ponds and *deep-well injection sites.*

In 1987 the area was selected as a *Superfund* site and prepped for a $2-billion cleanup to be performed by the U.S. Army and Shell. The cleanup primarily involved dealing with the contaminated *groundwater*, ensuring it did not spread, and removing contaminated *soil*. The *remediation* took over 10 years and a few billion dollars, but the cleanup has been deemed a success. [*hazardous waste disposal, military; toxic waste; Hanford Nuclear Reservation; LUSTs; Love Canal*]

Rocky Mountain Institute (RMI)

RMI is a well-respected *NGO* that concentrates on "the efficient and restorative use of resources." *Amory Lovins* is RMI's chief scientist. They use market-based solutions to transition business and industry toward efficiency and *renewable energy* sources in a cost-effective manner. RMI describes themselves as "practitioners, not theorists. We do solutions, not problems."

RMI was founded in 1982 by Amory Lovins and L. Hunter Lovins, who co-authored the book *Natural Capitalism*. [*negawatts; soft energy path; alternative energy sources*] {rmi.org}

rodenticide

A *pesticide* that kills rodents such as mice and rats. The most common rodenticide is warfarin, which when eaten by animals, causes internal bleeding and death. Warfarin is not selective, meaning it will kill or make seriously ill animals other than the targeted pests—whoever eats it. [*natural pesticides; insecticides*]

rookery

A breeding or nesting site for animals—usually birds. [*guano; penguins; Arctic National Wildlife Refuge; caliology*]

room and pillar method

A method used to provide structural support in *subsurface mining* for *coal*. [*overburden; spoil banks; mountain top removal mining*]

Roosevelt, Theodore

(1858–1919) If you have ever been in a *wildlife refuge*, *national park* or *forest* in the U.S., there is a very good chance you should thank our 26th president, Theodore Roosevelt, for its existence. He was our first and most important conservationist president, in office from 1901 to 1909. He created the *U.S. Forest Service*; 51 federal bird reservations, 4 national game preserves, 150 national forests, 5 national parks, and 18 national monuments. He protected about 230 million acres of *public land* for future generations to enjoy and savor. He signed into law the *Antiquities Act* and used it extensively.

He was a prolific writer of nature and of the wild west in particular. He was personal friends with *John Burroughs* and with *John Muir*, spending time with both in the *wilderness*, talking about saving lands for future generations. He was the founder and first president of the *Boone and Crockett Club* and helped establish the New York Museum of Natural History. His presidency has been called the Golden Age of *Conservation* in America.

After leaving office, Roosevelt continued his love of the outdoors by embarking on an expedition into the *Amazon* River of Doubt, where he came close to death during the ordeal. (The River of Doubt was later renamed Rio Roosevelt in his honor.) [*National Wildlife Refuge System; Forest Service, U.S.; Grand Canyon; Migratory Bird Act; preservation*]

r-strategists

r-strategists are usually small organisms that are short-lived, produce large numbers of offspring, and offer little or no parental care. These organisms expend a great deal of energy producing dozens, hundreds, are even thousands of eggs or live young. Their reproductive strategy is that there is safety in numbers—if enough offspring are produced, some of the young are likely to survive. Most *insects* and some reptiles, *amphibians*, and small *mammals* use this strategy.

The *population* size of these organisms is controlled by *density independent* limiting factors, factors that have little or nothing to do with the density of the existing population. For example, extremes in temperature, shortages of water, or unexpected population explosions of *predators* are *limiting factors* that control these populations. The population size of r-strategists does not usually reach the environment's *carrying capacity* as it does with *K-strategists*. [*population ecology*]

runoff

That portion of *precipitation* that is not absorbed by *soil* and washes away. Runoff collects in *surface waters*, and then typically makes its way to the *ocean*. Storm water runoff can contain numerous kinds of pollutants, including *gasoline*, *deicing road salts*, *plastic* products, and just about anything you see in a street, including *litter*. Agricultural runoff refers to water that washes off of *irrigated* land and usually contains large amounts of *fertilizers* and *pesticides*. This often results in *nutrient enrichment* of *aquatic ecosystems*. *Acid mine drainage* runoff comes from mining operations and carries harmful acids into bodies of water. [*water pollution*]

run-of-the-river hydropower

Also called "streaming hydropower," this relatively new technology generates power without the *dams* and reservoirs used by the traditional form of hydropower. Instead, it uses the natural movement of a river to turn turbines that generate electricity. It does not cause the environmental problems associated with damming a river and is viewed by many as a good way to use hydropower in a less damaging way.

However, it has some disadvantages. It does not provide *baseload power*, meaning the power is not continuous. Usually, river speed and volume is seasonal, which impacts the amount of power that can be generated. Also, the overall amount of power generated is small compared to systems with large dams. In most cases these facilities will only generate 30 *MW* or less of power. But this smaller, distributed method of producing energy is considered a possible path for the future. [*hydropower, traditional; hydropower, nontraditional*]

running water habitats

Freshwater ecosystems can be divided into *standing water habitats* and running water habitats, which include streams and *rivers*. Since the *water* is always moving, organisms attached to rocks and other substrate play the most important role instead of the floating *plankton* that are important in most other *aquatic ecosystems*.

Most moving bodies of water are usually shallow, so sunlight penetrates to the bottom, allowing plants and *algae* to grow and act as food for organisms such as *insects*, establishing a *grazing food chain*. *Predatory* insects and fish feed on these insects.

Nutrients are also added to the stream by dead and decaying organisms falling into the water. Leaves, twigs, and bark from trees fall in, and animal waste and other *organic* particles are washed in. This organic matter is attacked by *bacteria*, *fungi*, and microorganisms, which establish a decomposing *food chain*. Insects eat the bacteria and fungi that are attached to the decaying leaves. The decaying leaves and associated organisms flow downstream and eventually sink to the bottom to become a food source for many other organisms, especially aquatic insect *larvae*. [*running water habitats, human impact on*]

running water habitats, human impact on

Freshwater ecosystems can be divided into *standing water habitats* and *running water habitats*. Running water habitats include rivers and streams people have used for hundreds of years as toilet bowls, where *pollutants* and waste could be "flushed out of sight." The vast majority of rivers in the United States are polluted to some degree. Streams and rivers can cleanse themselves when low levels of waste are introduced, but the volume and speed with which humans produce these wastes cause irreparable harm.

Industrial water pollution pours huge quantities of *toxic waste* and byproducts into rivers. *Water-treatment facilities* often allow raw *sewage* to be poured into rivers and streams in vast quantities over short periods of time, changing the chemical makeup of the water and often eliminating the usual inhabitants. *High-impact agriculture* results in *runoff* of vast quantities of *pesticides* and *fertilizers* into rivers and streams, changing the chemical composition of the water and killing aquatic organisms. *Radioactive waste* has also been found in the *sediment* of streams and rivers. Water that is drawn from rivers to cool *nuclear power plants* is then returned to the river but at temperatures that destroy ecosystems, creating *thermal water pollution*. [*water pollution; rivers in decline; in-stream water use; Wild and Scenic River System*]

R-VALUE
See *thermal insulation, buildings*.

safe and just operating space
See *safe operating space.*

Safe Chemicals Act of 2013
See *Toxic Substances Control Act.*

Safe Drinking Water Act
Created in 1974, this law required the *Environmental Protection Agency* to establish maximum contaminant levels for all *drinking water* pollutants so they do not pose a danger. This includes chemicals, animal waste, *pesticides,* and human waste. Although the law has reduced the amount of microbiological contaminants, it does not adequately protect us from all forms of contaminated water. Today many cities deliver water not considered acceptable under this law.

New water problems are appearing across much of the United States. Old, outdated pipes *leach* out *lead* into the drinking water. Now traces of pharmaceuticals are also found in many water systems. These new contaminants are not tested or regulated in most municipalities. Even New York City, long able to claim some of the cleanest public drinking water in the country, is finding lead in some of its water. Experts in the field say that in 10 to 15 years, everyone (on public water) will have to have water-filtration systems in their homes to ensure clean drinking water. Some websites list the quality of water for most large cities. [*water pollution; domestic water pollution; freshwater springs; potable water; toilet-to-tap drinking water; water treatment; water use for human consumption; combined sewer systems; groundwater pollution and depletion*] {water.epa.gov/drink/local}

safe operating space
In a paper published in June 2009 in <u>Nature</u> magazine, scientists suggested a framework called the safe operating space that humans must live within if we are to survive

as a species on our planet. The phrase "safe operating space" has become a popular message about our planet's state of affairs.

This safe operating space can only be maintained if planetary systems remain within boundaries. They presented nine such planetary-system boundaries: *climate change*, *biodiversity loss*, disruption of the *nitrogen* and *phosphorus cycles*, *stratospheric ozone depletion*, *ocean acidification*, global *freshwater* use, changes in land use such as converting natural lands to farms and *urban* areas, *atmospheric* aerosol loading (*particulate matter*), and chemical *pollution*. They went on to give specific values as to where these boundaries lie. (With the exception of the last two.)

The scientists believe human activities have already pushed three systems beyond their boundaries: climate change, biodiversity loss and the nitrogen cycle (just the nitrogen portion). [*social boundaries; anthropocene geological epoch, tipping points*]

Sagebrush Rebellion

During the 1980s, many ranchers and miners felt too much land in the west was being set aside for environmental protection and initiated a crusade to stop these actions. This drew a battle line between ranchers, loggers, miners, and others against the Washington bureaucrats in a fight over the West's land, water, and mineral resources. This rebellion is often cited as the first anti-environmental movement in the United States. [*wise-use movement; environmental movement; buffalo; multiple use policy*]

salinity

The concentration of dissolved salts in water, usually measured in parts per thousand (ppt). For example, most oceans have about 35 ppt of salinity. The Dead Sea has the highest natural salinity in the world, with more than 330 ppt of salinity. *Freshwater* is not completely void of salts, but averages less than 0.5 ppt of salinity. [*aquatic ecosystems; marine ecosystems; freshwater ecosystems; brackish water; estuaries*]

salinization, soil

Salinization of *soil* is a harmful side effect of *agricultural irrigation*. When irrigation water washes over the surface of the soil, it dissolves and collects salts, making the water salty. When the water evaporates, it leaves behind increased concentrations of these salts in the soil. This process is called soil salinization.

In arid regions (where most irrigation occurs), evaporation occurs quickly and soil salinization progresses rapidly, stunting the growth of crops and eventually making the land useless. About one-fourth of the world's crops are believed to be growing at reduced levels because of soil salinization.

Methods of renewing salinized soil include flushing the region with excessive amounts of water to remove the salts, but this often results in *waterlogging* and excessive concentrations of salt in local *aquifers.* [*saltwater intrusion*]

saltatorial

Organisms that move by hopping, leaping, or bounding. For example, grasshoppers are saltatorial. [*niche, ecological; ecosystems; acoustic niche*]

saltwater intrusion

When *groundwater* in *aquifers* is removed faster than it can be replaced (recharged), the *water table* drops. This is called *water mining*. When this occurs in coastal areas, the aquifers may become saturated with saltwater from the ocean instead of being recharged with more freshwater. When saltwater replaces freshwater in aquifers, it is called saltwater intrusion. Saltwater intrusion has become a problem in many coastal areas such as California, New York, and Florida. [*water shortages, global; water pollution; domestic water pollution; domestic water use*]

sand

Particles of *soil* that do not adhere to each other and are between 2.0 and 0.0625 millimeters in size. Sand types are divided into very coarse, coarse, medium, fine, and very fine sand. [*soil texture; soil particles; soil types; topsoil; sedimentation*]

San Francisco and compostable waste curbside pickup

San Francisco always ranks at the top of the greenest U.S. cities, but when it comes to minimizing waste entering *landfills*, they are the undisputed leader. About 80 percent of all *municipal solid wastes* are diverted into various *recycling* programs and, most recently, even their food waste is picked up to be *composted*. About 650 tons of organic waste food is picked up in green bins every day. The compost is then available to town farms, orchards, vineyards, and other businesses in need of rich, natural, organic *fertilizer*. Today almost 100 other cities have also initiated some form of organic food waste pickups. [*composting, large-scale; sewage treatment facilities; sludge disposal; night soil - biosolids; toilet-to-tap drinking water; toilet paper, eco-; biosolids and wastewater; cities, ten best American green*]

saprophyte

A plant such as *fungi* (mushrooms) that obtains nutrients from dead, decaying *organic* matter. They are also commonly called *decomposers*.

satellite mapping

Many satellites orbiting Earth are equipped with cameras that take digital pictures of extraordinary resolution. These images are often used to create digital maps showing incredible clarity and allowing scientists to advance their knowledge. For example, recent satellite mapping of emperor *penguins* in *Antarctica* revealed twice as many individuals than originally estimated by traditional means. [*Landsat; Great Lakes Environmental Assessment and Mapping Project; drones, conservation; Wide-Field Infrared Survey Explorer*]

savanna

One of several kinds of *biomes*. The primary factors that differentiate biomes are temperature and precipitation. Savannas (also called tropical grasslands) are similar to *grasslands* but receive more precipitation, between 30 and 60 inches per year in a seasonal pattern.

The *environment* appears similar to that of grasslands but contains some trees. Fire often destroys these regions, so most of the trees are fire resistant. Some of these trees are capable of *nitrogen fixation*. Trees provide a *habitat* for animals that otherwise could not survive in the region. Most animals are *grazing mammals* or rodents, birds, *reptiles*, and *insects*. The larger mammals differ depending on the location of the savanna. For example, *Australia* has kangaroos, Africa has antelope, and South America has llamas. [*anthromes; ecosystems*]

Save the Redwoods League

Founded in 1918, the goal of this *NGO* is to acquire and protect *coast redwood* and giant sequoia *forests*. The boundaries of many existing redwood parks are inadequate to maintain the redwood *ecosystem*, so the *watershed* lands surrounding these parks must also be acquired. The League has purchased more than a quarter of a million acres of redwood lands located in state parks and monuments. It also works to prevent or at least limit commercial *logging* of redwoods. [*redwood forest destruction; Hill, Julia "Butterfly"; dual-use policy; Luna Redwood; ancient forests; forest health, global; douglas fir; timber, DNA sequencing*] {savetheredwoods.org}

scatology

The study of animal feces. [*biomass energy*]

scavengers

Animals that obtain nourishment by eating organisms that have recently died. They do not kill the animal and are therefore not considered *predators*. Scavengers usually are the first to arrive at an animal carcass, called carrion, and begin the process of decomposition, which is then taken over by the *decomposers*. Terrestrial scavengers include a wide range of organisms from fly larvae (*maggots*) to hyenas. Aquatic scavengers include snails and catfish. [*food web; saprophyte*]

schools and brownfields

A significant number of public schools end up being built on old, contaminated *brownfields*. This happens for two reasons in combination. First, few states have any regulations about investigating sites where schools are to be built. Second, many municipalities have little available space to find a location for a new school, so less than desirable sites are often chosen. Combine these two factors and you end up with schools on polluted grounds.

This issue came to light after a high-profile case in 2006, when a new public school named Carson-Gore Academy of Environmental Sciences was found to be located directly on an old chemical waste facility, still containing old buried tanks.

In this case, the danger was identified in advance of the school's opening and was cleaned up. However, this is not the norm, because in most cases the town would probably never have investigated and known any problem existed. The issue is addressed by an *NGO* called *The Center for Health, Environment & Justice*, founded by *Lois Gibbs* of *Love Canal* fame. The irony, of course, is the name and type of school that first addressed this environmental issue. [*cancer alley; cancer villages; POPs, dirty dozen; hazardous waste; toxic waste; toxic pollution; Superfund; remediation of hazardous waste; persistent organic pollutants*]

science

Science is the process of gaining and organizing knowledge. The step-by-step process used to gain new knowledge is called the scientific method. As science adds new knowledge to what is already known, it is organized into disciplines of study such as biology, botany, and bacteriology. So science is both the process of gaining new knowledge and the body of work where that knowledge is collected and organized.

The scientific method is based on a few fundamental assumptions. The most important is that events in nature have a specific cause. The scientific method is designed to link the cause to the event (a cause-and-effect relationship). If a direct correlation exists between cause and event, it should be reproducible, providing opportunities for other scientists to confirm, or contradict, what has been learned. This reproducibility is an important part of the scientific method. Once a cause-and-effect relationship has been established and confirmed, the results are considered a fact.

It is easy to confuse facts for opinions or just outright guesses, because many events appear to have a cause-and-effect relationship when they do not. A simple example makes the point. People often think trees in some regions lose their leaves in the fall because it gets colder. Certainly a correlation can be seen between its getting colder and a loss of leaves. However, a correlation doesn't make it a fact—just an opinion or a guess that someone might have. Scientists, using the scientific method, have determined that the true cause-and-effect relationship is the shortening of the days—not the temperature dropping.

Today science is fraught with people who believe a guess or an opinion is as good as a fact. Many people also don't know the difference between science and *pseudoscience*. It is science and the facts that we have learned that brought our civilization to where we are now. If it were not for science, we would still be living in the dark ages. [*opinions and science; argumentum ad ignorantiam; climate change deniers; environmental literacy; environmental education; pig-gate; Agenda 21 conspiracy*]

S-curve

If any one particular *population* of a *species* were allowed to grow unchecked, it could take over the entire planet. If you were to plot a graph of this unchecked growth, it would look something like the letter "J" and is therefore called a *J-curve*.

However, resources such as food are limited, and conditions such as temperature and moisture are never ideal. All of these limitations are called *environmental resistance* and the single most important factor is called the *limiting factor*. So in nature, most populations, when plotted on a graph, start by looking like a J-curve but then level off when they reach a maximum size the *environment* can support, called the *carrying capacity*. This creates a growth curve that looks something like the letter "S"—hence the S-curve of population growth. Almost all species follow this growth pattern—but not humans. [*tolerance range; population ecology; population explosion, human*]

seafood eco-ratings

Many *environmental rating systems* help consumers understand the environmental impacts of companies and their products and services. There are ratings for cosmetics, foods, wood products, and just about everything else. Seafood ratings are important because *overfishing* is wiping out *fish populations* globally (which is why half of our seafood now must come from *aquaculture* or fish farms). Seafood is one of the more difficult products to rate because they involve many variables.

Most seafood ratings are based on the following five factors: overfishing; *bycatch*; *habitat destruction*; aquaculture vs. *wild caught*; and whether it is *illegal, unregulated, or unreported (IUU).* Most of these factors have to do with how and where the seafood is caught. For example, wild-caught shrimp from the United States gets a far better rating than farm-raised shrimp from Thailand, which gets a poor rating.

Because most of us don't have time to spend figuring these things out, the most popular seafood eco-ratings is from the Monterey Bay Aquarium, with its color coding system. They typically are rated as follows: Green means the "best choice"—fish stocks are abundant and well managed; they can be wild or farmed. Yellow means a "good alternative"—there are concerns about factors mentioned above. Red means "avoid"—fish stocks are dangerously low or the way they are caught is harmful to the environment. [*food eco-labels; organic food labels; eco-labels; environmental certification programs*] {seafoodwatch.org}

sea gliders

Small, unmanned, robot-like submarines being used for research. For example, they have been used for studying underwater volcanoes, tracking the *BP Deepwater Horizon oil spill*, tracking radiation releases from the *Fukushima nuclear power plant disaster*, and inspecting *icebergs* and underwater cables. In many respects, they are the aquatic equivalent of *conservation drones*. Today about 400 gliders are in use. They are far less expensive than underwater submersibles that take people down to depths. Some sea gliders can go down to 20,000 feet, and they are so quiet they don't disturb wildlife, allowing for close-up investigations. [*beetlecam; robotic bees*]

sea level rise

One of the most dramatic results of *climate change* is the rise in sea level worldwide. Many low-lying island nations fear their entire population might be forced to migrate to new lands if climate change is not mitigated.

For 2000 years, sea levels remained unchanged. Then, between 1880 and 2011, the level increased an average of 0.07 inches per year. If you only look at the time between 1993 and 2011, the level increased 0.12 inches per year. The *Intergovernmental Panel on Climate Change Fifth Assessment Report* projects sea levels will rise between 3.3 and 9.8 feet by 2300. Even the low end of this range will put many coastline cities and islands under water.

sealing

Inhabitants of eastern Canada have hunted seals for thousands of years, and European settlers have done so for hundreds of years. Many countries have participated at one time or another, but most sealing today is done in Canada. By the late 19th century, many seal species were decimated. The 1960s brought protests from animal rights activists trying to stop harp seal hunts in Canada. It was not hard to demonstrate the cruel slaughter of adults and especially the young, called whitecoats.

The 1970s saw a consolidation of anti-sealing organizations and an upwelling of support from the public. These *NGOs* broadened their efforts to include applying economic pressure on the importing nations, as opposed to the source of the sealing. When Europe imposed bans to seal product imports, this method was successful temporarily. By the 1990s, regulations were again loosened, and today sealers are killing more seals than ever. This time, for some reason, there has been little public outcry. *China* and Russia, with an ever-increasing higher-income population, have become the main markets for seal coats. [*poaching; wildlife trade and trafficking; ivory trade; CITES; rhinoceros and their horns; elephant poaching; whaling; animal rights movement and animal welfare*]

seaport-related air pollution

Most of us think of *automobiles* when we think about *air pollution*. However, cities with large seaports can get as much, if not more, of their air pollutants from the seaports as from all the cars in the city. Many of the ships that import goods into the United States use a low-grade, highly polluting fuel that produces *smog*-forming chemicals and *particulate matter*, all contributing to nasty air. The exhausts of these ships make car exhaust pipes look clean.

As an example, in California, the Los Angeles and Long Beach ports combined make up the fifth largest port complex worldwide. Winds carry this polluted air across the state, where it impacts 80 percent of the entire state's population and produces significant public health risks.

But there is good news for those in both California and elsewhere. The Vessel Fuel Rules went into effect in 2012, after a long legal battle, requiring cleaner fuels to be used within 24 nautical miles of the coast. And in 2015, new international laws will require ships within 200 miles of the United States and Canada to use cleaner fuels as well. [*air pollution; airplane pollution; cruise-ship pollution; ship pollution; hazardous air pollutants; black yogurt*]

Sea Shepherd Conservation Society

A well-known *NGO* and probably the only one to ever have a successful television show about its exploits. "Whale Wars" on the Animal Planet network follows this group's acts of what many would consider extreme environmentalism. The society was founded and remains led by Captain Paul Watson.

Their mission "is to end the destruction of *habitat* and slaughter of *wildlife* in the world's *oceans* in order to conserve and protect *ecosystems* and species. Sea Shepherd uses innovative direct-action tactics to investigate, document, and take action when necessary to expose and confront illegal activities on the high seas."

They use aggressive tactics that few other environmental groups would consider. The following excerpt from their "brief history" web page describes how the Sea Shepherd first started and describes their methods. "Its first mission was to sail to the ice floes of Eastern Canada to interfere with the annual killing of baby harp seals known as *whitecoats*. In the same year, the Sea Shepherd hunted down and rammed the notorious prolific pirate whaler the Sierra in a Portugal harbor ending its infamous career as the scourge of the seas. Since those early days, Sea Shepherd has embarked on more than 200 voyages covering many of the world's oceans and defending and saving defenseless marine life all along the way." [*whaling; factory ship; pirate fishing; sealing*] {seashepherd.org}

seasonal foods

This refers to foods that are eaten during the season when they are locally grown. Before *high-impact agriculture*, globalized markets, and advances in *transportation* took hold, we ate foods that were locally grown when they were available. When certain foods were out of season, we simply did not eat them until the following year. Now, if a food can be grown anywhere on the planet, it can be eaten anywhere else on the planet. Seasons have little to do with mass-marketed food.

There are those trying to get back to eating seasonal foods with the *farm-to-table movement* and *locavores*. Eating seasonal foods accomplishes many things. It lessens your *ecological footprint*, typically offers better taste and healthier foods, offers variety throughout the year instead of eating everything all the time, lessens the likelihood of *food-borne illness*, and supports local farmers. [*tomatoes, tasteless but red; farm-to-school movement; organic farming and food; food miles; farmers' markets; farmers' markets, fake; slow food movement; fregan; imported food dangers; WWOOF; campus quad farms; community gardens; community-supported agriculture; heirloom plants; teaching farms; microgreens and urban gardening*]

sea turtles

A group of about 270 species of land-nesting turtles found in tropical or subtropical regions. It includes green turtles, loggerheads, leatherbacks, and Kemp's ridleys. Almost half of them are rare, *threatened,* or *endangered species,* mostly because, although they primarily live in the sea, they make themselves vulnerable to humans when they lay eggs on land. People have historically collected their eggs and killed the turtles for food. In less developed nations, these turtles might be an important source of protein for local residents. However, *poachers* also collect them just for their shells to make trinkets to sell. Just as with other examples of the *wildlife trade*, these animals are killed for one small part of their bodies.

They are also in decline from *ocean pollution* and are *bycatch* of *commercial fishing* fleets. In addition, the hatchlings are often confused by the lights of buildings along beaches as they emerge, because they naturally are attracted to what they think to be moonlight over the sea. Overall, sea turtles have not done well with humans. [*light pollution; biodiversity, loss of; CITES; endangered and threatened species lists; extinction, sixth mass; International Union for Conservation of Nature; whaling; seafood eco-ratings; ship ballast and invasive species; species recently gone extinct, ten; Treaty on Biological Diversity; whale and commercial ship collisions*]

secondary air pollutants

Burning *fossil fuels* to drive *automobiles* and to generate electricity produces five primary air pollutants responsible for *air pollution*. Secondary air pollutants are substances created when the primary pollutants combine with one another or with some other substance. These reactions are driven by radiant energy from the sun. For example, *sulfur dioxide*, a primary pollutant, reacts with oxygen and moisture in the atmosphere to produce *sulfuric acid*, a secondary pollutant. Also, *photochemical smog* is produced when *nitrogen compounds* (a primary pollutant) produce *ground-level ozone* (a secondary pollutant). [*hazardous air pollutants; atmospheric brown cloud; Clean Air Act; criteria pollutants; dust dome; fly ash; seaport-related air pollutants; thermal inversion*]

sedimentation

When *soil particles* are picked up and transported elsewhere, usually by water, they accumulate resulting in sedimentation. Sedimentation causes bodies of water (*aquatic ecosystems*) to gradually fill during the process of *succession*. The soil can also be moved by wind or by ice, as with *glaciers*. [*soil erosion; soil texture; soil types; dredging; estuaries; bogs*]

seed banks and vaults

Britain's Millennium Project Seed Bank and Norway's Svalbard Global Seed Vault (aka the *doomsday seed vault*) are huge repositories designed to collect and store endangered and economically important seeds from around the world as a backup in case of disaster. Their primary goal is to prevent a loss of *biodiversity*. This might be saving one species from *extinction* or many species of important crops if a natural or man-made disaster wipes them out.

Other types of repositories, such as Project Baseline in the United States, are designed primarily for the study of evolutionary biology. Millions of seed varieties have been collected and are stored in these facilities. [*seeds and climate change, Project Baseline; heirloom plants; vegetable seeds, heirloom*]

seeds and climate change, Project Baseline

Most *seed banks and vaults* are designed to protect seeds in case the original source is destroyed. For example, a nuclear accident (or war) wipes out a variety of seed found only in one region. The plant will not become extinct because a seed vault contains a backup supply. However, Project Baseline stores seeds for a different reason. With the advent of *climate change* and other environmental stresses such as *invasive species* and *habitat loss*, plants will evolve by *natural selection*, as they always have, but probably at an accelerated pace. Evolutionary biologists study these changes and feel this accelerated change can be monitored only if they have a baseline to work from in the future. How can they tell what changes have occurred if they don't know what the plants were like before the changes occurred?

This need will be fulfilled by a grant funded by the *National Science Foundation* that will sponsor biologists spending four years collecting seeds to be stored until needed for study, probably 10 to 50 years from now. [*pollination, wild; bats; heirloom plants; plant-oil fuels; quiescence; vegetable seeds, heirloom*] {baselineseedbank.org}

Seeger, Pete

(1919–) Seeger is a singer, songwriter, peace activist, and *environmentalist*. He and Woody Guthrie, are credited with creating the modern-day folk music genre. He wrote and recorded more than 100 songs but might be best known for "Where Have All the Flowers Gone" in 1961. During the 1970s Seeger became an active environmentalist working on many causes, but most notably he led the fight to clean up the terribly polluted Hudson River in New York. He helped create the environmental *NGO* called Hudson River Sloop Clearwater, which he still supports and where he continues to sing at their annual festivals that fund the river's cleanup and maintenance programs. [*environmental movement; environmental justice; ethics, environmental; philosophers and writers, environmental; jazz, Earth; hip-hop, green*]

seismic air guns, whales and

While searching for *oil* and *natural gas*, companies often use high-pressure seismic air guns to fire explosive impulses every 10 or so seconds while being towed through the water. These guns can be used persistently for days or even months at a time in certain areas, causing irreparable harm and even death to *whales*. Extensive use of these devices is being proposed along the U.S. Atlantic Coast. The U.S. government's own estimates state that over the next eight years, when it is proposed for use, well over 100,000 *marine* mammals, including the *endangered* North Atlantic right whales, will be harmed or killed. [*sonar testing, underwater; Natural Resources Defense Council*]

selective breeding (conventional breeding)

Long before *genetically modified organisms (GMOs)*—actually, thousands of years before—new varieties of crops and livestock were developed by selective breeding. (Now, usually called conventional breeding.) This involves selecting individuals (of both sexes) with traits that are wanted—avoiding individuals with traits not wanted—and cross-pollinating or mating them to produce individuals most likely possessing these positive traits. Repeating this process through successive generations produces new varieties of the species. In plants, these new varieties are called cultivars and in animals it is usually called a new breed. If two distinct varieties of plants or animals are used to develop the cultivar or breed, the resulting new variety is called a hybrid. For example, if one variety of a species of corn grows tall and another variety (of the same species) grows fast, crossing these two varieties results in a new hybrid variety.

A recent example of selective breeding is a cassava plant suited to grow in central Africa, where the soil and climate are marginal for this plant. It has excellent taste and is full of vitamin A. Another example is a sweet potato with high levels of vitamin A for farmers in Uganda and Mozambique, where malnourishment is common.

The *green revolution* of the 1960s occurred by using selective breeding methods. [*DNA sequencing technology; genome; synthetic biology*]

selective harvesting

A logging method in which only certain types of trees are removed from a forest—for example, harvesting a certain species of tree or only mature trees. Selective harvesting causes much less harm to a forest ecosystem than does *clear-cutting*. [*forest management, integrated; forest health, global; afforestation; reforestation; deforestation; deforestation and climate change; forest, sustainable; logging, illegal; Rainforest Alliance; rainforest destruction; timber, DNA sequencing of*]

self-concept and the human microbiome

The *human microbiome* is the collection of microbial organisms that live within each of us and outnumber our own "human" cells 10 to 1. Most of these microbes have co-evolved with humans, meaning they cannot live without us and we cannot live without them. Only recently have the magnitude of their numbers and the importance of their presence been realized. The discovery of the *human microbiome* has caused many to rethink the meaning of "I" and the perception of self-concept. Some believe we should consider ourselves an *ecosystem* instead of an individual organism.

semiochemicals

The best way to understand semiochemicals is to first understand hormones. Hormones are chemicals typically produced in one organ within an organism's body that then travel to another location to control or regulate other organs or cells within the body. For example, the pituitary gland secretes a growth hormone that travels throughout the body and regulates the growth of bones and muscle tissue. Insulin is produced in the pancreas, travels throughout the body, and lowers blood sugar levels.

Now back to semiochemicals. They can be thought of as "external hormones." Semiochemicals do much the same thing as hormones but are produced by one individual and travel—typically through the air—where they then influence the behavior of another individual. Some semiochemicals affect individuals of the same species and some affect individuals of another species. Of most interest to environmentalists are the ones that affect the members of the same species, called *pheromones.*

septic tank

About 24 percent of U.S. homes use underground septic tanks designed to accept, hold, and decompose the contents of *domestic wastewater. Bacteria* break down the waste, leaving an *organic sludge* that settles to the bottom of the tank to be removed on a regular basis. The remaining liquid waste, called *effluent,* flows out of the tank through underground pipes into the surrounding area, called the septic field. As long as the system is in good working order, it has little negative impact on the environment. [*sewage treatment facilities; water pollution; domestic water pollution*]

sequestration, budget cut

The budget sequestration cuts implemented by Congress on March 1, 2013, are gradually taking effect. If Congress allows these cuts to continue, they will cause significant impact on environmental protection. The most significant cuts are at the *Environmental Protection Agency,* where pollution-monitoring programs will be cut back and far fewer inspections will be done to ensure compliance of environmental regulations. Also, staff cuts are already taking place in *national parks* and historic sites, resulting in closures and fewer services. Efforts that fight *overfishing* will also be cut back.

Adding insult to injury, although environmental protection will get significant financial cuts, *fossil fuel subsidies* remain fully funded, doling out millions of dollars to the richest companies in the world—the ones that do the most polluting.

sere community

The *community* of organisms in an area gradually changes over time, beginning with a *pioneer community* and ending with a *climax community*. This process is called *succession*. Each community that forms during this transition is called a sere community and the entire sequence is called a sere.

sewage treatment facilities

These are facilities that treat wastewater that has been collected as sewage. This treatment renders the wastewater less harmful before it is released into the environment.

Sewage treatment is usually divided into three stages: primary, secondary, and tertiary. In the United States, facilities are required to use primary and secondary sewage treatment processes. However, the primary and secondary processes don't remove large amounts of some substances, such as *phosphates*, nitrates, *heavy metals*, and *hard pesticides*, so many municipalities now use a tertiary treatment process that does so, at least in part.

The primary process filters and settles out large particles, including everything from sticks to sand, but leaves the *organic* material in the water. *Bacteria* and other microbes begin to multiply and decompose the organic matter.

Secondary treatment holds the treated water until the microbes can complete their job of decomposing the organic matter. Secondary treatment encourages microbe growth by aeration because these microbes require oxygen. The microbes use the organic matter as food and incorporate it into their own bodies. The microbes are larger than the organic matter, so they settle out of the water and are removed. This mass of microbes (both alive and dead) is called *sludge*.

If the water is to be released following the secondary treatment, it is first disinfected to reduce the number of microbes remaining in the water. This is usually accomplished with *chlorination*. Some municipalities and industries use a tertiary treatment to reduce remaining pollutants and contaminants. Some industries are required to do so to remove contaminants specific to their industry. [*sewer and sewage; combined sewer systems*]

sewer and sewage

A sewer is a conduit system that takes *wastewater* from buildings or storm water from roads either to a *sewage treatment facility* or directly to a natural body of water. *Sewage* refers to the wastewater flowing through a sewer.

There are three basic types of sewer systems: A) Sanitary sewers direct waste that originates from either a domestic source such as residences or office buildings (called *domestic waste*) or from an industrial source such as manufacturing facilities (called *industrial waste*) and delivers the wastewater to a sewage treatment facility. B) A storm drainage system takes water that drains down a street sewer (called *storm sewage*) and carries it directly into a stream, river, or larger body of water. C) Some older, but still common, sewer systems are called *combined sewers,* because they are capable of carrying both forms of waste, combining the two.

Many environmental problems are associated with these sewers. Sanitary sewer treatment facilities make water much safer, but often it still contains many *pollutants* when released. Storm sewage contains *runoff* from the streets and surrounding areas (such as *pesticides* and other *toxic* substances, *fertilizers,* and many other synthetic products) and then dumps it directly into bodies of water, causing *water pollution, cultural eutrophication,* and other damage. Combined sewer systems cause the most harm, because raw, untreated sewage from homes and industry can be released untreated directly into bodies of water. [*beach water quality and closings; detergents, phosphates; drinking water; water treatment*]

sex pheromones (attractants)

Some organisms produce *pheromones*, which are substances produced by individuals of a species to communicate information to other members of the same species. Some of these pheromones act as sex attractants, used by the females to attract males. Most studies on sex attractants have been performed on *insects* and have unique applications.

An interesting alternative to using *pesticides* is to use synthetically produced sex attractants (pheromones) that mimic the real thing. These artificial sex attractants are sprayed in the infested area. The abundance of the attractant confuses the males searching for females—if not driving them "crazy." As a result, the males cannot find the females, so no young are produced and the pest population dwindles in size. Sex attractants are now available for about 30 insect pests. Some of these attractants can be detected by males almost two miles away.

A common residential scene in some areas of the country includes yellow gypsy moth traps, which contain a sex attractant under the brand name Gyplure. Male moths are attracted to the lure and become trapped. Female gypsy moths cannot fly, so they are doomed to remain outside the trap, while the males are stuck inside.

Many applications have proved successful to control pests without the negative impact caused by traditional pesticides. Sex attractants are most effective when used as part of a comprehensive *integrated pest management plan.*

On the business side, pheromones cannot be patented, because they are natural substances. Companies, therefore, concentrate on the delivery mechanism, which can be patented and is vitally important to the success or failure of the pheromone as a pesticide. Contrary to what some perfume companies might imply, no human sex pheromone has yet to be discovered. [*natural pesticides; biological control; biological control methods; biological pesticides*]

sex ratio

A *population's* birth rate and death rate are dependent, in part, on two factors: *age distribution* and sex ratio. The sex ratio is the number of males relative to the number of females. With humans, about 106 males are born for every 100 females. The ratio differs wildly among different types of organisms. Many *insect* populations, for example, are almost entirely female, as with aphids. [*population ecology; population dynamics; population stability*]

sexual reproduction

When new individuals are created by uniting two cells, called gametes, it is called sexual reproduction. In most plants and animals, the two cells come from two individual parents, but in some cases, both are supplied by the same individual. The gametes may unite by internal fertilization, as with humans, or may come in contact with each other externally. External fertilization can occur via the water (as with many fish) or the wind (as with many grasses). Most flowering plants must rely on insects to transport these cells for them [*pollination; asexual pollination; hermaphrodites; parthenogenesis; traumatic insemination*]

shale beds

One third of U.S. *natural gas* comes from vast geological formations of shale beds. [*fracking; Bakken shale bed; Marcellus shale bed*]

shale gale, the

An informal term referring to the U.S. boom in *unconventional natural gas extraction* from *shale bed* formations, which has resulted in low prices and boomtowns appearing where *fracking* operations are located. [*natural gas extraction*]

shallow ecology

Most often used in a comparison with *deep ecology*. The term has a negative connotation (such as a shallow person) when used in counterpoint to deep ecology. But the only people who appear to define shallow ecology are those who adhere to deep ecology. To use the deep ecology definition of shallow ecology, it views humans as separate from the rest of nature—as opposed to being one with nature. Shallow ecology works within the current economic worldview; we should produce less, recycle more, manage our waste more efficiently, conserve, and so on. Deep ecology, on the other hand, requires a new and very different worldview. [*ethics, environmental; biocentric; anthropocentric*]

shared use in national parks (U.S.)

National parks have long been a battleground between those wanting to protect natural resources and those wanting to remove them. The history of these parks is complex and often antagonist. The *Grand Canyon* remains open to *uranium* mining, and *coal mining* occurs in Bryce Canyon National Park.

The term "shared use" usually pertains to national parks, as opposed to *dual use* (and even *multiple use*), which usually pertains to *national forests*.

shareholders, corporate

If you own a publically held stock, you are the public and you have a say in what the company does. They don't have to listen to you, but if enough people want the same thing, they often do listen. It usually takes about 30 percent of all the shareholders voting yes on a particular issue to get the company to pay attention and possibly take the action.

Shareholders are also consumers and are often concerned about how a company does business. Do their products or the processes that manufacture the products harm the environment? Are they harmful to the health of those who make these products or to those who consume the products—someone like you?

There is a significant movement of people presenting shareholder resolutions that pertain to environmental or societal issues. Examples of shareholder resolutions include wanting *BPA* removed from soda cans or being against any involvement with *fracking* or *coal* burning and its accompanying pollution. One of the most popular resolutions pertains to making companies environmental reporting more transparent. [*green consumers; CERES; money market funds, environmental; college divestment campaigns; payroll deduction, environmental*]

shark finning

Shark fins are delicacies used in shark fin soup in *China* and Hong Kong —so much so they are one of the most expensive sea food products in the world, bringing a as much as $350 per pound. Shark fins are also used in everything from cookies to cat food and—in Japan—sushi. The global trade of this one animal part is estimated at half a billion dollars per year, resulting in the death of 73 million sharks per year.

The practice of finning is as inhumane as one can image. The fish are caught and their fins (they have four) are hacked off; then the live fish are thrown back into the sea, where they sink to the bottom to die. Realistic estimates show that roughly 50 million sharks are killed each year, just for their fins (although estimates go as high as 200 million). Overall, shark populations are declining six to eight percent per year. Much of this catch is illegal, so the total number of sharks taken is far greater than reported.

Because sharks are long-lived and produce few young, they are vulnerable to *overfishing,* putting populations in many parts of the world in jeopardy. Populations in the tropical regions of the Pacific are in the worst shape and are believed to have dropped 90 percent in the past 50 years. Large areas such as the Caribbean and the Mediterranean are now considered shark-free, and in other areas sharks are considered locally extinct.

At least 17 nations are taking action against shark finning, but it is usually *poaching* that causes the most harm. The *ecotourism* industry often joins the fight to protect sharks, because wildlife in many regions is more valuable to an entire country than dead sharks for a few poachers. The United Kingdom, Maldives, Ecuador, and Costa Rica are some of the nations that have banned finning, and Hawaii has become the first U.S. state to do so by making it illegal to possess, sell, or distribute fins. China has surprised many by banning shark soup at official banquets which is a small step in the right direction. *CITES* recently placed many species of shark on Appendix II, meaning the trade will be regulated. [*shark sanctuaries; wildlife trade and trafficking; biodiversity, loss of; endangered and threatened species lists; extinct in the wild; extinction, sixth mass; polar bear trade; seafood eco-ratings; species recently gone extinct; United Nations Convention on Biological Diversity; whaling; Wildlife Conservation Society; habitat loss; ivory trade; whaling and commercial ship collisions; pelagic marine ecosystems*]

shark sanctuaries

The Republic of the Marshall Islands created a 1.2-million-square-mile *marine protected area* specifically for sharks. Fishing for sharks and importing or exporting of shark products is prohibited in this area. Five other countries have created shark-protected areas as part of this plan, which will expand it overall to 2½ million square miles. [*shark finning; no-take zones, marine*]

sheet erosion

Erosion of a uniform layer of surface *soil* from a large area, caused by *runoff* water. [*soil erosion; soil conservation; topsoil*]

shelter belts

Trees planted around the perimeter of cropland to reduce wind and thereby reduce *soil erosion.* [*windbreaks; Great Green Wall of Africa, the; Green Belt Movement; conservation agriculture*]

ship ballast and invasive species

Ship ballast is any weight placed in the cargo hold of a ship to help stabilize it. If the ship is loaded with cargo, there is no need for ballast because the cargo itself adds weight. However, a ship travelling without cargo can be unstable and requires ballast. In the 19th century, ships used rocks and anything heavy for ballast. Once pumps were created, the simplest form of ballast became water.

Prior to a trip with the cargo hold empty, it would be filled with water (the ballast) directly from the body of water it sat in, meaning many forms of marine life commonly catch a ride with the ballast. When the ship reaches another port and gets ready to load cargo, the ballast is emptied, with all of its hitchhiking organisms. If some of these organisms are new to the region and begin reproducing, they become an alien and possibly an *invasive species.* The *zebra mussel* is an example of an invasive organism transported via ship ballast.

Estimates show that almost 60 invasive species have invaded the U.S. *Great Lakes* since 1950, and about half of those arrived probably via ship ballast.

Recent regulations stipulated by the United Nations International Maritime Organization and issued by the *Environmental Protection Agency* require all ships built after 2013 that enter the United States to have approved onboard ballast-treatment systems to reduce the release of organisms. The hope is this will reduce the number of invasive species. *Biofouling* is another method by which invasive organisms take hold. [*alien species; native species; biodiversity, loss of*]

ship-breaking industry

Also called ship recycling, this term refers to the business of dismantling large scrapped vessels and selling the scrap steel. The practice is scorned by environmentalists, because it typically causes *water pollution* as well as contaminating the land near where the process occurs. In addition, human rights advocates are concerned because it is a dangerous process, in terms of both accidents and inhaling of noxious fumes and *asbestos* by workers.

Bangladesh started the practice in the 1960s and continues today to be in the top five ship-breaking nations. Although they discontinued the practice a few years ago for environmental and health concerns, they recently restarted it out of economic concerns. [*Basel Convention; environmental justice; ship-sinking program; Reefs-to-Rigs Program; e-waste recycling and disposal*]

ship pollution

The shipping industry causes pollution, but probably more so than many people assume. For example, it is responsible for twice as much *carbon dioxide emissions* as is the entire aviation industry. It also emits about 15 percent of all the *nitrous oxide* emissions and almost 10 percent of all the *sulfur oxide* emissions. This is caused mostly by the use of bunker fuel, called *black yogurt* for its consistency, as the fuel source for these ships. When burned, it emits these pollutants with large amounts of *particulate matter*. In coastal cities with large ports, most of their *air pollution* can come from ship-caused pollutants.

However, the good news is the air should start to clear up soon. In 2012, the United States joined the European Union and a few other nations in requiring cleaner fuels for any ship entering national waters. The *Environmental Protection Agency* and the Coast Guard enforce these new regulations.

Cruise ship pollution poses a different set of problems at sea. [*seaport-related pollution; ship-breaking industry*]

ship-sinking program

The U.S. Navy has a program called SINKEX (sinking exercise program), where old, decommissioned military vessels are sunk to create artificial reefs. Artificial reefs have been successfully created in the past. A *Rigs-to-Reefs* program converts old, derelict oil rigs into reefs. Although this seems good at first, a serious problem has caused many environmentalists to come out against this program.

The problem is that large amounts of *PCBs* were used on these ships and remain in them when they go down. In time, the PCBs *leach* out of the ships into the local environment, wreaking havoc with the *ecosystem*. The first such ship sunk was in 2006. Even though the Navy cleaned out the aircraft carrier prior to sinking it 23 miles off the coast of Pensacola, Florida, it is reported that about 700 pounds of PCBs remained in the vessel when it went down. [*ship pollution; seaport-related pollution; cruise ship pollution; water pollution; Ocean Dumping Ban Act*]

short-lived climate pollutants (SLCPs)

Methane gas and *soot* (black carbon) both cause *global warming* but do not remain in the *atmosphere* for long periods of time and are therefore called SLCPs. This designation compares these pollutants to the most common *greenhouse gas, carbon dioxide,* which remains in the atmosphere for hundreds of years. Even if we could stop emitting carbon dioxide into the atmosphere today, it will continue causing global warming for decades to come. However, if we could stop emitting SLCPs today, the methane would be gone within 10 years and the soot gone in a few weeks. [*climate change*]

shrimp farms, aquaculture

Americans eat four pounds of shrimp per year—the most of any seafood, surpassing even canned *tuna*. The worldwide popularity of this *crustacean* has resulted in the decimation of their wild populations. Global populations are ranked as being depleted at best to fully exploited at worst, meaning there are not many shrimp left in the wild anywhere. This wild shortage has prompted an upsurge in shrimp *aquaculture*. Most of these farms are created along coastal regions, such as near *mangrove forests*. Many are in *China*, Thailand, and Vietnam, with the United States being the biggest importer.

Few if any regulations exist in most of these countries, and the impact of these farms is becoming a major environmental problem as natural *habitats* are being destroyed by these aquatic farms. An even bigger problem is that the primary feed for these shrimp farms is *forage fish*, which themselves are becoming overexploited and in jeopardy.

A big push is on to get shrimp aquaculture certified in a meaningful way so consumers can know what they are buying, much like the *seafood eco-ratings* system. The *Aquaculture Stewardship Alliance*, formed by the *World Wildlife Fund,* has proposed a meaningful certification program, but the industry has created its own weak certification program, touting it as sufficient. [*aquaculture, producing countries, ten biggest; tuna aquaculture; hydroponic aquaculture; fish populations and food; fishing, commercial; forage fish; genetically modified organisms-basics about; Aquaculture Stewardship Alliance; biofuels, second-generation; vertical farms; Magnuson Fishery Conservation Management Act*]

Sierra Club

The Sierra Club, a *NGO*, was founded by *John Muir* in 1892. Its first goal was to help preserve the beauty of the Sierra Nevada mountain range. Since that time, the Sierra Club has grown to 650,000 members and has played a significant role in the formation of America's *National Park and Wilderness Preservation Systems*, protecting more than 132 million acres of *public land*. They concentrate their *preservation* efforts on clean air and water, safe disposal of *toxic wastes*, and energy and *population* issues. It files lawsuits and lobbies agencies to further its goals. It has helped pass hundreds of pieces of legislation since its founding and played an important role in getting

Superfund legislation passed. The Sierra Club has numerous publications and is considered the world's largest conservation publisher. [*Group of 10*] {sierraclub.org}

silent killer
See *asbestos*.

Silent Spring
See *Carson, Rachel.*

Silkwood, Karen
(1946–1974) Silkwood worked as a technician in a *nuclear fuel* production plant in Oklahoma. She became a vocal activist critical of the lack of safety procedures at the facility. She died in a suspicious one-car accident while on her way to meet a New York Times reporter to provide information about the facility where she worked. She had high levels of radioactivity in her body at the time of her death. The speculation around her life and death was made into a book and a popular movie, Silkwood. [*nuclear reactor safety and problems; nuclear waste disposal; nuclear waste dilemma, the global; REM; Mendes, "Chico"; Fossey, Dian*]

silviculture
The practice of growing and tending *forests.* [*forest health, global; deforestation; forest management, integrated; arborculture; deforestation and climate change; agroforestry*]

single-stream recycling collection
In an effort to make *recycling* as easy as possible, over 60 percent of all towns and cities have switched to single-stream recycling, where all recyclable materials are thrown in one large bin. The idea is that more people will recycle more waste if it is as easy as possible, and it is easier to throw everything into one bin than to have to separate it into multiple bins: one for paper, one for glass, one for aluminum, etc. From that perspective, it typically works. Recovery rates (how much is placed into recycling bins) improve when this method is used.

However, recycling programs only work if there is a market for the recycled material. The problem with single-stream programs is they produce far less reusable material because of significant spoilage and cross-contamination. Separated recycling bins produce only about 2 percent spoilage, but single-stream recycling produces about 25 percent. Companies that buy these materials say it is like trying to unscramble an egg. The effort we save by throwing everything into one bin means the recyclers must spend a great deal of extra money separating them back out again. Paper-recycling companies say the paper is contaminated with glass materials, and the glass recyclers don't get enough clean glass. Some recycling companies refuse to buy recycled waste from single-stream facilities. Although the move to single-stream might have seemed good at the time, if it reduces the market for recycled materials, it will be a failure. [*paper recycling; glass recycling; plastic recycling; soda can recycling; bottle bills; disposal fee; municipal solid waste disposal; recycling, moving toward zero-waste; source separation; waste stream; source reduction*]

sinkholes

When *groundwater* is substantially reduced by *water mining* or drought, the once saturated Earth can collapse, forming a large depression in the land's surface called a sinkhole. This usually occurs in regions where *karst terrains* are common. Some sink-holes can be large enough to destroy entire homes.

In Florida there is even a "sinkhole season" when they are most likely to occur. In March 2013, a sinkhole outside of Tampa opened up, swallowing an entire house and, unfortunately, killing the occupant. Outside of Philadelphia in 2013, a sinkhole devoured a creek and a duck pond, although no duck deaths were reported.

Other notables include a 100-foot-deep by 70-foot-wide sinkhole in Guatemala City in 2011 and a sinkhole in Bayou Corne, Louisiana, that grew to over nine acres and a depth of 5000 feet. This one, however, was caused by a collapsing salt dome that had been mined for many years.

sinks, environmental

We are not referring to the sink in your kitchen or bathroom. Scientifically speaking, a sink is something that holds or stores something. For example, a heat sink stores heat. In the fall, as weather cools, the oceans act as a heat sink, retaining the heat from the summer much longer than the air. Another example is a *carbon sink,* which stores carbon. This has become an important issue because of *greenhouse gases,* of which carbon is one. Forests are considered carbon sinks because they store vast amounts of carbon. [*carbon sequestration storage*]

sixth sense in animals

People and most other animals get by with our five senses: smell, hearing, touch, sight, and taste. However, some animals have a sixth sense, the ability to detect geo-magnetism—the electromagnetic force produced by Earth. Birds and some fish, such as salmon, are known to follow their *migratory routes* thanks to this additional sense. Also, some *insects* use it to find their way. [*smell sensitivity in albatross; stereoscopic smell in moles; animals experiencing grief; bumblebee senses; camouflage, squid; communications between plants; hummingbird and flight; whistle names, bottlenose dolphin*]

skin bacteria

Bacteria on our skin are part of our *human microbiome*. Some bacteria occasionally cause problems, but most of the time they play an important role in protecting our bodies. Our skin cells produce a waxy secretion that surface bacteria feed on. The bacteria then produce a moisturizing film (yes, it's their waste products) that keeps the skin moist and supple, making it a perfect barrier that prevents other organisms from entering and causing infections. [*Human Microbiome Project; infectious disease; Lyme disease; maggot medicine; medical ecology; nanoparticles and human health; pathogens; probiotics; spillover effect; superbugs; antibiotics and the human microbiome; poop pills*]

Skin Deep, EWG's

See *lipstick and toxic substances*.

SLAPP

An acronym for Strategic Lawsuit Against Public Participation. The First Amendment of the U.S. Constitution guarantees an individual's right to participate in any peaceful, legal action to influence the government. This includes actions such as signing petitions, writing letters, reporting violations of the law and participating in public demonstrations, public hearings, boycotts, and the like. A SLAPP is a lawsuit filed by someone with opposing views, whose sole purpose is to stifle these activities—in other words, to make it very uncomfortable and expensive to speak out in opposition (or in favor of) some issue.

Hundreds of such suits have been filed in efforts to stifle environmentalists trying to express their opinions about environmental issues. People who have protested against developers, utilities, and private companies have been hit with SLAPPs. SLAPPs have also been used against environmental organizations such as the *Nature Conservancy* and the *Sierra Club*.

However, 28 states have now passed legislation protecting individuals and groups from SLAPPs. There is currently no federal anti-SLAPP legislation, but it has come up a few times and might still become law. [*farm animal cruelty legislation; law and the environment; environmental justice; SLAPP back*]

SLAPP back

In response to the threat of *SLAPPs*, many individuals and environmental organizations have successfully filed lawsuits "against" those who filed the original SLAPP, called a SLAPP back.

slash-and-burn cultivation

People in many *less developed countries* use an old practice called slash-and-burn cultivation to grow gardens and crops. A small area within a forest is cleared of vegetation and then burned. *Forest* soils are naturally poor in *nutrients* (the nutrients are all up in the trees), so the dead, burned vegetation acts as an *organic fertilizer*, enriching the soil so crops can grow. The cleared land, however, can only be used for a few years before the nutrients are depleted and the area abandoned. As long as small tracts are cleared, they are replaced with new growth, but larger expanses of forests cleared in this manner cause *soil erosion* and severe damage to the *forest ecosystem*. [*deforestation; forest fragmentation; forest health, global; clear-cutting; logging, illegal; rainforest destruction; Statement of Forest Principles; governments of the forest; subsistence farming*]

slow food movement

A movement created to counteract and resist fast foods. It was established in 1986 by Carlo Petrini in Italy to protest a fast food restaurant moving into his neighborhood. Their website states, "Slow Food is a global, grassroots organization. . .linking the pleasure of good food with a commitment to their community and the environment." They emphasize local and regional foods and advance projects on *sustainable agriculture*, *seed vaults,* and *heirloom* varieties to preserve the natural *biodiversity* of regions. They work to stop *genetically modified organisms*, synthetic *pesticides*, and *land grabs*. [*farm-to-table movement; farm-to-school movement; seasonal foods; food miles; organic farming and food; food's carbon footprint, the best and worst; doggie bags; food eco-labels; seafood eco-ratings; fregan; vegan and vegetarian diets and the environment; vegetarianism*]

slow money movement

Refers to investing money in the *slow food movement,* which includes small, local agricultural projects, the *farm-to-table* movement, and *organic farming* in general.

sludge

Sludge is a thick, gooey mass of microbes—a variety of *organic* matter—and other solids that are removed from *sewage treatment facilities* or septic tanks, before the water is released back into the environment. (It is also called floc.) The volume of sludge produced can be staggering. Cities can easily produce thousands of metric tons of sludge each day. Recent technologies can convert sludge into useful organic products, but no one would be interested in using anything called "sludge." Therefore, when sludge is deemed worthy of reuse, it is now called a biosolid. [*biosolids and wastewater; night soil – biosolids; compost; composting; composting, large scale; wastewater-to-energy power plants; sewer and sewage; sludge disposal*]

sludge disposal

Once *sludge* has been collected from *sewage treatment facilities*, it must be disposed of. Beginning in the 1920s and continuing through the late 1980s, the *ocean* was the sludge dumping ground. Hundreds of tons of sludge were dumped within a few miles of many shores, destroying the fish and shellfish industries in these areas.

Almost all sludge is now disposed of on land, in either *landfills* or *incineration* plants. However, more and more cities are turning their sludge into usable materials called *biosolids*. These biosolids are used most often as *compost* but also building materials such as brick, cardboard, and pavement. To convert sludge into some form of biosolids, it must go through additional treatments to remove *toxic waste* substances such as *heavy metals*.

slum

When a city or any urban area—especially in poor nations—cannot provide essential services such as safe *drinking water*, sanitation, and living space in a secure manner, they are often considered slums, a term used throughout most of the world. Estimates have the number of people living in slums increasing by six million per year globally, primarily because of increased *urbanization*. Slum populations are considered the most important issue pertaining to public health in this century.

However, many countries have successful slum reduction programs. From 2000 to 2010, improved sanitation and clean water access took over 225 million people out of what were previously considered slums. This was one of the accomplishments of the *United Nations Millennium Development Goals:* more than double the number of people targeted were lifted out of slums. [*hunger, world; brownfields; schools and brownfields; standard of living; environmental justice; United Nations Development Program; United Nations, global environmental issues and the*]

smallest vertebrate

A recent discovery in the New Guinea rainforest has identified the world's smallest *vertebrate:* a tiny frog with an average size of 7.7 mm, which is less than half the diameter of a dime. (The previous record holder was a tiny carp.) It lives in *leaf litter* and probably eats tiny *invertebrates*. Frogs appear to have found their *niche* by being tiny; numerous species use their diminutive size to live in the microhabitats formed by vegetation at the bottom of *tropical rainforests*. [*species on Earth*]

small house movement

One of the two biggest factors that determine your *ecological footprint* is where you live (with your *automobile* being the other). Large houses use much more energy and resources than smaller ones. People interested in reducing their footprint can join the small house movement by reducing the physical footprint of their homes. Over a dozen companies now specialize in building small to tiny homes. They look just like beautiful large homes, but in miniature, with most ranging from 500 square feet down to a Lilliputian 100 square feet. It is only for those willing to live with less, because there is no room for more.

Surprisingly, most cities and towns have few building codes about building large McMansions but many restrictions regarding small ones. Therefore these small homes and micro-villages popping up are often zoned as RV parks to get around these restrictions. [*Small Is Beautiful; conspicuous consumption; sustainable growth; smart growth; eco-cities*]

Small Is Beautiful

This is the title of a landmark book written by E. F. Schumacher in 1973. Schumacher follows Gandhi's philosophy advocating small-scale, decentralized, self-sustaining economies. Schumacher believes, and this book illustrates, that the most appropriate technology, often small on scale, will have the best results, use the least amount of energy to get the job done, and produce the least amount of pollution. [*growthmania; smart growth; eco-cities; smart energy; economics and sustainable development*]

smart cities

See *smart growth*.

smart energy

The use of new technologies to efficiently manage a home's or business's *energy use* to reduce overall energy demand and lower one's costs. These technologies include products such as smart meters that let you remotely or automatically control when certain electronic devices or equipment turn on or off, in concert with services such as *eco-apps* that indicate the cost of power at any given moment. Smart meters can usually be controlled with a smartphone. Coordinating power use with the fluctuations in the cost of the power is a good example of using smart energy. Smart energy means you don't need to wait to get a monthly bill to see your power use; instead a smart energy system lets you manage your power use in real time.

Some of these smart energy systems were first provided by companies such as Google and Microsoft working in concert with utility companies. However, in both cases, even though they were offered free, there was little consumer interest. It was probably a technology ahead of its time—but only a little ahead, because this technology will most likely take off as soon as people realize how much money can be saved. The big box retailers are getting into the act. Best Buy has opened Home Energy Learning Centers and sells smart energy products, and Lowe's is doing the same.

Some utility companies are trying to keep costs down by providing smart meters to all of their customers. Almost every home in the state of Oklahoma has a smart meter. The amount of electricity that could be saved by running everything more efficiently is believed to be enough to cut U.S. energy needs by 25 percent. [*light bulb – recent history and legislation; light bulb technologies, domestic; power grid; microgrids, power; energy-saving apps and programs; energy consumption, historical; energy (fuels); energy efficient states, top ten U.S.; energy, A very brief history of U.S.; soft energy path*]

smart farms

See *precision farming.*

smart growth

Cities, and all that comes with *urbanization*, used to be condemned by environmentalists, but today they are embraced as a possible solution to many of our problems. Well-planned, density-oriented urban planning is gaining momentum and is now called smart growth. The most important aspect of this growth is cities that are densely populated, where people do not need *automobiles* for *transportation* and can either *walk* or *bicycle* to work, shopping, and recreation or use *mass transit*. This planning requires a change in the way most of today's *zoning laws* work to keep sectors separate, keeping residential and business areas far apart when in fact they should be close together. [*eco-cities; infrastructure, green; small house movement; urban gardens; megacities; megalopolis; brownfields; handshake buildings; urban sprawl; smart energy*]

smart meters

See *smart energy*

smell sensitivity in albatross

Albatross, a *predatory* bird, can smell fish—their *prey*—from the air. Research has shown that these birds alter course toward the location of the fish even when the fish are out of visual range. The birds appear to be able to cover a one-mile-wide area of *ocean* with their sense of smell as they search for their prey. [*sixth sense in animals; stereoscopic smell in moles; animals experiencing grief; bumblebee senses; camouflage, squid; communications between plants; hummingbird and flight; whistle names, bottle-nose dolphin*]

Smithsonian Institution

Most people know of the Smithsonian in Washington D.C. for its wonderful museums and the National Zoo. In addition to educating people of all ages, the institution is also involved in *conservation* and environmental research. These studies are conducted in research facilities around the country and in many parts of the world.

The institution uses more than 6000 volunteers each year to teach at and staff museums and the *zoo*. They also publish educational materials for everyone from scientists to laypeople, as well as the colorful <u>Smithsonian Magazine</u>. The institute was founded in 1846 and has 2.6 million members. [*environmental education; environmental literacy; American Zoo and Aquarium Association; Student Environmental Action Coalition; Biosphere 2; nature centers; touchpools; Blue Marble photographs, original and updated; National Park and Wilderness Preservation System*] {si.edu}

smog

First used in the early 1900s, the term smog originally referred to the combination of smoke (from smokestacks) and fog. Today, however, it refers to any accumulation of pollutants that cause *air pollution*. Some pollutants, called primary pollutants, are emitted directly into the air and others—produced by chemical reactions under the influence of the sun—are called *secondary air pollutants*. [*photochemical smog; brown cloud; atmospheric brown cloud; acid rain; air pollution in Beijing, China; brick kilns; Clean Air Act; dust dome; gray air smog; personal portable air-pollution sensing device; seaport-related air pollution; thermal inversion*]

smoke detectors

Some smoke alarms are called "ionization" detectors because they use minute quantities of americium 241, a *radioactive* element. As long as the seal within the device is not broken, the radiation will not be released. The problem, however, is what happens to all the americium when these smoke detectors are disposed of. *Landfills* containing this radioactive substance, which is dangerous for hundreds of years, could leak into the soil and *groundwater supply*. If you use this type, the best way to dispose of it at the end of its useful life is to return it to the store where you purchased it (in most cases they will accept it), send it back to the manufacturer (they will accept it), or bring it to a *hazardous waste* pickup day in your community. What you don't want to do is just throw it into the *garbage*. [*hazardous waste disposal; toxic waste disposal; indoor air pollution; laser-printer ozone; inkjet and laser printer cartridges*]

snags
Old, dead, standing trees in a *forest*. [*dendrochore; ancient forests; old-growth forests; forest health, global; forests, pop-up; Luna Redwood*]

sneakers, green
Three hundred million pairs of sneakers are discarded each year. Because they are not typically *recycled* or even *upcycled*, they end up in *landfills*. Mostly made of *plastic* materials, they do not *decompose* readily. To make an attempt to be environmentally friendly when buying sneakers, look for products made from recycled, *post-consumer waste*, with nontoxic materials. Some even have additives embedded in them to encourage microbes to help out with the decomposition once the shoes are discarded. Some are advertised as being totally degradable, created from hemp, cotton, cork, and a biodegradable plastic product. One brand even states you can plant the sneakers and the sewn-in wildflower seeds will bloom. [*stonewashed jeans; blue-jean pollution; textile industry; flip-flops and preconsumer waste; green products; flat screen televisions, green; wine-bottle cork debate; wrapping paper alternatives; greenest cars, top ten; lipstick and toxic substances; eco-labels; consumer green; companies, top ten green*]

sniggling
See *eels and elver fishing*.

social boundaries
In recent years, scientists have defined what they call a *safe operating space* of our planet. As long as we stay within this safe space, Earth should be capable of sustaining life for generations well into the future. To remain within this safe space, we must not violate nine planetary boundaries that have been identified as the most critical to our planet's future. Passing these upper-end planetary boundaries will jeopardize life on our planet.

This model was extended in 2013 to also include social boundaries that define what is needed to ensure all people a minimum quality of life on our planet. Eleven social foundations were defined and lower-end boundaries established. If people remain above these minimum levels of social boundaries, they will find themselves in the safe and just operating space of our planet. (By adding this additional dimension of social boundaries to the existing planetary boundaries, the name was changed from a "safe" operating space, to a "safe and just" operating space of our planet.)

By combining both the upper-level limits of the planetary boundaries (environmental factors) with the lower-level limits of the social boundaries (social factors), the safe operating space becomes a useful tool to provide insight of what must be done to make Earth a safe and just place.

The eleven social boundaries are food security, income, water and sanitation, health care, education, energy, gender equality, social equality, voice, jobs, and resilience (to poverty). Although three of the nine planetary boundaries are believed to have already been surpassed, eight of the eleven social boundaries are believed to be below the desired levels. [*hunger, world; water shortages, global; population growth, limits of human; land grabs; standard of living; tipping point; United Nations Population Fund; Millennium Development Goals; United Nations Environment Programme; urbanization and urban growth*]

social media, environmental

An informal category of online services that foster communications to benefit the environment in some way. It can include texting, tweeting (*microblogging*), Facebook, *blogging, crowd sourcing, crowd funding,* and facilitating *online petitions*. Social media has helped mobilize people and provide a powerful outlet for the *environmental movement*. It has raised monies for environmental research and let thousands of people view environmental tragedies as well as wonders first hand.

In *Kenya*, crowd funding has paid for the first *conservation drones* in a wildlife preserve. In *China*, microblogging helped spread the word about some of the world's worst *air pollution* and taught its citizens the meaning of $PM_{2.5}$. Online petitions have helped win health-care services for victims of water poisoning on a U.S. army base and helped stop animal cruelty in Belarus. In 2013, on World Water Day, the organizing group sponsored a *tweet* campaign that raised more than a million dollars used to support more than 6000 clean water projects in 19 countries. From an educational perspective, streaming videos of wildlife have been viewed by thousands on Facebook, such as an osprey-cam that was part of a public outreach project gathering information on *toxic* threats to birds.

Social media has expanded the quality and quantity of environmentalism in much of the world.

sociobiology

The study of the biology of social behavior. [*Wilson, E.O.; animals experiencing grief; ant slaves; beehive sound research; honeybees, pollination, and agriculture; migration; sixth sense in animals; whistle names, bottlenose dolphin; anthropocene geological epoch; anthropologists, environmental; biosphere, shadow; community ecology; systems ecology*]

soda can recycling, aluminum

Aluminum is the most valuable material we routinely recycle. New aluminum is costly, so recycled aluminum is also valuable. Because *recycling* old aluminum into new takes far less energy and resources, the more we recycle, the less environmental impact on the planet. Using recycled aluminum takes one-eighth the energy compared to using new aluminum. Moreover, some problems with these cans might be corrected in the near future, resulting in more efficient recycling.

A soda can is made from two different types of aluminum; the sides are one type and the top and bottom another. This makes the recycling process more difficult and more expensive and, therefore, recycling more expensive. The industry is working to make an entire can out of one type of aluminum, which will result in less expense in the recycling process and make aluminum even more valuable, giving this recycled product a larger market. [*appliance recycling; refrigerator recycling; refrigerator doors; energy, ways to save residential; automobile recycling; mattresses, recycling; cigarette butt litter and recycling; downcycling; upcycling; motor oil recycling; paper recycling; recycling metals; plastic recycling; e-waste recycling; single-stream recycling*]

soft energy path

A prophetic phrase used back in 1976 by *Amory Lovins* of the *Rocky Mountain Institute* regarding energy strategies the United States and the world could take moving forward. A soft energy path means *conservation* and efficiency strategies along with *renewable energy* sources, as opposed to taking a hard energy path, which is business as usual based on *fossil fuels*. [*alternative energy sources; negawatts; alternative fuels; clean technology; smart energy*]

soft pesticides

Pesticides that decompose into harmless chemicals within a few hours or days. They are not *persistent pesticides* that remain in the environment for long periods of time. [*insecticides; biocides; toxic pollution*]

soil

Soil forms a thin layer over the Earth's surface. The word soil refers to the combination of the *topsoil* and the subsoil as described by horizontal layers called the *soil horizon*. Below these two layers is a layer of *parent rock*. As the parent rock breaks down over long periods of time, it forms subsoil above. When subsoil mixes with *organic* matter such as waste from animals and dead, decaying organisms, it forms the topsoil above. Soil also contains air and moisture. Topsoil provides the *nutrients* for plant life. Good quality soil consists of (by volume) about 45 percent minerals, 25 percent water, 25 percent air, and at least 5 percent organic material.

Soil is composed of a mixture of many types of *soil particles*. The soils form in different ways because of differences in the parent rock and surrounding formations, the climate, and the types of organisms that have inhabited the region. About 14,000 soil types are classified in the United States alone. [*geological time scale; soil erosion*]

soil conservation

More than one-third of the *topsoil* originally found on croplands in the United States is gone because of *soil erosion* by water and wind. Soil washes away and wind blows off topsoil from our croplands much faster than soil is created naturally. Soil conservation refers to the techniques and practices that help stabilize soil to reduce this erosion.

Soil conservation methods include the following: *conservation tillage, contour farming, strip-cropping, terracing, alley cropping, gully reclamation, windbreaks*, and *zoning*. These soil conservation methods are only used on about one-half of U.S. croplands.

A great deal of human-caused soil erosion, however, is from *high-impact agriculture*. Much of it comes from other factors such as *clear-cutting* forests, *urban sprawl* and its associated development, and mining practices such as *strip-mining*. [*Dust Bowl; Natural Resource Conservation Service; Conservation Reserve Program*]

soil erosion

The movement of *soil* (usually *topsoil*) from one place to another caused by water or wind is called soil erosion. *Erosion* occurs naturally as *runoff* water flows into streams and rivers but is exacerbated by human activities. The quantity of soil moved by erosion is immense. The Mississippi River transports about 325 million metric tons of soil from the middle of America to the Gulf of Mexico every year.

Most soils are protected from erosion by plants that stabilize the soil, but serious erosion occurs when human activities remove most of the plant cover and expose soil to the elements. Activities such as *deforestation, clear-cutting, high-impact agriculture*, construction, and off-road vehicles cause soil erosion.

Even though soil formation occurs naturally, human activities remove it much faster than it is created. It takes 200 to 1000 years in most regions to create about one inch of topsoil, soil that can be eroded away in a few days.

Areas that get little rain on a regular basis or are prone to drought may experience wind-caused soil erosion instead of water-induced erosion. During the *Dust Bowl* in the Great Plains of the United States during the 1800s, millions of acres of topsoil blew away in a matter of weeks.

soil horizons

The horizontal layers of *soil*. The upper horizons usually contain *organic* matter consisting of partially decomposed plants and animals, while the lower horizons consist almost entirely of *inorganic* matter. There are three main horizons, each one divided into more specific layers.

The top layers form the A horizon, called the *topsoil*; the middle layers form the B horizon, called the subsoil; and the lower layer forms the C horizon, composed of solid *parent rock*. The horizons are slowly formed from the natural *weathering* of the parent rock below and the accumulation of dead organisms that lived above. [*soil conservation; soil erosion; soil organisms*]

soil organisms

Soil provides *nutrients* and support for terrestrial plants to grow, but it also is a *habitat* teeming with life. A single ounce of rich soil can contain 75 billion *bacteria*, 15 million *fungi*, 900,000 protozoans (single-celled animals), and 50,000 *algae*.

The soil community is hard to see but a fascinating and very busy *habitat*. Bacteria, fungi, protozoans, *nematodes, earthworms*, and *insects* establish their own complex *food webs*. Bacteria and fungi are major players in the decomposition of dead organisms and waste material, breaking down *organic* molecules into simpler molecules so they can be reused by plants.

Some fungi attack and kill nematodes. Many protozoans are *parasitic* on other soil organisms. Nematodes, also called roundworms, play many roles, such as decomposition; some are an important food source for other soil inhabitants, while others are *predatory*. Earthworms condition the soil for plant growth. Although many soil insects are pests, many more are beneficial, acting as decomposers and soil conditioners.

Little is known about soil ecosystems. It is thought that only about one percent of all soil species have been identified. An effort to remedy this situation is called the

Global Soil Biodiversity Initiative. They hope not only to identify more species but also to determine their role in soil ecosystems and the global environment overall, including their impact on issues such as *climate change*.

soil particles

Soil particles are divided into three categories, based on size. From largest to smallest, they are sand (0.0625 to 2.0 millimeters), silt (0.002 to 0.05 mm), and clay (less than 0.002). Particles larger than 0.05 mm are called gravel. Soils that contain a lot of sand feel gritty when wet, soils with a lot of silt feel smooth like flour, and those with a lot of clay feel sticky when wet. [*soil texture*]

soil profile

A vertical section of soil that reveals the *soil horizons*.

soil texture

A soil's texture is determined by the varying mixtures of the three *soil particles*: sand, silt, and clay. For example, a *soil* with 40 percent sand particles, 40 percent silt particles, and 20 percent clay particles is called *loam*. For a soil to be called sand, it must contain at least 85 percent sand particles, with the remaining portion a mixture of silt and clay. For soil to be called clay, it must contain at least 40 percent clay particles, with the remainder a mixture of sand and silt. [*soil horizons; soil types; humus*]

soil types

There are thousands of soil types, but the most important soils can be grouped into two major categories: *grassland* soils called chernozem and *forest* soils called podzol. Grassland soil is formed in areas with limited amounts of rainfall. This keeps the *nutrients* in the top *soil horizon* (A horizon) from *leaching* down to the B horizon. The nutrients remain near the surface, so root systems are shallow. Forest soils, however, occur in areas with more rainfall, resulting in nutrients leaching downward, producing a deeper B horizon that supports large root systems.

solar cells

See *photovoltaic cell*.

solar charging of electronic devices

Wireless charging of electronic devices is just beginning to become popular. However, solar technology companies are already taking the next step. They offer devices that can charge your smartphone or other devices without the need for any electrical outlet at all. Instead, they use mini-solar panels that generate electricity for charging your device, usually through a USB port. [*solar power; solar cookstoves; photovoltaic cell; photovoltaic cell film; biotechnology; nanotechnology*]

solar cookstoves

In *developing nations*, small, open, primitive *wood-burning cookstoves* are used for cooking. Because they do not burn wood efficiently, they release smoky, *soot*-filled fumes that cause respiratory problems in millions of people. These primitive cook-stoves are considered one of the world's biggest health concerns because so many people use them.

They also contribute to *deforestation* because forests are cut down for the wood, and they produce *greenhouse gases*. If people had affordable alternatives, their health and the environment's health would improve.

Solar cookstoves and ovens collect sunlight and reflect the heat into a stove to heat water and cook foods. People in developing nations are getting these stoves through a public/private initiative. Their popularity has spread from the poor to others, because some people in the U.S. Southwest and southern California have started to use them as well. [*air pollution; air pollution in Beijing, China; brick kilns; smog; fireplaces, wood-burning*]

solar glitter

A still-on-the-drawing-board idea for *photovoltaic cell* technology where the light is collected, not in a traditional solar panel, but on almost anything that has this solar glitter applied to the surface. Theoretically, even materials such as clothes and bags could be covered with this glitter and used to generate electricity. [*solar power*]

solar panels

See *photovoltaic cell*.

solar ponds

Unique *solar power* plants that use the sun's energy to heat water. They consist of black plastic bags filled with saltwater, which are laid out to cover an area of at least one acre. The bags are held in a container-like structure or a cavity carved into the Earth. The water at the bottom contains the highest level of saline and the top layer, the least. This is an important aspect for efficient heat collection. As heat is collected within the water, it is used to produce steam that then turns turbines to produce electricity.

The largest solar pond facility is in Israel near the Dead Sea. About 60 such facilities can be found around the globe, with the majority providing power to heat industrial buildings. Many nations have plans to build these sites. *India* recently completed its first site. Some analysts believe solar ponds can supply significant amounts of pollution-free power in the future.

solar power

At present, only one percent of the world's energy supply comes from solar power. Some scientists, however, believe the sun to be the ultimate *alternative energy* source and the answer to most of our energy problems. The sun delivers far more energy than all the peoples of the world could ever need or use, but the problem is harnessing it. Less than five percent of all the radiant energy that strikes Earth is used by living organisms for *photosynthesis*.

Solar power can be divided into four methods: 1) *Passive solar heating systems* absorb the sun's energy as heat for immediate use. 2) *Active solar heating systems* use a solar collector device and pipes to transfer the heat to a target area. Passive and active systems are typically used to heat homes and produce hot water. 3) *Solar thermal power plants* use solar collectors to heat water or other liquid to then generate electricity. 4) The *photovoltaic cell*, currently the most popular method, converts sunlight directly into electricity.

None of these methods produce significant *air pollution* or *water pollution*; add *carbon dioxide*, a *greenhouse gas,* into the *atmosphere;* or destroy the land. [*pumped-storage hydropower; power islands*]

solar thermal power plants

Also called concentrated solar technology, solar thermal power plants use enormous mirrors, called heliostats, to focus sunlight onto a central heat-collection facility composed of pipes containing water or oil. The heat captured by this fluid is used to create steam that then spins turbines to generate electricity.

The three such *solar power* plants in the Mojave Desert (California) use thousands of acres of these mirrors and generate 377 megawatts of power—enough to meet the needs of almost 150,000 homes. The largest such facility is found in the *eco-city* called, Masdar City in the United Arab Emirates. It has 192 parabolic mirrors, each over a football field in length and 20 feet wide. Others even larger are under construction—one in Phoenix, Arizona, and another in northern Nevada. Many projects in various stages of development across the globe are planned to be four times the size of the one in Masdar City.

Most of the existing facilities use parabolic troughs to capture and concentrate the solar rays. However, newer technology is replacing these with clustered mirrors around the central tower because they can generate much higher temperatures.

These plants cause no *air pollution* or *water pollution* and are relatively inexpensive and quick to build when compared to other power plants. The cost for this electricity is becoming competitive and cheaper than *nuclear power*. The biggest problem is obvious—there must be sunlight—so only a few regions are candidates. It is also not considered a form of *baseload power*—meaning not available 24/7—and therefore must be supplemented with additional power by some other method.

solar tube lighting

A high-efficiency replacement for skylights in homes and other buildings. They are tubes that allow sunlight to enter from the roof that can be directed down, even around corners, to shed light in a room. They are much less expensive to install than skylights or roof windows and take much less time to do so. They can be from 10 inches to 20 inches in diameter and are so good at capturing light, they can bring in light from a bright, full moon.

The tubes have clear dome tops about the size of a basketball that capture sunlight (or moonlight). The inner surface is a highly reflective silver-coated metal. On the inside, at the other end, is a frosted glass diffuser that spreads the light out across the room. [*indoor ecology; Living Buildings; LEED certification; solar glitter*]

solid waste

See *municipal solid waste.*

sonar testing, underwater

Both the *oil* industry and the U.S. Navy use underwater sonar testing and plan to use much more in the future. Oil companies use a form called low-frequency active sonar as part of their exploration for new oil fields, and the Navy uses high-intensity mid-frequency sonar to detect enemy submarines.

This type of testing was first brought to light in 2002 when a dozen beaked whales beached themselves on the Canary Islands and later died while NATO forces were testing sonar devices that would be used to detect submarines. Since then, a great deal of research has shown that these types of sonar seriously harm and often kill *dolphins* and *whales.*

In Peru in 2010, 900 dolphins died and autopsies revealed damage caused by the low-frequency sonar. The Navy's use of mid-frequency sonar is known to injure and potentially kill whales and other marine *mammals.* The Navy plans to use this technology across 70 percent of the world's *ocean,* meaning the potential for harm is huge. By the Navy's own account, this activity will result in 1000 deaths, 5000 serious injuries, and millions of cases of temporary hearing loss (which will probably mean death) to these marine mammals. Surprisingly, the *National Marine Fisheries Services* has approved the Navy's proposal to move ahead. There are legal battles ensuing, trying to stop or at least minimize this use of sonar. [*seismic air guns, whales and; dolphins and porpoises, true; whaling; whaling and commercial ship collisions; International Whaling Commission; Ocean Conservancy; oceanic zone ecosystems; Marine Mammal Protection Act; sea gliders*]

soot

Also called black carbon, soot is a type of *particulate matter* composed of fine carbon particles. These particles are released by incomplete burning of most *fossil fuels* and are especially prevalent in burning of *coal, wood,* and *diesel* fuels.

Soot has long been known to cause health problems. It is of greatest health concern in *developing countries*, where primitive *wood-burning cookstoves* used for heating and cooking in homes produce soot. Fortunately, soot is not as bad a health problem in developed nations as it used to be. Cleaner-burning diesel engines and more efficient coal-burning *electric power plants* have improved the situation in nations with strict regulations.

In the United States, the *Environmental Protection Agency* recently lowered the acceptable standards of soot from 15 micrograms of fine particulates (soot) per cubic meter of air to 12. They estimate that every dollar spent on pollution control, such as reducing the levels of soot in the air, results in $30 of health-related savings.

Soot is also considered an environmental issue, contributing to *climate change*. In 2013, research demonstrated that soot probably contributes far more to *climate change* than previously thought. This occurs because soot, being black, absorbs heat, so at any given time, soot in the air contributes to increasing temperature. Add to that, soot found in frozen regions collects on the white snow that typically reflects most of the light, causing the darker sooty soil to reflect less light and therefore absorb more heat. It is now believed soot is second only to *carbon dioxide* when it comes to causing climate change and is even worse than *methane* gas. [*air pollution; brick kilns; cost/benefit analysis*]

soundscape

A soundscape encompasses all of the sounds of a *habitat*. It includes sound from three sources: the geophony, which is abiotic sounds such as running water and wind; the biophony, which is the biotic sounds made by all life other than humans; and the anthrophony, which is the sounds people make—often called just plain noise. [*acoustic niche; apps, eco-; beehive sound research; jazz, Earth; noise pollution and control*] {eoearth.org/view/article/179209/}

source reduction

A method of reducing the amount *of municipal solid waste* that usually ends up being dumped into *landfills* or *incineration* plants. Many people consider landfills and incinerators "bandaid fixes" for a much bigger problem: the production of too many disposable materials.

Source reduction means minimizing the use of a material that will have to be disposed of at a later time. Source reduction of excessive *product packaging* materials is one important area. *Plastic bottles* and aluminum *soda cans* have been reduced by as much as 35 percent of their weight since they were first introduced, a form of source reduction. CDs that used to have excessive packaging (so they could be seen on the older record racks) were reduced in size. Concentrated forms of detergent are often offered by companies as alternatives to larger bottles containing diluted solutions. [*waste minimization*]

source separation

Most municipalities have some form of waste *recycling*. Source separation means people must separate out the recyclable waste from the nonrecyclable waste. This improves recycling efforts by providing a *waste stream* containing only recyclable waste materials. [*single-stream recycling collection; municipal solid waste; waste minimization; disposal fee; bottle bill; recycling*]

soy ink

An alternative to the traditional petroleum-based (*oil*-based) printing inks, which contain a large amount of *VOCs*. Soy ink is a natural product extracted from soybean oil and has become popular with companies involved in printed materials and concerned about environmental issues. You will often see the soy ink logo for these printed materials. [*bioplastics; biomaterials; biomimetics*]

space agriculture

Recent research shows that, contrary to previous thought, plants can grow roots naturally, without the need for gravity, the force that was thought to guide the growth process. This is important because it means plants can be cultivated in weightless environments. Space agriculturists are thinking about greenhouses on Mars or possibly farms in space stations to feed the space residents. [*vertical farms; hydroponics; hydroponic aquaculture; integrated multi-trophic aquaculture; biofuels, second-generation; organic farming and food; hunger, world*]

space conditioning
See *thermal insulation, building.*

space debris
Hundreds of objects are shot into space each year and placed into orbit around Earth. These objects and the remnants of these objects have become so numerous as to pose a danger of a collision with a new rocket or satellite entering space. NASA estimates that circling Earth are more than 100 million tiny fragments of space junk at least one millimeter across, about half a million pieces of junk bigger than a marble, and more than 22,000 pieces bigger than a softball. Plans are being made to attempt a *space debris cleanup* program. [*near earth objects; Blue Marble photographs, original and update; Earthrise photograph; outer space; Spaceship Earth; extinction and terrestrial impact; panspermia*]

space-debris cleanup
The problem of *space debris* has been getting worse over the years. Sooner or later, someone (or some country) was going to think of a solution, and it turns out to be Switzerland. They plan to send up a spacecraft called CleanSpaceOne, designed to hone in on defunct manmade satellites in need of destruction. It will alter the satellite's orbit, forcing it to head earthward, where it will burn up on reentry. They plan to begin by taking down one of their own satellites and then, if that works, offer their services to other countries.

Spaceship Earth
In 1969, man landed on the moon and people saw the first pictures of our planet from *outer space.* These pictures are so iconic they have names such as *The Blue Marble* and *Earthrise.* It was then that Spaceship Earth became a metaphor for our one-and-only home. Thinking of Earth as a spaceship is an excellent way to frame our existence—fragile. The phrase was first coined by author *Kenneth Boulding,* a Quaker economist and educator in 1966. The analogy was also used by *Barbara Ward* in her book Spaceship Earth (1966 – quoted below) and by *Buckminster Fuller* in his book Operating Manual for Spaceship Earth (1969). Many believe the only way humans will survive for the long haul is by understanding and appreciating this analogy.

From Spaceship Earth: "The most rational way of considering the whole human race today is to see it as the ship's crew of a single spaceship on which all of us, with a remarkable combination of security and vulnerability, are making our pilgrimage through infinity. Our planet is not much more than the capsule within which we have to live as human beings if we are to survive the vast space voyage on which we have been engaged for hundreds of millennia—but without yet noticing our condition. This space voyage is totally precarious. We depend on a little envelope of soil and a rather larger envelope of atmosphere for life itself. And both can be contaminated and destroyed. Think what could happen if somebody were to get mad or drunk in a submarine and run for the controls. If some member of the human race gets dead drunk on board our spaceship, we are all in trouble. This is how we have to think of ourselves. We are a ship's company on a small ship. Rational behavior is the condition for survival."

speciation

When *natural selection* (along with other factors) results in the development of a new species, the process is called speciation. This can occur when individuals of one species *migrate* to separate areas with different conditions. For example, some individuals belonging to an early species of fox are believed to have migrated north, while others migrated south. Those that went north developed heavier fur (for warmth), shorter ears (to reduce heat loss), and a white coat to blend in with the snow, creating a new species called the arctic fox. Those that went south developed a thin coat, long ears, and a darker coat, establishing a new species called the gray fox.

The process of speciation probably can occur within a few hundred or thousand years with organisms such as *insects* that reproduce rapidly, but takes tens of thousands to millions of years in higher forms of life such as *mammals*. [*Darwin, Charles*]

species

A group of similar organisms capable of reproducing with one another. Species is the lowest (most specific) category of biological classification. [*speciation*]

species on Earth

See *biodiversity*.

species recently gone extinct, ten

The following species are just a sampling of organisms believed to have become *extinct* within the past few years (with the probable cause of their demise): the golden toad from Costa Rica (*chytrid fungus, air pollution,* and probably *global warming*); the Yangtze River *dolphin* from *China* (*overfishing,* boat traffic, *habitat destruction, pollution,* and *poaching*); the Hawaiian crow from Hawaii (possibly avian malaria); the

Pyrenean ibex from Spain and France (unknown); the Liverpool pigeon from Tahiti (unknown); the West African black *rhino* from Cameroon (poaching); the black-faced honey creeper bird from Hawaii (habitat destruction and disease); the Alaotra grebe from Madagascar (habitat destruction); and the Holdridge's toad from Costa Rica (chytrid fungus and probably global warming). [*endangered and threatened species lists; biodiversity, loss of; CITES; de-extinction; seed banks and vaults; species recently gone extinct; Lazarus Taxon; International Union for the Conservation of Nature; extinction and extraterrestrial impact; extinction, mistaken*]

speleology
The study of caves. [*karst terrain; sinkholes*]

spillover effect
The process of a *pathogen*—found in nonhuman *host* organisms—making a jump to human hosts. The scientific term for this is *zoonosis*. These are rare but dramatic events in human public health. It is believed that the HIV *virus* made this jump from primates in East Africa to humans. [*antibiotics and the human biome; bird flu virus; bubonic plague; geomedicine; maggot medicine; nanoparticles and human health; superbugs; virome; Quammen, David*]

spoil banks
The resulting environmental damage that occurs when *strip-mining* is not restored. [*surface mining; mineral exploitation (mining); acid mine drainage; gold mining; mountain top removal mining; overburden; room and pillar method*]

stand
A group of trees with uniform characteristics, such as all of the same *species*, age, or other condition. [*tree farms; afforestation; reforestation; clear-cutting; deforestation and climate change; logging, illegal; timber, DNA sequencing; ancient forests; monoculture*]

standard of living
The standard of living is similar to what many people call the "quality of life." It is difficult to quantify because needs and values differ among different peoples. The *United Nations* has designated countries as *more developed countries* (MDCs) and *less developed countries* (LDCs). These designations can also be used to distinguish countries with high standards of living versus low standards. *Population growth* appears to be inversely related to a nation's standard of living. Those countries with the highest

standards of living typically have the lowest growth rates, and vice versa. This is considered an important aspect of the *demographic transition* theory.

Certain indicators are often used to give a reading on a peoples' standard of living. The infant mortality rate is a good indicator of standard of living. In 2011, the United States (an MDC) had 6.1 deaths per 1000 births, while *Kenya* (an LDC) had 59 deaths per 1000 births. (Even though the United States is far better in this statistic when compared to Kenya, it has a poor rating when compared to other MDCs. In 2013, it came in 30th among other developed nations.) Life expectancy (at birth) is another good indicator of standard of living. In 2011, the United States had a life expectancy of 78 years while in Kenya it was only 57 years. [*safe operating space; environmental justice; land grabs; hunger, world; water shortages, global; push-pull hypothesis*]

standing water habitats

Freshwater ecosystems can be divided into *running water habitats* and standing bodies of water such as ponds, lakes, and reservoirs. If a standing body of water is deep enough, it has a top layer called the *euphotic zone* in which sunlight penetrates and a lower level that receives no light, called the *aphotic zone.*

In areas where light reaches the bottom of the water, rooted plants grow. This part of the euphotic zone is called the *littoral region*. It may contain flowering plants such as water lilies and cattails. These are called *emergent plants* because they rise above the water. Aquatic plants that remain entirely submerged are called *submergent.*

Algae also play an important role in these habitats. They form thick layers on the bottom or at the surface, attached to vegetation. The emergent and submergent plants, plus the algae masses, provide food and shelter for many species of fish, *insects*, crayfish, clams, and other aquatic organisms.

If the body of water is deep, the lower portion of the euphotic zone has no rooted plants and is called the *limnetic region*. These regions are in many ways similar to *marine ecosystems*. *Phytoplankton* floats in the sunlit area, performing *photosynthesis* and becoming the first link in many *food chains*. *Zooplankton* feed on the phytoplankton and are then eaten by small fish, which in turn are eaten by *predatory* fish.

The lower levels (aphotic zone) receives no light, so organisms must get their *nutrients* from dead and decaying organisms and waste that floats down from the surface regions, or they may travel to the surface to find food. The *organic matter* that continually falls to the bottom produces a nutrient-rich environment. Organisms that dwell in the organic *ooze* that accumulates at the bottom are mostly *anaerobic bacteria*, because little oxygen is available at this level.

Many terrestrial and flying organisms are often associated with these ecosystems, including dragonflies, turtles, frogs, mallards, and muskrats. [*nutrient enrichment; dams, hydroelectric; habitat loss; dams, removal*]

Statement of Forest Principles

One of five documents written for the *United Nations Conference on Environment and Development* (UNCED) in June 1992. This document contains 15 principal elements. The first states that in accordance with the Charter of the United Nations and the principles of international law, "States have the right to utilize, manage and develop their *forests* in accordance to their development needs and level of socioeconomic development." The statement also supports the need for financial resources to developing countries with significant forests who are establishing programs for their conservation. [*forest management, integrated; forest health, global; deforestation; forest, sustainable*]

steady-state economics

This was the first of many environmentally aware economic models that have proliferated in numbers but not in adoption. It was first alluded to by the philosopher John Stuart Mill, who stated that one day economies could become mature and reach a steady state. *Herman E. Daly* advanced this to a full-blown thesis in many of his books written in the 1970s.

A simplified explanation of such a model is that economies cannot grow forever because we live on a planet with finite resources and space and, sooner or later, a limited number of people, dictated by the finite resources and space. [*economy vs. environment; Beyond GDP; CERES; climate capitalism; companies, top ten green global; corporate social responsibility; corporate sustainability rankings; Sustainability Imperative; economics and sustainable development; Genuine Progress Indicator; green net national production; integrated bottom line; natural capital*]

STEM

A commonly used acronym for educational programs on science, technology, engineering, and math. For example, the U.S. National Science Foundation has made a commitment to improve STEM education. [*environmental education; environmental literacy; science; pseudoscience; opinions and science; law and the environment; volunteer research, scientific; geocaching*]

steppe

See *grassland*.

Stereocaulon volcani

Stereocaulon volcani is a *lichen*, but it is an iconic lichen. If you remember back to your basic biology class, think about *succession:* the process where a volcano erupts, spewing lava that then hardens and becomes totally barren rock with absolutely no life on it. Succession is the process of life colonizing a new area. The first form of life that appears, beginning the process of turning barren rock into a rich environment full of life, is the lichen, because it is one of the only life forms capable of surviving on pretty much nothing—but barren rock. Stereocaulon volcani is the species of lichen most likely to start this entire process of succession. Its spores float about in most parts of the planet, waiting until the right raw environment presents itself and it can begin the long process. [*succession, primary terrestrial; biodiversity*]

stereoscopic smell in moles

Recent research shows that the Eastern American mole can smell in stereo, providing a precise way to locate their *prey* in spite of being totally blind and hearing little. This is the first time stereoscopic smell has been identified in any wild *mammal* species. [*smell sensitivity in albatross; sixth sense in animals; animals experiencing grief; bumblebee senses; camouflage, squid; communications between plants; hummingbird and flight; whistle names, bottlenose dolphin*]

Stockholm Convention on Persistent Organic Pollutants

This convention was put into force in 2004 and has almost 150 signatories. Its purpose is to control *persistent organic pollutants (POP)*. Special emphasis is placed on the 12 worst substances, informally called the *POPs Dirty Dozen*. The convention works to restrict or ban their production, manage existing stockpiles, control their import and export, and promote replacements. This convention can be viewed as a continuation of the work started by *Rachel Carson* in the 1960s.

Stockholm Declaration

Properly called the Declaration on the Human Environment, it was the principal document created at the *United Nations Conference on the Human Environment,* held in 1972.

Stockholm Earth Summit of 1972

See *United Nations Conference on the Human Environment.*

stonewashed jeans

See *pumice.*

stool transplant

As disgusting as it sounds, this new medical procedure is saving many lives. After large doses of antibiotics, some people have their *human microbiome* disrupted, resulting in a population explosion of the bacteria Clostridium difficile. This causes a distension of the person's gut, which can be life threatening. Fourteen thousand people die from this in the United States each year.

Recent research has resulted in doctor's transplanting the human microbiome from a healthy person into the bowel of the sick person, resulting in the person's cure. Transplanting and otherwise manipulating a person's microbiome will probably be a new medical development in a few years. Research in this field is being advanced by the *Human Microbiome Project.* [*poop pills*]

stranded natural gas

See *natural gas as an energy source.*

straw-bale gardening

You can grow many types of plants, including vegetables such as *tomatoes*, with nothing more than a single bale of hay, available at any garden center for a few dollars. By adding some water and a little fertilizer to get the process started before the natural decay and decomposition process takes hold, you can get attractive flowers or edible produce from this small, container-like garden. A bale is about 24 by 42 by 18 inches. [*urban gardens; community gardens; rain gardens; microgreens and urban gardens; campus quad gardens; guerilla gardening; hydroponics; vertical farms; vegetable seeds, heirloom*]

strip-cropping

Soil erosion can be reduced by alternating rows of one crop such as corn with a cover crop such as alfalfa. This is a *soil conservation* technique called strip-cropping. The alfalfa covers the soil and protects it from erosion by catching *runoff* from the alternate crop (corn in this case). This method also helps minimize the spread of pests and plant diseases because there is no continuous stand of a single crop. The cover crop can be a *nitrogen-fixing* plant that also helps restore the soil's fertility. [*conservation agriculture; cover crops; gully reclamation; intercrop; low-input sustainable agriculture; regenerative agriculture; shelter belts; sustainable agriculture*]

strip-mining

There are two kinds of strip-mining (also called surface mining). Both methods are commonly used to mine *coal*, but the processes are similar to mining for other minerals as well.

The first kind, called area strip-mining, is used on flat terrain where the vein of coal is no more than 300 feet beneath the surface. The material above the coal is called the overburden. Heavy equipment removes the overburden and places it alongside the vein. The vein of coal is then removed and the overburden replaced to fill in the vein.

The second method, called contour strip-mining, is used on hilly terrain. In this method, the hill is cut into a series of terraces with the overburden from one dropping into the excavated terrace below.

In both methods, the land must be manually restored or *soil erosion* destroys the region. In area strip-mining, failure to restore the site results in a series of small eroded hills and valleys called *spoil banks*. In contour strip-mining, a wall forms that is quickly eroded, called a highwall. [*coal formation; acid mine drainage; fossil fuel; mineral exploitation (mining); Mining Law of 1872; mountain top removal mining*]

Student Conservation Association

This group's mission is to "build the next generation of conservation leaders and inspire lifelong stewardship of our environment and communities by engaging young people in hands-on service to the land." The association provides expense-paid, year-round, team-based conservation internships for students over 18.

Their Student Conservation Corps provides an opportunity to work in any of hundreds of projects that help protect and restore *national parks, marine sanctuaries*, cultural landmarks, and community green spaces all over the United States. [*Student Environmental Action Coalition; environmental literacy; environmental education; consumers, green; ecotravel; Environmental Protection Agency; United Nations, global environmental issues and the*] {thesca.org}

Student Environmental Action Coalition (SEAC)

Begun in 1989, this organization of student *environmentalists* began at the University of North Carolina. Hundreds of universities now have SEAC chapters. It is the largest student organization on environmental issues and includes high schools. They describe their group and mission as "SEAC is a student and youth run national network of progressive organizations and individuals whose aim is to uproot environmental injustices through action and education. We define the environment to include the physical, economic, political, and cultural conditions in which we live. By challenging the power structure that threatens these conditions, students in SEAC work to create progressive social change on both the local and global levels." [*Student Conservation Association; environmental justice; environmental literacy; environmental education;*

Environmental Protection Agency; United Nations, global environmental issues and the; consumers, green; ecotravel] {seac.org}

styropeanuts

The packaging "peanuts" you often find pouring out of a shipping box are called styropeanuts. Just like foam coffee cups, they are made from *polystyrene (PS)*, which accounts for about 11 percent of all *plastics*. Because they are inherently bulky, they take up a great deal of space in *landfills* and—like most plastic products—are non-biodegradable. In an effort to resolve the disposal problems caused by styropeanuts, many options now exist.

Some *recycling* programs reuse styropeanuts in their original form, but polystyrene does not lend itself well to recycling. (Many companies are moving away from using polystyrene products.) Alternatives to styropeanuts are springing up, including some that use organic materials such as various types of starch. *Bioplastics* made from natural products such as *fungi* and other substances are also beginning to appear. Most of these new types of *product packaging* materials readily degrade and are non-toxic. [*plastic recycling; plastic pollution; litter*]

suburbia

See *urban sprawl.*

suburbia and biodiversity

Many people think *wildlife* is only found in the wild, but this is not so. Most people are in much closer proximity to lots of wildlife than they might think. Eighty-two percent of all Americans live in cities and suburbs; two-thirds of all North American wildlife species live there as well. A study done a few years ago showed that 1200 rare or *endangered species* live in the 35 fastest growing metropolitan areas of one million or more residents, and over 500 of these species live only in these areas and nowhere else.

What does this mean? Two things: First, people in urban and suburban areas have an enormous opportunity to enjoy wilderness where they live, if they look in the right places, and second, these same people can help protect biodiversity right where they live. Many of us are becoming more aware of these facts and embracing them. Numerous books have been written about identifying and helping to sustain urban and suburban wildlife, and interest in urban wildlife research has increased. *NGOs* such as the *National Wildlife Federation* offer programs about creating your own *Certified Wildlife Habitat*, even in a city. [*biodiversity; biodiversity hotspots; easement, conservation; migratory bird protection; urbanization and urban growth; Alliance for a*

Paving Moratorium; deicing roads; eco-cities; greenways; megacities; smart cities; tract development; urban sprawl; urban gardens; urban farms; urban open space; urban livestock; white roofs; zoning, land use]

subsidies
See *fossil fuel subsidies*.

subsistence farming
Raising crops or livestock on a small scale, usually to meet the immediate needs of a family or small group, as opposed to *high-impact agriculture* where vast quantities of crop or product are produced for the masses. Subsistence farming was the typical state of agriculture before technological advances produced *monocultures* requiring *pesticides, fertilizers,* and mechanized harvesting equipment.

In many *less developed countries,* subsistence farming remains the norm. Although it is considered a far more environmentally friendly form of agriculture, it is responsible for some *deforestation,* because wood is the primary energy source on these types of farms.

Much of the *farm-to-table* movement is about "getting back" to subsistence farming methods of producing food. The food is less contaminated, the *soil* is less damaged, and the *water* is less polluted. [*permaculture; community-supported agriculture; green manure; organic farming and food; regenerative agriculture; sustainable agriculture*]

subsurface mining
When coal is too deep for *strip mining,* subsurface mining is used. A deep vertical tunnel is dug and a series of tunnels and rooms are blasted out so the coal can be hauled to the surface. The room and pillar method of subsurface mining leaves about half the coal in place to act as pillars that provide structural support to the maze. [*coal formation; acid mine drainage; fossil fuel; mineral exploitation (mining); Mining Law of 1872; mountain top removal mining*]

succession
While organisms live in an environment, they change that *environment,* making it less suitable for themselves and more suitable for other types of organisms. This process results in an area being inhabited by a collection of organisms in predictable stages (called a *sere*), with the simplest forms of life present early and more complex forms later. This series of changes in the *community* is called succession.

The first stage is called the *pioneer community* and the final stage, which remains relatively stable with little further change, is called the *climax community*. All the stages together are called a *sere*. The stable climax communities for land-based *ecosystems* are called *biomes*.

There are two major kinds of succession—primary and secondary. *Primary succession* occurs in an area that has never been colonized by organisms, such as barren rock created from a lava flow. This is called *terrestrial primary succession*. When the environment is a water habitat, it is called *aquatic primary succession*.

Secondary succession, which is more common, occurs when an existing community has been totally destroyed by events such as forest fires, floods, or *clear-cutting* a forest. Secondary succession differs from primary succession in that the land has not been reduced to bare rock.

succession, primary aquatic

All ponds and lakes are destined to become part of the land. This natural progression is called primary aquatic *succession*. This process begins as soon as the body of water is created and continues until the body of water is completely filled in and becomes land. The entire process can take hundreds or even thousands of years.

Aquatic succession results from the continual flow of *topsoil* and *organic matter* from the surrounding land into the water. As the *nutrient*-rich materials enter the water, more organisms establish themselves. When they die, they contribute even further to the increase in sediment and organic matter at the bottom. As the water becomes shallower, new species of plants establish themselves and emerge out of the water.

Once the sediment makes the body shallow enough so plants can thrive near the surface, the water begins to dry out creating a *wet meadow*. Once the body is dry, terrestrial organism move in with the process continues until a *climax community* is established. [*peatlands; bogs*]

succession, primary terrestrial

A barren terrestrial region will become inhabited by simple organisms, called a *pioneer community* at first and then gradually give way to more advanced organisms, until a relatively stable *climax community* develops. The process of gradual change in the makeup of a community is called *succession* and usually takes several hundred years to complete. When the process begins with barren rock, it is called primary terrestrial succession. (If it began with standing water it is called *primary aquatic succession.*)

In the past, large portions of the Earth were left barren when the glaciers receded at the conclusion of the last *Ice Age*. Today, barren areas are occasionally produced naturally due to volcanic action creating new lava formations, large scale mud slides, the creation of new sandbars, or other natural catastrophic events. Barren regions are also created by human processes such as *strip mining* that denudes the Earth's surface.

The successional stages of primary succession, beginning with bare rock are as follows. The pioneer community includes *lichens* that help establish a shallow layer of *topsoil* by breaking down the rock (along with *erosion*) and forming *humus* when they die and decompose. The lichens also create a *habitat* for some microbes. The soil allows mosses, some annual plants, small worms, and *insects* to establish themselves. As the soil becomes richer due to the decomposition of these organisms, the intermediate stages of succession occur. This includes the establishment of grasses, shrubs and often some shade intolerant trees, along with larger animals.

The final climax community, which may include shade tolerant trees and a full range of animals, begins to establish itself. The climax community is determined by many factors including the temperature range and amount of water for the area. Succession may begin in some areas with *sand* instead of rock. (Sand is simply pulverized rock.) [*plate tectonics; peatlands*]

succession, secondary

Secondary *succession* is similar to *primary succession* except for the initial condition of the environment. In primary succession, the land is barren and has never been colonized by life. In secondary succession, an area previously colonized with life is devastated by some natural or human-caused events. Examples of natural occurrences include forest fires and floods, and artificially produced events include *clear-cutting*, intense *water pollution*, *mineral exploitation*, or clearing woodlands for *agriculture* (and then abandoning the farm).

Topsoil or *sediment* and *organic nutrients* still exist in these environments, so the speed with which succession occurs is more rapid than primary succession, taking very roughly one or two hundred years, depending on the type of *climax community* that forms.

For an example of secondary succession, consider an abandoned farmland in the southeastern United States. The stages begin with an empty plowed field. The *pioneer stage* (the first couple of years) involves the establishment of annual grasses, such as crabgrass, when the dormant seeds that lay in the soil germinate. Over the next decade, other grasses and small perennial plants take hold, along with a few shrubs. Sedge is a dominant species during the early stages. The sedge stabilizes the soil enough for pine and spruce seedlings to take root. Over the next 10 to 20 years, the pine trees grow tall enough to shade the ground, eliminating the sedge. The succession now moves into the intermediate stages.

Competition for moisture and sunlight prevents new pine seedlings from surviving among the existing *stand* of trees. Hardwood trees such as oak begin to grow and force out the pines. Finally, shade-tolerant trees such as dogwood fill the *understory* in the *climax stage,* which becomes a relatively stable community.

sudd
A large floating mass of plants and debris that can clog *rivers* and *dams*. [*aufwuch*]

sugarcane to biofuel
See *bioethanol*.

sugar, sugar alternatives, and the environment
The sugar crop has a long history of environmental destruction. However, today some *organic* brands of sugar are grown sustainably. They are indicated as such by *eco-labels*, *eco-certification programs*, or *NGOs* approval of some sort. Other possibilities for those with a sweet tooth include some sugar alternatives that cause little environmental harm. Stevia is a sustainably grown herb in Latin America with zero calories. Others considered eco-friendly (but not calorie free) include brown rice syrup, agave nectar, maple syrup, and honey. [*sustainable yield; sustainable agriculture; organic farming and food*]

sulfur dioxide
Sulfur dioxide is one of five primary pollutants found in *air pollution*. It is released when *fossil fuels* that contain *sulfur* are burned. The sulfur is often present in fossil fuel because it was naturally found in the organisms that, over millions of years, became the fossil fuel.

Burning *oil* or *coal* that contains sulfur produces an odor and causes respiratory irritation. Sulfur can react with oxygen and moisture and produce acids that are called *secondary air pollutants*. These acids contribute to *acid rain* and can destroy lung tissues if in sufficient quantities.

Industrial smog (also called gray air smog), contains sulfur dioxide. Modern coal and oil burning facilities often use "scrubbers," devices that dramatically reduce these emissions. Some U.S. power plants and—especially—those in less developed countries still generate *sulfur dioxide* in quantities harmful to humans.

The best way to reduce sulfur emissions is to convert from using high-sulfur-content coal to low-sulfur-content coal, oil, or natural gas, which reduces emissions by a minimum of 65 percent. Another method is to remove the sulfur from the coal before it is used, but this process is expensive and rarely used. [*Clean Air Act; acid rain; hazardous air pollutants; criteria air pollutants; air pollution in Beijing, China; Mercury and Air Toxics Standards; photochemical smog*]

sunflower oil as a biodiesel fuel

Refined forms of sunflower oil (as well as soy and canola oil) have been used experimentally as an alternative form of *biodiesel* automobile fuel. Environmentally, this would provide many advantages over conventional *fossil fuels*. Because sunflower oil doesn't contain sulfur, it doesn't produce *sulfur dioxide* when burned. Tests also indicate that this fuel—when compared to traditional fuel—reduces soot emissions by 80 percent, carcinogens by 65 percent, visible smoke by 65 percent, and *carbon monoxide* by 30 percent. Whether it ever becomes feasible and economical remains to be seen. [*biofuels; biofuels, second-generation; biomass energy; alternative fuels*]

sunsetting

Gradually phasing *toxic* substances out of production and use. For example, sunsetting *mercury* could begin with its banishment from batteries and paints, followed by its removal from many other products and processes, until it is totally eliminated. [*leaded gasoline; toxic pollution; toxic waste; toxic cocktail; body burden; toxicology*]

superbugs

This does not refer to *insects*. The phrase refers to *bacteria* that have become resistant to most modern antibiotics. These superbugs are becoming more common, most likely because of the extensive use of *antibiotics on livestock* to keep them from getting sick in *CAFOs*. They are resistant to all of the standard antibiotics used, leaving few alternatives to stop infections.

Recent studies have found that a large percentage of the meat sold in the United States contains these superbugs. They were found in 81 percent of the ground turkey sampled, 69 percent of the pork chops, 55 percent of the ground beef, and 39 percent of the chicken breasts, wings, and thighs. [natural *selection; antibiotics and the human microbiome; medical ecology; nanoparticles and human health probiotics; skin bacteria; spillover effect*]

supercontinents

The continents as we know them did not always exist. Because of *plate tectonics*, the continents have always been moving and changing. If you go back about 300 million years before now, there was one supercontinent called *Pangaea*, which later broke up into today's continents. If you go back about a billion years before that in *geological time*, there was another supercontinent called Rodinia, which then broke up. And if you go back about 700 million years before that, a previous supercontinent called Nuna existed.

Scientists have determined that these supercontinents form and then break up in cycles of very roughly every half a billion years. They have even discovered the locations of these past continents. Because they have determined this to be a cyclical event, they are expecting the continents to once again come together and form a new supercontinent to be called Amasia, about 100 million years from today.

super insulation

Insulation used in buildings constructed with a variety of the most advanced technologies, including triple-glazed, gas-filled, low-emissivity windows and heavy insulating fiberglass with R-values of 40 to 70. Super insulation dramatically reduces energy costs in cold climates. [*buildings, green; homes, green; LEED certification; indoor air pollution; zero-energy buildings; Living Buildings; thermal insulation, building*]

Superfund

In 1980, Congress established *CERCLA* (the Comprehensive Environmental Response, Compensation and Liability Act) to begin cleaning up the worst *hazardous waste* sites in the United States. It set aside a large trust fund, commonly called the Superfund, to be managed by the *Environmental Protection Agency*. Sites to be cleaned up by using Superfund money are placed on the National Priorities List.

The act was designed to establish a comprehensive cleanup plan and make those responsible for the pollution pay for much of the cleanup. This act has been updated and funds increased, but it is still considered by many environmentalists to be more smoke than substance. Of the 1200+ sites on the National Priorities List, to date fewer than 60 have been cleaned up.

One of the reasons for this poor track record is that offenders find it more economical to fight the EPA in court than to actually clean up the wastes. More than five billion dollars of the Superfund's budget has been spent on legal fees instead of actual cleanup costs. [*Love Canal; superfund sites, U.S. counties with the most; schools and brownfields; remediation of hazardous waste; POPs, dirty dozen*]

superfund sites, U.S. counties with the most

The following U.S. counties contain the largest number of *superfund* sites: 1) Santa Clara, CA (23); 2) Los Angeles, CA (17); 3) & 4) Montgomery, PA (16) and Nassau, NY (16); 5) Burlington, NJ (15); 6) & 7) Middlesex, MA (14) and Middlesex, NJ (14); 8) Harris, TX (13); 9) & 10) Chester, PA (12) and Suffolk, NY (12).

super weeds

Weeds that have become resistant to *herbicides*, such as *glyphosate*. Especially those that grow among GMO crops. [*genetically modified organisms (GMOs)—resistance in; natural selection*]

surface mining

See *strip mining*.

surface water

The world's freshwater resources are divided into surface water and *groundwater*. Surface waters include streams, rivers, ponds, lakes, and human-created reservoirs. All freshwater that is not absorbed into the Earth (becoming groundwater) or returned to the *atmosphere* as part of the *water cycle* is considered surface water. Surface waters comprise only about 0.02 percent of all water on our planet. [*freshwater ecosystems; drinking water; water use for human consumption; domestic water use; water pollution*]

sustainababble

The prefix *eco-* and the word *green* were used meaningfully at first. After years of use, they became hijacked by marketing, advertising, and just about everything else as buzzwords for something that is supposed to be environmentally friendly but often is not. Now they are even used to describe just how useless they have become: *green scam*, *greenspeak*, and *greenwash*, for example. Overuse has made them questionable words, unless the consumer is aware of what is and what is not factual.

The same thing is happening to the word sustainable. In <u>State of the World 2013</u>, published by the *Worldwatch Institute*, an entire chapter describes how the word sustainable has also become hijacked and turned into sustainababble.

All of these terms have been stretched beyond their limits of usefulness. The only solution is to become knowledgeable so you can see through the babble, scam, speak, and wash and learn the facts about the *science* that is hidden beneath the words. [*environmental literacy; consumer, green; green marketing; green products;*

Sustainability Imperative

The premise that businesses must be socially responsible partners with people and the planet—not only because it is ethically the proper thing to do, but also because that's what consumers want and it is proving profitable. This idea was first stated in a paper called "The Sustainability Imperative" by David A. Lubin and Daniel C. Esty, published in <u>The Harvard Business Review</u> in May 2010.

The sustainability imperative implies that capitalism and sustainability are not opposed to one another but rather complementary; sustainability has become a key driver of corporate success. Many people believe sustainability constrains the economy, but the sustainability imperative states it is expanding the economy.

A former vice president of Exxon, Øystein Dahle, stated, "Socialism collapsed because it did not allow the market to tell the economic truth. Capitalism may collapse because it does not allow the market to tell the ecological truth." Capitalism is a long way from collapse, but the statement could be foreboding in the long run. [*sustainable development; sustainable growth; Sustainable Development Goals; CERES; climate capitalism; corporate sustainability rankings; decoupling; economy vs. environment; internalizing costs; growthmania; triple bottom line*]

sustainable agriculture

Many *conventional agricultural* practices used today contaminate the *soil* and *water* with *synthetic fertilizers* and *pesticides. Soil erosion, salinization*, and waterlogging are all harmful side effects of today's *high-impact agriculture.* Short-term productivity is paid for by long-term destruction of the land. Sustainable agriculture tries to establish an ongoing relationship with the land, resulting in long-term productivity.

Sustainable agriculture uses both the latest advances in technology and old-fashioned farming practices to assure continued use of the land and less harm to the environment. Ongoing research in the field of sustainable agriculture is not a "cute" idea but one that is economically feasible for the farmer as well as environmentally sound for everyone.

Many land grant universities have established low-*input sustainable agriculture* programs with federal research dollars. Some aspects of *sustainable agriculture* are similar to *organic farming* techniques, but on a larger scale. A combination of both old and new farming practices is used. *Crop rotation* and *strip farming* are examples of the former, and genetically engineered crops (*GMOs*) that resist pests and drought, examples of the latter. *Integrated pest management* is also used in sustainable agriculture to reduce the use of pesticides.

Even though these practices work and are gaining popularity, they are still not widely used when compared to conventional or high-impact agriculture.

sustainable development

Probably the most common phrase issued from anyone's lips when talking about our planet's future, the environment, our economy, or anything related. The idea was first discussed back in 1974 at a *United Nations Environment Programme* symposium. The *International Union for Conservation of Nature* first used the actual phrase in 1980. However, it was the book <u>*Our Common Future*</u> that brought it to the forefront of the environmental movement. That book's definition was simple, eloquent, and remains the standard. It states that sustainable development is "development that meets the needs of the present without compromising the ability of future generations to meet their own needs."

A fine but clear line exists between sustainable development and *sustainable growth*. These two phrases actually contradict one another, because infinite growth is not sustainable in a finite world. [*economy vs. environment; Beyond GDP; CERES; climate capitalism; economics, ecological; steady-state economics; United Nations Educational, Scientific, and Cultural Organization; United Nations Environment Program*]

Sustainable Development Goals (SDGs)

At the close of the *United Nations Conference on Sustainable Development* in 2012, there was hope that a new set of environmental goals would be created, called the Sustainable Development Goals. These goals would emphasize environmental issues such as setting targets for *green jobs*, curbing overconsumption, protecting *marine* life, providing food security, and others. In addition, it would continue the *Millennium Development Goals* that expire in 2015. However, the SDGs were never developed at that summit. Instead, they agreed that the SDI goals will become part of a "post-2015 Development Agenda" to be established sometime in 2014. Kicking the can down the road has become a central theme, both in the U.S. government and apparently the international community as well.

sustainable economy

See *economics and sustainable development*.

sustainable growth

An oxymoron to many, because our *biosphere* has limited resources, making infinite growth impossible—even with new technologies. As opposed to *sustainable development,* which is a reasonable and necessary goal.

sustainable management

Sustainability has become the key word for our planet's environmental health. If something is sustainable, it is good for the planet; if it is not, it is bad for the environment. Although this sounds like an oversimplification—and it is—it still has value. If any process, organization, or system is going to remain sustainable, it will only become so by being properly managed. In today's business acumen, managers must understand how to manage with sustainability a primary focus.

Sustainable management is the study of the intersection of management and the need for sustainability. A business based on a finite resource—and most resources are finite—cannot continue if the resource is depleted. The growing importance of sustainable management is underscored by the large number of advanced degrees offered in the field. Most large companies have high-level management positions with the term sustainable in their title. In many cases, these positions report directly to the president. Hopefully, they don't report to marketing. [*economy vs. environment; Beyond GDP; CERES; climate capitalism; economics, ecological; steady-state economics*]

sustainable marketing

See *green marketing.*

sustainable yield

Any type of harvest (fish, crops, trees, nuts, etc.) can be either sustainable or not. If not, the harvest will cease to exist, at least temporarily and often forever. If sustainable, enough of the product remains to reproduce and continue the species in this location indefinitely. The sustainable yield is that quantity that allows for the continuation of the crop. This is a balancing act that requires scientific expertise and research. Sustainable yields have been successfully pinpointed when left to science, with regulations in place to enforce maximum yields allowed. It rarely works when left to those whose livelihoods depend on the harvest or is voluntary, as is often the case. [*overfishing; fish populations and food; Magnuson-Stevens Fishery Conservation and Management Act; tragedy of the commons; overgrazing; overpopulation; hunger, world*]

swamp

A *wetland* type of *ecosystem* with woody plants such as trees and shrubs being the dominant forms. (This distinguishes them from *marshes* and other wetlands.) They remain partially covered by water all year long and are very productive, with great *biodiversity*. Swamps and other wetlands historically were considered wastelands because they could not be used for development and often bred mosquitos and other so-called pests. They were, and remain, *ecosystems* under threat of development. [*inland wetlands; coastal wetlands; eco-services*]

Swanson's law

Swanson's law is about *photovoltaic cells* used for most *solar power*. It states that the cost of photovoltaic cells drops 20 percent every time the number of these cells manufactured doubles. It is named after Richard Swanson, the founder of a large solar panel company. Although it might not be completely true, the point is that the cost has been dropping dramatically and appears able to continue to do so, making photovoltaic power more and more competitive, at least in sunny regions. [*alternative energy sources*]

symbiotic relationships

When two organisms of different species live together over long periods of time in close physical contact, with one or both benefiting in some way, it is called a symbiotic relationship. Symbiotic relationships can be divided into four categories: *parasitism*, *commensalism*, *mutualism*, and *parasitoidism*.

synecology

Synecology (also called community *ecology*) studies how groups of populations, called a *community*, coexist in the same *environment*. It studies why the number and mix of different species in a given area change over time and what controls the dispersion of a *population* in an area. [*invasive species; anthropogenic stress*]

synfuels

Coal or *biomass* that has been converted into either a gas or a liquid fuel is called synfuel, short for synthetic fuel. The name is misleading because it is not a synthetic product. It just refers to the fact that the coal or biomass has been converted, with human intervention, into a fuel.

A process called *gasification* converts the coal or biomass into a gas synfuel called *syngas* because it is similar to *natural gas*. By using a process called liquefaction, the syngas can be converted into a liquid fuel called *methanol*.

Both types of synfuels produce less *air pollution* than coal. Liquid fuels are more functional than solid fuels for heating homes and running automobiles and other forms of transportation. They can also be transported through pipelines, whereas coal must be shipped.

Synfuel facilities, however, are expensive to build and run when compared with a coal-burning *electric power plant*, even with a full complement of air-pollution control devices. Synfuels also have a lower heat content, meaning more must be used to produce the same amount of energy from synfuels than directly from coal. [*energy (fuels); alternative fuels*]

syngas

A gaseous fuel typically created from either *coal* or *biomass* by the process of *gasification*. The name is short for synthetic *natural gas* because it is similar to natural gas but usually has less available energy. Syngas can also be used for other purposes, such as solvents and *synthetic fertilizer*. [*energy (fuels); alternative fuels*]

synthetic biology

The field of biology that manipulates organism's *genomes* (their genetic makeup) to create new traits or even new life forms. Today it is most notably the science used to create genetically modified organisms. [*genetically modified organisms (GMOs)— basics about; DNA sequencing technology; timber, DNA sequencing of; cryptic species; de-extinction; recombinant DNA; transgenic crops*]

synthetic fertilizers

Organic fertilizers are natural substances made from plants, animals, or their waste. Synthetic *fertilizers* are manufactured substances. The most common are ammonium sulfate, ammonium phosphate, ammonium nitrate, and ammonium chloride. These substances are created by using free *nitrogen* from the air and combining it with hydrogen from *natural gas* at high temperatures. With the exception of *organic farms*, virtually all agriculture in the United States—called *high-impact agriculture*—uses these synthetic fertilizers produced from *fossil fuels* (natural gas).

These artificially created fertilizers get the job done by supplying the needed *nutrients*, but at the expense of the environment. They must be constantly replenished, they degrade the quality and the texture of the soil because they do not replenish the soil with organic matter, and they easily run off from the land into bodies of water where *nutrient enrichment* can occur. [*conventional agriculture; monocultures*]

synthetic natural gas (SNG)
See *synfuels*.

systems ecology
A branch of *ecology* that studies the relationships between all the components of an *ecosystem*, living and nonliving, and concentrates on how energy and chemicals flow through the system. A "system" can be as simple as one particular *habitat* (a pond or a *forest*) or as large as the entire *biosphere*. *Computer modeling* is often used to predict how an ecosystem might react to changes such as increased concentrations of pollutants or gases in the air.

tagging programs

Tagging or marking animals is a useful and accepted method of monitoring and studying *wildlife*. Tagging fish, banding bird legs, attaching radio transmitters to grizzly bears, and using delicate Mylar patches on *monarch butterflies* have all been used successfully. These programs help scientists study growth rates, *migration paths*, and breeding cycles. The geographical range of the *polar bear*, the wintering grounds of the Monarch butterfly, and the breeding grounds of some *sea turtles* have all been discovered thanks to tagging programs. [*ecological study methods; descriptive ecology; crowd research, scientific; robotic bees; nanotechnology; satellite mapping; Landsat; drones, conservation*]

taiga

One of several kinds of *biomes*. The primary factors that differentiate biomes are temperature and precipitation. The taiga is also called a northern coniferous forest or a boreal forest. With precipitation between 10 and 40 inches per year, these regions don't have a shortage of water, but they have long, harsh winters with short, cool summers. The dominant plants are coniferous trees, including spruce, fir, and larch, with needle-like leaves that minimize water loss because most of the water through the winter is frozen in snow and ice. Most birds are *migratory* and most *insects* become inactive in the winter. Many small *mammals* inhabit the taiga. Larger animals found in the taiga include deer, moose, and *wolves*. [*anthromes; ecosystems*]

tailings, uranium mine

After uranium ore is mined for use in *nuclear reactors*, it must be crushed for processing. The remaining, unused crushed rock, called tailings, releases low-level radiation and poses health risks. The fine particles can be blown by the wind, so the source of the *radioactivity* can move away from the original tailings pile and contaminate areas far from the source. Radioactive particles can also *leach* into the soil and enter the *groundwater* or simply be washed into rivers or streams, contaminating them.

Mine tailings, whether from radioactive or nonradioactive minerals, don't lend themselves to plant growth, often resulting in *soil erosion*. [*nuclear fuel cycle; nuclear fuel cycle hazards; nuclear power; cooling ponds and towers; yellowcake*]

target organism

Pesticides are meant to kill a specific pest—called the target organism. Any organism affected, other than the target organism, is called a non-target organism. Unfortunately, most pesticides do not know the difference between target and non-target organisms. For this reason, *Rachel Carson*, in her famous book <u>Silent Spring</u>, suggested calling pesticides *biocides*. [*insecticides*]

tar sands oil extraction

Oil is found in many forms. Some of it can be easily extracted, but much of it has been either technologically impossible to get at or economically impractical to extract. However, with advances in technology and the high price of oil, new methods of extraction are now used to get at oil previously considered inaccessible. It is called *unconventional oil* and includes tar sands oil and *shale bed* oil that requires *fracking*.

Tar sands is a form of oil that is very heavy and thick and embedded in *sand* or clay. As the name implies, it is difficult to extract and is considered, from an environmental perspective, to be one of the dirtiest types of oil.

Tar sands oil has come front and center in the news recently, because huge deposits are being extracted in Alberta, Canada, and the *Keystone XL pipeline* has been proposed to transport it across the United States. It would travel from Alberta south to the gulf coast of Texas, where it would be refined and then distributed for use throughout the United States.

Environmentalists are concerned about the danger of leaks while transporting this oil through the pipeline and also the use of such a polluting form of *fossil fuel*. And they are also concerned that the pipeline would commit the United States to the continued use of *fossil fuels* as opposed to *renewable, alternative energy* sources such as *wind* and *solar*.

Although the immediate concern is about tar sands in Canada, some companies are proposing to extract from tar sands in the United States, specifically in Utah, where there are large deposits of this oil.

Serious environmental problems have been associated with tar sands oil extraction. The common method is very water intensive (much like the fracking process). Two to five barrels of water are needed for every one barrel of oil processed. *Natural gas* is needed to run the refining equipment, which results in the inefficient need to burn one fossil fuel to produce another fossil fuel. This entire process generates about three times more *greenhouse gases* than *conventional oil* production. And finally, the process results in large quantities of a wastewater product mixed with sand, clay, oil residues, and other chemicals added to the water during extraction. This contaminated wastewater is then stored in tailing ponds—usually an environmental nightmare waiting to happen.

All forms of oil extraction involve some form of *habitat destruction* such as *strip mining*. Tar sands extraction is no different. It takes about two tons of tar sands removal to get one barrel of oil. The extraction of tar sands oil in Alberta, Canada, will destroy hundreds of thousands of acres of boreal forest (*taiga*).

Companies hoping to extract tar sands oil within the United States are well aware of these concerns and the opposition they will face. They state they have unique extraction techniques that are far less damaging to the environment. The fate of both the pipeline from Canada and the first U.S. tar sands mines remains up in the air. [*oil recovery or extraction; oil spills in U.S.; pipeline spills; oil spills in history, ten worst*]

Tasmanian tiger, extinct

Also called a thylacine, this interesting marsupial looked much like a dog in size and shape. This similarity between the two animals is an excellent example of *convergent evolution*. The tiger became *extinct* in the early 1900s, with the last survivor dying in a *zoo* in 1936. It was originally found throughout *Australia* and Tasmania, which were geologically connected at that time. Competition from the *predatory* dingo and government-imposed bounty to hunt them down resulted in the tigers' demise. [*biodiversity, loss of; passenger pigeon; dodo bird; species recently gone extinct; United Nations Convention on Biological Diversity; Animal Damage Control Program*]

taxonomy

The study of describing, naming, and classifying organisms. [*natural selection; speciation; species; kingdoms and domains*]

teaching farms

Many working farms have a primary goal to educate children (and adults) about farming. Their primary focus is teaching sustainable farming practices. The crops harvested and/or milk and eggs collected are often donated to food banks or used in school cafeterias as part of a *school-to-table program*. [*sustainable agriculture; low-input sustainable agriculture; campus quad farms; community gardens; community-supported agriculture; farmers' markets; farmers' markets, fake; organic farming and food; organic food labels; regenerative agriculture*]

telephone pole disposal

Of the approximately 120 million telephone utility poles in the United States, more than two million need replacement each year. To prevent decay and fire, the poles are treated with chemicals such as creosote, which the *Environmental Protection Agency* considers a *toxic* substance and is potentially *carcinogenic*. Because of these chemicals, the poles cannot be recycled and typically end up in *landfills*, where the substances can *leach* out and contaminate soil and possibly *groundwater*.

Research is ongoing to find methods of *bioremediation* that could render the toxic substances harmless and allow the poles to be recycled. [*recycling; tires, recycling / reprocessing*]

temperate deciduous forest

One of several kinds of *biomes*. The factors that differentiate biomes are temperature and precipitation. These forests receive more than 40 inches of rain, evenly distributed throughout the year. Seasonal temperature changes include cold winters during which the plants lose their leaves (*deciduous*) and become inactive until the following spring. Although life is abundant, most of these forests are composed of only a few species of trees. The forest floor contains many flowering plants that have access to sunlight in early spring before the leaves return to the trees. There are also many small shade-tolerant shrubs.

These regions contain numerous *insects*, small *mammals* such as mice and rabbits, and large *grazing mammals* such as deer. *Predators* such as foxes and badgers are also common. Most of the birds that live in these woods *migrate* south during the colder months.

The *topsoil* in a deciduous forest is fertile (primarily from the loss of leaves each season), and much of it has been cleared and converted to farmland or swallowed up by *urban sprawl*. In North America only 0.1 percent of the original deciduous forests remains. [*anthromes; anthropogenic stress*]

teratogens

Substances that cause birth defects. [*REMs; radiation; mutagenic; carcinogens; pathogens*]

terracing

Crops grown on steep slopes are prone to excessive *soil erosion*. Terracing is a *soil conservation* technique in which the slope is carved into a series of level terraces and crops are grown on each terrace. Water gradually flows from one terrace to the next in a cascade, thereby reducing erosion.[*conservation agriculture*]

terraforming

Describes the theoretical act of making another planet more earth-like. Often used in science fiction but also discussed by scientists discussing future possibilities. [*geoengineering; near earth objects; Blue Marble photographs, original and update; Earthrise photograph; outer space; Spaceship Earth; extinction and terrestrial impact; panspermia*]

terrestrial ecology

The study of organisms that live on land and their habitats. In the past this was the realm of field scientists using a great deal of manual labor to collect data to be analyzed. Today, a new initiative called NEON (the National Ecological Observatory Network) is bringing this traditional science into the 21st century by using high-tech scientific equipment.

Sixty sites around the United States are being fitted with 35,000 sensors to collect more than 500 types of data, such as precipitation, air pressure, wind speed, humidity, sunshine, *air pollutants* (including *ozone*), soil *nutrients*, vegetation growth, and even microbes.

All of these sensors will take consistent readings at the same time in the same place over long periods of time, something needed for *ecological studies*. By producing baselines across the country, it will be possible to make better predictions about how *ecosystems* are reacting to *climate change* or *invasive species* or any form of environmental stress. [*satellite mapping; Landsat; Biosphere 2; Amphibian Ark project*]

terrorism, eco-

Aggressive or violent acts designed to either protect the environment or use the environment as a pawn during a dispute. Trying to stop *deforestation* by driving spikes into trees, thus endangering the lives of loggers using chainsaws, would be an example of an aggressive act. Using environmental degradation as a weapon, such as occurred during the *Persian Gulf war,* is an example of violent eco-terrorism.

The FBI states that eco-terrorism has caused damage worth hundreds of millions of dollars over the past decade. [*monkeywrenching; ecotage; eradicating ecocide movement; tree-spiking; Earth First!; eco-warrior; environmental movement*]

textile industry

This industry is second only to *agriculture* when it comes to causing *water pollution*. Twenty percent of all *industrial water pollution* comes from textile mills, finishers, and the industry in its entirety. Most of this results from dyes, such as those used to make blue jeans blue. The industry produces *air pollution* and lots of solid wastes, as well.

However, some textile companies work toward reducing their environmental impact, and there are *eco-certification programs* that track how textile manufacturers are doing. For example, Bluesign is an independent company based in Switzerland that tracks a textile manufacturer's energy, water, and chemical use with goals of both protecting the environment and saving the company money; which often go hand and hand.

Some textile companies, such as Polartec, go out of their way to be green. They hire an independent group to audit energy, water, and chemicals used by textile mills they use so they know who they are dealing with and whether they are polluting their local environment. [*blue-jean pollution; stonewashed jeans; rivers in decline; consumer, green; dry cleaning; flip-flops and preconsumer waste; green products; sneakers, green; toilet paper, eco-*]

theme parks, greening of

Just like *baseball stadiums*, theme parks are going green, partly to make visitors happy but also to save money. Six Flags is using waste vegetable oil from its kitchens to fuel all of the parks' trains and vehicles. The Disney resorts have set a goal to *recycle* all of its waste and become *carbon neutral* by aggressive efficiency measures and, when needed, *carbon offsets*. The Disney park train runs on 100 percent *biodiesel* fuel. Tivoli Gardens is a theme park in Copenhagen, Denmark, that powers the entire park with *wind energy* it purchases. It plans to become *carbon neutral* by powering the park with an offshore *wind turbine*. [*NBA arenas, green; NFL stadiums, green; universities, top ten green; cities, ten best American; green gym movement, green; Facebook's carbon footprint*]

theoretical ecology

Theoretical ecology is one of the more recent methods used to study *ecology*. It uses mathematical equations and *computer modeling* on powerful computers to predict what will happen to *populations* of organisms and life on Earth. These equations and models are based on existing knowledge, but must also rely on assumptions. A typical example is the attempt to predict what will happen to life decades from now, if *climate change* or *deforestation* continues at current rates. [*Landsat; satellite mapping; Great Lakes Environmental Assessment and Mapping Project; ecological study methods; biotechnology; geoengineering; terraforming*]

thermal insulation, building

Fifty to seventy percent of the energy used in the average American home is used for space conditioning (heating and cooling). Proper thermal insulation dramatically reduces the amount of energy used for space conditioning by minimizing air leakage. The three most important areas requiring insulation are the attic space, under floors that are above unheated spaces, and around walls in basements and crawl spaces.

Thermal insulation is measured by the *R-value*, which indicates the resistance to heat flow. The *Department of Energy* has recommended R-values for each zip code in the United States. New construction buildings should meet the recommended R values throughout the building. For existing construction, insulation can almost always be upgraded to meet these recommendations.

There are many different types of insulation used. These include blankets or batts of mineral fibers, loose or blown fill made of fiberglass, rock wool or Cellulosic fiber, or Perlite or Vermiculite. Each type of insulation requires a different thickness to reach the desired R-value. For example, 9 to 10 inches of blown in rock wool offers an R value of 30, but you'll need 13 to 14 inches of fiberglass to attain the same R-value. [*heating, ventilation, and A/C system; air conditioning (A/C); weather stripping; super insulation; indoor air pollution; indoor ecology; LEED certification; buildings, green; homes, green*]

thermal inversion

Air pollutants produced in cities are usually dissipated when warm air at the surface rises and mixes with cooler layers of air above, carrying away the concentrated pollutants. However, cities within valleys or partially surrounded by mountains are prone to thermal inversions where a layer of warm air forms above the surface layer, creating a lid over the cooler air, trapping pollutants for extended periods of time and often resulting in *photochemical smog*. Thermal inversions and photochemical *smog* are problems in many cities including Los Angeles, Salt Lake City, Phoenix, Denver, and Mexico City. [*air pollution; dust dome; urban heat island; atmospheric brown cloud; brown cloud; criteria pollutants; industrial smog; ozone, ground-level; ground-level ozone, ten U.S. cities with the worst*]

thermal water pollution

One form of *industrial water pollution* is called thermal *water pollution*.

About 60 billion gallons of water are used each year by industry, with most of it used in *electric power plants*. (*Industrial water use* does not include *irrigation*.) About 200,000 million gallons of water per day are used in the United States just to generate electricity. These power plants convert water to steam, which drives turbines to produce electricity. The steam is then cooled by more water, which carries the heat away. If this heated water is released directly into natural bodies of water such as a river or stream, the *aquatic ecosystem* is affected, because all organisms have specific temperature and dissolved oxygen *tolerance ranges* in which they can survive. The damage done is called thermal water pollution. (Temperature controls, in part, the amount of dissolved oxygen in the water.) Changing the temperature just a few degrees, which often happens in these industrial processes, can change which organisms can and cannot survive.

Methods exist to reduce the water temperature before being released into natural bodies of water, such as using artificial ponds to dissipate the heat. [*nuclear reactor safety and problems*]

thermochemical conversion, biofuel

Biomass energy is produced by either a *biochemical conversion* process or a thermochemical conversion process. The latter produces synthetic natural gas, called *syngas*. Syngas is produced during a process called *gasification*, where a *biomass* such as wood is heated with little or no oxygen. As a result, the biomass breaks down into simpler substances, resulting in syngas, which can be used as fuel. Syngas can also be turned into a liquid fuel by a process called liquefaction, which creates *methanol*.

Thermodynamics, The Second Law of

When energy is converted from one form to another, some of the useful energy is lost, usually in the form of heat. For example, when *coal* (one form of energy) is burned to create electricity (another form of energy), large amounts of energy are lost as heat and becomes useless. Another example is the loss of useful energy that occurs between each of the *trophic levels* as illustrated in an *energy pyramid*.

Scientists are always looking for ways to minimize this "lost useful energy." The energy lost when generating electricity from coal can be recaptured in a process called *cogeneration*. And the lost energy between trophic levels is why you often hear that it is more efficient to feed people plants than meat. This is because when plants (one form of *biomass* energy) are eaten by animals (another form of biomass energy), about 90 percent of the energy originally stored in the plants is lost before it ever becomes incorporated into the animals. Scientists cannot change the Second Law of Thermodynamics but they do find ways to work with it.

third domain

An informal name for *Archaea*. [*kingdoms and domains*]

Third Pole, The

This is an informal name given to the ice sheets located in and around the Tibetan Plateau north of the Himalayas. It covers 40,000 square miles and consists of 46,000 glaciers, as well as over one million square miles of *permafrost*. Many of the sheets are hundreds of feet thick.

The *Arctic* (North Pole) and *Antarctic* (South Pole) are much larger, but the Third Pole is large enough to be another good indicator of *climate change*. There had been a great deal of controversy about whether the *glaciers* on the Himalayas are shrinking. In 2007, the *Intergovernmental Panel on Climate Change (IPCC)* stated the glaciers were receding so rapidly they might be gone by 2035. However, since then, the IPCC has retracted that statement.

Current research shows that although some of these ice sheets are advancing (increasing), the majority are receding, so that they are receding (decreasing) overall.

So all three poles (the North, South, and Third Poles), are receding, but not at the extremely rapid pace first announced. The melting of these ice sheets and glaciers has significant impact on the global environment. [*Arctic ice, changes in; melt ponds; glacial lake outbursts; glaciers, mountain; melt ponds; permafrost, Siberian; Petermann Glacier ice detachment*]

THM (trihalomethanes)

These are substances produced when water is chlorinated as part of a municipality's *drinking water* treatment process. When chlorine gas in the water comes in contact with *organic* matter also found in the water—such as leaves, grasses, and food particles—it reacts by forming THMs. Chloroform is a well-known THM.

Research has shown that some THMs are *carcinogenic* or *mutagenic*. For this reason, the *Environmental Protection Agency* has set an allowable limit of 80 parts THMs per billion parts of drinking water. Some municipal water treatment plants are having trouble meeting this standard, so many are looking for alternatives to replace chlorine gas. The most common alternative to chlorination is *ozone* sterilization, used in some countries but not as of yet in the United States. [*chlorination*]

Thoreau, Henry David

(1817–1862) Thoreau was a poet, writer, philosopher, and—some say—America's first *environmentalist*. Thoreau is best remembered for his book <u>Walden</u> and his essay "Civil Disobedience." Environmentalists appreciate how his works got people thinking about nature in our new land.

Thoreau graduated from Harvard, ran a school for a few years in Concord, Massachusetts, and then moved in with *Ralph Waldo Emerson* in the position of handyman. This is when he started writing and helped found (along with Emerson and Margaret Fuller) the Transcendental movement.

In 1845 Emerson allowed Thoreau to build his own cabin on Emerson's property in the woods, locally known as Walden Pond. It was there he wrote many of his books. He spent two years, two months, and two days living in this cabin in semi-solitude.

He believed cities were the evils of society and wilderness was our salvation. An interesting quote from Walden was prophetic: "Thank God man cannot yet fly and lay waste the sky as well as the Earth." Today Thoreau is considered one of the most important forces in the early American years of environmental writing. [*philosophers and writers, environmental; Muir, John; Burroughs, John; Whitman, Walt; airplane pollution*]

threatened species

Species that are likely to become *extinct* if some critical factor in their environment is changed (often called vulnerable species). In other words, the species now exists with some environmental factor at the minimum level sufficient for survival. For example, the *U.S. Fish and Wildlife Service* has listed the northern spotted owl as a threatened species under the *Endangered Species Act,* with its primary threat being *habitat loss* due mostly to *logging.*

The U.S. Fish and Wildlife Service maintains this list for species within the United States, and the *International Union for Conservation of Nature* maintains threatened and *endangered* lists for species on a global basis. [*endangered and threatened species lists; Convention on Biological Diversity, The United Nations*]

Three Gorges Dam

Completed in 2009 and located in the upper Yangtze River of *China*, this is the largest *hydroelectric dam* in the world, generating over 18,300 *MW* of power. When first put into operation, it provided an amazing 10 percent of China's energy needs, but because of the country's staggering growth and energy demands, this number is down to about 3 percent of the total needs.

A dam of this size cannot be without its detractors. Over one million people were relocated from their homes to fill the dam, and almost 75,000 acres of farmland were lost. Massive hydroelectric power plants such as this one—although a form of *renewable*, clean energy—are considered by most environmentalists as causing far more harm than benefit. However, others counter that had this dam not been built, most of the energy it now generates would have come from *coal*-burning *electric power plants*.

The environmental damage caused by this dam was projected to be huge—and it has not disappointed. Prior to the dam, the river and its upstream tributaries were drinkable, but are no longer because of *algal blooms* from *nutrient enrichment*. The waters now have increased levels of *heavy metal* pollution. There is severe *erosion* along much of the shores, far more than even anticipated. The infection rate from schistosomiasis, also called snail fever, has increased because the dam expanded the snail's *habitat*. The dam also destroyed many fish-spawning habitats and reduced the habitat of the *endangered* Yangtze finless porpoise and Chinese River dolphin.

Because of all of these environmental and health problems, China has begun a 10-year mitigation program to clean up the mess the dam created. But even in doing so, it appears they have learned little from these past lessons, because there are now plans for a dozen new dams in the upper Yangtze that will wreak further havoc. The river will become a series of giant dams, creating a vast amount of needed hydroelectric power but destroying what is left of the Yangtze *ecosystem*.

The new dams will even destroy a nature preserve that was created upstream to provide a protected habitat for the species harmed by the Three Gorges dam. The reserve is one of the few remaining habitats that protects almost 200 fish species; however, it will be gone as well, once these additional dams are built. [*alternative energy sources; dam, removal; Aswan High Dam; dams, ten largest; hydropower, traditional; hydropower, nontraditional; James Bay Power Project; pumped-storage hydropower*]

Three Mile Island (TMI) nuclear accident

This, the worst *nuclear power* plant accident in the United States, occurred on March 29, 1979, at the TMI nuclear power plant in Harrisburg, Pennsylvania. A series of equipment failures and human errors caused the reactor to lose its coolant and have a partial *meltdown* resulting in an unknown amount of *radiation* loss. An evacuation advisory was issued.

The government went to great lengths to ensure the public they remained safe. President Jimmy Carter went into the control room to demonstrate its safety. However, the accident started many in the United States to reconsider nuclear power as part of our power strategy. Completing the cleanup and closing the reactor took over 14 years and cost over one billion dollars. [*nuclear reactor safety and problems; nuclear waste disposal; nuclear waste dilemma, the global; Chernobyl; Fukushima nuclear power plant disaster*]

three R's

The environmental version of the three R's stands for Reduce, Reuse, and Recycle. This sounds overly simplistic, but it works if enough people follow it. As a quick example, "reduce" could refer to buying unpackaged or minimally packaged products, such as unpacked fruits and vegetables, or buying detergents that come as concentrated formulations requiring smaller packaging. For "reuse," the classic example is the reusable *water bottle* and a reusable-material shopping bag. And for "*recycling*," any program available counts, such as bottle return bins, *glass recycling*, and *e-waste recycling*. [*product packaging; upcycling*]

throwaway age
A phrase coined by Vance Packard in his 1960 book <u>The Waste Makers</u> to describe a society in which everything is disposable. [*conspicuous consumptions; growthmania; planned obsolescence; municipal solid waste disposal; economy vs. environment*]

tidal power
Twice each day the gravitational pull from the moon causes the ocean waters to flow in and out along the coasts, resulting in the tides. If the tides flow through narrow inlets, the water can be channeled through turbines which are turned by the moving water and generate electricity, an energy source called tidal power. Unlike other *renewable energy* sources, such as *wind power* and *solar power*, tidal power is consistent and works like clockwork.

A few of these tidal power plants exist—one in France and the other in Canada. The latest design is found at a site in Maine's Cobscook Bay that rests at the bottom of the bay, generating electricity as the tides move the water back and forth through the device. At the moment, it only generates enough electricity to power about 100 homes, but it is being used as a test site before twenty such turbines are built to provide power for about a thousand homes. It is run by the Ocean Renewable Power Company.

Although these power plants are an excellent source of clean energy, only a few sites are suitable for their construction. Some people, however, believe tidal power can be a major contributor to our power needs in the future.

Because tides are caused primarily by the moon, it is sometimes called moon power. [*hydropower, nontraditional*]

tight oil
See *unconventional oil and gas.*

timber, DNA sequencing of
Illegal logging is a global problem. Many countries, including the United States, levy fines on companies that use illegally harvested wood, but only if they can prove the wood was improperly taken. Many people want to be sure they are not purchasing wood products that might have contributed to *deforestation* or endangered the workers who cut the timber down, but this too is only possible if the source of the wood is known.

Furniture manufacturers and others that use wood products have found it difficult to track down and truly know where the wood came from, so not much could be done about illegal wood products. But now, *DNA sequencing technology* can determine the exact genetic makeup of a species, allowing it to be tracked to the source. It is the ultimate fingerprint. The process is new and therefore expensive, but many companies are signing up to use this technique, ensuring consumers they are purchasing environmentally friendly wood products. [*forests, sustainable; Forest Stewardship Council; cryptic species; de-extinction; genetically modified organism (GMOs)—basics about; recombinant DNA; transgenic crops; Franken-salmon*]

time shifting

Wind power and *solar power* have a big problem: inconsistency. The wind doesn't always blow and the sun, of course, is cyclical. Companies building these power facilities must figure out how to make energy available all the time and this is called time shifting—shifting to a time when it is available. They accomplish this by using power—while it is available—to store energy for use when it is not available. This is typically done with a process called *pumped-storage hydropower* or a new technology in development called *power islands*.

tipping point

Used in recent years by environmental scientists to illustrate how a gradual process of steady change, such as *global warming*, reaches a point where something happens that rapidly accelerates the process and makes it uncontrollable beyond that point. For example, some suggest that the melting of the ice caps is approaching a tipping point. Ice and snow are a light-color, so they reflect most of the radiant energy from the sun, keeping the region cold. As ice melts, the darker soil or water absorbs more heat, making it warmer and making even more ice melt. This is a negative feedback loop where a process becomes self-perpetuating. The end result could be a very different environmental state then previously existed; with profound repercussions.

Tipping points are now also discussed with a recent perspective about human survival called planetary boundaries or our *safe operating space*. If we go beyond one or more of these boundaries, as we have already done, we might pass a tipping point, resulting in a new accelerated state of decline. [*Arctic ice, changes in; sea level rise*]

tire, high-mileage

A great way to improve a car's miles per gallon is to use *tires* that reduce rolling resistance with the road. Many *automobile* manufacturers install high-mileage tires so they can advertise higher miles per gallon. Tires produce less resistance when they contain a powdered substance with strong chemical bonds. Historically companies used carbon black made from petroleum for this substance.

They are now using silica from sand that has been processed and modified to suit this purpose. However, there are now a few other green alternatives available. Most recently, the greenest alternative—both in production and in results—is to use a natural substance called phytoliths, produced by grasses to protect themselves from being eaten by *herbivores*.

Some tire companies, such as Pirelli, are getting these phytoliths from rice husks, which is an agricultural waste product. The husks are used to run small generators in some countries, such as *Brazil*. Once these husks have been burned to generate power, a *bottom ash* remains, once again becoming a waste product. Pirelli uses this ash as the source for phytoliths, which it then uses in its low-rolling-resistance, high-mileage tires. This source for green tires could become popular in any region that grows a lot of rice. [*CAFÉ; greenest cars, top ten; automobile propulsion systems, alternative; batteries, automobile; gasoline, saving on auto; idling your car*]

tire fires

Hundreds of millions of tires are discarded in the United States each year. Many end up in vast used-tire stockpiles. In 1983, a used-tire stockpile in Virginia burned 7 million tires and lasted for many days. In 1999, a fire in California lasted for a month. In both cases, there was significant *air pollution* and increased numbers of people with respiratory ailments during these events. [*tires, recycling / reprocessing; extended product responsibility; disposal fee*]

tires, recycling / reprocessing

More than 300 million tires are discarded in the United States each year. Twenty years ago, only about seven percent of tires were recycled. Today, about 75 percent are recycled or reprocessed and turned into products such as ground rubber products, fuels, and—most commonly—the base for roadways. The fact that there is a strong market for this type of recycled product has made it a *recycling* success story.

Even with these great strides in *recycling*, and because they cannot be placed into *landfills*, more than 275 million tires are in U.S. junkyards today (down from 2 billion 20 years ago). They pose a variety of health problems as well as a threat of huge *tire fires* spewing noxious fumes for miles. [*appliance recycling; refrigerator recycling; refrigerator doors; energy, ways to save residential; automobile recycling; mattresses, recycling; cigarette butt litter and recycling; downcycling; upcycling; motor oil recycling; paper recycling; recycling metals; plastic recycling; e-waste recycling; single-stream recycling*]

Title X of the Public Health Service Act

This is the only U.S. federally funded program devoted to family planning. Its budget is on an annual roller coaster ride with many ups and downs because Congress determines its funding levels and birth control is often a contentious issue. [*population growth, limits of human; baby boom; hunger, world; water shortages, global; China's One-Child Program; human population throughout history*]

Toilet Day, World

Many people in *developed countries* might think this a joke, but it is not. It should be considered a reality check. The United Nations General Assembly designated November 19 as World Toilet Day to focus the world's attention to the fact that 2.5 billion people do not have basic toilets. The United Nations states that 6 billion people have cell phones but only 4.5 billion have toilets. [*domestic water use; water shortages, global; hunger, world; night soil – biosolids; biosolids and wastewater*] {worldtoiletday. org}

toilet paper, eco-

Americans use 17 billion rolls of toilet paper per year, totaling over 160 million pounds of trash. The toilet paper *ecological footprint* is immense, and individuals can help reduce the impact in a few ways. Many manufacturers now offer tubeless toilet paper rolls that work just as well as those with tubes. Some also offer rolls made from *recycled paper* with as much as 90 percent from *post-consumer recycled waste*. Look for rolls that are bleach free and possibly dye and fragrance free as well. Another possibility is to use *bamboo* toilet paper. [*product packaging; waste minimization; source reduction; three Rs; municipal solid waste disposal*]

toilets and biosolids

See *biosolids and wastewater*.

toilet-to-tap drinking water

Although it sounds disgusting, it does work, it is safe, and you probably wouldn't know that's what you were drinking. As clean *drinking water* becomes scarce in many parts of the world, cities are looking for new ways to provide clean water. At the moment, a few municipalities perform this process and provide this product. Orange County Water District of California has been doing so since 2008, and some towns in Texas do so as well.

Although most *sewage treatment facilities* release the treated water into bodies of water to let nature perform the final cleansing, these water-recycling facilities go through many additional steps beyond the norm and then release the sterilized water into the *potable water* system to be enjoyed as liquid refreshment.

The initial steps are similar to the usual *water treatment*, where solids are removed and beneficial *bacteria* are added. However, then the *recycling* portion of the process is filtered by a series of membranes to remove solids, microbes, and chemicals. A small amount of hydrogen peroxide is added and UV light is used to finish the process. This might become a more common practice than you would expect or hope for in the future. [*water shortages, global; water grabs; Safe Drinking Water Act; water use for human consumption; wastewater-to-energy power plant; chlorination*]

tolerance range

The range of a specific factor such as temperature, in which an organism can survive. If the factor goes beyond the tolerance range (too high or too low), the organism will die. This then becomes the *limiting factor*. [*ecosystems; ecology; euroky; extremophiles; natural selection*]

tomatoes—tasteless but red

Tomatoes must be refrigerated when transported long distances, which destroys their taste and texture. Homegrown, fresh-picked tomatoes are far better, but many people agree tomatoes are not as tasty as they used to be. Researchers have learned why.

A few decades ago, a variety of tomato was discovered that was uniformly red when ripe—a nice trait for farmers wanting to harvest red tomatoes all at the same time, so this new, uniformly red variety was bred into almost all commonly grown varieties used today. (Previously, tomatoes grew red at varying times, with green or yellow patches.)

Scientists have learned that the same genes that made tomatoes red and were bred into most of the common varieties used today also dramatically reduced the production of sugars and aromas. So we have ended up with nice bright-red but tasteless tomatoes.

If you miss tasty tomatoes, you can look for heirloom varieties that retained their original genetic makeup. Some companies are even trying to put the original traits back into some varieties by using *conventional breeding* techniques. Another possibility is to look for the time when *genetically modified* tomatoes appear to possibly return tasty tomatoes to stores. [*heirloom plants; vegetable seeds, heirloom; seed banks and vaults; seeds and climate change, Project Baseline; organic farming and food; Non-GMO Project Verified label; rice, contaminated; seasonal foods; WWOOF; wine, organic and biodynamic*]

Tongass National Forest

The largest of all 156 U.S. *national forests* containing almost 17 million acres. It makes up about 80 percent of southeast Alaska and is 57 percent temperate rainforest. It contains trees 800 years old and 200 feet tall and has the world's largest concentration of *bald eagles* and grizzly bears.

In the 1950s, the *U.S. Forest Service* signed 50-year contracts with two large firms to log 90 percent of this forest. But beginning in the 1960s, the *environmental movement* began applying pressure to stop devastating *old growth* and *ancient forests*. During the late 1980s, *environmentalists* brought to the public's attention not only the loss of nature but also the loss of money these contracts had perpetrated on the public by practically giving away the rights to use this property for private enterprise.

Environmental *NGOs* worked relentlessly to get Congress to reform the Forest Service's management of Tongass. In 1990, Congress voted to halt timber sales by passing the Tongass Timber Reform Act. This act began the transition of the Forest Service's role of managing logging to one of fish and wildlife protection and restoring watersheds.

To balance lost jobs and income, the service began supporting job creation in areas such as fishing, *renewable energy*, *mariculture*, and *ecotourism*—instead of logging. [*dual-use policy; forests, sustainable; forest health, global; forest management, integrated; multiple use policy; deforestation and climate change; forest, pop-up; governments of the forest; logging, illegal; timber, DNA sequencing of; Forest Service Employees for Environmental Ethics; Forest Stewardship Council*]

ton of carbon dioxide look like?, What does a

Carbon emissions are always discussed in tons (or metric tonnes). When we hear the word ton, we think of weight, so visualizing a ton of a gas such as *carbon dioxide* is difficult. At standard pressure (which is important for a gas), one metric ton (2,205 pounds) of carbon dioxide gas would fill a balloon about 33 feet across, which is a pretty big balloon. To further the visualization, a car driven about 40 miles (getting 20 mpg) produces roughly one ton of carbon dioxide. {news.brown.edu/pressreleases/2013/04/earthweek}

topsoil

The uppermost layer of *soil*, which contains significant amounts of *organic* matter. It can typically range from an inch to about a foot in depth and stops where the soil becomes packed hard, with little organic matter, called subsoil. Rich topsoil has a great deal of decomposing organic matter as well as vast numbers of microorganisms including *bacteria, algae, protozoans, fungi, nematodes*, and *earthworms*, among others. Topsoil is where most plant life acquires its *nutrients*. [*soil horizons; soil organisms; soil profile; soil conservation; soil erosion*]

topsoil formation
See *soil erosion*.

torpor
Some mice, *bats*, and birds can dramatically reduce the body's physiological activity, creating a dormant or almost dead-like condition for short periods of time. For example, the ruby-throated *hummingbird* goes into a state of torpor each night when it drops its metabolism tenfold and then wakes up for another day of feeding. *Hibernation* is considered an extended form of torpor. [*estivation; cold resistance; quiescence; xenobiotic; adaptation; extremophiles*]

Torrey Canyon
An *oil tanker* that ran aground off the coast of Great Britain in 1967, causing the largest tanker *oil spill* up to that time and drawing worldwide attention. The ship broke apart, spilling roughly 35 million gallons of crude oil. The *oil* despoiled about 260 square miles, killing numerous seabirds and other wildlife. The incident helped fuel the burgeoning environmental protection movement. [*oil spills in U.S.; oil spills in history, ten worst; Amoco Cadiz oil spill; pipeline oil spills; Deep Water Horizon oil spill; Oil Pollution Control Act; oil pollution in the Persian Gulf*]

touchpools
Aquariums, like *zoos*, attract millions of visitors each year. Touchpools in an aquarium are the equivalent of the petting farm section of a zoo. Aquatic organisms that can be safely handled by visitors are placed into a shallow, easily accessible pool. Guides are usually present to urge visitors to handle the specimens and to explain their natural history. [*zoos, future; Zoo, Frozen; zoos, captive breeding programs in; nature centers; Smithsonian Institution; environmental education; volunteer research, scientific; voluntourism, environmental; WWOOF*]

tourism, eco-
See *ecotourism*.

toxic cocktail
An informal term loosely used to describe a mix of toxic substances found in a particular location but with an unknown impact. For example, it has been used to describe

the Ganges River in *India*, with its many pollutants. In this case, the exact extent of the harm caused to the river's *ecosystem* and to those that bathe in it is unknown.

It can be used with *body burden*—the mix of synthetic substances we put into the environment that ends up in our bodies—albeit in very small concentrations. No one knows for sure what the negative impact is of this toxic cocktail in our environment or within our bodies.

The primary concern is that the mix of toxic substances is likely to cause more harm than the sum of its parts. When used in this manner, the term—toxic cocktail— seems most appropriate. [*toxic pollution; toxic waste; hazardous waste; toxicology; Superfund; POPs, dirty dozen; cancer alley; cancer villages; Rachel Carson; persistent pesticides; volatile organic compounds*]

toxicology

The study of poisons and their effects on organisms. [*toxic pollution; toxic cocktail*]

toxic pollution

Any substance found in the environment that causes harm to an organism's normal functioning. These substances can be found in air, water, or soil, so toxic pollution is often a component *of air pollution* and *water pollution* as well as found in contaminated *soil*. In the United States alone, millions of tons of toxic pollutants are produced and disposed of as *toxic wastes* each year. The government has passed legislation is response to the public's concern about toxic pollution, including the *Resource Conservation and Recovery Act*, the *Toxic Substances Control Act,* and the *Superfund*.

Technological advances are enabling industry to use less of these substances without having a negative effect on their business and competitiveness, but the problem remains a serious one. Cleaning up existing *toxic wastes* and *hazardous waste disposal* sites may prove as difficult as preventing future damage. [*toxic cocktail*]

Toxic Substances Control Act (TSCA)

This piece of legislation, enacted in 1976, supposedly gave the *Environmental Protection Agency* the authority to require testing of new chemicals to see if they were deemed safe—before they can be made available on the market. In truth, most environmentalists and many health experts agree it is entirely inadequate and endangers the health and well-being of Americans, as well as our *environment*. The first problem with TSCA was that it grandfathered in about 60,000 chemicals already in existence— meaning they have never even been looked at, let alone tested for safety, but continue to be sold.

TSCA works like this. Companies must notify the EPA they plan to introduce a new chemical, but they don't need to provide any safety information. Instead, it is up to the EPA to find relevant scientific information to prove that the new chemical will pose a risk. The EPA can only "ask" the company for information or require testing if it first proves there is a potential risk, which is hard to do without the company's data. The logic is clearly upside down, putting the burden of proof on the EPA, who knows nothing about the chemical, and the manufacturer, who need not provide the information to the EPA. The absurdity of this law can be seen by looking at a few numbers. Since 1976, when about 60,000 chemicals were grandfathered in, the number of untested chemicals has grown to about 85,000. How many new chemicals has the EPA been able to actually declare unsafe and require regulation of some sort? That would be five.

This logic is contrary to what is done in Europe, where they use the *precautionary principle:* A substance is assumed harmful until proven safe, and the burden rests on the manufacturer.

There has been a battle in Congress in recent years to update TSCA with a new law called the *Safe Chemicals Act* of 2013, which would overhaul TSCA and make major improvements. It would require all chemicals to be proven safe to human health and the environment. If the manufacturer could not prove the chemicals to be safe, they could be removed from the market or prevented from entering the market. This new act would also encourage green chemistry, meaning safer alternatives would become available, as opposed to the current model where manufacturers have no incentive whatsoever to improve safety.

Possibly an indication of the importance of this act as it pertains to our health can be seen by those who endorse it, such as The American Academy of Pediatrics and dozens of other groups represented under the umbrella of an organization called Safer Chemicals, Healthy Families. The Safe Chemicals Act currently sits and languishes in a congressional committee, with no action as of yet.

Even without this new act, consumers are becoming more knowledgeable about toxic substances in products and the environment. They have become *green consumers* by demanding action from manufacturers and their local governments. For example, many stores stopped carrying *plastic bottles* containing *BPA* before it was even banned. You don't need to wait for Congress—you can take action into your own hands. {saferchemicals.org}

toxic waste

One type of *hazardous waste*. Toxic wastes include any substance that negatively affects the health of organisms. Some toxic substances are harmful or even lethal if an individual is exposed to a single large dose. This is called *acute toxicity*, which is measured by the *LD50* of the substance. The chemical released at the plant in *Bhopal* was acutely toxic. Substances such as *mercury* cause acute toxicity. Some toxic substances cause harm because they are carcinogens (cause cancer), teratogens (cause birth defects), or mutagens (cause genetic defects). Other toxic substances are harmful in low doses over long periods of time, which is referred to as *chronic toxicity*. [*toxic pollution*]

trace elements
Elements that occur in organisms in minute amounts but are essential to life. Examples are copper and zinc. [*nutrients, essential*]

tract development
A type of *urban sprawl* in which homes or other residential units are built over large areas, usually on land originally used for agriculture. [*urbanization and urban growth; suburbia; smart cities; high market value farming counties, U.S.; zoning, land use*]

tragedy of the commons
The phrase originated in biologist *Garrett Hardin's* essay "The Tragedy of the Commons," published in <u>Science</u> in 1968. In the essay, the commons was pastureland in England. Today, however, it has become a metaphor for our entire *biosphere*. The tragedy refers to the overuse of a common resource, resulting in the degradation or depletion of that resource. For example, fish are often harvested until the fishing grounds are empty, or land is *irrigated* until the *groundwater* is gone and the *topsoil* turns to dust. The tragedy is defined by the belief that "if I don't use it, someone else will." [*overfishing; overgrazing; deforestation; desertification; groundwater mining; environmental justice; environmentalist*]

Trans-Alaska pipeline
See *Alaska pipeline*.

transgenic crops
An early name, used about 30 years ago, for what are now called genetically modified crops. Companies producing these crops prefer to use this term instead of GMOs, because it meets with less opposition. [*genetically modified organisms (GMOs)—basics about; DNA sequencing technology*]

Transition Towns
A new movement consisting of people whose primary mission is to reduce *fossil-fuel* use and mitigate *climate change* by what they call "relocalizing." Relocalizing

means shifting all forms of production, products, services, and just about everything from the norm—that is, from global sources to local sources. Part of this is generating and using local energy sources and food supplies. The goal is to create strong communities, called Transition Towns, where people work together so they can face the anticipated tough times ahead. It has been described as a large-scale social experiment that appears to incorporate many characteristics of both survivalists and utopians.

Transition Towns is not a small movement. About 360 official Transition Towns exist in 31 countries, of which 85 are in the United States. The movement began in England and has only been in existence for a few years, so time will tell of its successes or failures. [*farm-to-table movement; community-supported agriculture; organic fertilizer; seasonal foods; slow food movement; food miles; locavores; Beyond GDP; alternative energy sources; alternative fuels; energy independence, United States*] {transitionus.org/transition-town-movement}

transpiration
Water loss from organisms, typically plants, through stomata, which are porthole-like openings in their leaves. Transpiration is an important part of the *water cycle*, which is one type of *biogeochemical cycle*.

transportation
Transportation is one of the three biggest requirements for *energy use* in *developed countries*. For this reason, it is also one of the biggest areas to look for energy savings, with alternative forms of transportation such as *hybrid cars* and *electric vehicles* and *mass transit*. Some people even consider *virtual transportation* to be an alternative. [*automobile; bicycles and bicycling; walking; rail transportation*]

transposon
See *hyperparasite*.

traumatic insemination
See *bedbugs and remedies*.

Treaty on Biological Diversity
This treaty signed at the *United Nations Conference on Environment and Development* held in *Brazil* in 1992 required participating countries to take inventories of plants and animal species and continue to protect *endangered species*. It also required participating countries to share the research, profits, and resulting technologies with

countries whose resources are used in the research. [*biodiversity; endangered and threatened species lists; Encyclopedia of Life; Census of Marine Life*]

tree farms

Farms that grow trees for commercial use. They usually contain only one species of tree and are therefore similar to a *monoculture* typically used for agricultural crops. They use species that grow quickly so they can be readily harvested. The practice is designed to produce a *sustainable yield*. [*sustainable agriculture; forest, sustainable; energy plantations; deforestation; afforestation; reforestation; bamboo*]

tree-spiking

The dangerous practice of driving spikes into trees about to be harvested. If a chain-saw hits the spike, it can kick back and harm or kill the operator. This practice was used back in the 1980s by a radical group of *environmentalists* belonging to an organization called *Earth First!* At that time some people believed civil disobedience and *ecotage* were legitimate forms of environmental activism. [*terrorism, eco-; monkey-wrenching; environmental movement*]

triple bottom line

Typically, businesses measure economic success by using the traditional bottom line of profits and losses. Recently businesses have realized they should take other factors into consideration. The triple bottom line refers to also looking at social and environmental values as part of the economic measurement of success.

This is one of many new business practices that expand the standard bottom line, including *corporate social responsibility* and the *integrated bottom line*. The triple bottom line is similar to the integrated bottom line approach; however, it uses three separate balance and income sheets instead of only one. [*CERES; Sustainability Imperative; climate capitalism; economics and sustainable development; economy vs. environment; polluter pays principle; extended product responsibility*]

trophic levels

See *energy pyramid*.

tropical fish tanks

Three hundred million dollars is spent on tropical fish for fish tanks each year. These fish can either come from the wild or from captive fish breeders. *Freshwater* fish are one of the most *threatened* groups of species. Most people believe it is more environmentally prudent to breed these fish so as to not decimate the wild populations. This is the generally accepted model and about 90 percent of all tropical fish in tanks come from fish farms.

However, others believe these fish can be caught in a sustainable manner and foster not only stewardship of nature but a strong economic advantage to the local region. For example, along the Rio Negro River in *Brazil*, an amazing 40,000 fish are caught and sold each year, yet the operation is considered sustainable and an economic boom to the local communities. This program, called Project Piaba, is endorsed by some *environmentalists* and the *Association of Zoos and Aquariums*. [*permaculture; integrated multi-trophic aquaculture; governments of the forest*]

tropical grassland

See *savanna*.

tropical rainforest

One of several kinds of *biomes*. Tropical rainforests receive 80 to 200 inches or more of rain per year. They are found near the equator where the temperatures remain warm year-round. The *biodiversity* of these biomes is truly amazing. Hundreds of species of trees may live close to each other. This diversity can cause three or four layers of *canopy*. The canopies are thick and provide a unique *habitat* that can be studied as an entire *ecosystem* unto itself. The plants living on the ground must be shade tolerant, because little light filters through all the trees. Ferns, moss, flowering plants, and vines are also abundant.

Animals include numerous birds, rodents, *insects*, snakes, and lizards. Many animals are tree dwellers ranging from *amphibians* to *mammals*. Even though tropical rainforests take up a fraction of the Earth's surface, they contain more plant and animal species than all the rest of the planet combined. About 25 percent of all pharmaceuticals are derived from plants that grow in tropical rainforests.

These regions are being harvested for timber, mined for minerals, and cleared to make way for crops and rangeland for beef cattle at staggering rates. [*deforestation; biodiversity, loss of; deforestation and climate change; governments in the forest; forests, sustainable; anthromes*]

tropics

The region of the Earth lying between the Tropic of Cancer, which is the northern limit where the sun appears directly overhead at noon, and the Tropic of Capricorn, the southern limit. The term is also used to refer to warm climates in general. [*tropical rainforest; biodiversity*]

tub bath vs. shower

A typical tub bath uses about 30 gallons of water. A typical shower runs at about six gallons per minute of shower time. This means a five-minute shower uses about the same amount of water as a tub bath. A ten-minute shower uses 30 more gallons of water compared with a tub bath. Showerhead flow controllers or low-volume showerheads can save water by substantially reducing the volume of water used in the shower. [*bathroom water-saving techniques; energy, ways to save residential; energy-saving apps and programs*]

tubeless toilet paper

See *toilet paper, eco-.*

tuna

Tuna are one of the world's most popular commercial fish. It is a common name for several large fish of the mackerel family, including albacore, yellowfin, and bluefin. They are widespread across the globe and important to the economies of many countries. This commercial importance has led to a drastic reduction in many tuna populations. Many countries have quotas to limit *overfishing* in efforts to save remaining populations. Estimates have populations of Atlantic tuna reduced by 90 percent since the 1970s.

Another environmental issue concerning tuna is *bycatch*. While fishing for tuna, many other species of marine life are captured and killed, including *dolphins*. This resulted in the development of so-called *dolphin-safe tuna* products in the 1970s.

Ocean pollution is another problem associated with many of these species. Because tuna are high up in *food chains*, they can have high levels of *toxic substances* in their bodies as a result of *bioaccumulation* and *biological amplification*. These substances impact their health and are reducing populations. Because tuna is a favorite food for humans and we are even higher up the food chain, it impacts our health as well. [*fishing, commercial; fish populations and food; longlines; tuna aquaculture; Flipper Seal of Approval*]

tuna aquaculture

As the populations of wild bluefin *tuna* continue to decline dramatically due to *overfishing*, commercial interests are looking for alternatives to keep feeding a public that loves their tuna. Tuna *aquaculture* is more difficult than raising other farm fish because tuna migrate and are warm-blooded and *predatory*. The first attempts at raising tuna involved capturing wild young fish and raising them to sell once large enough. However, this method never allowed the captured fish to breed, thus reducing wild populations even further. Now, a new species of tuna has been developed in Japan, specifically for tuna farms, to be raised and bred within the farm. The future of farmed tuna, however, remains in doubt. [*shrimp aquaculture*]

tundra

One of several kinds of *biomes*. Tundra (also called polar grasslands) get about the same amount of precipitation as deserts (10 inches or less), but the conditions are very different. Tundra is characterized by a permanently frozen layer of *soil* called *permafrost*. These regions occur at the extreme northern latitudes, north of the *taiga*, and at high altitudes where similar conditions exist, called alpine tundra. During a few months in the summer, just the top of the soil thaws and a few small types of plants quickly grow. The ground is always frozen, so water remains at the surface, creating many shallow bodies of water. *Migratory birds* fly in for a brief period. *Insects* become prevalent in the brief summer, also. A few hardy *mammals* can survive this environment, including mush oxen, caribou, and the arctic hare. [*permafrost, Siberian; archeologists, ice-patch; peatlands; anthromes*]

turtle excluder device (TED)

Prior to the implementation of TEDs, *shrimp* trawlers in the Gulf of Mexico and off the southeastern shores of the United States routinely trapped roughly 45,000 unwanted *sea turtles* each year in their long cone-shaped shrimp nets. More than 10,000 of these turtles died of suffocation, because they were trapped beneath the water for long periods of time and could not breathe. Federal legislation that began in 1987 mandates the use of turtle excluder devices (TEDs) that allow almost all of these creatures to escape from these nets (without releasing the shrimp).

The TED is a metal or nylon grid that is inserted into the middle portion of the shrimp trawl cone. The grid is large enough to allow shrimp to pass through and remain netted. But larger organisms, such as sea turtles, bump into the grid and are stopped. The grid is placed at an angle, so the turtles slide toward the side of the net where they are released through an opening.

In 1989, the United States passed a new law commonly called the "shrimp-turtle law." It requires all countries exporting shrimp to U.S. markets to also use TEDs. Unfortunately, the use of TEDs is not well-enforced outside the United States because local authorities in many less developed countries do not have the resources to enforce this law.

The World Trade Organization has challenged the legality of the shrimp-turtle law, but so far it still supports a U.S. ban on all shrimp imports from countries that do not agree to require TEDs. [*bycatch; overfishing; fish populations and food; bycatch; biodiversity, loss of*]

tweets, eco-
See *microblogs*.

tweetstorm, red panda on the loose
In June of 2013 at around 8 AM, a red panda escaped from the National Zoo in Washington D.C. CNN tweeted about the loss, asking for assistance. Within an hour, it was retweeted over 3000 times and #redpanda became the news headline of our nation's capital for a few hours. Around 1 PM, local residents were tweeting about Rusty's (that's his name) location and sending pictures to assist zoo personnel in their quest to find him. Within an hour Rusty was back home at the zoo. The zoo thanked everyone for their assistance. [*microblogs, green; blogs, green; social media, environmental; crowd funding, green; crowd research, scientific; crowdsourcing, eco-; tweetstorm, Rio+20*]

tweetstorm, Rio+20
During the *Rio+20 Earth summit* meetings in 2012, nearly 100,000 people sent tweets asking the United Nations negotiators at the meeting to end *fossil fuel subsidies*. #endfossilfuelsubsidies was sent for a 24-hour period and included tweets from Robert Redford, Nancy Pelosi, and many other notables. Although the blizzard of tweets was impressive, it did not result in any formal action being taken about ending these subsidies. [*microblogs, green; blogs, green; social media, environmental; crowd funding, green; crowd research, scientific; crowdsourcing, eco-; tweetstorm, red panda on the loose*]

Ugly Animal Preservation Society

Concern is growing that too much emphasis is placed on saving only those species that invoke human affection and concern. Most save-the-such-and-such campaigns pertain to *charismatic megafauna*—large mammals people think are cute or intelligent or have some other human characteristics. These *threatened species* deserve attention, but the concern is that they are not necessarily the most important species worth saving. Many small, nondescript, or even outright ugly species might be more important to the overall health of an *ecosystem*, and they are usually ignored. For example, you don't hear about a save-the-*blobfish* campaign for this incredibly ugly but *threatened species*.

The Ugly Animal Preservation Society is a European group that makes this point in a tongue-in-cheek fashion during comedy routines. However, the society's name is perfect for this serious problem. Most of the *Group of 10* environmental *NGOs* take this problem seriously and work to educate people that cuteness is not the most important criterion when it comes to *endangered species*. [*flagship species; biodiversity hotspots; CITES; extinction, sixth mass; Lazarus taxon; Lonesome George; species recently gone extinct; Treaty on Biological Diversity; Wildlife Conservation Society*] {uglyanimalsoc.com}

unburnable carbon

To understand this phrase, you must first understand two others: The first is *fossil fuel reserves,* which is the amount of *fossil fuels* known to exist and believed to be extractable. The value of an *oil* or *natural gas* company is based in large part on the quantity of reserves they own; the more, the better. The second phrase is *carbon budget*, which refers to the estimated maximum amount of *carbon dioxide* humans can emit into the *atmosphere* between now and the year 2050 before causing a catastrophic level of *global warming*.

A recent report states the carbon budget to be about 1000 *gigatons*, but companies already have close to 3000 gigatons of fossil fuel reserves—three times the carbon budget. If governments get serious about stopping global warming and finally pass legislation that places strict limits on carbon dioxide *emissions*, oil companies will find they have far more fossil fuel than can ever be burned. This extra amount of fossil fuel is called unburnable carbon.

Not only do these companies already own more fossil fuels than could be burned, but they continue to look for more. This contradiction leads many to believe oil and gas companies really do not believe governments will ever truly act on climate change. [*climate change; ton of carbon dioxide look like? What does a*]

unconventional oil and gas
Extracting *oil* and *natural gas* with new technologies that go beyond the so-called conventional methods. These new technologies include *fracking* and horizontal drilling to access and extract natural gas and oil from *shale beds* and *tar sands.*

Unconventional-extraction oil and gas takes a great deal of additional effort and causes far more environmental harm when compared to conventional resources. Not only is it more difficult to extract unconventional resources, but also it produces a poorer quality fuel. The *energy return on investment* (EROI) to produce liquid fuels such as *gasoline* shows conventional oil has an EROI rating of 16, while unconventional tar sands oil has an EROI rating of 5 (higher is better). [*energy (fuels); energy consumption, historical; energy independence, United States; energy sources, historical; energy, A very brief history of U.S.*]

understory
The layer of vegetation between the *canopy* and the *groundstory* in a *forest*. Usually formed by shade-tolerant trees of medium height. [*tropical rainforest; overstorey; forests; eocsystems; deforestation; biodiversity, loss of; deforestation and climate change; anthromes*]

unintentional pollutants
See *dioxins*.

Union of Concerned Scientists
This *NGO* was founded in 1969 by some faculty at MIT University. Today their mission is to combine citizen's action with scientific knowledge to promote a cleaner and safer environment. They are leaders in the field of science communications, helping experts explain their research to the general public. They produce excellent topical publications. Some of their programs are on *climate change*, global security, clean energy and vehicles, food, and *invasive species*. There are more

than 400,000 members. [*environmental education; risk assessment, environmental*] {ucsusa.org}

United Nations, global environmental issues and the

Discussing global environmental issues and initiatives is impossible without including the United Nations. The three most important conferences have been the *United Nations Conference on the Human Environment* in 1972 (also called the Stockholm Conference), the *United Nations Conference on Environment and Development* in 1992 (also called the Earth Summit or Rio Summit), and the *United Nations Conference on Sustainable Development* in 2012 (also called the *Rio+20 Summit*).

The United Nations has many global environmental management programs. These include The *United Nations Environment Programme*, the *United Nations Development Programme*, the United Nations Food and Agriculture Organization, and the *United Nations Population Fund*.

In addition, the United Nations plays host and sets agendas for meetings to establish international treaties and agreements. Some are preliminary meetings where parties get together to discuss issues, laying out a plan for future action. Possibly the best example is the 1983 *World Commission on Environment and Development*, responsible for generating, a few years later, the <u>*Our Common Future*</u> report where *sustainable development* first made a name for itself.

The United Nations makes its biggest impact by providing forums where action plans, including international rules and regulations, are put in place and *ratified*. There have been hundreds of such agreements, including the *United Nations Framework Convention on Climate Change*, the United Nations Convention on International Trade in Endangered Species of Wild Fauna and Flora (*CITES*), the *United Nations Convention to Combat Desertification*, the Ramsar Convention on *Wetlands*, the *United Nations Convention on Biological Diversity,* and the *Stockholm Convention on Persistent Organic Pollutants*.

Finally, the United Nations is active in *environmental education,* providing numerous materials and raising awareness with activities such as *World Earth Day.* [*signing vs. ratifying; convention vs. treaty*]

United Nations Convention on Biological Diversity (CBD)

One of two *conventions* created at the *United Nations Conference on Environment and Development.* It produced the *Treaty of Biological Diversity*, which became the first global agreement that attempted to conserve *biodiversity*. In general terms, its goal is to facilitate *sustainable development.* It advocates the *precautionary principle* and more recently has condemned some forms of *geoengineering.* [*biodiversity, loss of*]

United Nations Conference on Environment and Development (UNCED)

In 1992, the world's attention focused on this conference. UNCED, often called the Rio Summit or Earth Summit, was held in *Brazil* in June of 1992, which was the 20th anniversary of the first summit—*the United Nations Conference on the Human Environment* held in Stockholm. At that time it was the largest conference ever held, with 178 delegations represented. Most notable, this summit produced the *Rio Declaration on Environment and Development* and *Agenda 21*.

The Rio Declaration contains 27 principles about the rights and responsibilities of nations regarding the environment and economic growth, with significant changes made to the original document created back in 1972. An important change was that it read more favorably about the rights of a nation to their economic growth. *Agenda 21* was a long action plan that laid out a blueprint for *sustainable development*.

In addition to these documents, the conference opened two *conventions* for signature: the *United Nations Framework Convention on Climate Change* and the *Convention on Biological Diversity*. It also issued a nonbinding resolution on principles of *forest management*.

Much like the 1972 conference, this conference displayed the great divide that existed between the developing and *developed countries* of the world.

Comparing what was promised then to what has been accomplished since, is not hopeful. Most of the commitments made were never met. Since the time these commitments were made in 1992, global *carbon emissions* have increased almost 50 percent and *deforestation* has claimed 300 million hectares of *forests*.

United Nations Conference on Sustainable Development (UNCSD)

Commonly called Rio+20, which references the original Summit held in Rio twenty years earlier, this was a three-day environmental summit meeting held in Rio de Janeiro during the summer of 2012. Well over 100 countries sent some 50,000 representatives to discuss and hopefully agree to a document that covered almost 300 topics, called The Future We Want. The previous Rio summit 20 years earlier resulted in significant commitments, such as global agreements on *climate change* and protecting *biodiversity* (although accomplishments never met the commitments).

Rio+20, in general, promised little and accomplished less. Final opinions varied, but most were not positive. They ranged from the *NGO,* CARE's representative stating "nothing more than a political charade" and *Greenpeace* calling it "a failure of epic proportions" to the Oxfam CEO saying, "Rio will go down as the hoax summit."

Here are some highlights: Efforts were made to create an environmentally friendly replacement for the *Millennium Development Goals (MDGs)* that expire in 2015. The hope was to develop a set of *Sustainable Development Goals* (SDGs) to replace the MDGs. However, no one could agree on what they should include, so the best they could do was create a working group of 30 nations that will decide on what these goals should be. The hope is that these SDGs will replace the MDGs when they expire in 2015. Also, an effort to reduce *fossil fuel subsidies* was all but abandoned, and efforts to manage our global oceans were postponed three years, primarily by the United States.

Even though large-scale agreements did not occur, 700 voluntary commitments worth more than half a trillion dollars were made by governments, businesses, and NGOs for environmental causes and projects. Some examples include the island of Aruba pledging to transition to 100 percent *renewable energy*, the Bank of America pledging $50 billion over the next decade toward renewable energy projects and energy access, *Germany* pledging to transition to 80 percent of its electricity needs from renewable sources by 2050, and Microsoft pledging to be a *carbon neutral* company by the end of 2013.

The bottom line as far as international agreements is that they are getting even harder to come by; in fact, even when agreements are made, such as with the first Rio Summit, they are often not kept. Both *global warming* and *deforestation*, two major points from the first summit, are far worse now than they were then.

United Nations Conference on the Human Environment

Held in Stockholm in 1972 and therefore commonly known as the Stockholm Conference, this was the first major global environmental meeting for leaders and the forerunner of the Earth summits that came later. Delegates from 113 countries attended and many *NGO* conferences were running simultaneously.

This meeting exposed serious disagreements over environmental issues between *developed* and *developing countries*. The developed nations were more concerned with *pollution* and *conservation* (issues found primarily in these nations), and the developing nations were more concerned with environmental problems produced by poverty, such as lack of clean *drinking water* (found primarily in these nations). The disagreements also centered around who should pay to resolve these problems.

This conference produced many documents and agreements, including an Action Plan with 109 recommendations and the *Declaration on the Human Environment*, commonly called the Stockholm Declaration, which included 26 principles. Both were nonbinding agreements, meaning they agreed to disagree. But it was a start. Possibly the best contribution that came out of this conference was that it led to the creation of the *United Nations Environment Programme*, opening in 1973 and still the major driver of global environmentalism. [*eradicating ecocide movement*]

United Nations Convention on the Law of the Sea (UNCLOS)

Also called the Law of the Sea Treaty, the UNCLOS was adopted in 1982 but took 15 years of negotiating before finally coming to fruition. The treaty provides a comprehensive legal framework for governing the *oceans* and includes definitions and provisions such as those for territorial waters, international waters, and the *exclusive economic zone*. It did not take effect until 1994. As of 2012 it had 164 signatories. The United States is one of the few countries that has not signed the treaty.

United Nations Convention to Combat Desertification (UNCCD)

This *Convention* specifically addresses arid, semi-arid, and dry sub-humid areas, known as "drylands," where some of the most vulnerable *ecosystems* and peoples are found. Their 10-Year Strategy (2008–2018) was adopted in 2007 and guides their mission, which is "to forge a global partnership to reverse and prevent *desertification/ land degradation* and to mitigate the effects of drought in affected areas in order to support poverty reduction and environmental sustainability."

There are 195 parties involved. The specific goals include improving the living conditions for people in these drylands, to maintain and restore land and soil productivity, and to mitigate the effects of drought. The UNCCD uses a bottom-up approach, encouraging participation of local people to combat desertification and land degradation.

The UNCCD works closely with the other two Rio Conventions, the *Convention on Biological Diversity* and the *United Nations Framework Convention on Climate Change*.

United Nations Development Program (UNDP)

A United Nations agency created in 1965 to help *developing countries* achieve sustainable human development goals. Efforts include increasing literacy, creating jobs, reducing poverty, and increasing transfers of technology between industrialized and nonindustrialized nations. One of their targets is to help countries achieve the *Millennium Development Goals.*

United Nations Educational, Scientific, and Cultural Organization (UNESCO)

This United Nations agency works to create holistic policies to address social, environmental, and economic aspects of *sustainable development*. One part of their mission is to identify *World Heritage Sites*. UNESCO works to preserve and protect these sites with the assistance of the host governments.

United Nations Environment Programme (UNEP)

The United Nations Environment Programme was conceived at the 1972 *United Nations Conference on the Human Environment* and is considered the environmental conscience of the United Nations. Its original function was to raise the level of environmental action and awareness on a worldwide basis. In the 1990s, however, this was broadened to include implementation of *sustainable development*. UNEP is funded primarily by member nations, but also by *NGOs* and some businesses.

The program has been successful at coordinating international environmental negotiations and *conventions* on the most pressing issues, including *climate change* and *biodiversity loss*.

United Nations Framework Convention on Climate Change

Held in 1993, this was a follow-up conference to the *United Nations Conference on Environment and Development* held a year earlier. This second meeting officially stated that the results of the first conference were inadequate and called for the establishment of specific targets and deadlines to accomplish cuts in *greenhouse gases*. The result was the creation of the *Kyoto Protocol,* which took effect in 2005.

United Nations Man and the Biosphere Program

See *Biosphere Reserves.*

United Nations Population Fund (NFPA)

This was originally called the United Nations Fund for Population Activities but changed its name to the United Nations Population Fund in 1987. However, it retains the use of its original acronym, NFPA. This is the world's largest source of international funding, including both government and nongovernment sources for family planning. It supplies about 140 *less developed countries* with family planning and maternal and child health care. The fund encourages birth control measures when deemed appropriate, so it has been a target of many who oppose birth control. For example, The White House dramatically cut back the U.S. contribution to this fund between 1982 and 1992 to show its dismay about birth control. [*population growth, limits of human; population explosion, human; hunger, world; water shortages, global*]

universities, top ten green

Many institutions rank colleges and universities based on environmental criteria. TheDailyGreen.com ranked American universities by criteria such as the amount of green space, electricity consumption, *transportation*, *water use*, waste management, and many others. Their top ten are 1) College of the Atlantic, 2) Warren Wilson College, 3) Evergreen State University, 4) Oberlin College, 5) Middlebury College, 6) Berea College, 7) University of California System (10 campuses), 8) Harvard University, 9) Duke University, and 10) California State University, Chico. [*cities, ten best American green; countries, ten greenest; countries, ten most energy-efficient; companies, top ten green global; greenest cars, top ten; ecotourism, ten best countries for; energy efficient states, top ten U.S.; bicycling cities, ten best; GMOs in your food, the five most common*]

upcycling

A variation of *recycling*, upcycling means recycling old products or materials but turning them into products of more value than their first use. For example, some clothier stores take donated denim and turn it into designer clothes. Another company creates evening dresses from unused, waste swimsuit material. Other examples include converting old cane chairs into modern denim chairs and vintage leather jackets into high-end leather bags. Individuals do their share of upcycling as well. The craft site Etsy.com illustrates how popular the idea has become.

There is only a subtle difference between upcycling and recycling. Ballpoint pens are being marketed as being made from two plastic water bottles. Does this count as recycling or upcycling? It doesn't really matter, because the important thing is these materials do not become part of the *waste stream* and do not enter *landfills* or get *incinerated*. [*three R's; downcycling; municipal solid waste disposal; wedding dresses, reused; green products*]

upwelling

Off the western coasts of many continents, constant trade winds blow offshore. This wind pushes the surface water further out to sea, which in turn is quickly replaced with water from the deep that rushes up to fill the void. This process is called upwelling. The floor of most coastal regions of the ocean, called the *neritic zone*, receives sunlight and therefore is teeming with life and *nutrients*. When the cold deeper water rushes upward, it brings with it these nutrients. The nutrients help establish complex *food chains*, including *plankton*, fish, and *predatory* seabirds.

Upwelling occurs in only 0.1 percent of the total *marine* environment but contributes significantly to the overall productivity of the ocean. These regions are also important for humans because many commercial fisheries are found in these areas. They include the tuna catch off of the west coast of the United States, the anchovy catch off of Peru, and the sardine catch off of Portugal. [*ocean currents; gyres; fishing, commercial; oceanic zone ecosystems*]

urban

Urban means different things to people in different countries. In general terms, it refers to densely populated areas, typically called cities, and the areas surrounding the cities. Most Americans are surprised to learn that in the United States, urban refers to any village, town, or city with populations greater than 2500. That is not a lot of people. However, Spain defines urban as any place with more than 10,000 people, and Japan only calls an area urban if it has more than 30,000 people. So when reading about urban areas and urbanization, it might be more realistic to think of urban as the opposite of rural—areas simply more densely populated than in the country. [*urbanization and urban growth; megacities; megalopolis; smart cities; slum; suburbia; urban sprawl*]

urban farms

In recent years, there has been a farming revolution, but it's not in the wide open *farm belt*—it is in cities. Small, often tiny islands of *topsoil* in the so-called concrete jungle. Almost any open lot can and is being converted into a green space where people plant fruits, vegetables, herbs, and flowers. They do so for many reasons, such as exercise and being outdoors. Interest in *organic foods* and the *farm-to-table movement* are probably major motivators. Neighborhoods providing areas for local citizens to farm help build community pride and improve neighborliness. Some city schools provide areas for students to grow a harvest, with the added benefit of exercise, also a school requirement. Often called the *farm-to-school movement*. [*urban forests; urban gardens; microgreens and urban gardens; urban livestock; urban open space; zoning, land use; gardens, rain; straw-bale gardening; community gardens; community-supported agriculture; farmers markets; food miles*]

urban forests

These forests either in or adjacent to urban areas are also called community forests. They can include tree-lined streets, *greenbelts,* and city parks. They provide numerous benefits to numerous people. For example, they generally reduce the temperature during summer within cities and help mitigate the *urban heat island effect,* reducing the overall utility costs of cooling homes and buildings. They help control stormwater *runoff* from entering bodies of nearby water and help filter *air pollution*. They offer a *habitat* for some species and have been reported to have positive human health effects, as well. [*urban farms; urban gardens; vertical farms; forests, pop-up*]

urban gardens

Growing plants within and around cities—but with more to it than first appears. The plants are grown within a city with the intent of complementing the urban environment—making the city more natural. The two are not contradictory. For example, *greenbelts* can cool the air and help prevent *soil erosion, rain gardens* prevent water *runoff* that typically occurs in heavily paved areas, and *community gardens* and *urban farms* can reduce *food miles* and thereby reduce the *ecological footprint* of the local residents. [*microgreens and urban gardens; campus quad gardens; gardens, rain; guerilla gardening; hydroponics; vegetable seeds, heirloom; urban farms*]

urban heat island effect

Cars, factories, furnaces, and people in urban areas generate lots of heat. All the asphalt, concrete, steel, and other construction materials absorb and retain this heat. This tendency to generate and absorb heat causes cities to be 5–10 degrees F warmer than the surrounding countryside in the summer. This is called an urban heat island effect. This heat often creates a bubble to form over the city with its own microclimate that traps pollutants, dramatically increasing the level of *air pollution*. [*urban farms; urban forests; urban gardens; Alliance for a Paving Moratorium; handshake buildings*]

urbanization and urban growth

Urban refers to cities and their surrounding areas. Urbanization describes an increase in the percentage of individuals who live in these urban areas. Urban growth simply refers to an increase in the size of an urban population.

Globally, roughly 3.5 billion people live in urban environments. The number of people living in cities has grown from about 15 percent of the total population in 1900 to roughly 30 percent in 1950 and over 50 percent in 2011; it is projected to reach almost 70 percent by 2050.

Because urban areas of most *developed countries* have already expanded to the maximum, most of this urbanization and growth will occur in developing nations—especially Asia and Africa. Some of these cities have expanded so rapidly as to be called *megacities*.

People move to urban centers for many reasons, including the shift from a rural farming society to an industrial society, meaning the jobs are in the city. Today, however, many cities are popular because of the lifestyle they offer.

On the negative side, urban centers often come with a reduction in the *standard of living* for many individuals throughout the world—especially in less developed countries. Increasing the density of a population makes finding or providing food, shelter, services, and jobs more difficult. When areas within cities cannot provide these essential services, they are typically considered *slums*. Increased density and industrialization are often accompanied by increased levels of *air pollution* and *water pollution* and a general decrease in environmental quality.

However, urban areas that have been properly planned, called *smart cities*, can be one of the most environmentally friendly ways for people to live. People in well-planned cities have a small *ecological footprint,* primarily because they drive less or don't need a car at all because they can use *mass transit* and they live in smaller living spaces, such as apartments, and don't have *lawns*. [*eco-cities; infrastructure, green; interim use; megalopolis; mass transit; air conditioning; smart energy; sewage treatment facilities; San Francisco and compostable waste curbside pickup*]

urban livestock

Urban farms or city farms have become popular. Most people think about vegetables and flowers when they think about a farm in the city; however, a growing number of people also want to keep chickens and even goats on their little plot of city space. Backyard chickens can provide fresh eggs and goats can provide fresh milk—as locally produced as you can get—But, unlike plants, livestock comes with potential problems.

Keeping backyard chickens has been banned in some cities, primarily due to noise and sometimes smell. Other towns allow a maximum number of birds, while still others have no problem with as many as you want. Some communities are even overturning previous bans against these birds. Proponents say properly cared for animals are quieter than dogs and won't smell, but neighbors might disagree.

Other reasons exist for banning urban livestock, similar to why you are urged not to buy a bunny at Easter. Many people think creating a small urban farm is a great idea but then abandon their plans—along with their livestock—when they realize the work involved. There are groups that try to help farm animals that have lost their homes, and they say business is booming. Some of these homes for wayward urban farm animals are overwhelmed with chickens and goats from what were probably well-intentioned people who decided urban livestock was a mistake. [*egg farms; dairy cow output; CAFO; factory farms; farm animal cruelty legislation; rendering*]

urban miners

Old electronic equipment—called *e-waste*—such as computers and cell phones comprise the fastest growing segment of *municipal solid waste* on the planet. In 2009, the United States alone disposed of almost 50 million computers. These devices contain many valuable *rare earth metals*. Recycling e-scrap has become big business. It can be done correctly, by trained, qualified employees in a high-tech facility, using solvents and furnaces under controlled conditions and yielding 95 percent of the valuable metals to be used again.

Or, it can be done the wrong way. Old electronic equipment is often illegally shipped off to developing nations, where it is dumped and manually picked through by urban miners—poor locals who get paid close to nothing to go through the junk, picking out components of value for a low-tech removal of metal. These urban miners risk burns, inhale poisonous fumes, and come in contact with carcinogens—a process harmful to the urban miner's health and the environment as well. [*e-waste product life cycle; e-waste and you; e-waste recycling and disposal programs; Basel Convention; beryllium in computers; gold fingers*]

urban open space

Areas in urban settings that have been set aside for *outdoor recreation* are called urban open spaces. This property would often be much more valuable if it were to become developed, so city planners must see the intrinsic need for these spaces and their value to the city's residents. Some of the world's biggest cities planned for open spaces while still young. For example, New York's Central Park was created in the late 1800s and consists of 500 acres of some of the world's most valuable real estate. [*urbanization and urban growth; greenways; National Park and Wilderness Preservation System; national parks (global); nature centers; Nature Conservancy; zoning, land use; suburbia and biodiversity*]

urban sprawl

This phrase is actually a misnomer. Urban sprawl refers to individuals moving out of a city, into surrounding areas, creating what is commonly called suburbia or the suburbs. So it is not really a sprawl, which would mean the city is stretching outward; it is the creation of suburbs, which are entirely different from a city.

In more *developed countries*, many individuals have the wherewithal to move out of urban centers into surrounding areas that are less densely populated and more environmentally pleasing. However, the suburbs with their large lots, large *lawns*, large homes, and required *automobiles* to get back to the city, create a large *ecological footprint*—far greater than that of anyone living in the city.

Today, suburbia and its urban sprawl is considered one of the least environmentally friendly places to live, whereas people living in smart cities, that come about with *smart growth*, are considered the most environmentally friendly.

Some cities try to prevent urban sprawl, but few succeed. One that has succeeded is Portland, Oregon, which has an urban growth boundary to prevent the development of suburbs. [*urbanization and urban growth*]

USDA Organic label
See *food eco-labels*.

U.S. Green Building Council (USGBC)
Founded in 1993, the USGBC is probably the best-known and most respected green building organization in the world—a private, nonprofit, membership-based group that promotes sustainability in building design. They are best known and almost synonymous with their green building *environmental rating program* called *LEED*. They also hold an international exposition on *green buildings*. [*homes, green; eco-cities; small house movement; white roofs*] {ucsusa.org}

utilitarian environmental ethic
An ethic that believes the environment should be used for the greatest good for the greatest number of people. The belief that trees are best used for wood, prairies are best used for *agriculture*, and rivers are best used for *hydropower*. This was the original, primary mission of the *U.S. Forest Service*—not to save *forests*, but to use them on a sustainable basis. [*wise-use movement; dual-use policy; multiple use policy; conservation*]

Valdez Principles
See *CERES*.

vampire energy
Also called phantom energy or load, this is the electricity your electronic devices continue to use even when they are supposedly turned off. Televisions, computers, printers, and many other electronic devices are in a standby state that draws low levels of power even when off. Devices are available that track this use and some even turn the device off automatically. You can lower your electrical bills by monitoring which devices use vampire energy and turning off as many as possible. [*negawatts; energy, ways to save residential; energy-saving apps and programs; domestic water conservation; bathroom water-saving techniques; Energy Star rating; Christmas tree lighting*]

vector
An organism that carries a *pathogen* from one organism to another, causing illness. For example, mosquitoes and ticks are vectors of many human diseases. They pick up pathogens such as a *virus* or *bacteria* by "biting" one individual (called a *host*), and when they bite another individual, they transmit the infection. [*Lyme disease; bedbugs and remedies; bird flu virus; bubonic plague; spillover effect; superbugs; antibiotics for livestock; infectious disease; medical ecology; Millennium Development Goals*]

vegan and vegetarian diets and the environment
People are vegans or vegetarians for lots of reasons, some of which are related to environmental issues. *Meat production* and feeding the world's seven-plus billion people and the energy it takes to do so is one such issue. Almost five pounds of grain must be fed to cattle (instead of directly to humans) to produce one pound of beef for people to eat. A 10-acre farm can provide food for 60 people by growing soybeans, 24 people by growing wheat, 10 people by growing corn, and 2 people by raising cattle. [*meat, eating; vegetarianism; hunger, world; human population throughout history, energy*

pyramid; fruit waxing; pesticide residues on food; animal rights movement and animal welfare]

vegetable seeds, heirloom

The diversity of the foods we once ate was much greater than now. Today, *high-impact agriculture* uses *monocultures* that require one variety of plant. For example, almost 90 percent of all U.S. crop varieties used during the past 100 years are no longer commercially available. More specifically, in 1903, more than 400 pea varieties were for sale in the United States; today there are about 25, and only 2 are routinely cultivated. With monocultures come monotonous foods.

The few varieties we are left with are either created by *selective breeding* or are *genetically modified organisms*. Finding the original, tasty varieties of fruits and vegetables is difficult. But things are changing, because consumers are demanding more choice, and home gardening seed companies are responding by offering the older, vintage, nonhybridized, non-GMO varieties of plants once again. They are now typically called heirloom seeds. Burpee sells about 400 heirloom seed varieties, with *tomatoes* being the most popular. [*heirloom plants; pollination, wild; seeds and climate change, Project Baseline; doomsday seed vault*]

vegetarianism

If someone is a vegetarian, you could call that person a human *herbivore,* a person who does not eat meat. One of the first and probably most famous vegetarians was the philosopher Pythagoras. People who did not eat meat, all the way up to the mid-19th century, were even called Pythagoreans, and they included some of the rich and famous, such as Leonardo de Vinci, Ben Franklin, and Albert Einstein. The term vegetarian was coined in 1847 after a small movement in England and it has stuck since then.

People opt for vegetarianism for many reasons, including health, ethical reasons for the animals and the planet, taste, and simply cost. There are roughly 4.5 million vegetarians in the United States, which is about 2.5 percent of the population. Women are more likely to be vegetarians than men and, Republicans are more likely than Democrats. [*vegan and vegetarian diets and the environment; meat, eating; farm animal cruelty legislation; hunger, world; human population throughout history, energy pyramid; fruit waxing; pesticide residues on food*]

vehicles, hydrogen fuel-cell (FCV)

Hydrogen gas is an experimental source of energy that theoretically has the potential to provide all our energy needs. It burns cleanly, producing only water vapor, and has

2.5 times the energy content of *gasoline*. So if this is the case, why aren't we all driving around in hydrogen cars?

The problem is that this process requires pure hydrogen. Even though hydrogen is abundant, it is not readily available in free form because it is typically combined with other elements to form compounds such as *water* (H_2O). The only way to collect pure hydrogen is to go through some process that separates the hydrogen from the other elements, and this requires energy. If more energy is needed to create a fuel than can be supplied by that fuel, it is impractical as an energy source. To make free hydrogen useful, we must find efficient methods of creating it.

People have been talking about perfecting this form of energy for decades and it always seems to be just around the corner. Previous research focused on an efficient type of *solar cell* that breaks water down into hydrogen and oxygen molecules. However, at the moment, a process called steam reforming looks most promising. It uses *natural gas*, not water, as the source of hydrogen, so we must use a *fossil fuel* to create the hydrogen fuel. Although not a perfect solution, the end result will still be three times more energy efficient than a regular gas-burning *automobile*.

Some of the major auto manufacturers have rolled out FCVs on a test basis in significant numbers. An entirely new infrastructure will have to be developed to accommodate FCVs, such as hydrogen refueling stations. At the moment, the total number of hydrogen refueling stations in the entire United States stands at less than 100. *Electric vehicles*, however, have similar infrastructure problems. [*automobile propulsion systems, alternative; ammonia as a car fuel; natural gas vehicles*]

vertebrates

Animals with a backbone and internal skeleton made of bone or cartilage, as opposed to *invertebrates*, which do not have backbones. [*alligators; amphibian decline; Asian carp; bald eagle; bats; buffalo; California condor; circus elephant ban; dolphins and porpoises, true; elephants; Florida panther; forage fish; gorillas; hummingbirds and flight; lemurs; mammals; manatee deaths and algae blooms; northern spotted owl; passenger pigeon; penguins; peregrine falcon; primate research; sea turtles; sealing; shark finning; smell sensitivity in albatross; stereoscopic smell in moles; tuna; whales; whistle names, bottlenose dolphin; wolves, endangered*]

vertical farms

This refers to one specific type of *urban farm*—nontraditional farms that grow produce and rear fish within the confines of buildings that can be located anywhere, without the need of land. These farms typically consist of *hydroponically* grown plants but can also include indoor *aquaculture*. One such vertical farm found in Chicago, called The Plant, contains a 9000-gallon water system where wastewater from tilapia tanks flows into hydroponic beds to *irrigate* and feed lettuce. The lettuce roots filter the water, which is then returned to the tilapia tanks.

Similar vertical or urban farms are taking root elsewhere. The environmental benefits include reduced *farm-to-table* costs. This technology is only in its infancy, but the potential to help feed the masses over the next decade is worth a great deal of research. [*integrated multi-trophic agriculture; permaculture; low-input sustainable agriculture; sustainable agriculture*]

vineyards and assisted migration

An interesting way to look at the impact of *climate change* is to consider that crops such as wine grapes, with their very particular requirements, will need to move to new regions as the climate warms.

Assisted migration means people, not natural processes, will do the moving. Some reports show that regions now famous for their vineyards will no longer be suitable to grow wine grapes, and new regions, where one would never imagine vineyards, will become the norm.

The report shows that almost half of Chile's wine regions and 60 percent of California's wine-growing regions will no longer be suitable for wine. California's wine-growing region might have to migrate north to the United States/Canadian border. [*assisted migration; oysters and climate change; climate change and biodiversity; climate change and marine life; deforestation and climate change; natural selection; biodiversity, loss of*]

virgin habitat

A *habitat*, such as a forest, that have not been affected by human intervention; for example, a virgin forest. [*wilderness; ancient forests; old-growth forests; Arctic National Wildlife Refuge; Biosphere Reserves; Nature Conservancy; land acquisition, wilderness; land trusts*]

virome

The population of *viruses* found within the *human microbiome*. In most cases, these viruses specialize in infecting the *bacteria* that live within us. Even people with healthy microbiomes have a slew of viruses attacking bacteria. It appears to be the norm. [*hyperparasites; Human Microbiome Project*]

virtual transportation

This term has been used for applications such as Skype and many telepresence programs with which people have meetings—including verbal and visual communications—without the need for actual travel. Of course, this can be considered a form of communications, but it also is a form of *transportation*: virtual transportation. This is especially important when considered from an environmental perspective. Most forms of transportation have a huge negative impact on the environment; virtual transportation eliminates almost all of these impacts.

viruses

All viruses are *parasites* that live inside another organism's cells. They reproduce by forcing their *host* (the cell they live in) to manufacture replicas of the original virus. Many viruses cause disease in humans, including measles, yellow fever, rabies, small pox, AIDS, HPV, Ebola, and influenza. [*hyperparasite; virome*]

visible light spectrum

That part of solar radiation that is visible to the human eye—wavelengths between about 400 and 800 nanometers. Other animals are capable of seeing wavelengths beyond what humans can see. *Insects* can see down to 300 nanometers, meaning they can see ultraviolet light, and rattlesnakes can see up to about 850 nanometers, meaning they can see into the infrared range of light. [*photosynthetically active radiation*]

visual pollution

Visual *pollution* is one form of *aesthetic pollution*—highly subjective forms of pollution. Almost everyone would agree that an open *landfill* or *garbage* strewn on a *beach* is *visual pollution*. Most would call billboards lining an otherwise beautiful countryside visual pollution, but not the people advertising their products or services on the boards. Local and state governments often regulate what is and is not a form of visual pollution—hopefully, based on the preferences of the constituents who must live with the sights.

vitrification

Also called glassification, a technology that vitrifies *nuclear wastes* (turns them to glass) with heat and fusion. This is the planned method to dispose of the *nuclear waste* materials located at the *Hanford Nuclear Reservation*. The facility was originally scheduled to open in 1999. It has yet to begin operations as of 2013. [*nuclear waste disposal*]

viviparous

Animals that produce live young from the body of the maternal parent. Humans are viviparous. [*ovoviviparous; r-strategists; K-strategists*]

vivisection

Performing surgical procedures on live animals, usually for research. In 2010, more than 21,000 primates, mostly monkeys, were imported into the United States for scientific research involving vivisection. Most of them die after their usefulness in the lab has ended. Many people believe this procedure to be inhumane and unnecessary, and advocate for legislation to stop the practice. Some major research laboratories have begun reducing their use of vivisection. [*primate research; farm animals and animal cruelty legislation; dissecting frogs in schools; animal rights movement and animal welfare*]

volatile organic compounds (VOC)

A collective name given to *organic* pollutants that are gases at room temperature. These gases are emitted from certain liquids or solid substances. Many VOCs are emitted from common household and garden products, such as paints, lacquers, paint removers, cleaning products, *pesticides*, carpets, copiers, printers, and many glues and adhesives. Some of the most common VOCs include *formaldehyde*, benzene, and xylene. These products are primarily used indoors, so they are often responsible for *indoor air pollution*.

Electrical equipment and some *plastics* emit another type of VOC, called polychlorinated biphenyls, or *PCBs* for short. If improperly disposed of, these products release these PCBs. The PCBs then can either enter water supplies, causing *water pollution,* or become airborne and enter the body, where they accumulate in the fatty tissues in a process called *bioaccumulation.*

Some VOCs cause symptoms similar to a common cold, including some respiratory ailments. Others, such as PCBs are *carcinogenic* and a serious threat. Many VOCs have never been thoroughly tested; therefore, little is known about their inherent risks.

Alternatives depend on which product we are talking about. Many of the indoor products can be found in "VOC-free" forms or will specifically note the type of VOC—such as formaldehyde-free. [*persistent organic pollutants; POPS, dirty dozen; hazardous air pollutants; body burden*]

volcano

Openings, usually called vents, in the Earth's crust where lava, gases, and ash escape, often with enormous force. They typically form along the boundaries of the Earth's tectonic plates. One such boundary, called the Ring of Fire, is the most active region for volcanic activity. Volcanos are classified as active, dormant, or extinct. [*plate tectonics; supercontinents; Pangaea; aerosols; carbon cycle; nanoparticles; sea gliders*]

volunteer research, scientific

This is similar in some ways to crowd research, but more formal. Anyone can participate in crowd research if you are in the right place at the right time. However, scientific volunteer research pertains to scientists volunteering their time to help out other scientists. Usually these are quick and simple experiments with simple, straightforward data collection needs, so volunteers need not spend a great deal of time getting up to speed before being able to help out another scientist.

For example, a collaboration network called NutNet has scientists from 12 countries volunteering at 68 sites to run field experiments and collect data on how *grasslands* respond to *climate change*. Data collection is done through an online database, making the entire process quick, easy, productive, and free of funding needs. [*crowd research, scientific; voluntourism, environmental; WWOOF*] {nutnet.umn.edu}

voluntourism, environmental

People who want to combine their vacations with some *environmental* cause or some form of ecological research can go on voluntours. For example, volunteers joined scientists on a trip to Malaysia to monitor the health of the *coral reefs*, fish, *sea turtles*, *sharks*, and *dolphins*. They have a good time and benefit our planet as well. [*ecotourism; ecotravel; volunteer research, scientific; crowd research, scientific; WWOOF*] {ecotourism.org/voluntourism}

vulnerable species

See *threatened species*.

walking

An alternative form of *transportation* and a form of exercise. The statistics will surprise few people, but if one in every ten Americans started a regular walking program, the United States could save $5.6 billion in health care costs. And walking is not only good for our health—it is also good for the *environment*. Every trip made by walking instead of burning *fossil fuels* reduces *air pollution* and *carbon emissions*.

While most of us can always find a reason for not walking, part of the problem is a lack of places to walk. Some regions are better than others. *Greenways* and trails of all sorts have become popular, improving the desirability of communities. Advocacy groups work to increase the number and accessibility of places to walk. [*bicycles and bicycling; bicycling cities, ten best; Rails-to-Trails Conservancy; smart growth; fregan; zero carbon*]

Ward, Barbara

(1914–1981) A British economist and author, Ward was one of the first to discuss the concept of *sustainable development* and used the metaphor *Spaceship Earth*. She published Spaceship Earth (1960) and Only One Earth (with *Rene Dubos*), written for the 1972 *United Nations Conference on Human Environment*.

warming hole

From 1912 to 2011, the United States showed a general gradual warming trend. But a few states appeared to warm considerably more slowly than the rest of the country. This region has been called the warming hole and includes Arkansas, Georgia, and Alabama. One theory is this was caused by an *afforestation* project in the mid-1990s that might have resulted in more *carbon dioxide* being absorbed, lessening warming trends. Scientists point out that whatever the cause was has stopped working and the states in the warming hole are now warming at the same pace as the others. [*climate change*]

waste-conscious product development

See *extended product responsibility*.

Waste Isolation Pilot Program (WIPP)

Carlsbad, New Mexico, was designated by Congress to be the first location for a permanent *nuclear waste disposal* site. It consists of a network of caverns carved into the geological formations of salt found deep beneath the surface. The site accepts only transuranic nuclear wastes, which are solid materials such as tools, instruments, and building materials that are *radioactively* contaminated. Most of these wastes come from *nuclear weapons* production facilities around the country. The site began accepting nuclear wastes in 1999 and has received 9000 shipments, totaling about 2.5 million cubic feet of wastes.

This facility does not accept nuclear waste disposal of spent fuel rods, which is still in need of a permanent repository. However, WIPP is viewed as the only permanent nuclear waste disposal site that exists on Earth, so it is the model for how it can be done. [*Yucca Mountain; cooling ponds and towers; plutonium pit disposal; nuclear waste dilemma, the global*]

waste minimization

Reducing the volume of waste prior to its release into the environment. Instead of worrying about cleaning up wastes after they have been created (*end-of-pipe technologies*), waste minimization refers to reducing the need to use raw materials in the first place. It typically includes *source reduction* (using less to begin with) and *recycling*.

An example is using the least amount of chemicals possible during a manufacturing process and also collecting and reusing these chemicals for the same or possibly another process. The emphasis of waste minimization is to use *clean technologies* in the first place, so there is less need for end-of-pipe technologies—cleaning up after the environmental mess is made. [*recycling; disposal fee; bottle bill*)

waste stream

The flow of solid waste from its inception of becoming waste, such as being placed into a *garbage* container, to its final resting place in a *landfill* or an *incinerator,* is called a waste stream. About one-third of all U.S. *municipal solid waste* is comprised of *paper* and paper products. Waste destined for *recycling* or *composting* is still considered part of the waste stream. Another possibility for the final destination of the waste stream is to be used in a *waste-to-energy power plant*.

waste-to-energy power plant (WTE)

Power plants that incinerate waste products such as *municipal solid waste* to create electricity are called waste-to-energy power plants. Using solid waste, which is one type of *biomass energy*, reduces the need for *landfills* and converts the energy available within the waste into electricity. State of the art WTE facilities use intense heat to burn the waste thoroughly, destroying many of *the toxic substances*, often found in waste. They have sophisticated *air pollution* devices to minimize the contaminants that might be released with the *emissions*.

There are more than 800 WTE power plants operating in about 40 countries. But this remains only 11 percent of all the municipal solid waste generated worldwide. Japan has been a leader in this technology with more than 300 WTE power plants and consumes about 40 percent of their solid waste in this manner. WTE facilities are also common in Europe. The United States has about 87 of these facilities generating about 2500 megawatts of power, but this is only about 0.3 percent of our total energy supply. [*electric power plants; landfill gas-to-energy power plants*]

wastewater-to-energy power plants

Theoretically, *sewage treatment facilities* could generate power at the same time they clean our wastewater. If the facility could extract out the *organic* matter and the *nitrogen* found in wastewater, these materials could be sent to a separate part of the facility and used to generate power. Making the process work is not as difficult as making it economically feasible. [*electric power plants; biosolids and wastewater*]

water

Water is our planet's most abundant and important resource. Estimates place the total amount of water in all forms in the *biosphere* at 360 billion, billion gallons; it is no surprise that Earth is called the water planet.

More than 97 percent of all water is found in the *oceans*, about 2 percent in the ice sheets, 0.5 percent in *groundwater*, a mere 0.02 percent is in all of the *surface waters* (streams, rivers, ponds, lakes, reservoirs), 0.01 percent is in the *soil*, and even less is located in the *atmosphere*.

Water covers more than 70 percent of the Earth's surface and is a major force in controlling the climate by storing vast quantities of heat. It is essential to all forms of life; most of which are composed primarily of water.

Water passes through the hydrologic or *water cycle*, which constantly replenishes the *freshwater* supply that most life depends on. In spite of the vast quantity of all water, freshwater is in short supply in much of world including many parts of the United States where it is used heavily for *agricultural irrigation*.

The small amount of the water available for human consumption is found in groundwater and surface water and both are being used to their limits in much of the world. Groundwater *aquifers* supply much of the world's freshwater. However, it is being removed far more quickly than it can be replaced—a process called *groundwater mining*. *Water pollution* of the surface waters and contamination of many groundwater aquifers has made much of the freshwaters unfit for human consumption. Even though we live on the water planet, freshwater—the essence of life—is scarce and becoming more so. [*water shortages, global; water conservation; groundwater pollution and depletion; running water habitats, human impact on; drinking water; Kenya water find; Ogallala aquifer; water use for human consumption*]

water conservation

Water pollution and *water shortages* are considered by many to be one of our biggest environmental and public health challenges. The U.S. population has doubled over the past 50 years, but our use of water has tripled and at least 36 states now have water shortages. It is a far worse of problem in other regions of the world, where they have less water to begin with when compared to the United States.

Water conservation—saving the water we have available to us—must become a major commitment made by nations, states, the agriculture sector, and individuals.

People use water in one of four ways, each with its own set of potential water conservation measures. *Domestic water* use has its own set of *domestic water conservation* measures that anyone can use to lessen their water *ecological footprint*. *Irrigation* (agricultural use), however, uses the largest amount of water and requires the most attention from a global perspective. There is also *industrial water use* and *in-stream water use*.

water cycle

Water passes through a *biogeochemical cycle* as it passes through the *biosphere*. The energy from the sun drives this cycle by causing water from the Earth's surface to enter the *atmosphere*. It does this by evaporation—rising of vapor from bodies of water and the moisture in the soil—and also by *transpiration*—moisture in plants exiting into the air. Warm air carries the moisture until the air cools, causing the moisture to form water droplets that return to the Earth as precipitation. The water falls back into bodies of water, is absorbed by soil and reabsorbed by plants, or percolates into *groundwater*. If the soil cannot absorb the moisture, it runs off into streams and *rivers* and finally reaches the *ocean*. The water cycle acts like a filtration system, purifying and removing salts from the water to produce the three percent of all water that is fresh and available for organisms to consume.

water diversion

The flow of *water* may be physically diverted from an area with an abundance of water to an area where water is in short supply. Many of these projects are successful and pose little environmental threat compared to the benefits they provide. However, water is scarce in many agricultural regions, so legal battles are fought over whose water gets diverted elsewhere. For example, Northern and Southern California have feuded for many years over who gets the water that flows through the state's many aqueduct systems. Water diversion projects have also resulted in environmental disasters, such as the old Soviet Union's *Aral Sea*. [*water shortages, global; Colorado River*]

water grabs

See *land grabs*.

waterlogging and soil salinization

When soils become salty due to *soil salinization*, farmers often flush the land with water in an effort to remove the excess salts and make the soil productive again. However, in areas where layers of soil are exceptionally impermeable and the water cannot drain properly, the soil becomes waterlogged. The saline water remains, flooding the roots of plants and killing them.

water mining

See *groundwater mining*.

water pollution

Water *pollution* occurs when the natural quality of water is degraded in some way. This degradation results in damage or destruction of *aquatic ecosystems* and/or makes the water resource unfit for human consumption. Both *surface waters* and *groundwater* can become polluted. Water pollution is caused in numerous ways.

It is caused by chemicals and substances that are dumped into bodies of water, such as *heavy metals* and acids from *industrial waste*. Nitrates and *phosphates* from synthetic *fertilizers* and toxic substances from *pesticides* that *run off* into bodies of water, cause water pollution. Many oil-based products, such as *gasoline* and *plastic* products that make their way into bodies of water, also cause water pollution.

It can be caused by organic wastes—including *food waste* or human waste that makes its way from *sewage* into the water supply. This food or raw sewage harbors and encourages the growth of disease-causing microbes. Organic waste dramatically increases the abundance of *bacteria,* which use up all the available oxygen in the water and cause *hypoxia* and *dead zones*.

Radioactive wastes can also cause water pollution. When radioactive materials are produced, used, or disposed of, they can enter bodies of water, killing aquatic organisms or causing other long-term effects. Water used to cool manufacturing processes and *electric power plants* cause *thermal water pollution*. The water becomes intensely heated and when re-released, it damages ecosystems by changing the normal temperatures and lowering the oxygen levels.

Finally, excessive *sedimentation* from human-caused *soil erosion* damages or destroys aquatic ecosystems and makes the water less suitable for human consumption. [*groundwater pollution; toxic pollution; plastic pollution; hazardous waste; e-waste; persistent organic pollutants; acid mine drainage; biosolids and wastewater; sewage treatment facilities; CAFO; combined sewer systems; effluent; Clean Water Act; geothermal energy; industrial water pollution; landfill problems; ocean pollution; rivers in decline; running water habitats, human impact on; ship-breaking industry; textile industry; urbanization and urban growth; deep-well injection sites*]

watershed

The land surrounding a lake or river, also called catchment area or drainage basin. This area is responsible for most of the water entering the lake or river. Precipitation falls on the watershed, which in turn delivers the water to the lake or river as *runoff*. The region's size, shape, and vegetation are responsible for the amount and type of water to enter the body of water. For example, watersheds in forested regions deliver water that is rich in *organic* matter, but only after the trees have soaked up their share first. Bare rocks on a mountain side will deliver water with few organic nutrients, but rapidly as the water rushes over the rocks through the watershed into the body of water. [*freshwater ecosystems; rivers in decline; running water habitats, human impact on; urbanization and urban growth*]

water shortages, global

About 1.2 billion people live in areas where there is a physical water scarcity, meaning a shortage of clean water on a regular basis. Another 1.6 billion live in areas where there is an economic water scarcity, meaning even if the water is in the area, infrastructure does not exist to deliver the water to these people. Regions with the least available water are the Arab and sub-Saharan nations.

The vast majority of all water consumption—over 70 percent—is for *agricultural irrigation*. Much of this water is used inefficiently. Irrigation diverts *surface waters* and causes *groundwater mining*. Much of this agricultural water use is to grow crops—not for people but for livestock to then feed to people— an incredibly wasteful process. It takes fifteen times as much water to produce one pound of beef as it does to produce one pound of corn. Or, put another way, with the same amount of water, you could feed 15 pounds of corn to people but only one pound of beef to people.

The future of water use does not look good. A rather frightening statistic shows water consumption has been growing twice as fast as *human population growth* over the past 100 years. [*water pollution; water use for human consumption; water conservation; domestic water conservation; drinking water; groundwater pollution and depletion; meat, eating; Ogallala aquifer; comprehensive water management; desalination; aquifers; Australia; China, People's Republic of; Kenya, water find; drip irrigation; toilet-to-tap drinking water; water grabs*]

water table

The upper portion of the *zone of saturation* that is readily accessible for human consumption. [*water; aquifers; groundwater; groundwater mining; groundwater pollution and depletion; drinking water; water use for human consumption*]

water treatment

The process of making water suitable for human consumption. Several methods are used: with *sedimentation*, solids settle out; with *flocculation*, masses called floc accumulate and are separated out; with filtration, contaminants are filtered out by passing the water through sand or similar material; and with disinfection, chemicals such as chlorine are used to kill pathogens. [*drinking water; water use for human consumption; chlorination; sewage treatment facilities; water shortages, global*]

water use for human consumption

Water is used for human consumption in four ways: 1) *domestic water use*, 2) *agricultural irrigation*, 3) *industrial water use*, and 4) *in-stream water use*. About 70 percent of the water used worldwide is for irrigation.

watt and kilowatt hours

Electric power is measured in units called watts. A watt is equal to one *joule* per second. The total generating capacity of an *electric power plant* is measured in kilowatts (Kw) for 1000 watts and megawatts (MW) for one million watts. For example, the Grand Coulee Dam produces electricity by *hydroelectric power* and creates more than 6100 MW of power as compared to some small *wood-burning* power plants (*biomass power*) that produce only about 250 MW of power.

The ongoing quantity of electricity produced is measured in kilowatt hours (kWh) and is often used to compare the costs of various electricity production methods. For example, *nuclear power plants* produce electricity at about 12.5 cents per kWh, *coal* at about 5 kWh, *natural gas* at about 4 kWh, *wind power* at about 5 kWh, *hydroelectric power* between 5 and 10 kWh, and *solar power* between 15 and 30 kWh.

wave power
A few small experimental wave-power plants have been built that capture the energy from the motion of waves. Although limited to coastal regions, this form of power shows promise for further development. A test site is located off the coast of Tuscany, Italy, and others are planned for *India* and the United Kingdom. [*alternative energy sources*]

weapon dumps, abandoned
Abandoned landmines and weapon dumps litter many countrysides after wars have ended. Landmines have received a great deal of attention since the end of the Vietnam war, but weapon dumps are prevalent and more deadly. They have killed 4600 people since 1995, when record-keeping of such things began. In 2012, 442 people died from 46 explosions of abandoned weapons dumps in the Middle East, Africa, *India*, and across the old USSR states, among others. The dumps are accidentally set off by inno-cent people unfortunate enough to come upon them. They also pose a risk that terror-ist organizations might gain access and control of weapons that are still in working order. [*explosive remnants of war; Agent Orange and the Vietnam War*]

weathering
The physical, chemical, or biological wearing of rock that contributes to *soil* forma-tion. [*erosion; soil erosion; topsoil*]

weather stripping
Most old homes and some new homes have cracks, holes, and spaces in the struc-ture that allow warm air to escape in the winter and cool air to escape in summer. These air leaks account for about 35 percent of the total lost energy from a typical house. The easiest and most cost-effective method to minimize this lost energy is with weather stripping. Weather stripping consists of narrow bands of metal, vinyl, rubber, felt, or foam that is applied to the openings. It is usually applied to joints between var-ious structures in the house where the leaks occur, such as window and door joints, wall joints near the foundation, joints around a fireplace, attic openings, and open-ings around electrical outlets and service boxes. [*thermal insulation, building; R value; heating, ventilation, and A/C system; air conditioning (A/C); weather stripping; super*

insulation; indoor air pollution; indoor ecology; LEED certification; buildings, green; homes, green]

wedding dresses, reused

Almost anything can fit into the *three Rs* adage of reduce, reuse, recycle. *Recycling* gets most of the attention, but reuse is becoming more popular—even for products you would never think about. If websites are any indication of popularity there is a boom in the reuse of wedding dresses. Recently, about a dozen websites specializing in gently used wedding dresses have sprung up. Besides the obvious environmental advantages of things being reused, there is the cost savings.

However, reusing wedding dresses is simply an example of how products of all sorts can be reused or repurposed and how it can play a role in *waste minimization* and the overall *solid waste* disposal problem. [*upcycling; planned obsolescence; cradle-to-cradle; wrapping-paper alternatives; hotels, green; styropeanuts; zero-waste*]

weed

Any plant growing where humans don't want it to grow. One person's weed might be found cultivated in another person's garden. [*pest; pesticides; herbicides; genetically modified organisms (GMOs)—you routinely eat; microgreens and urban gardening; rain gardens; urban gardens; community gardens; straw-bale gardens; Dirty Dozen – fruits and vegetables; heirloom plants*]

weird life

There is a general belief among scientists that all life on our planet had a single common ancestor and *natural selection* took care of the rest. That "last universal common ancestor," called LUCA, was probably some sort of microbe that started life about 3.5 billion years ago. Every form of life that evolved thereafter was made of the same stuff. A few scientists, however, believe life possibly formed on our planet more than once. This double-genesis belief could mean that living among us are some species with totally different building blocks, or genetic matter. A few even believe an entire, shadow *biosphere* could exist that consists of weird life.

If so, the obvious question is why haven't we found them. The thought is that possibly they live in regions so remote and inhospitable that even *extremophiles* would dare not tread. They may live in areas near *hydrothermal vents* no one has yet explored. Only time and science will tell. [*genome; DNA sequencing technology*]

wet deposition

See *acid rain*.

wetlands

Wetland is a collective term for a number of *habitats*, including *marshes*, *swamps*, and *bogs*. Wetlands are divided into two types—*inland wetlands* and *coastal wetlands*. At least one of the two can be found in all 50 U.S. states. These habitats contain some of the most productive and useful *ecosystems* found anywhere. Wetlands are to life, what farms are to man. They produce huge volumes of food in the form of plants, both alive and more importantly dead and decomposing as *detritus*. This *biomass* acts as the first level of a *food chain* that feeds everything from small *invertebrates* to large *predatory* fish such as striped bass.

Acting like filters, wetlands maintain the quality of water in rivers and streams by removing and retaining *nutrients*, and helping to decompose waste. They also reduce the amount of *sedimentation* that occurs in rivers and streams.

Wetlands also minimize the effect of flooding by storing huge volumes of excess water, which protects all forms of downstream wildlife. They also buffer the shoreline from *erosion* and protect against *coastal flooding*.

Unfortunately, wetlands were viewed as wasteland, of little value, as late as the early 1970s. Because of this misconception, vast swaths of our wetlands were destroyed within the last few decades. Less than half the number of wetland acres exist today than in the 1600s. From the mid-1950s to the mid-1970s, more than 11 million acres of wetlands were destroyed, with most drained for agricultural use. The remaining wetlands are now protected under sections of the *Clean Water Act* but have been under attack once again in recent years by proponents of development.

The *biodiversity* of wetlands are among the most abundant in the world, similar in *net primary production* with *tropical rainforests*. Many species are unique to these habitats. Inland wetlands are inhabited with numerous species of freshwater fish and wildlife. Ducks, geese, and many songbirds feed, nest, and raise their young in wetlands and most recreational fish spawn in wetlands. Many *mammals* also feed and live in inland wetlands.

Coastal wetlands are habitats for estuarine and marine fish, shellfish, waterfowl, and many birds and mammals. Most commercial and game fish use coastal marshes and estuaries for spawning. [*estuary and wetlands destruction; Convention on Wetlands*]

whale and commercial ship collisions

The leading cause of *whale* deaths worldwide appears to be colliding with commercial ships. The whales sustain severe injuries, usually from the ship's large propellers. Whales' migratory paths often cross shipping lanes, resulting in disaster for the *mammals*. Photos of blue whales with their tail fins almost entirely cut off illustrate their sorry fate. Sri Lanka appears to be one of the worst locations for such kills. Fifteen miles off its coast is one of the busiest shipping lanes in the world, and it is heavily populated with whales. It is thought that the popularity of whale observation boats is forcing whales to feed farther from shore, directly in the shipping lanes. [*whaling*]

whale earwax

Scientists are using interesting but strange techniques to study our environment's recent past. We are familiar with the use of tree rings to tell us a story about the past of an old tree. The earwax found in *whales* is being treated much like tree rings to study the presence of toxic chemicals in our *oceans*.

Earwax removed from recently found dead whales is being used to determine the presence and concentrations of toxic substances in the water during the life of the whale. As the wax builds over the years, toxic substances are embedded in the wax. Cross-sections of the wax plug, much like cross-sections of a tree trunk, can reveal the amount of these substances chronologically. In a recent study, the *persistent pesticide chlordane* was found in a four-inch-long wax plug, showing a gradual decline over the years, probably because of chlordane's being banned in the United States and other nations. They also found *PCBs* present and consistent throughout the whale's lifespan. PCBs were not banned internationally until 2004. [*toxic pollution; persistent organic pollutants; timber, DNA sequencing of*]

whales

A group of marine *mammals* that belong to the order Cetacea, with flippers and horizontal tails called flukes. They are believed to be highly intelligent organisms, similar to *dolphins*. They include humpbacks, right whales, grays, and bowheads, among others. Whales have been hunted for hundreds of years for their food, bone, oils, and other products of value to humans. There have been international agreements to stop or limit *whaling* and there are *NGOs* dedicated to saving whales.[*whale earwax*]

whaling

Hunting *whales* for food and oil has been practiced for hundreds of years. The Inuit, Europeans, Americans, and Japanese have been heavily involved at one time or another. Long ago, the numbers hunting and the crude devices used caused little harm to whale populations. But increased numbers of those in the pursuit and better technology began driving these *mammals* towards extinction in the 20th century.

The first attempts to control whaling started in the 1930s but had little impact. In 1946, the International Whaling Commission (IWC) was formed and pushed for a sustainable catch of whales, but it to could not control the wanton destruction—mostly by Japan. In the 1960s, Japan killed more than 25,000 whales each year.

Illegal hunting was also pervasive and when exposed, helped the public see the true magnitude of the problem. By the late 1970s, many environmental and activist groups, including *Greenpeace* and later the *Sea Shepard Society* captained by Paul Watson, went on an all-out attack to stop the practice. But even with this anti-whaling activism, many species' numbers continued to decline.

In the early 1980s, the IWC realized the dire state of affairs and changed their position from that of sustainability to one of outright protection for whales. They imposed a global ban on whaling in 1986 that exists to this day. However, Japan, Iceland, and Norway have always defied the ban. But some of the worst whaling continues with the blessing of the ban. The ban allows whales to be killed for scientific research, which has been used as an excuse by Japan to continue killing on a large scale.

The Sea Shepard Society continues to thwart and expose whaling with a popular Discovery Channel show called Whale Wars.

Not all species of whale are experiencing population declines. There has been a push to remove some species from the ban. However, others believe that lower populations is not the main issue and, for ethical reasons, the killing of such a magnificent, intelligent animal should not be allowed. [*dolphins and porpoises, true; factory ship; marine ecosystems; fishing, commercial; Marine Mammal Protection Act; marine protected areas; seismic air guns, whales and; sonar testing, underwater; whale and commercial ship collisions; animals experiencing grief; biodiversity, loss of; sealing*]

whistle names, bottlenose dolphin

Research shows that bottlenose dolphin appear to recognize one another by unique whistles. These whistles can be thought of as the individual's name. The dolphins appear to remember those they have not seen in 20 years by their whistle. [*smell sensitivity in albatross; stereoscopic smell in moles; animals experiencing grief; bumblebee senses; camouflage, squid; communications between plants; hummingbird and flight; whistle names, bottlenose dolphin*]

White, Gilbert

(1720–1793) White's most famous book, <u>The Natural History of Selborne</u>, published in 1789, has taken on cult-like status among some environmentalists—especially in his homeland of England. He was one of the first to observe the relationships between all species, large and small, and the importance of each. He is viewed as someone who led the way for the next generation of nature writers, all of whom helped set the stage for environmentalism. [*philosophers and writers, environmental*]

whitecoats

See *sealing*.

white-nose syndrome (bats)

Bats have been behaving oddly in recent years—leaving their caves when they should be *hibernating*—burning up their stored food supplies and succumbing to the elements. It has become what some call a bat plague. In 2006, cavers in New York found bats with their noses covered in a white *fungus*. Since then, 5.5 million bats are believed to have died from this disease, called white-nose syndrome after the fungus that grows around the nose, mouth, wings, and body. It has now spread to 19 states and Canada and has infected seven species of bat.

Bats do not do well on the "cute scale" and have never garnered much sympathy from people. However, they are integral parts of many *ecosystems*, responsible for keeping *insect* populations in check. One bat eats 1000 insects per hour and the equal of its total body weight each evening. Without bats, much more of the food we eat would be eaten by insects. Plus, bats are responsible for a great deal of *pollination*. Estimates already have the value of the crops lost due to a reduction in bats and an increase in insect pests at $3.7 billion dollars. [*flagship species; Ugly Animal Preservation Society; chytrid fungus; Amphibian Ark project; colony collapse disorder; manatee deaths and algae blooms*]

white roofs

The *urban island heat effect* makes cities considerably hotter than outlying areas in the summer. This occurs for many reasons, but the most obvious one is heat absorption by the many dark colors found on city pavements and rooftops, as opposed to natural settings with far more plants and soil. This seems obvious, but most office building rooftops are black, increasing the temperature outside and making the *air conditioning* units work harder inside. Studies show painting roofs white reduces the average temperature by over 40 percent in the immediate area.

If less air conditioning is needed to cool, less *greenhouse gases* are produced, helping cities meet new, stricter emission standards. It is strange that it took research dollars to figure this out. [*eco-cities; urbanization and urban growth; megacities; smart growth*]

Whitman, Walt

(1819–1892) One of America's greatest poets. His collection, <u>Leaves of Grass</u>, published in 1855, is considered one of America's finest works of literature. However, at first he considered it a failure. His spirits were bolstered when *Ralph Waldo Emerson*— who strongly influenced Whitman—praised the work. Whitman continued to publish his collection, adding and dropping poems each time. His poems celebrate the individual, freedom, and the natural world.

During the Civil War, Whitman became a nurse for the Union Army in Washington, D.C., and spent most of his time in hospital wards helping the wounded. Here he met *John Burroughs*, beginning a long friendship. Burroughs wrote the first biography of Whitman. Many of Whitman's poems at that time, such as "Drum-Taps," turned to war and death. By the end of his career, he had attained respect as a great literary figure. [*philosophers and writers, environmental*]

Wide-Field Infrared Survey Explorer Satellite (WISE)

A satellite put into orbit in 2009 by NASA that scans the heavens for objects from 300 feet across to about a half-mile across. This helps scientists identify and track almost all of the near-earth asteroids. The hope is that if we know our planet is going to be hit by a large object, we would have time to do something about it—not that anyone at this time has any idea what to do. [*near earth objects; panspermia; extinction and extraterrestrial impact; outer space; space debris; Blue Marble photograph, original and update; Earthrise photograph; Landsat*] {wise.ssl.berkeley.edu}

Wild and Scenic Rivers System

A 1968 act created the Wild and Scenic Rivers System to protect rivers or portions of rivers deemed to have wild and/or scenic values. The act is managed by the *Bureau of Land Management*. Rivers in this system are to be kept free of development of any kind, with the only activities allowed to be camping, swimming, nonmotorized boating, sport hunting, and fishing. It encompasses 69 wild and scenic rivers found in 7 states, including more than 2400 river miles. Rivers are divided into categories of wild, scenic, and recreational. [*Blueways, National; national parks (global), National Park Service; National Park and Wilderness Preservation System*]

Wild Bird Conservation Act

This act, managed by the U.S. *Fish and Wildlife Service*, created laws to protect wild populations of exotic birds from the international pet trade; however, it only protects birds listed by *CITES*. This means many other species continue to be captured and sold, resulting in hundreds of species threatened with *extinction*. For example, several species of parrots are endangered because they are not on the CITES list. [*endangered and threatened species lists; biodiversity, loss of; wildlife trade and trafficking*]

wilderness

The United States Congress defined wilderness in the *Wilderness Act* of 1964 as "an area where the Earth and its community of life are untrammeled by man, where man himself is a visitor and does not remain." [*National Parks and Wilderness Preservation System*]

Wilderness Act

This act was created in 1964 to provide permanent protection to millions of acres of *wilderness*. Once designated, these areas cannot have any permanent structures or roads. Protecting these natural *habitats* protects the flora and fauna of these areas. Today 110 million acres, or roughly 5 percent of the nation, is set aside as wilderness, with most in Alaska and western states. In addition to the obvious advantages to wildlife, these lands are guaranteed to remain wild so future generations can enjoy them for camping, hiking, and other recreational activities. [*National Wildlife Refuge System; National Park and Wilderness Preservation System; national parks (global); outdoor recreation; Arctic National Wildlife Refuge; Bureau of Land Management; National Petroleum Reserve – Alaska*]

Wilderness Society

The Wilderness Society's mission remains the same since it was founded in 1935 by the famous conservationist and writer *Aldo Leopold,* who fostered the belief that land must be viewed, not as a commodity, but as a valuable resource. This *NGO* uses science, analysis, and advocacy to protect *wilderness*. Their track record is admirable, having led the effort to permanently protect more than 100 million acres of wilderness in 44 states. One of their signature accomplishments was helping to pass the 1964 *Wilderness Act*. The society has 400,000 members. {wilderness.org}

wildlife

All the uncultivated vegetation and nondomesticated animals living in an area. [*wilderness; wildlife management; natural resources*]

Wildlife Conservation Society

This *NGO* is headquartered in and closely associated with the Bronx Zoo in New York City. Founded in 1895, it is one of the oldest *conservation* societies. Its conservation efforts extend around the world in over 50 countries. {wcs.org}

wildlife management

Maintaining or promoting the survival of *wildlife*. Management might pertain to one species or to all the species in an area. If it pertains to all wildlife in an area, the goal might be to maintain natural populations against human intervention. However, it might pertain to managing one species of animal for a specific purpose, such as hunting.

Wildlife management involves studying and understanding all aspects of the organisms' *habitat*—food and water requirements, types of protection or cover from the elements and from *predators*, and many other factors. It also includes understanding the overall *population dynamics* of the area.

Once these studies are complete, decisions can be made about how best to manage the wildlife. The methods used can range from simply leaving the natural populations alone to introducing additional numbers of some species or reducing the numbers of other species.

Managing wildlife often includes habitat management. This may involve building structures that provide cover from predators, or destroying some part of a habitat to encourage or discourage a certain population. An example would be cutting mature trees to allow new saplings to grow and provide food for deer. Habitat management of some animals, such as *migratory birds,* has required the establishment of international agreements, because many of these birds can have habitats in three countries each year as they migrate along their *flyways*. [*forest management; integrated; comprehensive water management planning; conservation; fish populations and food; Magnuson Fishery Conservation Management Act; National Marine Fisheries Service; sustainable management; wildlife trade and trafficking*]

wildlife refuge

See *National Wildlife Refuge System*.

wildlife trade and trafficking

Wildlife trade is the sale of any wild plant or animal, or parts of either, but it usually refers specifically to the international sale of exotic pets. The term includes both legal and illegal transactions. However, wildlife trafficking specifically refers to the illegal trade of wildlife. For example, the *ivory trade* is a form of wildlife trafficking commonly called *poaching*.

The legal wildlife trade is a multibillion-dollar business. Sales just to the United States in 2009 included almost 150,000 mammals, almost a million *reptiles*, more than three million *amphibians*, close to 200,000 birds, and more than 165 million fish.

The numbers for illegal trade are hard to come by but assumed to be far greater than for the legal trade.

Illegal trafficking obviously has no restrictions, and the damage done to wildlife is rampant. But even the legal wildlife trade is poorly enforced and results in serious damage to *ecosystems* as well. The wildlife trade and trafficking are major contributors to *biodiversity loss* and the dramatic rise in the number of *endangered species*.

Regulating the U.S. wildlife trade is piecemeal at best, because different agencies regulate different aspects of the problem. The Centers for Disease Control is in charge of any aspect that impacts public health. The Department of Agriculture is responsible for livestock, poultry, and wild birds. The *U.S. Fish and Wildlife Service* handles the smuggling of wildlife such as endangered species and exotic pets. This patchwork causes inefficiencies, resulting in poor enforcement.

On a global scale, *CITES* is the primary international agreement trying to control the wildlife trade. [*whaling; sealing; polar bear trade; elephant poaching; rhinoceroses and their horns*]

Wilson, E. O.

(1929–) A world-renowned scientist (*entomologist,* to be exact), previously at Harvard University, who has won two Pulitzer Prizes. He is best known for his studies about evolutionary biology—particularly *sociobiology*—a discipline he brought to the forefront of science in 1971 with his work <u>The Insect Societies</u>. He extended the study of sociobiology to include humans and shared it with the world in his <u>On Human Nature</u> in 1978, for which he won a Pulitzer. He is noted for his extensive work on *biodiversity loss* and outspoken about what must be done to stop it. In his book <u>Biophilia</u>, he proposed the *Biophilia Hypothesis*. [*philosophers and writers, environmental*]

windbreaks

When *soil erosion* is caused by wind instead of water, windbreaks are used as a *soil conservation* technique. Windbreaks, also called *shelterbelts*, are long rows of trees that break the wind. They are usually used in wide-open, flat lands such as the Great Plains, where wind often causes erosion. [*conservation agriculture; regenerative agriculture; sustainable agriculture; Great Green Wall of Africa*]

wind-energy areas

Areas where *wind farms* could be installed. These are areas that are conducive to the technology but also avoid environmentally sensitive areas. Some environmentalists only favor offshore wind farms, which they believe would cause less environmental harm. [*floating wind turbines; Global Wind Day; wind power, brief history; wind power transmission line*]

wind farms

Building just one or two *wind turbines* to generate power is not economically feasible, so most plans are for large installations of dozens or hundreds of them in what are called wind farms. Many of these wind farms are offshore, with the United Kingdom leading in this area and the United States far behind.

However, there are plans for the United States to catch up. The *Cape Wind* facility has been approved and is being built off the coast of Cape Cod, Massachusetts, and others are planned in the Great Lakes. The U.S. government has started selling leases for wind farms off the coasts of Rhode Island, Massachusetts, and Virginia, with more planned. Estimates state that each of these leases could generate power for about 700,000 homes. [*wind power*]

wind power

Wind is a *renewable energy* resource that has been used for hundreds of years around the world. Wind power is *air* and *water pollution* free and does not emit any *greenhouse gases*. It is one of the most energy efficient technologies, converting more than 90 percent of the available energy into usable energy.

Wind power globally increased fourfold between 2005 and 2012. *China* has more than 26 percent of the total global capacity, followed by the United States, *Germany*, and then Spain. Within the Unites States, Texas has the largest capacity followed by California and Illinois. In 2012, wind power added more electricity production to the U.S. supply than any other power source. However, only about 3.5 percent of our current total energy supply comes from wind.

The growth of wind occurs in spurts due to the spurious nature of tax credits, such as the *Production Tax Credit* that helped encourage initial development. But wind power is already cost competitive with other sources of energy and is the least expensive of all renewable energy sources.

Most existing U.S. production occurs in California where more than 1300 MW are produced by 15,000 wind turbines. This supplies about one percent of the state's total needs—enough to power all of San Francisco. Although wind could theoretically supply all U.S. energy needs, experts believe it can realistically supply about 20 percent of our total energy within the next 20 years.

In spite of its many environmental advantages as an alternative to *fossil fuels*, many people oppose wind power. Proposals for *wind farms* have met with intense *NIMBY* resistance. These turbines are typically immense, and some feel cause *noise pollution.* They have been known to cause bird kills, but the biggest NIMBY complaint is *aesthetic pollution.* An entire landscape of these high-tech windmills is not considered attractive to many. *Cape Wind*—a wind farm off of Cape Cod, Massachusetts—spent years in court before being approved. [*wind power brief history*]

wind power brief history

Windmills have probably been in existence since before 1000 AD, most likely to bring up water or to mill flour. It wasn't until the late 1800s when there were attempts to generate electricity with wind power. In 1941 Vermont became the first location to build *wind turbines* to generate electricity that connected to the *power grid*. Then in 1980, New Hampshire was the first to open a true *wind farm*. In 2009, the first full-scale floating wind farm was constructed off the coast of Norway and in 2013, the largest offshore wind farm opened off the coast of England.

Wind turbine technology is fast improving with the most notable change being the size of the wind turbines and the height of the towers. Since 1990 the turbine blades are now 90 percent larger and the height is 45 percent greater. And most important, a turbine today generates about 17 times more power than one did in 1990.

Wind power is now a booming business, even though it has experienced growing pains. In 2012 alone, 6700 additional wind turbines were installed in the United States. About 3.5 percent of all electricity in the United States is supplied by wind and estimates are that by 2030 it will be about 20 percent. Most U.S. growth is expected in the Midwest and Plain states where much of their power already comes from wind. Iowa and South Dakota both already get about 20 percent of their electricity from wind.

As with many *renewable energy* sources, the U.S. government's interest in providing support and incentives comes and goes as Congress comes and goes. Tax credit programs such as the *Production Tax Credit* has helped build the industry but is often on the chopping block for funding.

wind power states, ten biggest

The following are the ten states that generated the largest percentage of their total power needs from *wind power* in 2011. They are South Dakota (22.3%), Iowa (18.8%) North Dakota (14.7%), Minnesota (12.7%), Wyoming (10.1%), Colorado (9.2%), Kansas (8.2%), Idaho (8.2%), Oregon (8.2%), and Oklahoma (7.1%).

wind power transmission line

At the moment, whenever a new offshore *wind farm* is built, a separate energy transmission line must be run between the wind farm and the land to tie the power into the *power grid*. There is a proposal to build a backbone main line in a corridor between southern Virginia and northern New Jersey. This would allow any offshore wind farm to quickly tie into the power grid, thereby reducing the number of individual lines needed. Google is a major investor in this project.

Getting the power from a wind turbine to its destination for use is a technical challenge. The United States is far better at this than other countries. For example, even though *China* produces a large amount of wind power, it has problems actually using it. Estimates are only a small percent of the total power produced by China's wind farms ever becomes electricity and is used. [*floating wind turbine; wind turbine*]

wind-speed vacuum

Wind farms have been shown to alter wind speed. Turbines downwind from other *wind turbines* are known to produce less power because they receive less wind. This reduction in wind speed is called a wind-speed vacuum. With this knowledge, newer wind farms are being designed so this effect does not impact the overall efficiency of the farm.

wind turbine

Wind power is one of the most promising *alternative energy* sources. The days of the quaint wind mills of yesteryear are gone and replaced by hi-tech wind turbines. Most of these turbines consists of two or three blades, a rotor, a transmission, an electric generator, and controls, all mounted on a tower that is anywhere from about 50 feet to well more than 300 feet high. The rotor is capable of turning to align itself with the wind. Most newer blades are made of fiberglass, while older models were aluminum. The larger the turbine, the more efficient the device. This makes them expensive and therefore more practical for large utility companies as opposed to small power plants or individual use.

Wind speed is a critical factor in the efficiency of wind turbines. A ten mph wind can generate 65 kilowatts of power but only a five mph increase to 15 mph bumps that up to 300 kilowatts. Wind turbines located at high wind locations—with an average wind of 15 miles per hour—are already highly efficient. However, most U.S. sites are considered moderate sites with winds averaging between 12 and 15 miles per hour, or low wind sites with averages below 12 miles per hour. Technological advances are needed to make these low and moderate sites more economically feasible before wind power flourishes in the United States. [*floating wind turbine; wind farms*]

wine-bottle cork debate

A debate has been ongoing about which type of wine bottle cork has the least environmental impact: natural cork, synthetic cork, or aluminum screw-caps. Cork oak trees grow in Europe and North Africa and do not have to be cut down to provide the needed cork—they are stripped by hand. Once a cork oak tree is 25 years old, half of the thick bark has been stripped off, leaving the other half to continue growing. This process is repeated every nine years and these trees can reach 250 years of age, so this is a sustainable practice.

Research has shown that using plastic corks generates 10 times more *greenhouse gases* than natural cork. And the aluminum screw-caps produce 24 times as much as real cork. In both cases, the manufacturing process makes them far less attractive than using the natural product. Many believe that in addition to using natural cork, the best way to go is to recycle *wine bottle corks.*

wine bottle corks, recycling

Natural *wine bottle cork* can be recycled. A cork oak, like any plant, acts as a *carbon sink*, storing *carbon dioxide*, a *greenhouse gas*, and therefore keeping it out of the atmosphere. The cork itself is composed of 50 percent carbon, and recycling them is a great way to reduce carbon emissions. Some companies specialize in recycling corks and turning them into consumer products such as flooring tile, building insulation, automotive gaskets, and even sandals.

Some wineries add nonrecyclable materials to the cork, which is problematic for *recycling.* Green social media has been helping out with this cause by promoting an *online petition* asking these companies to use natural cork, without any other materials. [*appliance recycling; refrigerator recycling; refrigerator doors; energy, ways to save residential; automobile recycling; mattresses, recycling; cigarette butt litter and recycling; downcycling; upcycling; motor oil recycling; paper recycling; recycling metals; plastic recycling; e-waste recycling; single-stream recycling*]

wine, organic and biodynamic

Wines produced from conventionally grown grapes are loaded with *pesticides*, ranking #7 as the most contaminated food on a recent list. A study in 2008 showed 100 percent of the wines tested contained *pesticide residues*, with numerous different pesticides in each bottle and a total of 24 contaminants in all, five of which were classified as *carcinogenic, mutagenic,* or endocrine *hormone disruptors.*

Organic food labels can help. Those labeled 100% USDA Organic have only organically grown ingredients. Those labeled USDA Organic are 95 percent organic, and those labeled Made with Organic Grapes or Ingredients must contain at least 70 percent organic contents.

The Biodynamic label is a private labeling that goes above the USDA standards. They avoid pesticides and synthetic *fertilizers* and attempt to grow the grapes on *sustainable agriculture*. Bottles with this label are mostly grown in California or Oregon. [*food eco-labels; pesticide dangers*]

wireless charging of electronic devices

How many devices do you own that must be charged? Most of us have many. And it doesn't take long before the device and the charger end up in a *landfill*. Because most devices contain *toxic* substances such as *mercury* and *lead*, they should be placed in special *hazardous waste* landfills but typically are not. About 100,000 tons of these devices are disposed of each year in the United States alone. Once in a landfill, they release their toxic substances, which then *leach* into the *soil* and possibly *groundwaters*. The waste produced by electronic devices is called *e-waste*.

Solutions to the need for charging these devices, such as universal chargers, have been slow in coming and will probably be eclipsed by a newer technology, anyway; wireless charging has arrived.

Wireless charging uses induction to transmit a charge. A device has a charging base that is plugged into a power outlet and a receiver that is built into the device in need of a charge. The charging base contains a metal coil that produces an electromagnetic field. The receiver has a built-in metal coil as well. When the device to be charged is placed on the base, an induction field passes the charge from the base coil to the receiver coil, charging the device.

Today, these wireless charging devices produce 86 percent of the power that wired charges do, but this is rapidly improving. Devices that do not come with built-in wireless receiver coils can be retrofitted. For example, some cell-phone cases have a receiver coil embedded. However, it is only a matter of time before most manufacturers of our electronic gadgetry put coils in all of these devices routinely. The future might be interesting, because receivers can be built into almost anything. Plans include inserting them into car dashboards and even furniture. [*solar charging of electronic devices*]

wise-use movement

Based on a phrase originally used by *Gifford Pinchot*, who believed in *dual-use conservation* practices of *forest public lands*. However, today it is a catchall term for a loosely knit movement of loggers, farmers, miners and property holders and promoted by some conservative political and religious groups, who are opposed to the mainstream environmental movement. They advocate *natural resources* as a benefit to those that live on the land. The phrase "wise-use" is thereby inherent in the eyes of the beholder. [*Sagebrush Rebellion; buffalo; multiple use policy; environmental movement*]

wolves, endangered

In Wyoming, wolves were recently removed from the *endangered species list,* meaning they can now be hunted and killed in most of the state. The state laws allow wolves to be shot in almost any manner, including from helicopters and in their dens. Court battles are ongoing to prevent the state from allowing the species' demise. [*endangered and threatened species lists; buffalo; Endangered Species Act; biodiversity, loss of*]

women farmers

In the United States, people don't necessarily connect women with farming. However, on a global basis, more than half of the world's food supply is produced by women farmers, and they make up over 40 percent of the total farming workforce. In some regions, such as sub-Sahara Africa, 85 percent of the farmers are women. The women's farming role varies country to country. For example, in *Indonesia* women work mostly in rice paddies doing labor-intensive, low-valued work; however, in the United States, more women than men are involved in organic and small farming.

Other regions of the world have a gender gap in *agriculture*. Even though women play a big part in farming, numbers show how inequities limit women's participation in much of the global agricultural economy. For example, in *Bangladesh*, only 3 percent of rural women get paid for their work as opposed to 24 percent of rural men. Also, women produce about 70 percent of the food in developing nations but own less than 2 percent of the land.

Reports show that nations with more progressive gender-sensitive business practices typically have more abundant, more nutritious, and more affordable food supplies. Of course, nations with these more progressive business practices probably also have more progressive educational policies, which gets to the heart of the problem. Almost 70 percent of the world's one billion illiterate are women. [*ecofeminism; environmental justice; Women's Earth Alliance; Green Belt Movement*]

Woman's Earth Alliance (WEA)

The WEA grew out of a project developed in the *David Brower* Center in Berkeley, California. It supports women-led community groups throughout the world that focus on environmental projects. Their mission states, "When women thrive, communities, the environment and future generations thrive." They offer training and resources on projects about water, land, food, and climate change. At the moment, the emphasis is in North America, *India*, and Africa. In India, the program fosters local women's groups that promote traditional knowledge, food sovereignty, and the rights of *women farmers*. [*ecofeminism; environmental justice; Women's Earth Alliance; Green Belt Movement*] {womensearthalliance.org}

wood-burning cookstoves, primitive

Most people in *developed nations* would be surprised, but the leading environmental cause of death in the world are primitive wood burning cookstoves. It contributes to almost two million deaths a year. While people in wealthy nations don't even know what they are, it is the most common method to heat a home and cook meals for half the world's population – which is poor.

These are inefficient, open fire stoves that burn primarily wood but also any *biomass* such as leftover crops, animal dung, or charcoal. These stoves burn the wood inefficiently, filling homes with dense smoke filled with *soot*, which is then routinely inhaled. The soot is believed to cause about 400,000 deaths, mostly of women and children (who spend most of their time in these homes) in south and east Asia. It also contributes to childhood pneumonia, emphysema, lung cancer, bronchitis, and cardiovascular disease.

A global effort is underway to replace these cookstoves with cleaner burning devices. The United Nations has a Global Alliance for Clean Cookstoves with a target called "100 by 20," which stands for having 100 million homes with cleaner burning stoves by 2020.

These stoves cause environmental harm as well, including *deforestation*, because the wood for these stoves must come from somewhere. A single-room home in the Andean Highlands of Peru burns about 3.5 tons of wood a year in these stoves.

Other possible solutions include a new innovation called *solar cookstoves*. [*energy poverty; less developed countries; air pollution; indoor air pollution; brick kilns; fuel wood; particulate matter*]

World Business Council for Sustainable Development (WBCSD)

This is a Swiss based business association of CEOs from about 200 huge transnational corporations. Their latest report, called <u>Vision 2050</u> is an agenda that lists the following "must haves": a) Incorporating the costs of *externalities*, starting with *carbon*, *eco-services* and *water*, into the structure of the marketplace; b) Doubling *agricultural* output without increasing the amount of land or water used; c) Halting *deforestation* and increasing yields from planted forests; d) Halving *carbon emissions* worldwide (based on 2005 levels) by 2050 through a shift to low-carbon energy systems; and e) Improved demand-side energy efficiency, and providing universal access to low-carbon mobility.

Exactly how much of this is true action towards sustainability as opposed to just promotion, depends whom you ask. Many *environmentalists* believe the group is mostly posturing with little substance, while others work closely with them to advance corporate sustainability. It includes companies from 20 major industries located in 35 countries across the globe.

The WBCSD played an active role at United Nations *Conference on Sustainable Development*. The president of this group commented about those people who are leery about business' role in supporting our planet as opposed to degrading it. His statements include: "The time for creating awareness is behind us. If you are not aware today than you are not a global leader in business..." "...it is time to kick into action." "The 20 percent of the really bad guys we need to regulate out of existence."

Critics quickly point out that while this group is correct in stating the importance of business—something that few people can deny— their preferred methods are not about regulation and not in line with most environmentalists. The group advocates voluntary actions—not mandatory regulation; something that many people believe will be necessary to get businesses to where they need to be if the planet is to reach true sustainability. [*CERES; climate capitalism; corporate social responsibility; economics and sustainable development; economy vs. environment; integrated bottom line; polluter pays principle; industry self-regulation*] {wbcsd.org}

World Commission on Environment and Development

In 1983, the United Nations General Assembly established this commission with the purpose of preparing and planning for our future and the environment we will live in. The result of this commission was the release, in 1987, of a report called *Our Common Future* (also called the Brundtland Report). This seminal report provided the most commonly used definition of *sustainable development*, still used today. The report is a balanced approach to both protecting our environment and growing the economy. Most businesses and many environmental organizations have embraced many of the key points; however, some more radical environmental groups oppose the need for economic growth and consider it a business-as-usual endorsement.

World Environment Day

The *United Nations Environment Programme* (UNEP) established this day in 1972 and it is celebrated on June 5th each year in a host country, with over a hundred countries participating. The thrust is to push for global political action on environmental issues. As the UNEP states, "World Environment Day is also a day for people from all walks of life to come together to ensure a cleaner, greener, and brighter outlook for themselves and future generations."

In 2013 the theme was "Think.Eat.Save."—an anti-food waste and food loss campaign. It encouraged everyone to reduce their "foodprint," which is an interesting twist on the term *ecological footprint*. The host country was Mongolia. [*United*

Nations, global environmental issues and the; environmental movement; Earth Day; Environmental Protection Agency] {unep.org/wed}

World Heritage Sites

The *United Nations Educational, Scientific, and Cultural Organization (UNESCO)* has a World Heritage Committee that consists of 21 member states that, on a rotating basis, award this status to locations around the globe that have "outstanding universal value" to humanity. Currently 759 cultural sites, 193 natural sites, and 29 in combination of these two are located in 160 countries.

This idea began 40 years ago by a few dozen countries wanting to protect the most important natural places on Earth from destruction for any number of reasons. World Heritage sites typically benefit from this designation, but in recent years some have also been harmed by becoming so high profile and attracting too many visitors—damaging the very sites they wish to save. This has become such a problem that UNESCO now announces World Heritage Sites in Danger.

Some World Heritage sites include The *Great Barrier Reef*, Belize Barrier Reef *Marine Reserve System*, The Great Wall of China, Sichuan Giant Panda Sanctuaries, Coffee Cultural Landscape of Colombia, the Leaning Tower of Pisa in Italy, Lake Baikal in Russia, the Pantanal Conservation Area and Brasilia in *Brazil*, the Prehistoric Pile dwellings around the Alps in *Germany*, the Acropolis in Athens, Masada in Israel, and the Canadian Rocky Mountain Parks.

Of 21 sites in the United States, a few are *Yellowstone*, the *Grand Canyon*, the Florida *Everglades*, Independence Hall, the Statue of Liberty, Carlsbad Caverns, Great Smoky Mountains National Park, Chaco Culture, Taos Pueblo, and Papahanaumokuakea.

Forty-four sites are now considered in danger. The purpose of this designation is "to inform the international community of conditions which threaten the very characteristics for which a property was inscribed on the World Heritage List, and to encourage corrective action." A few of these sites are the Belize Barrier Reef Marine Reserve System, the *Tropical Rainforest* Heritage of Sumatra, the Rainforests of the Atsinanana in Madagascar, Everglades National Park, and the Birthplace of Jesus: Church of the Nativity in Bethlehem. [*Biosphere Reserves; biodiversity*] {whc.unesco.org}

Worldwatch Institute

Worldwatch Institute is a world leader in providing global environmental information to decision-makers and the public. This *NGO* was founded in 1975. Their analysis of global issues and the advice they render are respected by world leaders yet the public can understand them. "The Institute's top mission objectives are universal access to *renewable energy* and nutritious food, expansion of environmentally sound jobs and development, transformation of cultures from consumerism to sustainability, and an early end to *population growth* through healthy and intentional child bearing."

Their annual State of the World and Vital Signs publications have become the premier references for environmental policy decision-making. Anyone needing any kind of information about anything related to the environment can find it within their publications. {worldwatch.org}

World Wildlife Fund (WWF)

One of the world's largest *NGO* environmental organizations, its primary mission is the preservation of *habitat* and *biodiversity* throughout the world. The WWF is credited with helping to establish over 500 *national parks* and preserves globally. It provides research and assistance to others to support and facilitate these efforts. [*Group of 10*] {worldwildlife.org}

wrapping-paper alternatives

The giving season produces far more *municipal solid waste* than any other time of year. Between Thanksgiving and Christmas, Americans generate 25 million "extra" tons of *holiday waste* from products such as wrapping paper, packaging, and discarded products. *Landfills* and *incineration* plants must deal with this influx, and the planet must deal with the aftereffects.

The most obvious alternative to most standard wrapping paper is to use paper that has already been *recycled* and contains 100 percent *post-consumer waste* material. Or look for recyclable wrapping papers, so at least you can recycle the new paper. Some of these recyclable wrapping papers use *soy based inks*. Other unique alternatives include companies that convert old LP records and other materials such as plastic bottles and cloth into colorful, reusable wrapping materials.

WWOOF (Worldwide Opportunities on Organic Farms)

WWOOF stands for Worldwide Opportunities on Organic Farms, which is a network of national groups that help travelers looking for volunteer work on *organic farms*. In exchange, travelers typically get free room and board during their stay. It started in 1971 in Great Britain but now has over 1700 host farms in over 50 countries. [*volunteer research, scientific; crowd research, scientific; voluntourism, environmental*] {wwoofusa.org}

xenobiotic

A foreign *organic* substance, such as an organic *pesticide*, found in *drinking water*. [*domestic water use; domestic water pollution; Safe Drinking Water Act*]

Xerces Society

This *NGO* organization, founded in 1971, is dedicated to preventing the *extinction* of *invertebrates*, including *insects*, caused by human intervention. The Xerces Society has many programs, including conservation science, education, and public policy. [*colony collapse disorder; honeybee sperm bank; honeybees, pollination, and agriculture; Monarch butterfly decline; neonicotinoid pesticide and honeybees; biodiversity, loss of*] {xerces.org}

xeriscape

A garden or landscape consisting of drought-resistant plants. These landscapes survive and flourish with a fraction of the water typically used in an artificial landscape. In some municipalities where water is scarce, they are mandatory or recommended for homes. When recommended, they often are accompanied with tax breaks to urge use. For example, water utilities in most cities in Nevada offer cash rebates to anyone who converts a green lawn into a xeriscape. [*tolerance range; limiting factors; biomes; xerophyte; cacti; extremophiles*]

xerophyte

A plant adapted for life in a *desert*. [*biomes; anthromes; xeriscape; extremophiles*]

Yale Environment 360 blog

This publication of the Yale School of Forestry & Environmental Studies is one of the best environmental blogs to be found. It provides excellent reporting, providing both brief articles and indepth reports. Besides delivering the facts, it offers analysis and opinion, and it is always packed with great photographs. If you want to remain current on the important environmental issues of the day, this is one of your best resources. [*blogs, green; microblogs, green; tweets, eco-; magazines, eco-; news literacy; environmental literacy; environmental education; science; pseudoscience; opinions and science*] {e360.yale.edu}

yellowcake

Yellowcake is produced during the *nuclear fuel cycle*. Low-grade uranium ore is mined to produce fuel for *nuclear reactors*. The ore goes through a milling process that involves crushing and treating the ore with solvents to concentrate the uranium. The resulting mixture, called yellowcake, then goes through an enrichment process to prepare it for use in a nuclear reactor. [*nuclear power*]

Yellowstone National Park

President Ulysses S. Grant made Yellowstone the first official *national park* in the United States in 1872. It is the largest park in the lower states and is found in parts of Wyoming, Montana, and Idaho. Almost all of the park remains undeveloped land.

Its most famous features are the active hot-spring geysers such as Old Faithful. Over 300 species of animals, 18 species of fish, and well over 200 species of birds reside in the park. *Endangered species* have been brought into the park in efforts to revitalize populations, including Peregrine falcons, *bald eagles*, and *whooping cranes*. When the population of wild bison (*buffalo*) in the west was in dramatic decline, herds were relocated into Yellowstone to afford protection.

It has been designated as a *World Heritage Site* and a *Biosphere Reserve* site. Over three million people visit the park each year. [*National Park System; national parks, global; Hetch Hetchy; Colorado River; Yosemite National Park*]

YIMBY FAP

Acronym for "Yes, in my backyard for a price." Waste disposal firms have found they can possibly overcome the *NIMBY* syndrome by offering large financial inducements to towns (or countries) that are then willing to allow large *landfills*, *incineration* plants, or the waste sites. Today, YIMBY FAP might be used during negotiations between landowners and companies interested in *fracking*.

Yosemite National Park

Yosemite was set aside for protection by President Abraham Lincoln in 1864 but did not become an official *national park* in 1890. It was our third national park, following *Yellowstone* and Sequoia. *John Muir* wrote the first proposal to convert the area into a national park.

Yosemite includes about 750,000 acres, is home to two of the world's highest waterfalls, giant sequoias, and mountain peaks reaching over 13,000 feet. *Wildlife* includes mountain lions, black bears, and about 200 species of birds. It also blossoms with over 1300 species of flowering plants. Ninety-four percent of the park is designated as *wilderness* and therefore is off limits to road building or other development. About 3.5 million people visit Yosemite each year.

Shortly after becoming an official national park, environmentalists, including *John Muir*, became embroiled in a conflict to convert a large portion of its newly protected land, called *Hetch Hetchy,* into a dam to supply water to *San Francisco*.

Yucca Mountain

See *nuclear waste disposal* and *nuclear waste dilemma, the global*.

zero carbon

Zero anything has become a mantra for many people, especially companies. In their efforts to become *green*—or at least sound as if they are—they often use terms such as *zero waste* and zero carbon. Zero carbon is the elimination of all *carbon emissions*. This would require many things, but most important, it requires the complete transition from *fossil fuels* (that produce *carbon dioxide*) to *renewable energy sources* that do not.

Companies can also produce zero carbon by using *carbon offsets* that negate—at least on paper—the carbon they do emit. Copenhagen, Denmark, is one of the cities working toward zero carbon by using all renewable energy and getting most of their citizens to ride environmentally friendly *mass transit*, *walk*, or ride *bicycles*. [*climate change; alternative energy sources*]

zero-energy buildings

Buildings built so efficiently they can maintain a comfortable indoor environment without needing external energy. Many countries have zero-energy building projects primarily for demonstration purposes, with the goal to make these types of buildings commercially available in the future. [*Living Buildings; U.S. Green Building Council; eco-cities; thermal insulation, building; buildings, green; homes, green*]

Zero Population Growth (ZPG)

See *populationconnection.org*.

zero waste

The goal of many municipalities, companies, and even individual households is to reuse or recycle all of their *municipal solid waste* so no waste needs to be sent to *landfills* or *incineration* plants or shipped elsewhere. Zero waste has not been achieved in any large municipalities, but some are coming close. *San Francisco*, for example, reuses 78 percent of all its incoming waste. Some companies, such as General Motors, have declared that certain manufacturing plants are zero waste, and some individuals declare they have attained this feat. There are new zero-waste supermarkets where customers must bring their own reusable shopping bags—no *plastic shopping bags* and no *paper shopping bags*—and all food that cannot be sold is donated or recycled in some way.

In addition to reusing and *recycling*, an important component is to reduce the use of products that would turn into waste. This is called *source reduction* and *waste minimization*. [*municipal solid waste disposal; three R's*]

zombie bees

There is a form of *parasitism* called *parasitoidism* where one organism (the parasitoid) lays eggs in another organism (the *host*). In this particular case, a certain type of fly lays its eggs in adult honeybees. The eggs hatch within the honeybee and begin feeding on the insides of the bee. In most parasitoid relationships, the host is weakened and usually killed when the parasitoid burrows its way out.

However, the honeybee hosts exhibit bizarre behavior before dying, such as flying erratically at night even though bees are daytime fliers. Because of this behavior of the host bees, they have been dubbed zombie bees. [*honeybees, pollination, and agriculture; colony collapse disorder; honeybee sperm bank; bumblebee decline*]

zones of life

Dating back to the turn of the previous century, scientists have used a variety of methods to define the life that exists in different regions of our planet. Different names have been given to these various methods over time.

Formations specifically refer to the plant life that exists in major regions and includes regions such as *desert* and coniferous forest. *Realms* address the distribution of animals in regions of the world and includes regions such as Nearctic and Neotropical.

The original Life Zones system, described around the turn of the 20th century, attempted to include both plants and animals into one scheme and divided the Earth east and west into transcontinental bands. A more recent theory was the Holdridge Life Zone System—a complex system taking into account numerous variables including altitude as well as latitude. Probably the most accepted term for describing zones of life is the *biome*. Biomes use plant formations and the animal life associated with these plants to describe distinct zones of life on the planet.

However, most recently a newer zoning system has been developed that uses *anthromes* or anthropogenic biomes to describe life zones. This method takes into account the changes humans have made to our planet.

zone of saturation

That portion of the Earth's crust saturated with *water*, called *groundwater*. Surprisingly, the zone of saturation within the Earth contains about 40 times as much water as all the *surface waters* (ponds, lakes, streams, and rivers).

zoning, land use

The specific use of land is often regulated by local or regional zoning laws. Land can be zoned, for example, as agricultural, commercial, industrial, recreational, or residential. The individuals who decide on the zoning regulations should, in theory, be qualified professional planners, with access to environmental consultants. Those individuals responsible for zoning regulations are responsible for the economic well-being of the area and should also be concerned and knowledgeable about the area's environmental well-being. [*smart growth*]

zoo

Zoos were originally built to satisfy human curiosities about wild animals, so most zoos were virtual prisons to their inhabitants. With few people opposing this type of facility, zoos remained in this state for decades. In the 1960s, animal rights organizations began protested how zoos treated their prisoners. The public began to listen and so did the zoos. Instead of being treated as curiosities, animals were viewed as part of a *habitat* and treated as the most fascinating aspect of that habitat.

In most large zoos, animals are not confined to cages and are treated humanely. Although still a debatable issue, most people believe the educational advantages and the heightened public awareness raised by zoos makes them worthwhile.

In addition—and possibly as important—zoos have become a major force in protecting *endangered* and *threatened* species from *extinction* with breeding and *conservation* programs. For example, the *Smithsonian Institution*, through their National Zoo, has successfully reintroduced the golden lion tamarin to the *Brazilian rainforests*. Many other zoos are playing important roles in saving species from extinction with captive breeding programs.

Zoos also raise money and contribute to conservation efforts. However, as with most publically funded institutions, money is often limited and many zoos can barely cover expenses for their live animals and breeding programs for endangered species. [*zoos, future; Zoo, Frozen; zoos, captive breeding programs in; animal rights movement and animal welfare*]

zoogeography
The study of where animals live and, furthermore, why they live where they do. A subdivision of biogeography. [*ecology; biomes; ecotone; experimental ecology; anthromes*]

zoology
The study of animals. [*caliography; cryptozoology; environmental science; ecology; oology; terrestrial ecology*]

zoonotic diseases
Diseases that are transmitted from animals to humans (called zoonosis), which are believed to account for 75 percent of all emerging infectious threats globally. It is estimated that 60 percent of all human *pathogens* are zoonotic in origin, probably arising from areas where humans infiltrated wild areas and increased the contact between the humans and *wildlife*. The most high-profile such outbreak was HIV in the 1980s.

The only time the public pays attention to this issue is when there is an outbreak such as HIV, West Nile virus, or monkeypox. Public health agencies are always working to reduce the likelihood of these disease transmissions. In 2012 a report stated that 2.7 million people die each year from zoonotic diseases. [*bird flu virus; infectious disease; spillover effect*]

zooplankton
Crustaceans, rotifers, and protozoans, the microscopic animals found in *plankton*. These animals usually fill the role of primary *consumers* in many *aquatic ecosystems*.

zoos, captive breeding programs in
It is sad, but out of necessity, *zoos* are often the last stand for species on the brink of total *extinction*. Many species have been declared *extinct in the wild*, meaning the few remaining individuals of that species can only be found in breeding programs within zoos. As zoos begin to move from educational entertainment to *conservation* efforts, their role in saving species is coming to the forefront. Of course, the hope is to save species in the wild, but when that fails, as it appears to be happening more and more often, then if not in zoos, where can species be saved? Some believe this effort must be broadened beyond saving *flagship species* such as lemurs, pandas, and rhinos (also called *charismatic fauna*) to also saving any species in danger of extinction, including

reptiles, *amphibians, insects,* and even those like the *blobfish.* [*biodiversity, loss of; Ugly Animal Preservation Society*]

Zoo, Frozen

The San Diego Zoo has a specialized type of zoo, called the Frozen Zoo, which is an archive of DNA samples and cell lines of thousands of species. This is one method of trying to ensure the preservation of at least the genetic makeup of species in case they become *extinct.* The hope is that someday these species might be brought back to life or at least, their genetic makeup remains available. The idea has caught on and there is now over 20 other zoos all participating in this new initiative, called The Frozen Ark. [*zoos, captive breeding programs in; zoos, future; Doomsday Seed Vault; biodiversity, loss of*] {frozenark.org}

zoos, future

Zoos play a vital role educating people about our *biosphere.* They have evolved from prison-like exhibits to modern-day institutions of *conservation,* education, and protection. A recent conference looked into the future of how zoos might continue to advance their missions. Some ideas were obvious, such as more natural, open-terrain exhibits for free-roaming animals. Others were more high-tech, such as vertical, high-rise zoos in congested urban areas. Some downright fantastic ideas went beyond saving breeds from extinction through captive breeding programs to using stored genetic material to clone and "re-create" *extinct* species like the *Dodo bird* or woolly mammoth. [*Zoo, Frozen; zoos, captive breeding programs in; de-extinction; conservation triage*]

℘

Source Material

The following periodicals and website/blogs were used extensively for original material and to continually update existing entries:

Discover. Waukesha, WI: Kalmbach. Print.

Website/blog. *Dot Earth.* New York Times. Web.

The Economist. New York: Economist Newspaper Limited. Print.

Website/blog. *Earthtalk.* Earth Action Network. Web.

E-The Environmental Magazine. Norwalk, CT: Earth Action Network. Print.

Website/blog. *Green: A Blog about Energy and the Environment.* New York Times. Web. (discontinued)

New Scientist. New York: Reed Business Information. Print.

The New York Times. New York: New York Times. Print.

OnEarth. New York: Natural Resources Defense Council. Print.

Popular Science. New York: Bonnier Corporation. Print.

Science. Washington, D.C.: American Association for the Advancement of Science. Print.

Scientific American. New York: Scientific American. Print.

Website/blog. *YaleEnvironment360.* Yale School of Forestry & Environmental Studies. Web.

Additional sources:

Ariely, Dan, and Tim Folger. *The Best American Science and Nature Writing 2012.* Boston, MA: Houghton Mifflin Harcourt, 2012. Print.

Assadourian, Erik, Michael Renner, and Linda Starke. *State of the World 2012: Moving toward Sustainable Prosperity: A Worldwatch Institute Report on Progress toward a Sustainable Society.* Washington D.C.: Island Press, 2012. Print.

Assadourian, Erik, Thomas Prugh, and Linda Starke. *Is Sustainability Still Possible?* Washington, D.C.: Island Press, 2013. Print.

Assadourian, Erik. *Vital Signs 2006-2007: The Trends That Are Shaping Our Future.* New York: Norton, 2006. Print.

———. *Vital Signs 2007-2008: The Trends That Are Shaping Our Future.* New York: Norton, 2007. Print.

Bates, Marston. *Man in Nature.* Englewood Cliffs, NJ: Prentice-Hall, 1964. Print.

Berry, Thomas, and Mary Evelyn. Tucker. *Evening Thoughts: Reflecting on Earth as Sacred Community*. San Francisco: Sierra Club, 2006. Print.

Berry, Thomas. *The Dream of the Earth*. San Francisco: Sierra Club, 1988. Print.

———. *The Great Work: Our Way into the Future*. New York: Bell Tower, 1999. Print.

Brinkley, Douglas. *The Wilderness Warrior: Theodore Roosevelt and the Crusade for America*. New York: HarperCollins, 2009. Print.

Brown, Lester R. *Plan B 2.0: Rescuing a Planet under Stress and a Civilization in Trouble*. New York: W. W. Norton, 2006. Print.

———. *Plan B 3.0: Mobilizing to save Civilization*. New York: W. W. Norton, 2008. Print.

———. *World on the Edge: How to Prevent Environmental and Economic Collapse*. New York: W. W. Norton, 2011. Print.

Callenbach, Ernest. *Ecology: A Pocket Guide*. Berkeley: University of California, 1998. Print.

Carson, Rachel. *Silent Spring*. Boston: Houghton Mifflin, 1962. Print.

Clapham, W. B., Jr. *Natural Ecosystems*. New York: Macmillan, 1983. Print.

Collin, P. H. *Dictionary of Environment & Ecology*. London: Bloomsbury, 2004. Print.

Dauvergne, Peter. *The A to Z of Environmentalism*. Lanham MD: Scarecrow, 2009. Print.

DesJardins, Joseph R. *Environmental Ethics: An Introduction to Environmental Philosophy*. Belmont, CA: Wadsworth, 1993. Print.

Ecosystems and Human Well-being: Synthesis. Washington D.C.: Island Press, 2005. Print.

Enger, Eldon D., and Bradley F. Smith. *Environmental Science: A Study of Interrelationships*. New York: McGraw-Hill, 2013. Print.

Finch, Robert, and John Elder. *The Norton Book of Nature Writing*. New York: W. W. Norton, 1990. Print.

Friedman, Thomas L. *Hot, Flat, and Crowded: Why We Need a Green Revolution-- and How It Can Renew America*. New York: Farrar, Straus and Giroux, 2008. Print.

Fuller, R. Buckminster. *Operating Manual for Spaceship Earth*. Carbondale: Southern Illinois UP, 1969. Print.

Gardner, G., Linda Starke, and Lyle Rosbotham. *Vital Signs 2011: The Trends That Are Shaping Our Future*. Washington, D.C.: Worldwatch Institute, 2011. Print.

Gardner, Gary T., Thomas Prugh, and Linda Starke. *State of the World 2008: Innovations for a Sustainable Economy: A Worldwatch Institute Report on Progress toward a Sustainable Society*. New York: W. W. Norton, 2008. Print.

Goleman, Daniel. *Ecological Intelligence: How Knowing the Hidden Impacts of What We Buy Can Change Everything*. New York: Broadway, 2009. Print.

Golley, Frank B. *A Primer for Environmental Literacy*. New Haven: Yale UP, 1998. Print.

Hardin, Garrett James, and John Baden. *Managing the Commons*. San Francisco: W.H. Freeman, 1977. Print.

Hawken, Paul, Amory B. Lovins, and L. Hunter Lovins. *Natural Capitalism: Creating the next Industrial Revolution*. Boston: Little, Brown and Company, 1999. Print.

Howard, Albert. *An Agricultural Testament*. Goa: Other India, 1996. Print.

Howat, John K. *The Hudson River and Its Painters*. New York: Viking, 1972. Print.

Hughes, Holly, and Mark Weakley. *Meditations on the Earth: A Celebration of Nature, in Quotations, Poems, and Essays*. Philadelphia, PA: Running Press, 1994. Print.

Humes, Edward. *Garbology: Our Dirty Love Affair with Trash*. New York: Avery, 2013. Print.

Kanze, Edward. *The World of John Burroughs*. New York: H.N. Abrams, 1993. Print.

Leopold, Aldo. *A Sand County Almanac. With Other Essays on Conservation from Round River*. New York: Oxford UP, 1966. Print.

Lovins, L. Hunter, and Boyd Cohen. *Climate Capitalism: Capitalism in the Age of Climate Change*. New York: Hill and Wang, 2011. Print.

———. *The Way Out: Kick-starting Capitalism to save Our Economic Ass*. New York: Hill and Wang, 2011. Print.

Makower, Joel, Nancy Tienvieri, and Cathryn Poff. *The Nature Catalog*. New York: Vintage, 1991. Print.

Marris, Emma. *Rambunctious Garden: Saving Nature in a Post-wild World*. New York: Bloomsbury, 2011. Print.

Mastny, Lisa. *Vital Signs 2005: The Trends That Are Shaping Our Future*. New York, NY: W. W. Norton &, 2005. Print.

McGraw-Hill Dictionary of Environmental Science. New York: McGraw-Hill, 2003. Print.

McKeown, Alice. *Vital Signs 2009: The Trends That Are Shaping Our Future*. Washington, D.C.: Worldwatch Institute, 2009. Print.

———. *Vital Signs 2010: The Trends That Are Shaping Our Future*. Washington, D.C.: Worldwatch Institute, 2010. Print.

McKibben, Bill. *American Earth: Environmental Writing since Thoreau*. New York, NY: Literary Classics of the United States, 2008. Print.

Meadows, Donella H., Jørgen Randers, and Dennis L. Meadows. *The Limits to Growth: The 30-year Update*. White River Junction, VT: Chelsea Green Pub., 2004. Print.

Meadows, Donella H. *The Limits to Growth*. London: Universe, 1972. Print.

Mongillo, John F., and Linda Zierdt-Warshaw. *Encyclopedia of Environmental Science*. Phoenix, AZ: Oryx, 2000. Print.

Muir, John. *Muir, Nature Writings*. Library of America: New York, NY, 1997. Print.

———. *A Thousand-mile Walk to the Gulf*. San Francisco: Sierra Club, 1991. Print.

Oreskes, Naomi, and Erik M. Conway. *Merchants of Doubt: How a Handful of Scientists Obscured the Truth on Issues from Tobacco Smoke to Global Warming*. New York: Bloomsbury, 2010. Print.

Orr, David W. *Earth in Mind: On Education, Environment, and the Human Prospect*. Washington D.C.: Island Press, 2004. Print.

———. *Ecological Literacy: Education and the Transition to a Postmodern World*. Albany: State University of New York, 1992. Print.

———. *The Nature of Design: Ecology, Culture, and Human Intention*. New York: Oxford UP, 2002. Print.

Our Common Future. Oxford: Oxford UP, 1987. Print.

Owen, David. *The Conundrum: How Scientific Innovation, Increased Efficiency, and Good Intentions Can Make Our Energy and Climate Problems Worse*. New York: Riverhead, 2011. Print.

Park, Chris C. *A Dictionary of Environment and Conservation*. Oxford England: Oxford UP, 2008. Print.

Pollan, Michael. *The Omnivore's Dilemma: A Natural History of Four Meals*. New York: Penguin, 2006. Print.

Poole, Robert. *Earthrise: How Man First Saw the Earth*. New Haven CT: Yale UP, 2008. Print.

Quammen, David. *Natural Acts: A Sidelong View of Science and Nature*. New York: W. W. Norton, 2008. Print.

Renner, Michael, Eric Anderson, and Linda Starke. *Vital Signs.* Washington, D.C.: Island Press, 2013. Print.

Renner, Michael, Linda Starke, and Lyle Rosbotham. *Vital Signs 2012: The Trends That Are Shaping Our Future*. Washington, D.C.: Worldwatch Institute, 2012. Print.

Renner, Michael. *Vital Signs 2003: The Trends That Are Shaping Our Future*. New York: W. W. Norton, 2003. Print.

Reynolds, Heather L. *Teaching Environmental Literacy: Across Campus and across the Curriculum*. Bloomington: Indiana UP, 2010. Print.

Rhodes, Frank Harold Trevor. *Earth: A Tenant's Manual*. Ithaca: Cornell UP, 2012. Print.

Roach, Mary, and Tim Folger. *The Best American Science and Nature Writing 2011*. Boston, MA: Houghton Mifflin Harcourt, 2011. Print.

Schreurs, Miranda A., Elim Papadakis, *The A to Z of the Green Movement*. Lanham, MD: Scarecrow, 2009. Print.

Shapiro, Judith. *China's Environmental Challenges*. Cambridge, U.K.: Polity, 2012. Print.

Sheehan, Molly O'Meara, and Linda Starke. *State of the World 2007: Our Urban Future: A Worldwatch Institute Report on Progress toward a Sustainable Society*. New York: W. W. Norton, 2007. Print.

Silver, Cheryl Simon., and Ruth S. DeFries. *One Earth, One Future: Our Changing Global Environment*. Washington, D.C.: National Academy, 1990. Print.

Starke, Linda. *State of the World 2009: Into a Warming World: A Worldwatch Institute Report on Progress toward a Sustainable Society*. New York: W. W. Norton &, 2009. Print.

Stone, Michael K., and Zenobia Barlow. *Ecological Literacy: Educating Our Children for a Sustainable World*. San Francisco: Sierra Club, 2005. Print.

Tal, Alon. *Speaking of Earth: Environmental Speeches That Moved the World*. New Brunswick, NJ: Rutgers UP, 2006. Print.

Walker, Charlotte Zoë. *Sharp Eyes: John Burroughs and American Nature Writing*. Syracuse, NY: Syracuse UP, 2000. Print.

Ward, Barbara. *Spaceship Earth*. New York: Columbia UP, 1966. Print.

Warren, James Perrin. *John Burroughs and the Place of Nature*. Athens: University of Georgia, 2006. Print.

Wells, Edward R., and Alan M. Schwartz. *Historical Dictionary of North American Environmentalism*. Lanham, MD: Scarecrow, 1997. Print.

Wilson, Edward O. *In Search of Nature*. Washington, D.C.: Island, 1996. Print.

Yaeger, Bert D. *The Hudson River School: American Landscape Artists*. New York, NY: Smithmark, 1996. Print.

Zimmerman, Michael. *Science, Nonscience, and Nonsense: Approaching Environmental Literacy*. Baltimore: Johns Hopkins UP, 1995. Print.

Zokaei, A. Keivan, L. Hunter Lovins, Andy Wood, and Peter Hines. *Creating a Lean and Green Business System: Techniques for Improving Profits and Sustainability*. Boca Raton: CRC, 2013. Print.

Zuboff, Shoshana, and James Maxmin. *The Support Economy: Why Corporations Are failing Individuals and the next Episode of Capitalism*. New York: Penguin, 2002. Print.

છ૭

About the Author

H. Steven Dashefsky is a Professor at Norwalk Community College in Connecticut, where he has taught environmental science and business technologies for the past 14 years. He also taught biology at Northern Virginia Community College in Alexandria and The Greenhill School in Dallas. He worked as an Environmental Protection Specialist at the Environmental Protection Agency in Washington, D.C., and spent many years in corporate technical training and support.

Steven has over a dozen books published by Random House and McGraw-Hill that simplify science and technology for the layperson and student. He has degrees in biology and entomology.

Acknowledgments

I want to thank Keith Dashefsky for his assistance performing online research for this book. Also, my thanks to Jeffery Morse for his help with some entries and my wife for her help preparing the manuscript.

Further Information

If you have any questions or comments about this book, please send a message to: Info@E-Literacy.com. (See the "Crowd Suggestions" page at the beginning of this book as well.)

BT
10/6/14
29.95

Made in the USA
Charleston, SC
19 September 2014